"REALMS OF GOLD":

A CATALOGUE OF MAPS IN THE LIBRARY

OF THE AMERICAN PHILOSOPHICAL SOCIETY

"REALMS OF GOLD":

A CATALOGUE OF MAPS IN THE LIBRARY OF THE AMERICAN PHILOSOPHICAL SOCIETY

Murphy D. Smith

American Philosophical Society
Independence Square Philadelphia
1991

MEMOIRS OF THE
AMERICAN PHILOSOPHICAL SOCIETY
Held at Philadelphia
For Promoting Useful Knowledge
Volume 195

Library of Congress Catalog Card No: 91-55271
International Standard Book Number: 0-87169-195-7
ISSN: 0065-9738

Cover design: Colophon from James Thackera's map of the city of Washington, 1792. Collection of the American Philosophical Society, #1078.

Much have I travell'd in the realms of gold,
And many goodly states and kingdoms seen;
Round many western islands have I been
Which bards in fealty to Apollo hold.

John Keats,
"On First Looking into Chapman's Homer"

CONTENTS

List of Illustrations xv

Foreword xvii

Acknowledgments xviii

Introduction xix

User's Guide xxiii

Part I

*Manuscript Maps (1-52)**

Part II

Printed Maps (53-1635)

Space (53-61)

World (62-87)

Eastern hemisphere (87a-87b)

Western hemisphere (88-91)

Arctic (92-100)

Antarctic (101-104)

Europe (105-118)

England and Wales (119-143)

Ireland (144-147)

Scotland (148-158)

Sweden (159-163)

Norway (164-165)

Denmark (166-185)

Low Countries (186-193)

The Netherlands (194-213)

Belgium (214-216)

**Entry numbers*

Luxembourg (217)
Switzerland (218-244)
France (245-275)
Central Europe (276-287)
Germany (288-306)
Poland (307-311)
USSR (Russia) (312-320)
Finland (321-322)
Mediterranean area (323-325)
Spain and Portugal (326-340)
Italy (341-358)
Turkey (359-362)
Asia (363-370)
Near East (371-377)
Far East (378-380)
Japan (381-386)
China (386a-398)
India (399-401)
Indochina (402-403)
Indonesia (404-407)
The Philippine Islands (408-420)
Africa (421-426)
Egypt (427-433)
West Africa (434-442)
Central Africa (443-446)
South Africa (447-448)
North America (449-470)
Canada (471-482)
Eastern Canada (483-491)
Nova Scotia and New Brunswick (492-496)
Quebec (497-503)
Ontario (504-514)
Manitoba (515-526)
Saskatchewan (527-530)
Alberta (531-536)
British Columbia (537-544)

Northern Canada (545-549)

United States (550-623)

Eastern U.S. (624-625)

Northeastern United States and Canada (626-633)

New England (634-636)

Maine (637-650)

New Hampshire (651-660)

Vermont (661-668)

Massachusetts (669-696)

Boston (697-710)

Rhode Island (711-718)

Connecticut (719-749)

Middle Atlantic States (750-782)

New York (783-843)

New York City (844-857)

New Jersey (858-894)

Pennsylvania (895-986)

Philadelphia (987-1040)

Delaware (1041-1052)

Maryland (1053-1077)

District of Columbia (1078-1080)

Southern States (1081-1093)

Virginia (1094-1138)

West Virginia (1139-1144)

North Carolina (1145-1156)

South Carolina (1157-1169)

Georgia (1170-1183)

Florida (1184-1199)

Alabama (1200-1208)

Mississippi (1209-1214)

South Central United States (1215)

Louisiana (1216-1224)

Texas (1225-1235)

Kentucky (1236-1253)

Tennessee (1253a-1262)

Arkansas (1263-1264)

Oklahoma (1265-1268)
Old Northwest (1269-1273)
Great Lakes area (1274-1301)
Ohio (1302-1317)
Indiana (1318-1320)
Illinois (1321-1328)
Mississippi River (1329-1343)
Michigan (1344-1353)
Minnesota (1354-1357)
Wisconsin (1358-1373)
Central States (1374-1380)
Iowa (1381-1382)
Missouri (1383-1388)
North Dakota (1389-1390)
South Dakota (1391-1397)
Nebraska (1398)
Kansas (1399-1401)
Mountain States (1402-1407)
Colorado (1408-1421)
New Mexico (1422-1424)
Western States (1425-1429)
Montana (1430-1432)
Wyoming (1433-1446)
Idaho (1447-1448)
Utah (1449-1453)
Arizona (1454-1456)
Nevada (1457-1460)
The Northwest (1461)
Pacific Coast (1462-1465)
Washington (1466-1473)
Oregon (1474-1478)
California (1479-1494)
Hawaii (1495)
Alaska (1496-1500)
Central America (1501-1502)
Mexico (1503-1507)

Guatemala (1508)
Honduras (1509-1510)
Belize (British Honduras) (1511-1512)
Panama (1513-1516b)
West Indies (1517-1553)
South America (1553a-1562)
Central South America (1563-1564)
Colombia (1565-1568)
Venezuela (1569)
Peru (1570)
Bolivia (1571)
Brazil (1572-1574)
Chile (1575-1577)
Argentina (1578-1580)
Australia (1581-1610)
South Australia (1581-1583)
Western Australia (1584-1585)
Queensland (1586a-1605)
Victoria (1606)
New South Wales (1607-1610)
Oceans—World (1611-1615)
Pacific Ocean (1616-1618)
Polynesia (1619-1620)
Atlantic Ocean (1621-1633)
Indian Ocean (1634-1635)

Part III
Atlases (1636-1771)

Part IV
Globes and Model (1772-1774)

Index

LIST OF ILLUSTRATIONS

William Bartram. The Great Alachua-Savana, in East Florida, above 60 miles in circumference.

G.J. de Bois St. Lys. Carte de la partie Française de l'isle de St. Domingue. 1800.

Samuel Fahlberg. Special map, representing Man of War Shoal . . . 1801.

James Wilkinson. A sketch of the Muscle Shoals of the Tenessee [sic] River.

Charles Willson Peale. [New York, showing British and American troop positions.]

Henry Popple and C. Lempriere. Map of the British empire in America with the French and Spanish settlements adjacent thereto. 1733.

Samuel Dunn. A map of the British Empire in North America. 1774.

Henry de Bernière and H. Dearborn. Sketch of the action on the heights of Charles Town 17 June 1775, between his majesty's troops under the command of M. Genl. Howe and a large body of American rebels. 1818.

FOREWORD

The Andrew W. Mellon Foundation, seeking to increase the scholarly usefulness of our collections and enrich the intellectual life of the Library, made a grant in 1982 to the American Philosophical Society establishing a fellowship program in bibliography, research scholarship, and staff development. The major thrust of the program is to provide fellowships for scholars who will prepare bibliographic studies on topics well represented in the Library's collections. To date, five bibliographies have been published through the Mellon Fellowship program, providing informative and authoritative guides to our collections in the transfer of science, technology, and medicine from Europe to the United States; natural history in the New World; anthropology and archaeology; genetics; and the modern life sciences.

The Society's rich collection of nearly 3,000 maps, charts, and surveys clearly merited similar treatment, and after Associate Librarian Murphy D. Smith retired in 1984, I appointed him for that purpose to a Senior Mellon Fellowship in Bibliography for 1985-86. Mr. Smith was well qualified for his task, having served the Society with distinction for thirty-two years. He spent the majority of his career in charge of the manuscript department, and came to be recognized as one of the most knowledgeable and helpful historical research librarians in the country. He has published a number of useful guides and bibliographies, most recently *Historical American sketches: an illustrated guide to sketches in the manuscript collections of the American Philosophical Society* (G. K. Hall, 1984).

The impressive catalogue of the Society's maps contained within these pages bears testimony not only to the wealth of the Library's manuscripts and imprints, but also to the indefatigable labors of its author. In retrospect, however, it became evident that a one-year fellowship was hardly sufficient to do justice to such an enormous project, and the following members of the staff generously pitched in to help bring the work to final publication: Beth Carroll-Horrocks, Manuscripts Librarian; Marian Christ, Cataloguer; David J. Rhees, Assistant Librarian for Research and Programs; and Hildegard G. Stephans, Associate Librarian.

The length of the catalogue, combined with the desire to insure its wider distribution, prompted the decision to publish it as a *Memoir* of the Society rather than in the soft-cover format of the other Mellon bibliographies.

We would like to acknowledge the generous support of The Andrew W. Mellon Foundation, which made possible this publication.

Edward C. Carter II
Librarian
American Philosophical Society

ACKNOWLEDGMENTS

Thanks are due to many people for help received in preparing this catalogue. The librarians of the Academy of Natural Sciences of Philadelphia, the Free Library of Philadelphia, the Library Company of Philadelphia and the Library of Congress were most generous with their help, and I thank them. I also thank Dr. James E. McClellan III for information pertaining to the maps of St. Domingue (Haiti). The most thanks are due to the entire staff of the American Philosophical Society, but especially to Librarian Edward C. Carter II for his continued interest in this project; and to Beth Carroll-Horrocks, Manuscript Librarian; Marian Christ, Cataloguer; David Rhees, Assistant Librarian for Research and Programs; and Hildegard Stephans, Associate Librarian. At times I thought I must have worn out my welcome with the latter five, but they remained unfailingly pleasant and helpful. Much of the design of the format for these entries was the result of their interest in my work and in the future use of this volume.

INTRODUCTION

This volume is a catalogue of the rich and extensive collection of maps in the Library of the American Philosophical Society in Philadelphia. It contains information on some 1,750 printed maps, over 1,000 manuscript maps, 136 atlases, two globes, and one model.[1]

This project was begun in 1985 shortly after my retirement as Associate Librarian of the Society, when Librarian Edward C. Carter II named me Andrew W. Mellon Senior Research Fellow. The map collection had been catalogued by Mrs. Doris E. Broomall, but it was decided by Associate Librarian Hildegard Stephans and Cataloguer Marian Christ that more complete descriptions of the maps and a far more comprehensive index were essential for the better use of the collection. We determined that all maps in the Manuscript Collection as well as all printed maps, certain atlases, globes, and the terrestrial model would be listed, but that no microform maps would be included. Restrikes, reprints, and facsimiles of maps also are included. Occasionally maps which once belonged to a volume had been removed and placed in the map collection, and these, too, are listed in this catalogue.

The only maps in published works which are included in this catalogue are those listed in James Clement Wheat and Christian F. Brun's *Maps and Charts Published in America before 1800. A bibliography* (New Haven: Yale University Press: 1969). Finally, I have included in the catalogue the three engraved copper plates owned by the Society. One is the copper plate of the first published map of the Lewis and Clark Expedition (*History of the Expedition under the Command of Lewis and Clark* . . . Philadelphia: Bradford and Inskeep; New York: Abrm. H. Inskeep, J. Maxwell: 1814), which was recently repulled for the first time (see no. 566). Also, there are two huge copper plates pertaining to the port of Philadelphia which were used in the publication of the *Atlantic Neptune* (see no. 1659).[2]

The origin of the Library's map collection may be traced ultimately to Benjamin Franklin, who founded the American Philosophical Society, the oldest learned society in the United States, in 1743. In that year Franklin published a prospectus for the Society, *A proposal for promoting useful knowledge among the British plantations in America,* which specified that there be at least seven members from Philadelphia, one of them a geographer. Further, he added that the subjects of correspondence among the members should include

> Surveys, Maps and Charts of particular Parts of the Sea-coasts, or Inland Countries; Course and Junction of Rivers and great Roads, Situation of Lakes and Mountains, Nature of the Soil and Productions; &c.

Franklin himself was keenly interested in maps, and in 1785 on his return to America he made thermometrical observations and sketched a chart of the Gulf Stream, which was published in the second volume of the Society's *Transactions* in 1786 (see nos. 1622-1626).

[1]The number of manuscript maps is perhaps misleadingly large because it includes maps, however small, from notebooks in the collections of J. Peter Lesley (ca. 300 maps), Richard Joel Russell (40), Elsie Clews Parsons (37) , and Robert Cushman Murphy (35). Aside from these, there are some 600 manuscript maps described in this catalogue.

[2]Numbers given for maps are entry numbers. Manuscript maps are denoted by two numbers: an entry number for the collection, followed by a number in parentheses for the individual map within that collection.

Few, if any, maps were acquired in this early period, for by 1746 the American Philosophical Society had lapsed into inactivity. In 1769, however, the Society was revived and united with the American Society for Promoting Useful Knowledge to form the "American Philosophical Society, held at Philadelphia, for Promoting Useful Knowledge." Two years later the Library of the Society was begun when the Society published its first volume of *Transactions* and commenced a system of exchange of publications with sister institutions which continues to this day. Initially, maps were considered part of the Cabinet of Curiosities, or Museum, rather than the Library. However, by 1850 maps had become part of the Library and were no longer included as part of the Cabinet. The collection grew through gifts, purchases, and exchange of publications.

The bulk of the American Philosophical Society Library's map collection is more or less evenly distributed among four fifty-year periods: 1751-1800 includes 532 maps; 1801-1850 includes 326 maps; 1851-1900 includes 472 maps; and 1901-date includes 430 maps.

The earliest individual map (excluding facsimiles) is *Le pais de Brie* by Guiljemus Blaeu, published ca. 1644 (no. 245). Even older maps may be found in the atlases, such as the *Theatrum orbis terrarum* by Ortelius, ca. 1569-70. The most recent map is the *General highway map of Oconee County, South Carolina,* published in 1983 (no. 1169a).

For the period subsequent to 1850 many of the maps acquired were government publications, both state and federal. Two prominent members of the Society, Alexander Dallas Bache and Ferdinand Rudolph Hassler of the U. S. Coast and Geodetic Survey (as it came to be known), forwarded many maps as they were issued, as did some of the outstanding geologists as they explored the West. The geology maps continue to arrive to this date, but most of them are not included in this catalogue, for they are in book form or in hard boxes for storage, and are filed with the publications they illustrate.

Not surprisingly, North America is the principal geographical area represented in the collection. Maps of North America, principally the United States, make up two-thirds of the printed maps. Over one-half of the manuscript maps are of North American locations, of which three-fourths are Pennsylvania sites. The next best represented area of the world is Europe, which accounts for about 15 percent of the printed maps and 8 percent of the manuscript maps.

Of course, quantity is not necessarily a measure of inherent interest or uniqueness. For instance, while there are only 33 manuscript maps of the Arctic region, they include maps from the papers of Elisha Kent Kane (1820-1857; APS 1851), who made two voyages in the 1850s searching for the lost Arctic explorer Sir John Franklin [nos. 23 (1-12) and 32 (79)]. Another small but interesting group of Arctic maps are those made or gathered by the noted anthropologist Franz Boas (1858-1942; APS 1903) during his studies of Baffin Island Eskimos in the early 1880s [no. 6 (1-6)]. There is also a stunning series of 38 colored manuscript maps, dated 1800-1801, of Cuba, St. Domingue (Haiti), and Puerto Rico, made by Georges Joseph de Bois St. Lys and others [no. 32 (25-29, 32-53, 55-64)].

Indeed, there is a wealth of manuscript maps in the APS Library, of which the following are a few of the outstanding examples:

* Although poorly delineated, the map of the Appalachian Mountains made by John Bartram ca. 1750 is noteworthy for showing the areas where fossil sea shells had been found; it was presented to the Society by Benjamin Franklin, who wrote on the back: "Mr. Bartram's Map very curious" [no. 32 (3)].

* By contrast, one of the loveliest eighteenth-century maps in the collection was drawn by John Bartram's son, William Bartram. It is of "The Great Alachua Savana, in east Florida," showing the drainage of the "Savana" [no. 4 (1)].

* All the maps in the journals from the Lewis and Clark expedition are of great historical interest, but the one of the Great Falls and Portage of the Missouri River is particularly beautiful [no. 28 (2)].

* A map of the British fortifications of Boston Neck is of importance because it was presented with the Richard Henry Lee Papers [no. 25 (1)].

* A map drawn by Thomas Jefferson is based on his survey of 400 acres of Virginia land which devolved upon him through his wife. His conveyance of the land to Nicholas Lewis is with the map [no. 22 (1)].

* A map by Frederick Ridgely, "An Eye-Draught of the Mammoth-Cave in Warren County, Kentucky," was done by eye, for "the [compass] needle does not traverse" the cave [no. 2 (5)].

* A map of New York City drawn ca. 1776 by Charles Willson Peale shows the British and American army positions [no. 38 (2)].

* The War of 1812 found the nation unprepared, so General Jonathan Williams (nephew of Benjamin Franklin, the first Superintendent of West Point, and an active member of the American Philosophical Society) ordered William Strickland to make a map of "the country nine miles west of the city of Philadelphia" for the "sub-committee of defence" [no. 32 (68)]. Strickland, a famous architect, also was a member of the Society.

* APS member James Wilkinson sketched the "Muscle Shoals of the Tenessee [sic] River" in 1802 [no. 32 (66)].

* Sebastian Bauman made three preliminary drafts, dated 22-28 October 1781, of the "Plan of Yorktown, Virginia, depicting the armies when Cornwallis surrendered" [no. 32 (9)]. Although the pencil sketches are blurred, they are magnificent.

* General Henry Dearborn and Henry de Berniere made a map of the "action on the heights of Charles Town 17 June 1775, between his magestys troops under the command of M. Genl. Howe & a large body of American rebels," and the "parts in red are corrections of the original by Maj. Gen. Dearborn" [no. 32 (6)].

Among the many treasures found in the printed maps are the following:

* One of the oldest maps in the collection (and certainly the largest, measuring approximately eight feet by eight feet) is Henry Popple and Clement Lempriere's 1733 "Map of the British empire in America with the French and Spanish settlements adjacent thereto" (no. 449). Presented to the Society in 1834, it is probably one of the great maps which hung in Independence Hall in July 1776. The map has been deacidified, repaired, and remounted, and was a prominent feature of the 1976 bicentennial exhibition, *A Rising People,* organized by the Library Company of Philadelphia, the Historical Society of Pennsylvania, and the American Philosophical Society. It was also displayed in the 1988 *Legacies of Genius* exhibition organized by the Society and fifteen other members of the Philadelphia Area Consortium of Special Collections Libraries.

* Another printed map with unique associations is the "Plan of the city of Washington in the territory of Columbia, ceded by the states of Virginia and Maryland to the United States of America," published in 1792 by James Thackara (no. 1078). It was presented by George Washington on 22 April 1793 to the Earl of Buchan, an avid Scots republican and a member of the American Philosophical Society, who in turn gave it to the Society on 17 July 1793.

The two most outstanding maps the American Philosophical Society ever published were:

* The 1771 map by Thomas Fisher (reprinted 1789) of a proposed canal across the Delmarva Peninsula connecting the Chesapeake and Delaware bays (nos. 760 and 760a), and

* William Maclure's geological map of the United States, published in 1818 (no. 570). Both maps appeared in the Society's *Transactions.*

The uses to which the maps have been put have varied over the years. The most common is reference and research. Some maps have been borrowed for exhibitions, such as the Popple map noted above. Perhaps the most important use of the maps concerned the Northeast Boundary Dispute with Great Britain of 1828-29. Two maps were borrowed by Albert Gallatin and were bound in the volume to be placed before the arbiter, the king of The Netherlands (nos. 453 and 550). They were not returned until 1852.

Over the centuries a few maps have been lost and never recovered. For the sake of completeness, and with the hope that they someday may be found, they are listed as follows:

1. On 20 July 1769 a draft from actual surveys of the Mississippi River to Fort Chartres was received, as was another draft of the Mobile River to Fort "Tombeebe, a length of 96 leagues, taken from the survey made by Philip Pilman in 1767." These were presented by D. Clark through Hugh Williamson.

2. John J. Hawkins sent a sketch of a proposed city which was received 19 September 1800. The description runs four large pages of small script, but the plat has disappeared. The members of the Society took this sketch seriously, for towns were being founded all along the frontier, and the committee's report was comprehensive.

3. Ambroise Tardieu presented "Cartes des États Unis" in four sheets, the same in smaller size, and "Carte des Indes Occidentales," both of which arrived 4 November 1808.

4. Around 5 November 1824 Gaspard Deabatte presented a map of Turin, Italy.

5. Prior to 1826 Henry Schwenk Tanner presented his map of Mexico which he had published the previous year.

6. P. W. Sheafer sent on 7 April 1865 a manuscript sketch showing the tidewater area "relative to the various coal and oil regions of the U.S." Eli Bowen used this "in publishing his late work on coal and oil fields."

7. Archibald Campbell gave on 20 September 1867 two sheets of a photographic copy of the Northwest Boundary Survey.

8. The preliminary map of Ohio, prepared by the chief geologist of the state, J. S. Newberry, was received 16 December 1870 as a gift from the State of Ohio.

9. The (California?) Commission on Irrigation presented 6 November 1874 a "fine map of California" and a map of the delta of Canvery(?).

USER'S GUIDE

The catalogue is divided into four parts: **Manuscript Maps** (nos. 1-52), **Printed Maps** (nos. 53-1635), **Atlases** (nos. 1636-1771), and **Globes and Model** (nos. 1772-1774). Each part has its own format, given below.

When the issue date of a map is unknown, I have used the earliest known date and have bracketed the date. A case in point is the *West India-Pilot,* an atlas of the West Indies by Joseph Smith Speer (nos. 1145, 1170, 1503, 1509-13, 1520-34, 1565-67). As I worked on the map collection, these maps, individually catalogued and filed, seemed to me to have been issued as an atlas which had never been bound. I checked them against the bound atlas in the Library Company of Philadelphia and therefore was able to use the title page imprint for the date of issue. Another example was the material presented by Mathew Carey in 1803 which he used for William Guthrie's *The general atlas for Carey's edition of his geography improved* (1795). I assumed that he included a series of such maps which were "made for Chamber's edition of Guthrie" as is engraved on some of the plates (nos. 128, 153, 160, 192, 195, 233, 306, 308, 350, 361, 367, 424, 1554, 1555). There was never an edition by Chambers, and the maps were the same size and format of the Carey edition, so I appended the donation note to them. Manuscript maps were included by Carey, also, in this gift [nos. 32 (18a-19, 21-21a, 23-24)].

Facsimiles and reprints are listed by the date of the original map.

Titles of maps generally have been transcribed as they appear on the map. Idiosyncrasies of spelling and punctuation likewise have been preserved, although a few silent changes were made in order to insure clarity. Supplied titles have been bracketed.

The call number of each collection, map, or atlas was included for ready reference by any researcher who wishes to see the map in question or acquire additional information from it. There is also a note following the call number: Small, Large, Oversize, or Extra-oversize. Since many maps are trimmed, mounted on linen, or folded, it is important to remember that such notes as Small, etc. refer to the location of an item, *not* to the size of the map (which is listed elsewhere in the entry and is given in centimeters).

Manuscript Maps

The manuscript maps are ordered alphabetically by name of collection, and chronologically within each collection. Manuscript maps have two numbers: the collection entry number, followed by the number of the map within the collection set in parentheses. For instance, the maps in the Lewis and Clark journals (entry number 28) are 28 (1), 28 (2), 28 (3), etc. Manuscript maps are described as follows as the information warrants:

Name of Collection
Date (date of reproduction is used for facsimiles)
Title or description (supplied titles are bracketed)
Number of pieces (if more than one)
Scale
Cartographer
Size of paper
Colored
Provenance

Note (publication, cross-references, bibliographic references, etc.)
Call number

Printed Maps

The printed maps are arranged in the order in which they were catalogued, which for the most part follows the Dewey Decimal area classification system. Within a given area or country they are arranged chronologically. "Wheat numbers" are given for maps cited in James Clement Wheat and Christian F. Brun's *Maps and charts published in America before 1800. A bibliography.* The printed maps are described as follows as the information warrants:

Date (date of reproduction is used for facsimiles)
Title or description (supplied titles are bracketed)
Number of pieces (if more than one)
Scale
Cartographer(s)
Engraver(s)
Size (neat line, plate impression, or size of paper; unless otherwise noted, measurements are taken from neat lines)
Inset(s)
Coloring
Provenance
Note (cross-references, bibliographical references, etc.)
Wheat number
Call number

Atlases

Atlases are listed in chronological order. They are described as follows as the information warrants:

Date (date of reproduction is used for facsimiles)
Title or description
Cartographer(s)
Engraver(s)
Size (height at spine)
Provenance
Note (cross-references, bibliographical references, etc.)
Call number

Globes and Model

The globes and model are entered as follows as the information warrants:

Date
Description
Cartographer(s)
Engraver(s)
Size (diameter of globe)
Colored
Provenance
Note (cross-references, bibliographic references, etc.)

Index

The Index is one alphabet with listings of names under cartographers, donors, engravers, lithographers, and publishers, thereby making cross-references simple. I have tried to make the index as complete as possible with the full name and dates of the persons cited and a simple notation of each person's profession. For this I have relied heavily upon the *Dictionary of American biography; A catalog of books represented by Library of Congress printed cards*; James C. Wheat and Christian F. Brun's *Maps and charts published in America before 1800. A bibliography*; Ronald Vere Tooley's *Directory of mapmakers*; and several bibliographical dictionaries of artists and engravers: Emanuel Banezet, *Dictionnaire critique et documentaire des peintres, sculpteurs, dessinateurs et graveurs de tous les pays par un groupe d'écrivains specialistes français et étrangers.* Nouvelle édition. Libraire Grund: Paris: 1976. 10 vols.; George C. Groce and David H. Wallace, *The New-York Historical Society's dictionary of artists in America, 1564-1860.* New Haven: Yale University Press: 1957; William Young, *A dictionary of American artists, sculptors and engravers from the beginnings through the turn of the twentieth century.* Cambridge, Mass.: William Young & Co.: 1968; Shearjashub Spooner, *A biographical history of the fine arts; or, memoirs of the lives and works of eminent painters, engravers, sculptors, and architects. From the earliest ages to the present time.* New York: J.W. Bouton: 1865. 2 vols.; Michael Bryan, *Dictionary of painters and engravers, biographical and critical.* London: George Bell & Sons: 1889. 2 vols.

PART I

Manuscript Maps

1. AMERICAN PHILOSOPHICAL SOCIETY. Archives.

The Archives of the American Philosophical Society are amazingly complete, spanning the period from 1758 to date. The few maps in the collection are arranged chronologically.

1 (1) 1800

[Chart of a voyage across the Atlantic Ocean from Philadelphia to Cork, Ireland]. By "Hiram." 1800.

Size: 49.5 x 19.7 cm.

"Hiram" offered this small volume entitled "Navigation made easy or Mariners complete guide" using his "newly invented time pieces," for consideration for the Society's Magellanic Premium. The volume is illustrated "with a drawing of the Voyage, distinctly pointing out, each days sail, the nearest way, and number of miles from Land to Land." The volume is "Dedicated to Thomas Jefferson Esquire, President of the American Philosophical Society."

1 (2) 1816

Sketch of the plan of an ancient work, three miles to the southeast of Lexington [Kentucky]. By Charles Wilkins Short. [31 August 1816].

Cartographer: Charles Wilkins Short.

Size: 19.7 x 24 cm.

Short's description was read on 4 October 1816 to the Society's members. It was published, with a small engraving of the drawing, in the American Philosophical Society *Transactions*, n.s., vol. 1: pp. 310-312, plate IX, fig. 3.

1 (3) 1840

[Great Indian mound near Washington, Adams County, Mississippi]. By Caleb Goldsmith Forshey. September 1840.

Cartographer: Caleb Goldsmith Forshey.

Size: 24.7 x 19 cm.

This map accompanied Forshey's article which was read on 18 September 1840 before the Society. It describes the mound and its contents in seven large pages. This great mound was 84 feet high. An abstract of the article, but not the map, was published in the American Philosophical Society *Proceedings*, vol. 1 (1838): pp. 305ff.

(APS Archives)

2. AMERICAN PHILOSOPHICAL SOCIETY. Archives. Manuscripts Communicated to the A.P.S.

These scientific communications, about 560 in number, were sent to the APS between 1748 and 1837 by members of the general public anxious to gain approval of and support for their ideas and inventions. The topics include mechanics, engineering, trade, navigation, agriculture, medicine, natural history, education, mathematics, and astronomy. Several papers were submitted as entries in APS-sponsored competitions, and many were published in the Society's *Transactions*.

2 (1) [1805]

The nitre caves of Kentucky. By Samuel Brown. [1805].

Cartographer: Samuel Brown.

Size: 37.6 x 22.9 cm.

This map was accompanied by an article by Brown published as "A description of a cave on Crooked creek, with remarks and observations on nitre and gun-powder." American Philosophical Society *Transactions*, vol. 6 (1809): pp. 235-247.

2 (2) 1805

Plan of the nitre cave described by Dr. Samuel Brown. By John James Du Four. 1805.

Cartographer: John James Du Four.

Size: 41.3 x 51.4 cm.

The cartouche reads: "A Survey of the Great Salt Petre Cave on Crooked Creek Madison County, Ky. by John James Dufour. 1805."

2 (3) 1810

[Imaginary river to illustrate the movement of Chapman's "new-invented Steam Boat"]. By Isaac A. Chapman. 20 March 1810.

Cartographer: Isaac A. Chapman.

Size: 19 x 30.5 cm.

Chapman sent this communication to member James Mease to be presented to the Society for the possible award of a premium.

2 (4) 1807

Sketch of course of Mississippi River up to Natchez & of the country bordering. By William E. Hulings. 1807.

Cartographer: William E. Hulings.

Size: 44.5 x 34.3 cm.

The meridian of New Orleans is given. On the reverse is a note to John Vaughan from Hulings: "I have marked the situation of Mr. [William] D[unbar]'s place on the second Creek, but am unable to do it with Geometrical precision."

2 (5) 1811

An Eye-Draught of the Mammoth-Cave in Warren County, Kentucky. By Frederick Ridgely. 15 March 1811.

Cartographer: Frederick Ridgely.

Size: 40.6 x 25.4 cm.

Colored.

The covering letter addressed to Benjamin Rush states that "the needle does not traverse" the cave, therefore the "Eye-Draught." (Arch. III, 1)

3. ANDREWS, EMMA B. Journal.

This journal was kept by Mrs. Andrews during seventeen trips up the Nile River. One of her relatives was Theodore M. Davis, who excavated in the Valley of the Kings. Three maps pertain to royal tombs. Note: page numbers are repeated several times in each volume.

3 (1) 1903

Plan of the tomb of Thotmes IV. By Mrs. Andrews. 3 February 1903. Vol. 2, p. 136.

Cartographer: Emma B. Andrews.

Size: 8.3 x 8.9 cm.

3 (2) 1904

[Part of the plan of the tomb of Thotmes I]. By Mrs. Andrews. 4 February 1904. Vol. 2, p. 31.

Cartographer: Emma B. Andrews.

Size: 6.3 x 8.9 cm.

3 (3) 1907

[Sketch of the plan of the tomb of Queen Tyi]. By Mrs. Andrews. 19 January 1907. Vol. 2, p. 20.

Cartographer: Emma B. Andrews.

Size: 8.3 x 5 cm.

(916.2: An2)

4. BARTON, BENJAMIN SMITH. Papers.

Barton was an outstanding Philadelphia physician and naturalist. His major interests were medicine, American flora and fauna, and the American Indian. He published widely in these fields.

Within the Barton Papers are drawings by William Bartram, an eminent naturalist, who was perhaps the best American draughtsman of botanical specimens of his time. With his father, John, he toured parts of the southern colonies and published *Travels through North & South Carolina, Georgia, East & West Florida* . . . (Philadelphia: 1791). This map may have been made for inclusion in that work.

4 (1) [ca. 1774]

The Great Alachua-Savana, in East Florida, above 60 miles in circumference. Near 100 miles W. from St. Augustin, & 45 miles W. from the River St. Juan. By William Bartram. [ca. 1774].

Cartographer: William Bartram.

Size: 40.6 x 32.4 cm.

(B: B284.d)

5. BECK, RICHARD. Journal

Beck was a tourist from England who described his visit to the United States in a journal which also contains sketches, maps, and letters.

5 (1) 1880

[Map tracing the Cunard ship's track from Ireland to New York City]. By Richard Beck. 1880.

Cartographer: Richard Beck.
Size: 46.3 x 26.7 cm.

5 (2) 1880

Philadelphia center city street plan. By Richard Beck. 1880.
Cartographer: Richard Beck.
Size: 20.3 x 16.5 cm.

(917: B38)

6. BOAS, FRANZ. Papers.

Boas, known as the "father of modern anthropology," spent some time early in his career on Baffin Island where he studied the natives and made geographical observations. He sailed on the polar research schooner *Germania* for Cumberland Sound, Baffin Island. He charted 250 miles of coastline and used Eskimo maps and descriptions; he also identified two large lakes on the island. Some of these maps were published in Boas's *The Central Eskimo* in the *Sixth annual report . . .* of the Smithsonian Institution's Bureau of Ethnology (1888).

6 (1) 1874

Chart of Niantilic harbour by William A. Mintzer, U.S. Navy, from observations by Lieut. Wilkins, U.S. Navy, navigating officer of the U.S. St. Tigress on Polaris search. 1874.
Scale: ½ inch = 1 mile.
Cartographers: W. A. Mintzer and Lt. Wilkins.
Size of paper: 25.9 x 20.6 cm.

(178: 1874: M667nia Small)

6 (2) [ca. 1883]

[Hudson Bay, Baffin Bay and Greenland area]. By Franz Boas. [ca. 1883].
Cartographer: Franz Boas.
Size of paper: 35.9 x 35.6 cm.
Colored.
Printed map with travel routes marked in red and blue pencil.

(B: B61p)

6 (3) [ca. 1883]

[Sketch of area from Arsebemiling to Ipiakdiuak]. By Franz Boas. [ca. 1833].
Cartographer: Franz Boas.
Size: 30.8 x 30.5 cm.
On the verso is a map of Niantilik harbor.

(B: B61p)

6 (4) [ca. 1883]

[Route from American Harbor to Kingawa and return]. By Franz Boas. [ca. 1883]. 2 copies.

Cartographer: Franz Boas.
Size: 31.8 x 28.6 cm.
Colored.

(B: B61p)

6 (5) [ca. 1884]

[Map made by Boas tracing his trip to Baffin Island]. By Franz Boas. [ca. 1884].
Cartographer: Franz Boas.
Size: 34.6 x 37.1 cm.
Colored.

(B: B61)

6 (6) [ca. 1884]

[Baffinland]. By Franz Boas. [ca. 1884].
Cartographer: Franz Boas.
Size: 71.6 x 95.25 cm. (irregular).
Colored.

(B: B61 Large)

All these maps were presented by Mrs. Helene Boas Yampolsky, 1961-62.

7. BROWN, DAVID J. Papers.

Brown was a Scottish geologist who was active during the latter half of the nineteenth century.

7 (1) 1867

[Geological map of the plain of Cumberland, England, showing the Pennine fault and the Eden River]. By Joseph Lowthian. 11 April 1867.
Cartographer: Joseph Lowthian.
Size: 39.4 x 25.4 cm.

7 (2) 1871

Western Peebles-Shire showing the calcareous beds of the Wrae Hill, Glencotho and Kilbucho. By Charles Lapworth. 24 April 1871.
Cartographer: Charles Lapworth.
Size: 14 x 19.7 cm.
Colored.

7 (3) 1871

Catchment basin of Saint Mary's Loch [Scotland]. By Charles Lapworth. April 1871.
Cartographer: Charles Lapworth.
Size: 20.3 x 23.5 cm.

7 (4) n.d.

[Geological map of the neighborhood of Galashiels, Scotland]. By David J. Brown. n.d.
Cartographer: David J. Brown.
Size: 14 x 20.3 cm.

7 (5) n.d.

North Esk reservoir. By David J. Brown. n.d.
Cartographer: David J. Brown.
Size: 20.3 x 14 cm.
Colored.

7 (6) n.d.

[Moorfoot Hills, Scotland]. By David J. Brown. n.d.
Cartographer: David J. Brown.
Size: 37.5 x 28.6 cm.
Colored.

(B: B813)

8. CLARK, WILLIAM. Journals.

William Clark and Meriwether Lewis made their epochal expedition to explore the Louisiana Purchase territory from 1804 to 1806. Later, Clark served as an army officer who fought the Indians in the Old Northwest, an Indian agent, and then was made governor of the Missouri Territory in 1813. He drew most of the maps for the Lewis and Clark expedition. More maps from the expedition are described below under entry no. 28 (Lewis and Clark Expedition Journals).

8 (1) 1806

[River basin]. By William Clark. 1806.
Cartographer: William Clark.
Size: 10.8 x 16.5 cm.

(917.3: L58cl)

8 (2) 1808

[Fort Osage (draft)]. By William Clark. 1808.
Cartographer: William Clark.
Size: 10.2 x 16.5 cm.

(917.3: L58c)

9. CLYMER, GEORGE. Papers.

Clymer was a signer of the Declaration of Independence, an outstanding Philadelphia merchant, and a land speculator.

9 (1) 1838

J. C. Fisher's rear land. By S. A. Law. 28 May 1838.

Cartographer: S. A. Law.

Size: 35.6 x 30.5 cm.

Colored.

Annotated: "J. C. Fisher's Front Patent Lot all disposed of: And, of his Rear Division, on *hand,* only the Two Remnants, in *Yellow* Shade, of . . . 86¾ acres. 1 May 1838: 28 May 1838."

9 (2) 1839

[Authorized donations from Meredith and Clymer Land. 1839].

Size: 28.6 x 25.7 cm.

Annotated: "The line of the North End of No. 22 - can make nothing of - from the orig[ina]l Survey & conclude cannot but be a blunder! [The part of No. 5 and 22] called, say, a hundred acres, is the Reserved Piece or Parcel, appropriated for a Donation . . . to the Town, for the purpose of encouraging Schooling, or litterary, Instruction: and which, as yet, has never been so granted & conveyed."

(B: C625)

10. COUCH, JONATHAN. Papers.

Couch was an English naturalist whose family had been long resident in Polperro.

10 (1) n.d.

Sketch of a portion of . . . the coast of Cornwall and Devonshire as they were to be fortified in 1588 against the landing of any enemy. By Jonathan Couch. n.d.

Cartographer: Jonathan Couch.

Size: 33.3 x 20.9 cm.

Copied from an engraving made by T. Pine, 25 March 1740. As a local historian, Couch wrote a *History of Polperro,* published after his death. His comments on this map are specific, for Couch had to correct the geography.

(B: C831)

11. DARWIN, CHARLES ROBERT. Papers.

11 (1) [1837]

[Area of ocean showing the limits of coral islands]. By Charles Robert Darwin. [19? December 1837].

Cartographer: Charles Robert Darwin.

Size: 23.5 x 18.4 cm.

This map is in a letter from Darwin to Charles Lyell of the above date.

(B: D25.L)

12. DAY, SHERMAN. Ephraim Dyer IV Collection of the Sketches of Sherman Day.

Day was the grandson of Signer Roger Sherman and the son of Jeremiah Day, president of Yale College. He travelled in Europe and North America and wrote *Historical collections of the state of Pennsylvania* (Philadelphia: 1843). This work was lavishly illustrated with woodcuts made by John Warner Barber from sketches by Sherman Day. He travelled about the state and gathered stories and reminiscences from the older residents. Barber had written and illustrated several volumes on histories of other states, and Day used them as the prototype for his *Historical collections.*

12 (1) 1840

Fort Pitt in 1795 - copied from the large drawing at the Monongahela House. By Sherman Day. 1840.

Cartographer: Sherman Day.

Size: 19.7 x 12.7 cm.

(917.48: D33, #D)

12 (2) 1840

[Susquehanna Valley of central Pennsylvania]. By Sherman Day. 1840.

Cartographer: Sherman Day.

Size: 27.9 x 26 cm.

(917.48: D33, #C)

13. DUHAMEL DU MONCEAU, HENRI-LOUIS. Papers.

Duhamel du Monceau was an eighteenth-century French philosophe who wrote sixty memoirs. He is most famous as a botanist and his botanical works dealt with the physiology and anatomy of plants.

13 (1) 1722

Plan de la Rue St. Jean. . . . By Robert Heroguel. 8 April 1722.

Cartographer: Robert Heroguel.

Size: 28.5 x 202 cm.

Colored.

Plan for the building or rebuilding of the church of St. Jean.

(B: D87, #30)

13 (2) 1761

Verdun. 1761.

Size: 55.5 x 85 cm.

(B: D87 Extra-oversize)

13 (3) 1776

[L'Isle St. Louis, Paris]. 1776.

Size: 17.8 x 23.7 cm.

(B D87, #2)

14. FOUGEROUX DE BONDAROY, AUGUSTE DENIS. Collection.

Fougeroux de Bondaroy was a plant physiologist and archaeologist who traveled extensively in France and Italy. He studied plant physiology under his uncle, Henri-Louis Duhamel du Monceau, and worked with him on the projected revision of his most famous book, *Traité des arbes et arbustes.* Fougeroux de Bondaroy published *Recherches sur les ruins d'Herculanéum* (1769), *Art du tunnelier,* and *Mémoires sur la formation des os* (1763).

In this collection are two volumes which pertain to his tour to Rome and Naples in 1763; there is one map in each volume.

14 (1) [1763]

Port de Civita Vecchia. [1763].

Size: 29.2 x 20.9 cm.

Colored.

14 (2) [1763]

Environs de Naples. [1763].

Size: 47.3 x 21.6 cm.

(B: F8245)

15. FRANKLIN, BENJAMIN. Papers.

Franklin, perhaps Philadelphia's most important citizen during the eighteenth century, was in England at the time his house was erected in Philadelphia; he directed its construction and furnishing through the post. His involvement in the development of the city of Philadelphia and its environs is evident in the numerous sketches and maps within his papers.

15 (1) [ca. 1732?]

[State House Square (later Independence Square), Philadelphia]. By Benjamin Franklin. [ca. 1732?].

Size: 31.8 x 38.1 cm.

Endorsed on the verso: "B— F—."

(974.811: P53.1)

15 (2) 1743

Plan of a tract of land belonging to Lawrence Growden and Langhorne Biles. Situate in the county of Bucks as the same was divided into lots. By Nicholas Scull. 11 November 1743.

Cartographer: Nicholas Scull.

Size: 31.7 x 20.3 cm.

(B: F85, vol. 66, fol. 21)

15 (3) [1765]

[Plan of Franklin Court property in Philadelphia. 1765].

Size: 37.1 x 23.5 cm.

(B: F85, vol. 69, fol. 106)

15 (4) [1765]

[Plan of Franklin Court property in Philadelphia. 1765].

Size: 37.1 x 23.5 cm.

(B: F85, vol. 69, fol. 106)

16. FRANKLIN, WILLIAM. Papers.

William Franklin, like many others, invested in the vast territories which lay beyond the settled boundaries of New York and Pennsylvania. The maps in this collection pertain to such speculation.

16 (1) [1770?]

Plan of Franklin Township containing 31,723 a[cres], 1 rod 34 p[erches] of land divided among the Proprietors into twelve lots; viz, the River-lots containing 1,860 a[cres] . . . each, and the back or rear lots 3,426 a[cres] each. Copied from Robert Lettis Hooper, Jr., plan dated in November 1770.

Scale: 1 inch = 100 chains.

Cartographer: Robert Lettis Hooper, Jr.

Size: 24.1 x 35.6 cm.

(B: F861.27)

16 (2) [ca. 1770?]

A Plott of a tract of 69,000 acres of land (granted Feby 3d 1770 by letters patent to Charles Reade and others) in the county of Tryon, . . . New York. [ca. 1770?].

Size: 47 x 37.5 cm.

The land is situated along the Susquehanna River. Annotated: "NB All the Lots on which no Quantity is mentioned contain 1000 Acres each with the usual allowance for High-Ways."

(B: F861.27)

17. GAULD, GEORGE. A general description of the sea-coasts, harbours, lakes, rivers &c. of the province of West Florida.

Gauld was the British Admiralty surveyor of West Florida in 1769. One of his associates was Thomas Hutchins, British surveyor and Assistant Engineer in the British Army in North America. Hutchins surveyed areas ranging from Florida to the Indiana region, and published some of his maps.

17 (1) 1772

A sketch of the Middle River & Yellow River in West Florida . . . By Thomas Hutchins. 1772.

Cartographer: Thomas Hutchins.
Size: 26.3 x 22.5 cm.

(917.59: G23)

18. GOODRICH, B. G. Survey notebook and maps.

Goodrich was a surveyor who flourished during the middle of the nineteenth century. He worked in Wayne County, Pennsylvania.

18 (1) 1854

Draft of land sold by Richard Doney to Chauncy Demming. Surveyed April 12th, 1854.
Surveyor: B. G. Goodrich.
Size of paper: 30.5 x 20.1 cm.

18 (2) 1857

Lot of lands which Hugh Connor purchased of Abrm Brink. 15 May 1856.
Cartographer: B. G. Goodrich.
Size: 21 x 16 cm.

18 (3) 1857

McMullen survey for Aldens. July 1857.
Cartographer: B. G. Goodrich.
Size: 23 x 20 cm.

18 (4) 1859

Draft of Oliver Stevenson's land and mill pond in Mount Pleasant. May 1859.
Cartographer: B. G. Goodrich.
Size: 58 x 43 cm.

18 (5) 1859

Harvey D. Williams tract. Scott Township. 1 October 1859.
Cartographer: B. G. Goodrich.
Size: 31 x 20 cm.

18 (6) 1863

Bate Saw Mill. 28 February 1863.
Cartographer: B. G. Goodrich.
Size: 31.5 x 20 cm.

18 (7) n.d.
David C. Boyd road. n.d.
Cartographer: B. G. Goodrich.
Size: 31 x 19.5 cm.

18 (8) n.d.
John Mung's land. n.d.
Cartographer: B. G. Goodrich.
Size: 17 x 13 cm.

18 (9) n.d.
Joseph Bodie's home farm. Dyberry Township, Wayne Co., Pa. n.d.
Cartographer: B. G. Goodrich.
Size: 27.5 x 22 cm.

18 (10) n.d.
Land sold by D. W. Manning to Christian Hartung. n.d.
Cartographer: B. G. Goodrich.
Size: 32 x 20 cm.

18 (11) n.d.
Patterson Hartshorn land. n.d.
Cartographer: B. G. Goodrich.
Size: 43 x 35 cm.

(526.92: G62)

19. HARE-WILLING FAMILY. Papers.

Thomas Willing and Robert Hare were prominent Philadelphians of the eighteenth and nineteenth centuries. Willing was president of the Bank of North America and Hare was perhaps the most famous of all Philadelphia chemists.

19 (1) 1852
[Hare Street in Philadelphia, between the Schuylkill River and Darby Road]. By C. Hare. 21 June 1852.
Cartographer: C. Hare.
Size: 20.3 x 31.7 cm.

(Ms. Coll. #6, vol. 18)

19 (2) 1872
[Sault Ste. Marie]. By Horace Binney Hare. 15 July 1872.
Cartographer: Horace Binney Hare.

Size: 20.9 x 13.3 cm.

(Ms. Coll. #6)

20. HENRY, MATHEW SCHROPP. English-Lenni Lenape dictionary.

Henry was interested in the American Indian and corresponded with members of the Society on the subject. Also, he compiled an English-Lenni Lenape, Lenni Lenape-English dictionary. He drew the following maps, and carefully cited the original maps from which these were taken. He then inserted Indian place names, etc. on them.

20 (1) n.d.

[Chesapeake Bay area]. By Mathew S. Henry. n.d.
Cartographer: Mathew S. Henry.
Size: 37.5 x 47 cm.

20 (2) n.d.

[Delaware Bay]. By Mathew S. Henry. n.d.
Cartographer: Mathew S. Henry.
Size: 21.6 x 19.7 cm.

20 (3) n.d.

Delaware Bay and River. By Mathew S. Henry. n.d.
Cartographer: Mathew S. Henry.
Size: 47.6 x 21.6 cm.

20 (4) n.d.

[Delaware River]. By Mathew S. Henry. n.d.
Cartographer: Mathew S. Henry.
Size: 22.9 x 19.7 cm.

20 (5) n.d.

Discoveries and expeditions of Sir Walter Raleigh in America. By Mathew S. Henry. n.d.
Cartographer: Mathew S. Henry.
Size: 15.9 x 20.3 cm.

20 (6) n.d.

[Long Island, New York]. By Mathew S. Henry. n.d.
Cartographer: Mathew S. Henry.
Size: 25.4 x 40 cm.

20 (7) n.d.

[Lower Delaware River]. By Mathew S. Henry. n.d.

Cartographer: Mathew S. Henry.

Size: 19.7 x 19 cm.

(497.33: H39)

20 (8) n.d.

Northampton County [Pennsylvania], showing Lenni Lenape names. By Mathew S. Henry. n.d.

Cartographer: Mathew S. Henry.

Size: 68.6 x 55.9 cm.

(649.757: 1752: N86efd)

21. HORSFIELD, TIMOTHY. Papers.

Horsfield, a justice of the peace at Bethlehem, Pennsylvania, was active in fighting the Indians in 1756. He worked closely with Benjamin Franklin and the government at Philadelphia.

21 (1) 1742

[Delaware River between Philadelphia and New Jersey]. By Robert Longshow. July 1742.

Cartographer: Robert Longshow.

Size: 38.6 x 31.7 cm.

(974.8: H78, vol. 1, no. 5)

22. JEFFERSON, THOMAS. Papers.

Jefferson was the third president of the American Philosophical Society (1797-1814).

22 (1) 1790

[Land surrounding] R. Carter's Pat[ent] for 400 acres. By Thomas Jefferson. 17 October 1790.

Cartographer: Thomas Jefferson.

Size: 20.3 x 30.5 cm.

Annotation: "The line which divides Rob. Carter's 400 a[cre]s between Mr. [John] Wayles and Carter H. Harrison (who purchased the moiety) was run by Col. Carrington from Carter & Shelton's corner red oak to Willis' Creek S. 65 W. 272 po."

Jefferson sold his share of this land, which came to him through his wife, Martha Skelton, to Nicholas Lewis, and the indenture which conveyed the land is with the map.

(B: J35.7)

23. KANE, ELISHA KENT. Papers.

A massive collection of Kane family papers was presented to the Library by descendants in 1967. Part of this collection was the corpus of the papers of Elisha Kent Kane, who made two voyages searching for the lost English Arctic explorer, Sir John Franklin. Sir John was not found, but Kane became a great hero to the American public due to his part in these searches.

23 (1) [1853-55]

[Melville Bay and surrounding area]. By Elisha Kent Kane. [1853-55].
Cartographer: Elisha Kent Kane.
Size: 60.4 x 73 cm.

(172.3: 1853-1855: G895exp Large)

23 (2) [1853-1855]

[Base of Sylva Mountain, observatory and brig]. By Elisha Kent Kane. [1853-1855]. Draft.
Cartographer: Elisha Kent Kane.
Size: 44.8 x 63.4 cm.

(172.3: 1853-1855: G895exp Large)

23 (3) [1853-1855]

Sylva Mountain, position of brigg and observatory. By Elisha Kent Kane. [1853-1855].
Cartographer: Elisha Kent Kane.
Size: 18.3 x 20.8 cm.

(172.3: 1853-1855: G895exp Large)

23 (4) [1853-1855]

[Unidentified]. By Elisha Kent Kane. [1853-1855].
Cartographer: Elisha Kent Kane.
Size: 44.8 x 63.4 cm.

(172.3: 1853-1855: G895exp Large)

23 (5) n.d.

Bolivia and Paraguay. By Elisha Kent Kane. n.d.
Cartographer: Elisha Kent Kane.
Size: 21.6 x 22.2 cm.
Colored.

(B: K132)

23 (6) n.d.

Bon Secour Bay. By Elisha Kent Kane. n.d.
Cartographer: Elisha Kent Kane.
Size: 25.4 x 19.7 cm.

(B: K132)

23 (7) n.d.
British Isles. By Elisha Kent Kane. n.d.
Cartographer: Elisha Kent Kane.
Size: 19.7 x 22.9 cm.
Colored.
(B: K132)

23 (8) n.d.
Map of Baffin's Bay. By Elisha Kent Kane. n.d.
Cartographer: Elisha Kent Kane.
Size: 24.1 x 19.7 cm.
(B: K132)

23 (9) n.d.
[Northernmost penetration of the first Kane expedition]. By Elisha Kent Kane. n.d.
Cartographer: Elisha Kent Kane.
Size: 32.4 x 50.5 cm.
(B: K132)

23 (10) n.d.
Mobile Bay, Alabama. By Elisha Kent Kane. n.d.
Cartographer: Elisha Kent Kane.
Size: 25.4 x 19.7 cm.
(B: K132)

23 (11) n.d.
Mobile Bay, Alabama. By Elisha Kent Kane. n.d.
Cartographer: Elisha Kent Kane.
Size: 25.4 x 19.7 cm.
(B: K132)

23 (12) n.d.
[Six unidentified maps]. By Elisha Kent Kane. n.d.
Cartographer: Elisha Kent Kane.
Sizes: various sizes.
(B: K132)

24. LECONTE, JOHN LAWRENCE. Papers.

LeConte was a physician who never practiced medicine. He became a great American authority in the field of entomology.

24 (1) n.d.

[Unidentified surveyor's exercise]. n.d.

Size: 53.9 x 33 cm.

This map contains such notes as: "Beginning at a Small rock & Black Oak," "Iron bolt in a rock where a mark'd red oak stood opposite the old Mill," "marked hickory near the side of the creek," etc., with the "Turn Pike" and the stream clearly marked.

(B: L493)

25. LEE, RICHARD HENRY. Papers.

Lee, of the famous Lee family of Virginia, was a statesman, patriot, member of the Continental Congresses, and an avid worker for colonial rights. Lee was the mover of the resolution for independence for the United States and later was a signer of the Declaration of Independence.

25 (1) n.d.

[Fortifications of Boston Neck by the British]. By John Trumbull (?). n.d.

Cartographer: John Trumbull (?).

Size: 39.5 x 32.1 cm.

Provenance: Presented by Richard Henry Lee II, 1825.

(B: L51 Oversize)

26. LESLEY, J. PETER. Papers.

Lesley (1819-1903), a native Pennsylvanian and a graduate of the University of Pennsylvania, was an outstanding geologist. He worked on the First and Second Geological Surveys of Pennsylvania where he earned the approval of Henry Darwin Rogers. Although ordained a minister, he retired and undertook the supervision of the Second Pennsylvania Geological Survey. He wrote and published many papers on various scientific subjects and was an original member of the Academy of Natural Sciences of Philadelphia. His interest in the American Philosophical Society was deep; he worked actively in his roles of Secretary and Librarian of the Society.

There are ca. 300 maps in this collection, generally small, and in journals and field notebooks. Since they are so numerous and small (pertaining to a discrete aspect of a geological problem), they are not separately listed in this guide.

(B: L56)

27. LEWIS, MERIWETHER. Journal.

Lewis was the leader of the famous Lewis and Clark expedition which explored the Missouri River, across the Rocky Mountains, and the Columbia River to the Pacific Ocean. After President Jefferson appointed him leader of the expedition, Lewis travelled on the eastern seaboard to prepare for the trip. This journal of his trip down the Ohio River where he joined William Clark contains some sketches and two small maps.

27 (1) 1803

[Form of an island]. By Meriwether Lewis. 20 November 1803.

Cartographer: Meriwether Lewis.

Size: 20.3 x 12 cm.

27 (2) 1803

[Form of river islands]. By Meriwether Lewis. 21 November 1803.

Cartographer: Meriwether Lewis.

Size: 20.3 x 12 cm.

(917.3: L58p)

28. LEWIS, MERIWETHER, AND WILLIAM CLARK. Journals.

The most famous of the United States exploring expeditions was the one led by Meriwether Lewis and his associate, William Clark. Thomas Jefferson had long hoped to have an exploration made of the trans-Mississippi River West and projected this trip to go up the Missouri River, across the Rocky Mountains, to the Pacific Ocean. When the Louisiana Purchase was made the need for such a trip became even more important. The expedition left St. Louis in May 1804 and returned in September 1806. The importance of this well-executed and well-planned expedition was enormous: it opened vast territories to the United States and its influence on the West is incalculable.

The original journals of the expedition are in the Library.

28 (1) [1804-1805]

[Mississippi River and St. Peters River confluence. 1804-1805].

Cartographer: William Clark.

Size: 20.3 x 12 cm.

(917.3: L58 Codex C, p. 255)

28 (2) 1805

Draught of the Falls and Portage. [Missouri River, Great Falls]. July 1805.

Cartographer: Meriwether Lewis.

Size: 40.6 x 12 cm.

Colored.

(917.3: L58 Codex E, pp. 132-33)

28 (3) 1805

No. 1. Great Falls of River. October 1805.

Cartographer: William Clark.

Size: 20.3 x 12.7 cm.

Colored.

(917.3: L58 Codex H)

28 (4) 1805

[Columbia River]. 18 October 1805.

Cartographer: William Clark.

Size: 12 x 20.3 cm.

Colored.

(917.3: L58 Codex H, p. 33)

28 (5) 1805

No. 3 [Columbia River, Great Rapids]. October 1805.
Cartographer: William Clark.
Size: 20.3 x 12 cm.
Colored.

- (917.3: L58 Codex H, p. 4)

28 (6) 1805

No. 2. A Sketch of the Long and Short Narrows of the Columbia River. October 1805.
Cartographer: William Clark.
Size: 40.6 x 12 cm.
Colored.

(917.3: L58 Codex H, pp. 2-3)

28 (7) 1806

[Cape Disappointment, Washington]. January 1806.
Cartographer: William Clark.
Size: 20.3 x 12 cm.
Colored.

(917.3: L58 Codex I, p. 152)

28 (8) [1805]

Point Adams. [December 1805].
Cartographer: Meriwether Lewis.
Size: 16.5 x 9.5 cm.

(917.3: L58 Codex Ia, p. 7)

28 (9) 1806

[Mult-no-mah River and confluence]. 4 April 1806.
Cartographer: Meriwether Lewis.
Size: 20.3 x 12 cm.

(917.3: L58 Codex K, p. 28)

28 (10) [1806]

[Chopunnish Indian sketch map of Snake River. 1806].
Cartographer: William Clark.
Size: 40.6 x 12 cm.

(917.3: L58 Codex M, pp. 1-2)

28 (11) 1808

[Clark's River to the Missouri-Medicine confluence]. September 1808.
Cartographer: William Clark.
Size: 10.6 x 12 cm.

(917.3: L58 Codex N, pp. 149-150)

28 (12) n.d.

[Multonomah River (fragment)]. n.d.
Cartographer: William Clark.
Size: 35.6 x 9.5 cm.

(917.3: L58 Vol. VIII)

28 (13) [1808]

[Fort Osage (draft)]. [1808].
Cartographer: William Clark.
Size: 10.2 x 16.5 cm.

(917.3: L58c)

28 (14) [1806]

[River basin. 6-10 January 1806].
Cartographer: William Clark.
Size: 10.8 x 16.5 cm.

(917.3: L58.c1)

29. LINDSAY, JOHN, earl of Crawford. Military journals and papers.

These maps are in four volumes of the papers of British soldier John Lindsay, earl of Crawford and Lindsay (1702-1749). A Scotsman, he fought all over Europe. He was a captain in the British army and was permitted to join the Imperial army under Prince Eugene. In 1738 he received the rank of general from the Czarina Anna and fought against the Turks. He returned to the imperial forces and continued to fight until the Peace of London of 1748. Much of these manuscript volumes, other than the maps, were reproduced by Richard Rolt, *Memoirs of the life of the Right. Hon. John Lindsay, Earl of Crawford and Lindsay* (London: 1753) and an 1843 three-volume edition contains the same. Many of the maps are not reproduced in these volumes, due doubtlessly to the expense involved.

The maps were drawn by Henry Kopp, secretary and draughtsman for Lord Crawford. They are inscribed to George II, king of England; Lord Loudoun, etc. The volumes are titled: 1. Miscellaneous papers relating to the wars in Europe, 1681-1737; 2. Journal of a campaign with the Russian army against Turkey, 1739; 3. Journal of a voyage from the Thames to Russia, and of campaigning with the Russian army, 1738-1739.

The actual dates on which the maps were drawn is conjectural. These volumes were most probably prepared for Earl Crawford between 1743 and 1747 so they could be better seen and studied in the family's muniment room.

The volumes were once the property of Benjamin Franklin and were purchased at the sale of his library in 1803.

VOLUME 1. Account of some campaigns of the British Army from 1689 to 1712, and journal of a campaign under Prince Eugene on the Upper Rhine.

29 (1) [1743-47]

Plan & situation du Rhin et de ses environs jusque'au [sic] Coblenz. . . . [1743-47].

Cartographer: Henry Kopp.

Size: 175.9 x 85.7 cm.

Colored.

(940: M68)

29 (2) [1743-47]

Plan & situation du Camp . . . entre les Villages Sertzenich & Trierweiller. [1743-47].

Cartographer: Henry Kopp.

Size: 33.6 x 23.5 cm.

Colored.

(940: M68)

29 (3) [1743-47]

Plan & situation du Camp de l'Infant; Imper[ator]; et Auxil; près d'Ering. . . . [1743-47].

Cartographer: Henry Kopp.

Size: 23.5 x 31.1 cm.

Colored.

(940: M68)

29 (4) [1743-47]

Plan & situation du Camp de Bicon. [1743-47].

Cartographer: Henry Kopp.

Size: 35.6 x 25.4 cm.

Colored.

(940: M68)

29 (5) [1743-47]

Plan & situation du Camp des Imper[ator]: le long de la Rivier Salme entre les villages Ribnich & Salmerohr. [1743-47].

Cartographer: Henry Kopp.

Size: 34.3 x 24.1 cm., with insert: 19 x 17.1 cm.

Colored.

(940: M68)

29 (6) [1743-47]

[De Vauban's design for fortification. 1743-47].

Cartographer: Henry Kopp.

Size: 22.9 x 36.8 cm.

Colored.

(940: M68)

29 (7) [1743-47]
[Baron de Cochorn's design for fortification(?). 1743-47].
Cartographer: Henry Kopp.
Size: 22.9 x 36.8 cm.
Colored.
(940: M68)

29 (8) [1743-47]
[Imperial troops near Mayence. 1743-47].
Cartographer: Henry Kopp.
Size: 38.7 x 27.3 cm.
Colored.
(940: M68)

29 (9) [1743-47]
Plan & situation du Camp de Bingen. . . . [1743-47].
Cartographer: Henry Kopp.
Size: 39.4 x 26 cm.
Colored.
(940: M68)

29 (10) [1743-47]
Plan & situation du Camp des Imper[ator]: près de Bretzenheim sur La Nahe. [1743-47].
Cartographer: Henry Kopp.
Size: 35 x 24.8 cm.
Colored.
(940: M68)

29 (11) [1743-47]
Plan & situation du Camp des Imper[ator]: près de Durrenbach. [1743-47].
Cartographer: Henry Kopp.
Size: 35 x 23.5 cm.
Colored.
(940: M68)

29 (12) [1743-47]
Plan & situation du Camp des Camp des Allies près de Simmern. [1743-47].
Cartographer: Henry Kopp.
Size: 34.2 x 24.8 cm.
Colored.
(940: M68)

29 (13) [1743-47]
Plan & situation du Camp des Imper[ator]; près de Hirschfeld. [1743-47].
Cartographer: Henry Kopp.
Size: 35 x 22.9 cm.
Colored.
(940: M68)

29 (14) [1743-47]
Plan & situation du Camp près de Monsfeld. [1743-47].
Cartographer: Henry Kopp.
Size: 33 x 22.9 cm.
Colored.
(940: M68)

29 (15) [1743-47]
Plan & situation du Camp près de Muhlheim. [1743-47].
Cartographer: Henry Kopp.
Size: 35.6 x 25.4 cm.
Colored.
(940: M68)

VOLUME 2. [Journal of a campaign with the Russian army against Turkey, 1737-1739].

29 (16) [1743-47]
[Battle formation outside Pelegrad]. [1743-47].
Cartographer: Henry Kopp.
Size: 26.7 x 19.7 cm.
Colored.
(947: J82)

29 (17) [1743-47]
Plan von der Ordre de Bataille wie solche bey Krozka den 30 Juny 1737 . . . [1743-47].
Cartographer: Henry Kopp.
Size: 36.2 x 24.1 cm.
Colored.
(947: J82)

29 (18) [1743-47]
Lager bey Kollar. [1743-47].
Cartographer: Henry Kopp.
Size: 22.2 x 36.8 cm.
Colored.
(947: J82)

29 (19) [1743-47]

Lager zwischen Kirsna und Losanitz . . . [1743-47].
Cartographer: Henry Kopp.
Size: 22.2 x 36.8 cm.
Colored.

(947: J82)

29 (20) [1743-47]

Lager bey Lipoda. [1743-47].
Cartographer: Henry Kopp.
Size: 22.2 x 36.8 cm.
Colored.

(947: J82)

29 (21) [1743-47]

Lager bey Lapova. [1743-47].
Cartographer: Henry Kopp.
Size: 22.9 x 35.6 cm.
Colored.

(947: J82)

29 (22) [1743-47]

[Battle formation (?) outside Pataschin]. [1743-47].
Cartographer: Henry Kopp.
Size: 26.7 x 19.7 cm.

(947: J82)

29 (23) [1743-47]

[Battle formation outside Jagodina]. [1743-47].
Cartographer: Henry Kopp.
Size: 26.7 x 19.7 cm.

(947: J82)

29 (24) [1743-47]

[Encampment]. [1743-47].
Cartographer: Henry Kopp.
Size: 25.4 x 19.7 cm.

(947: J82)

29 (25) [1743-47]

Lager bey Schapeljack. [1743-47].
Cartographer: Henry Kopp.

Size: 22.2 x 36.8 cm.
Colored.

(947: J82)

29 (26) [1743-47]
[Morara River near Zworneck. 1743-47].
Cartographer: Henry Kopp.
Size: 19 x 26.7 cm.
Colored.

(947: J82)

29 (27) [1743-47]
Lager bey Wresina. [1743-47].
Cartographer: Henry Kopp.
Size: 23.5 x 36.8 cm.
Colored.

(947: J82)

29 (28) [1743-47]
Gros Vizier Berg oder Konigsfeld . . . [1743-47].
Cartographer: Henry Kopp.
Size: 25.4 x 19.7 cm.

(947: J82)

29 (29) [1743-47]
[Fort. 1743-47].
Cartographer: Henry Kopp.
Size: 26.7 x 20.9 cm.

(947: J82)

29 (30) [1743-47]
[Town of Nissa (?). 1743-47].
Cartographer: Henry Kopp.
Size: 23.5 x 36.2 cm.
Colored.

(947: J82)

29 (31) [1743-47]
[Road. 1743-47].
Cartographer: Henry Kopp.
Size: 36.8 x 47 cm.

(947: J82)

29 (32) [1743-47]
Lager bey Pesniza. [1743-47].
Cartographer: Henry Kopp.
Size: 22.9 x 36.2 cm.
Colored.
(947: J82)

29 (33) [1743-47]
Plan von der Palanka Krujowatz. [1743-47].
Cartographer: Henry Kopp.
Size: 22.9 x 36.8 cm.
(947: J82)

29 (34) [1743-47]
Lager bey Desternick. [1743-47].
Cartographer: Henry Kopp.
Size: 25.4 x 20.3 cm.
(947: J82)

29 (35) [1743-47]
Feld Lager bey Ozaizack . . . 1737. [1743-47].
Cartographer: Henry Kopp.
Size: 22.9 x 36.2 cm.
Colored.
(947: J82)

29 (36) [1743-47]
Feld Lager bey Lupit, 22 September 1737. [1743-47].
Cartographer: Henry Kopp.
Size: 27.3 x 19.7 cm.
Colored.
(947: J82)

29 (37) [1743-47]
Situation nebst der Palancka von Czazack. [1743-47].
Cartographer: Henry Kopp.
Size: 26.7 x 20.9 cm.
(947: J82)

29 (38) [1743-47]
Lager bey Sabatz. [1743-47].
Cartographer: Henry Kopp.

Size: 22.9 x 36.2 cm.
Colored.

(947: J82)

29 (39) [1743-47]

Plan der action zwischen Cornia and Carischa, 23 June - 4 July 1738. [1743-47].
Cartographer: Henry Kopp.
Size: 88.9 x 40.6 cm.
Colored.
Attached is: Plan von der Schantze zu Mehadia . . .

(947: J82)

VOLUME 3, PART 1. Journal of a voayage from the Thames to Russia and of campaign with the Russian army [spine title].

29 (40) [1743-47]

[Theatre of war between Russian Imperial army and the Turks and Tartars, 1736 and 1737 (Crimea)]. [1743-47].
Cartographer: Henry Kopp.
Size: 191.8 x 47.6 cm.
Colored, on printed map.

(947: L64, pt. 1)

29 (41) [1743-47]

Ordre de bataille de l'Armée Principate de sa Majesté Impériale de toutes les Russies. [1743-47].
Cartographer: Henry Kopp.
Size: 67.3 x 45.1 cm.
Colored.

(947: L64, pt. 1)

29 (42) [1743-47]

Plan of the operations of the Russian Army upon the banks of the Dneister . . . 1738. [1743-47].
Cartographer: Henry Kopp.
Size: 143.5 x 50.8 cm., with 8.9 cm.-high title piece.
Colored.

(947: L64, pt. 1)

29 (43) [1743-47]

[Town of Essek (?)]. [1743-47].
Cartographer: Henry Kopp.
Size: 72.4 x 52.1 cm.

Colored.

(947: L64, pt. 1)

29 (44) [1743-47]

Plan von Peterwardein. [1743-47].

Cartographer: Henry Kopp.

Size: 105.4 x 63.5 cm.

Colored.

(947: L64, pt. 1)

29 (45) [1743-47]

Order of battle of His Imperiale Majestys Army in Hungary commanded by . . . Konigseck. [1743-47].

Cartographer: Henry Kopp.

Size: 173.4 x 46.4 cm.

Colored.

(947: L64, pt. 2)

29 (46) [1743-47]

[Belgrade. 1743-47].

Cartographer: Henry Kopp.

Size: 52 x 35.6 cm.

Colored.

(947: L64, pt. 2)

29 (47) [1743-47]

Demolition de Panzowa (Pancevo). [1743-47].

Cartographer: Henry Kopp.

Size: 24.8 x 43.8 cm.

Colored.

(947: L64, pt. 2)

29 (48) [1743-47]

Plan van Temesewar. [1743-47].

Cartographer: Henry Kopp.

Size: 69.2 x 56.5 cm.

Colored.

(947: L64, pt. 2)

29 (49) [1743-47]

Plan und project von Temesewar. [1743-47].

Cartographer: Henry Kopp.

Size: 69.2 x 56.5 cm.
Colored.

(947: L64, pt. 2)

29 (50) [1743-47]
Plan von Banjalucka. . . . [1743-47].
Cartographer: Henry Kopp.
Size: 74.3 x 22.9 cm.
Colored.

(947: L64, pt. 2)

29 (51) [1743-47]
Plan von der Haupt Festung Commoren. [1743-47].
Cartographer: Henry Kopp.
Size: 94.6 x 64.1 cm.
Colored.

(947: L64, pt. 2)

29 (52) [1743-47]
Plan der Festung Ratsche. [1743-47].
Cartographer: Henry Kopp.
Size: 38 x 26 cm.
Colored.

(947: L64, pt. 2)

29 (53) [1743-47]
Plan der Kay Vestung und Insul Orsova. [1743-47].
Cartographer: Henry Kopp.
Size: 216.5 x 66.7 cm.
Colored.

(947: L64, pt. 2)

29 (54) [1743-47]
Plan de redoute der Insul Orsova gegenuber. [1743-47].
Cartographer: Henry Kopp.
Size: 64.8 x 47 cm.
Colored.

(947: L64, pt. 2)

29 (55) [1743-47]
Vestung Bender. [1743-47].
Cartographer: Henry Kopp.

Size: 58.4 x 45.7 cm.
Colored.
(947: L64, pt. 2)

29 (56) [1743-47]
Plan de la vielle et nouvelle fortification de la ville d'Oczakov. [1743-47].
Cartographer: Henry Kopp.
Size: 71.8 x 49.5 cm.
Colored.
(947: L64, pt. 2)

29 (57) [1743-47]
Project pour fortifier la ville de Belgrade. [1743-47].
Cartographer: Henry Kopp.
Size: 89.5 x 62.9 cm.
Colored.
(947: L64, pt. 2)

29 (58) [1743-47]
Projet de Monsieur le General de Wutquinaw pour fortifier Belgrade. [1743-47].
Cartographer: Henry Kopp.
Size: 45.1 x 29.8 cm.
Colored.
(947: L64, pt. 2)

29 (59) [1743-47]
Profil von Belgrade. [1743-47].
Cartographer: Henry Kopp.
Size: 72.4 x 50.2 cm.
Colored.
(947: L64, pt. 2)

29 (60) [1743-47]
Plan of the battle of Krotska in the year 1739. [1743-47].
Cartographer: Henry Kopp.
Size: 89.5 x 57.8 cm.
Colored.
(947: L64, pt. 2)

30. LYMAN, BENJAMIN SMITH. Papers.

Lyman studied geology and mining engineering in France and in Germany. He worked with J. Peter Lesley in the Second Geological Survey of Pennsylvania, and later worked on the

Iowa Geological Survey. Lyman was surveyor of the coal fields of Cape Breton Island and Nova Scotia and of the gold fields of California, and he worked on many other geological and topographical surveys in the United States. He worked for several years in Japan, having been appointed general geologist and mining engineer for the empire in 1873.

There are approximately forty maps in this huge collection, and most of them date from the time of Lyman's work in Japan. The few maps that can be identified pertain to the Tootoomi oilfield, the Ani mines, and the Magama area.

Provenance: Deposited by the Academy of Natural Sciences of Philadelphia, 1942.

(B: L982)

31. MALESHERBES, CHRETIAN GUILLAUME DE LAMOIGNON DE. *Voyage en Angleterre,* 3 April-27 May 1785.

Malesherbes was a French statesman and politician who was interested in agriculture and botany. In a diary kept during a tour of England in 1785 he wrote an eight-page description of the Blenheim Palace grounds, which includes this map.

31 (1) 1785

[Blenheim Palace grounds]. 30 April 1785.

Size: 24.8 x 18.4 cm.

(B: M291, p. 108)

32. INDIVIDUALLY CATALOGUED MANUSCRIPT MAPS (NOT IN COLLECTIONS)

Although most of the Library's manuscript maps are parts of collections, there are almost one hundred individual manuscript maps which were catalogued and filed with the Library's collection of printed maps.

32 (1) 1747

Plot of adjusted survey [of] Moravian road: Authorized: Court of Quarter-Sessions, New Town [Pennsylvania], June 11, 1747. By Thomas Craig, George Gray, and Robert Greeg. 1747. Photograph, July 1963.

Cartographers: Thomas Craig, George Gray, and Robert Greeg.

Size: 26 x 34.4 cm.

Provenance: Presented by John Robert Connelly, December 1964.

Annotated: "From the King's road near *Bethlehem* to the *Mahoning Creek* passing near the *Healing Waters* beyond the Blue Mountain."

See: John Robert Connelly, "Independent confirmation of the magnetic declination 1737-1738" (typescript, 2 pp., 1963).

(649: 1747: P376mor Small)

32 (2) 1752

Northampton County [Pennsylvania]. 1752.

Scale: 1 inch = 5 mi.

Size: 66 x 68 cm.

Note: This map is actually an 1854 copy of a 1752 map. Includes also the following counties: Lehigh, Schuylkill, Columbia, Carbon, Monroe, Pike, Luzerne, Wyoming, Bradford, Susquehanna, and Wayne. See: A. D. Chidsey, Jr., *The Penn patents in the forks of the Delaware.* Easton, Pa.: John S. Correll: 1937 (929.4: H39).

(649.757: 1752: N816efd Large)

32 (3) [1750s]

[Middle Atlantic states, showing rivers and mountains and location of sea shells on the tops of the mountains]. By John Bartram. n.d.

Cartographer: John Bartram.

Size: 31 x 38 cm.

Provenance: Presented by Benjamin Franklin.

As early as 1741, John Bartram sent some fossil sea shells to Sir Hans Sloan; other shells were sent to his London friend Peter Collinson in 1742. He wrote to Collinson in 1743 or 1744 that he had observed such fossils everywhere, "even on the top of the mountain that separates the waters of Susquehanna and St. Lawrence." (William Darlington, *Memorials of John Bartram and Humphry Marshall* [Philadelphia: 1849], p. 169.) Bartram was used to making rough maps of his travels and he made no pretense of being a competent surveyor. He apologized to Collinson for a map which he said was "Clumsily done, —having neither proper instruments nor convenient time," since he was drawing by the early light of dawn or by candlelight.

Franklin wrote his friend Jared Eliot on 16 July 1747 of Bartram's discoveries:

The great Apalachian Mountains, which run from York [Hudson] River back of these Colonies to the Bay of Mexico, show in many Places near the highest Parts of them, Strata Sea Shells, in some Places the marks of them are in the solid Rocks. 'Tis certainly the *Wreck* of a World we live on! We have Specimens of those Sea shell Rocks broken off near the Tops of those Mountains, brought and deposited in our Library [the Library Company of Philadelphia] as Curiosities. If you have not seen the like, I'll send you a Piece. *(The Papers of Benjamin Franklin* [New Haven: Yale University Press: 1961], vol. 3, p. 169.)

The endorsement on the back reads, in Franklin's hand: "Mr. Bartram's Map very curious."

(650: [ca.1750s]: At61mvc Small)

32 (4) [ca. 1765]

Pennsylvania. [ca. 1765]. Photostat in 4 pieces.

Size: 86.1 x 102.4 cm.

(649: [ca.1765]: P376pmj Small)

32 (5) [ca. 1772]

West Florida, part of E. Florida, Georgia, part of So. Carolina including . . . Chactaw, Chickasaw & Creek nations with [ ] Pensacola through ye Creek nation 16 August & Charles Stuart . . . superintendent of Indian affairs. [ca. 1772]. 30 sectional photostats.

Cartographers: Bernard Romans and David Taitt.

Size: 120 x 90 cm.

Provenance: From original in William L. Clements Library, 1940.

(636.5: [ca.1772]: St97fLa Book map)

32 (6) 1775

Sketch of the action on the heights of Charles Town 17 June 1775, between his magestys troops under the command of M. General Howe & a large body of American rebel's. 1775.

Cartographers: Henry de Berniere and H. Dearborn.

Size: 34.7 x 51.2 cm.

Colored.

Annotated: "The parts in red are corrections of the original by Maj. Gen. Dearborn."

The signature of J. V. N. Throop is on the map. Printed version is in the same folder.

(644: 1775: D325 chm Small)

32 (7) [ca. 1775-1777]

[Maps of Lake Superior and the Canadian Lakes. ca. 1775-1777]. Four maps on one folio sheet.

Size: 37 x 22.8 cm.

The sheet contains the following maps (No. 1 is not present):

"No. 2. From map of North America by Eman Brown, geog. to his majesty, 1775; in Jefferey's atlas of 1776."

"No. 3. Map of North America from d'Anville published by Sayer & others. London 1775; in Jefferey's atlas of 1776."

"No. 4. Map of the British empire by Samuel Dunn; published by Robert Sayer, 1774, London in Jefferey's atlas."

"No. 5. From a map of the British colonies in North America; engraved by William Faden, 1777."

(626.2: 1777: Su77mcp Small)

32 (8) 1777-1782

A collection of plans &c. &c. &c. in the province of New Jersey. By John Hills. 15 maps. Reproductions.

Cartographers: J. Hills, Grant, A. Sutherland, B. Morgan, A. Dennis, Rue, A. B. Dunhan, Taylor, Skinner, I. Williams, T. Milliadge and J. Fisher.

Sizes: Various.

Reprinted from the originals in the Library of Congress for the Portolan Press by the Meriden Gravure Company, 1976.

(648: 1777-1782: H55c Small)

32 (9) 1781

[Plan of Yorktown, Virginia, depicting the armies when Cornwallis surrendered]. 22-28 October 1781. By Sebastian Bauman. 1781. 3 pieces (preliminary drafts).

Cartographer: Sebastian Bauman.

Sizes of paper: 44.3 x 56.8 cm; 31.7 x 43.5 cm; 30.5 x 43.7 cm.

Colored.

Provenance: Presented by Richard Randolph, 7 October 1831.

The disposition of the troops and fleets are carefully delineated. This map was published in Philadelphia in 1782.

(654: 1781: B321ytv Small)

32 (10) 1784

Les deux parties française et espagnol de St. Domingue réduites en triangles, lieues marines carrées et carraux de cent pas sur cent d'après la carte gravée sur une plus grande échelle . . . Madrid. By Juan Lopez. 1784.

Scale: 50 Lieues marines de 2851½ toises chacune ou 4888 2/7 Pas.

Cartographer: Juan Lopez.

Size: 32.6 x 37.9 cm.

Colored.

Annotated: "Explication des base, perpendiculaire et surface de chaque triangle: Calcul des deux parties de St. Domingue et islots adjacents: Résultat en lieues marines carrées et en carrouse de cent pas carrés."

(724: 1784: L885fed Small)

32 (11) 1785

Mapa de las cerc de Mexico que comprehende todos sus lugares y rios, las lagunos de Tescuco, Chalco, Xochimiles, Ste. Christobal Zumpago . . . By Juan Lopez. 1785.

Scale: Leguas de une hora de camino a de 20 al grada.

Cartographer: Juan Lopez.

Size: 37.2 x 40 cm.

This map contains the names of the lagoons, rivers, mountains, towns, etc.

(707: 1785: L885mmr Small)

32 (12) [1786?]

[Parts of Glynn and McIntosh counties, Georgia, showing the Alatamaha and Mud Rivers, Buttermilk Sound and Hampton Creek. 1786?].

Size of paper: 38 x 48.1 cm.

(658.751: 1786?: G293arg Small)

32 (13) 1793

Maryland. Taken principally from Fry and Jefferson. 1793.

Scale: 69½ miles to a degree.

Cartographers: Fry and Jefferson.

Size: 32.6 x 41.9 cm.

This map contains far more settlements and better definition of streams than the 1751 map by Joshua Fry and Peter Jefferson. The inset shows Alleganey ridge.

(652: 1793: M365mfj Small)

32 (14) 1793

The [state of] New Jersey according to the best authorities. 1793.

Scale: 1 degree = 69½ British miles.

Size: 59 x 21.5 cm.

A note in the folder reads: "Nota Bene. This map should be used with discretion for the following reasons. . . . " See also information furnished by Lewis M. Haupt, 16 March 1917 (typescript with the maps).

(648: 1793: N466suj Large)

32 (15) 1795

Carta esférica de las costas de la Nueva Galicia, reconocida desde el surgidero de Mazatlan asta el Cava Corrientes. By José Maria Narvaez. 1795.

Cartographer: José Maria Narvaez.

Size: 50.8 x 35.5 cm.

Narvaez was "Primer Pilato de la Marina Nacional."

(706.2: 1795: N176ngm Small)

32 (16) [ca. 1795]

[Map showing parts of the intendencias of Vera Cruz, Yucatan, Chiapas, Oaxaca, etc., reaching from the Gulf of Mexico to the Gulf of Tehuantepec. ca. 1795].

Scale: 2 cm. = 5 leagues.

Size: 38.3 x 43.4 cm.

Annotated: "Los grados de Longitud y Latitud se han Pueste seguin el calculo de Baron de Homboldt."

(707: ca.1795: M575tco Small)

32 (17) [ca. 1795]

[Map showing the area between Guadalaxxara and Salamanca, Mexico. ca. 1795].

Size: 40 x 43.5 cm.

(706: [ca.1795]: M575gsm Small)

32 (18) [ca. 1795]

[Mexico: showing parts of the states from Vera Cruz to Toluca, with the bishoprics of Oaxaca to the south and the archbishopric of Mexico and the bishopric of Puebla to the west: also, "Partidos" of Antiqua, Zalape, Mizantla, Cordova, Cosamaluapan, Tehuacan, Haujapa, Tepixi de la Seda, Tepeaca, Orizava, Los Llanos, Tezuitlan, Telela y Xonotla and Zacatlan. ca. 1795].

Scale: Escala de [docc?] leguas.

Size: 44.6 x 63.5 cm.

(708: [ca.1795]: M575vms Large)

32 (18-a) [1795]

State of Massachusetts. Compiled from the best authorities. By Samuel Lewis. [1795].

Scale: 1 degree = 69½ miles.

Cartographer: Samuel Lewis.

Size: 37 x 49.4 cm.

Provenance: Presented by Mathew Carey, 18 October 1805.

See: M. Carey, *Carey's American atlas.* Philadelphia: Carey, 1796, no. 5.

The American Philosophical Society *Transactions* (vol. 6) state that Carey presented the "Materials from which Guthrie's Geography were compiled." This is one of those maps. See: William Guthrie, *The general atlas for Carey's edition of his geography improved.* (Philadelphia: Carey: 1795).

Wheat: 213

(644: [1795]: L585smc Large)

32 (19) [1795]

The state of Virginia from the best authorities. Engraved for Carey's edition of Guthrie's geography, improved. [Philadelphia: Carey: 1795].

Scale: 1 degree = 69½ American miles.

Cartographer: S. Lewis.

Size: 34.4 x 53.1 cm.

Provenance: Presented by Mathew Carey, 18 October 1805.

Adapted from the map of Virginia in the general atlas for *Carey's edition of Guthrie's geography improved* (Philadelphia: Carey, 1795). See also: *Notes on the state of Virginia,* by Thomas Jefferson. 3rd American edition. (New York: M.L. & W.A. Davis, for Furman & Loudon: 1801).

The American Philosophical Society *Transactions* (vol. 6) state that Mathew Carey presented the "Materials from which Guthrie's Geography was compiled," on 18 October 1805. This is part of that gift.

Wheat: 568

(654: [1795]: L585vcg Large)

32 (20) [ca. 1795]

The Waldo patent. Copied by Osgood Carleton. [ca. 1795].

Scale: 1 inch = 3 miles.

Cartographer: Osgood Carleton.

Size: 105.7 x 73.1 cm.

Colored.

(641: [ca.1795]: W149wpm Large)

32 (21) [1795]

Map of the Tennassee government, formerly part of North Carolina. Taken chiefly from surveys by Gen. D. Smith & others. [1795]. 2 pieces.

Scale: 1 inch = 14 miles.

Cartographer: David Smith.

Size: 33.7 x 83.7 cm.

Provenance: Presented by Mathew Carey, 18 October 1805.

Knoxville and Nashville are shown, as well as public roads, Indian boundaries, Indian towns, etc. The width of the river is in yards.

The American Philosophical Society *Transactions* (vol. 6) state that Mathew Carey presented the "Materials from which Guthrie's Geography was compiled," on 18 October 1805. This is part of that gift.

(662: [1795]: Sm5.3 Small)

32 (21a) [1795]

A map of the Tennasee government, formerly part of North Carolina, taken chiefly from surveys by Gen[era]l D. Smith & others. [1795].

Scale: 1 inch = 22 miles.

Cartographer: D. Smith.

Size: 24.4 x 53.5 cm.

Colored.

Provenance: Presented by Mathew Carey, 18 October 1805.

See also an engraving of the above by "J.T. Scott, Sculp" that was "engraved for Carey's American edition of Guthrie's geography improved." There are numerous manuscript notations on this printed map.

The American Philosophical Society *Transactions* (vol. 6) state that Mathew Carey presented the "Materials from which Guthrie's Geography was compiled," on 18 October 1805. This is part of that gift. See: William Guthrie, *The general atlas for Carey's edition of his geography improved.* (Philadelphia: Carey, 1795).

Wheat: 650

(662: [1795]: Sm5.1 Large)

32 (22) 1796

[Map of western Pennsylvania and eastern Ohio showing the locations of early Indian paths & towns of the Christian Indian missions . . . between 1772 & 1787 . . . by John G. E. Heckewelder in 1796.] Reproduced and copyrighted by the Western Reserve Historical Society, Cleveland, Ohio: 1968.

Cartographer: John G. E. Heckewelder.

Size: 33.4 x 59.7 cm.

Facsimile of an original manuscript.

(649: 1796: H365paoh Large)

32 (23) [ca. 1796]

The state of Pennsylvania. Reduced (with permission) from Reading Howell's large map. By Samuel Lewis. [ca. 1796].

Scale: 1 degree = 69½ American miles.

Cartographers: Reading Howell and Samuel Lewis.

Size: 29.6 x 47.1 cm.

Provenance: Presented by Mathew Carey, 18 October 1805.

The American Philosophical Society *Transactions* (vol. 6)state that Mathew Carey presented the "Materials from which Guthrie's Geography was compiled," on 18 October 1805. This is part of that gift.

Wheat: 448

(649: [ca.1796]: L585rhm Large)

32 (24) [1796]

The state of Rhode Island; compiled from the surveys and observations of Caleb Harris. By Harding Harris. [1796].

Scale: 1 inch = 4½ miles.

Cartographers: Caleb Harris and Harding Harris.

Size: 35.4 x 24.3 cm.

Provenance: Presented by Mathew Carey, 18 October 1805.

The American Philosophical Society *Transactions* (vol. 6) state that Mathew Carey presented the "Materials from which Guthrie's Geography was compiled," on 18 October 1805. This is part of that gift.

Wheat: 250

(645: [1796]: H243ri Small)

32 (25) [1800]

Carte de la partie Française de l'isle de St. Domingue. [1800].

Scale: 5 cm. = 10 toises.

Cartographer: G. J. Bois St. Lys.

Size: 44 x 55.2 cm.

Colored.

This map was acquired with the purchase of a large collection of maps about St. Domingue and other Gulf of Mexico areas from Henry Schwenk Tanner on 14 December 1835.

Information about G. J. Bois St. Lys and the significance of these maps may be found in James E. McClellan, *Colonialism and science in the old regime: the case of Saint Domingue* (forthcoming).

(725: [1800]: Sa27fsd Small)

32 (26) 1800

Carte de l'isle de Cuba, comprenant les jurisdictions de Philippine, de la Havane, des quatre Bourgs, de la ville de St. Marie du Port au Prince, de Bayamo et de St. Yago do Cuba, desinée d'après celle de Dn Juan Lopez, . . . par G. J. Bois St. Lys. 1800.

Scale: 20 lieues marines au dégré.

Cartographers: Juan Lopez and G. J. Bois St. Lys.

Size: 49.2 x 89.6 cm.

Colored.

This map was acquired with the purchase of a large collection of maps about St. Domingue and other Gulf of Mexico areas from Henry Schwenk Tanner on 14 December 1835.

(723: 1800: S27phs Large)

32 (27) 1800

Carte de l'isle de la Gonave et de la côte depuis la baye de l'Artibonite jusqu'à celle des Baradaires dans l'isle de St. Domingue. By G. Bois St. Lys. 1800.

Cartographer: G. Bois St. Lys.

Size: 57.7 x 88.9 cm.

Colored.

This map was acquired with the purchase of a large collection of maps about Saint Domingue and other Gulf of Mexico areas from Henry Schwenk Tanner on 14 December 1835.

(725: 1800: Sa27arf Large)

32 (28) 1800

Carte de l'isle de Puerto Rico. Réduite d'après celle de Dn. Thos. Lopez . . . par Georges Jh. Bois St. Lys. 1800.

Scale: Echelle de dix lieues Marines, de vingt au dégré.

Cartographers: Don Thomas Lopez and Georges Jh. Bois St. Lys.

Size of paper: 54.3 x 100 cm.

Colored.

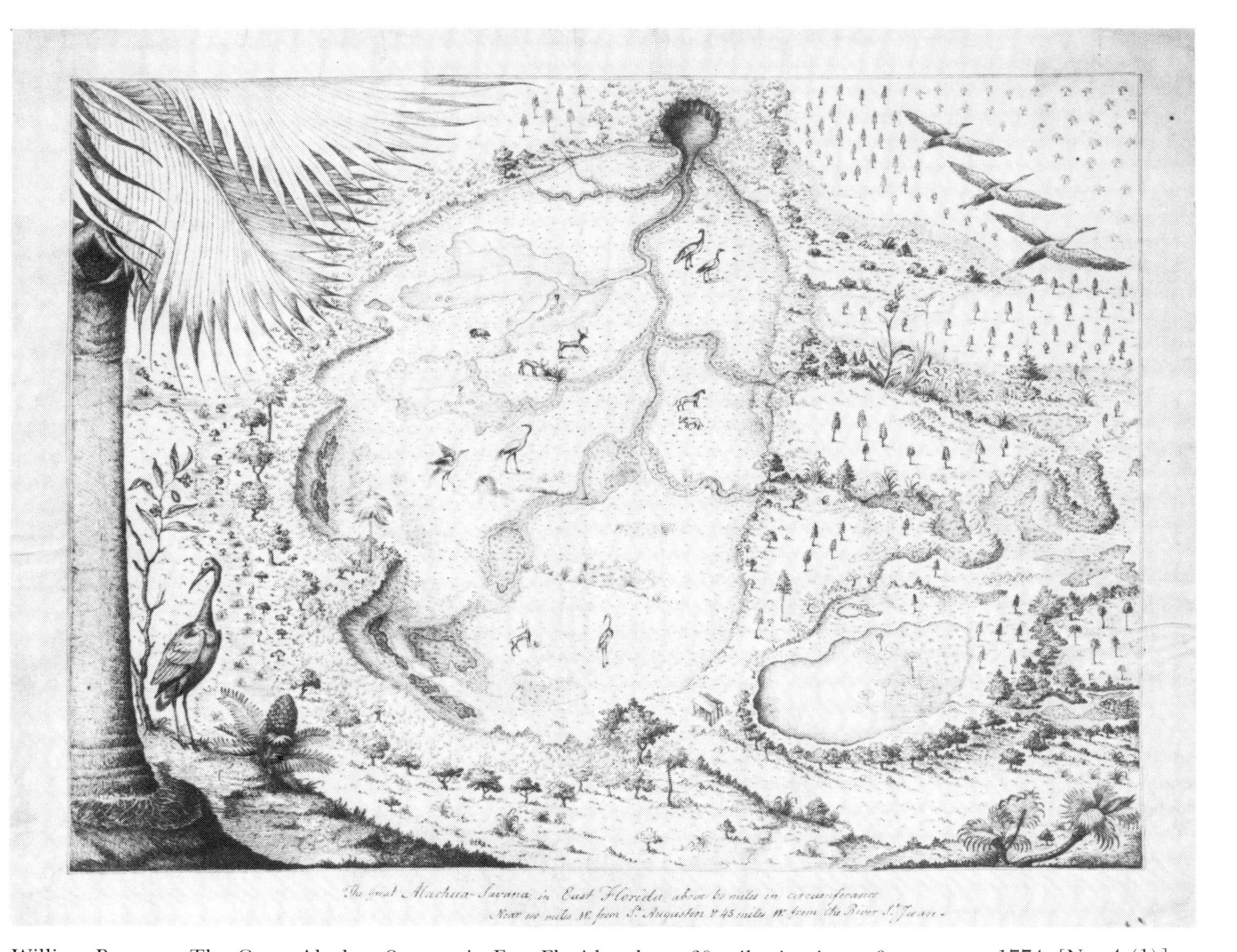
The great Alachua-Savana in East Florida above 60 miles in circumferance
Near 100 miles W. from St. Augustine & 45 miles W. from the River St. Juan.

William Bartram. The Great Alachua-Savana, in East Florida, above 60 miles in circumference. ca. 1774. [No. 4 (1)]

G.J. de Bois St. Lys. Carte de la partie Française de l'isle de St. Domingue. 1800. [No. 32 (25)]

This map was acquired with the purchase of a large collection of maps about St. Domingue and other Gulf of Mexico areas from Henry Schwenk Tanner on 14 December 1835.

(728: 1800: Sa27cd Large)

32 (29) 1800

Carte de l'isle de St. Domingue. Par G. J. Bois St. Lys. 1800.

Scale: Lieues marines de vingt au dégré.

Cartographer: G. J. Bois St. Lys.

Size: 49.6 x 90.3 cm.

Colored.

This map was acquired with the purchase of a large collection of maps about St. Domingue and other Gulf of Mexico areas from Henry Schwenk Tanner on 14 December 1835.

(724: 1800: S27std Large)

32 (29-a) [ca. 1800]

[Chart of the coast of southern Florida, the Bahamas and Cuba]. [ca. 1800].

Cartographer: J. J. de Ferrer.

Size: 44.2 x 66.3 cm.

Colored.

(659: [ca. 1800]: F373cfc Large)

32 (30) [ca. 1800]

A map of part of Onondaga Lake, and its vicinity, in the state of New York. From actual survey. [ca. 1800].

Scale: 1 inch = 765⅙ yards.

Size: 27.5 x 40.5 cm.

Colored.

(647: [ca. 1800]: On66ol Small)

32 (31) [ca. 1800]

[Pennsylvania. ca. 1800].

Scale: 1 inch = 15 miles.

Size: 30.6 x 51.8 cm.

Annotated: "Explanation: County towns, villages, forges & mills, houses, roads & portages, roads to be opened & improved on the straightest line practicable."

Northumberland and Allegany Counties occupy a disproportionately large area of the state.

(649: [ca. 1800]: P376pen Large)

32 (32) 1800

Plan de la baye de Dame Marie dans l'isle de St. Domingue. Fait et réduit par G. J. Bois St. Lys . . . pour M. E. Stevens. 1800.

Scale: 2.9 cm. = 100 toises.

Cartographer: G. J. Bois St. Lys.

Size of paper: 53.3 x 64.7 cm.

Colored.

This map was acquired with the purchase of a large collection of maps about St. Domingue and other Gulf of Mexico areas from Henry Schwenk Tanner on 14 December 1835.

(725: 1800: Sa27dmd Large)

32 (33) 1800

Plan de la baye de l'Acul dans la partie du nord de l'isle de Saint Domingue. Faite et réduit par G. J. Bois St. Lys . . . pour Mr. E. Stevens, Consul G[ener]al. 1800.

Scale: 2.4 cm. = 300 toises.

Cartographer: G. J. Bois St. Lys.

Size: 53.3 x 63.9 cm.

Colored.

This map was acquired with the purchase of a large collection of maps about St. Domingue and other Gulf of Mexico areas from Henry Schwenk Tanner on 14 December 1835.

(725: 1800: Sa27acu Large)

32 (34) 1800

Plan de la baye de Samana dans l'isle de St. Domingue. Fait par A. P. Advenier de Breuilly. 1800.

Scale: Cinq lieues de trois milles chacune.

Cartographer: A. P. Advenier de Breuilly.

Size of paper: 55.3 x 99 cm.

Colored.

This map was acquired with the purchase of a large collection of maps about St. Domingue and other Gulf of Mexico areas from Henry Schwenk Tanner on 14 December 1835.

(725: 1800: Ad11brd Large)

32 (35) 1800

Plan de la baye des Irois dans l'isle de St. Domingue. Fait et réduit par G. J. Bois St. Lys . . . pour M. E. Stevens. 1800.

Scale: 2 cm. = 100 toises.

Cartographer: G. J. Bois St. Lys.

Size of paper: 53.5 x 64.5 cm.

Colored.

This map was acquired with the purchase of a large collection of maps about St. Domingue and other Gulf of Mexico areas from Henry Schwenk Tanner on 14 December 1835.

(725: 1800: Sa27iro Large)

32 (36) 1800

Plan de la baye d'Ocoa dans l'isle de St. Domingue. Fait par Advenier de Breuilly. 1800.

Scale: 1.9 cm. = 400 toises.

Cartographer: A. P. Advenier de Breuilly.

Size of paper: 59.4 x 95.3 cm.

Colored.

This map was acquired with the purchase of a large collection of maps about St. Domingue and other Gulf of Mexico areas from Henry Schwenk Tanner on 14 December 1835.

(725: 1800: Ad11sdo Large)

32 (37) [ca. 1800]

Plan de la baye et du bourg de Tiburon dans l'isle de St. Domingue. By Georges Bois St. Lys. [ca. 1800].

Scale: 7 cm. = 500 toises.

Cartographer: Georges Bois St. Lys.

Size: 39.7 x 56.5 cm.

Colored.

This map was acquired with the purchase of a large collection of maps about St. Domingue and other Gulf of Mexico areas from Henry Schwenk Tanner on 14 December 1835.

(725: [ca.1800]: Sa27bbt Small)

32 (38) 1800

Plan de la baye, ville et forts de St. Yago dans l'isle de Cuba. Fait et réduit par G. J. Bois St. Lys. 1800.

Scale: 2.2 cm. = 200 toises.

Cartographer: G. J. Bois St. Lys.

Size: 48.1 x 63.4 cm.

Colored.

This map was acquired with the purchase of a large collection of maps about St. Domingue and other Gulf of Mexico areas from Henry Schwenk Tanner on 14 December 1835.

(723: 1800: S27dsy Large)

32 (39) [1800]

Plan de la ville des Cayes dans l'isle de St. Domingue. [1800].

Scale: 1.3 cm. = 50 toises.

Cartographer: G. J. Bois St. Lys.

Size: 41.7 x 50.1 cm.

Colored.

This map was acquired with the purchase of a large collection of maps about St. Domingue and other Gulf of Mexico areas from Henry Schwenk Tanner on 14 December 1835.

(725: [1800]: Sa27pvc Large)

32 (40) 1800

Plan de la ville de Jeremie et de ses environs. [1800].

Scale: 2.8 cm. = 50 toises.

Cartographer: G. J. Bois St. Lys.

Size: 46 x 47.3 cm.

Colored.

This map was acquired with the purchase of a large collection of maps about St. Domingue and other Gulf of Mexico areas from Henry Schwenk Tanner on 14 December 1835.

(725: [1800]: Sa27pvj Large)

32 (41) [ca. 1800]

Plan de la ville de St. Marc dans l'isle de St. Domingue. [ca. 1800].

Scale: 2.6 cm. = 560 toises.

Cartographer: Georges Bois St. Lys.

Size: 43.7 x 60.1 cm.

Colored.

This map was acquired with the purchase of a large collection of maps about St. Domingue and other Gulf of Mexico areas from Henry Schwenk Tanner on 14 December 1835.

(725: [ca.1800]: Sa27pvm Small)

32 (42) 1800

Plan de la ville de Santo Domingo. Réduit d'après celui de Dn. Lopez par G. J. Bois de St. Lys. 1800.

Scale: 1.8 cm. = 100 toises.

Cartographers: Don Lopez and G. J. Bois St. Lys.

Size: 55.5 x 71.6 cm.

Colored.

This map was acquired with the purchase of a large collection of maps about St. Domingue and other Gulf of Mexico areas from Henry Schwenk Tanner on 14 December 1835.

(726.971: 1800: Sa27pvs Large)

32 (43) [ca. 1800]

Plan de la ville, des rades et des environs du Port au Prince dans l'isle de St. Domingue. [ca. 1800].

Scale: 5.5 cm. = 200 toises.

Cartographer: Georges Bois St. Lys.

Size: 44.4 x 61.5 cm.

Colored.

This map was acquired with the purchase of a large collection of maps about St. Domingue and other Gulf of Mexico areas from Henry Schwenk Tanner on 14 December 1835.

(725: [ca.1800]: Sa27ppd Small)

32 (44) [1800]

Plan de la ville du Cap Français dans l'isle de St. Domingue. [1800].

Scale: 1.8 cm. = 50 toises.

Cartographer: G. J. Bois St. Lys.

Size: 48.8 x 62 cm.

Colored.

This map was acquired with the purchase of a large collection of maps about St. Domingue and other Gulf of Mexico areas from Henry Schwenk Tanner on 14 December 1835.

(725: [1800]: Sa27cfid Large)

32 (45) [1800]

Plan de la ville et de la baye du Fort Dauphin dans l'isle St. Domingue. By Georges Bois St. Lys. [1800].

Scale: 1 cm. = 100 toises.

Cartographer: Georges Bois St. Lys.

Size: 41.5 x 60.3 cm.

Colored.

This map was acquired with the purchase of a large collection of maps about St. Domingue and other Gulf of Mexico areas from Henry Schwenk Tanner on 14 December 1835.

(725: [1800]: Sa27fds Large)

32 (46) [1800]

Plan de la ville et de la rade du Petit Goave dans l'isle de St. Domingue. [1800].

Scale: 2.1 cm. = 100 toises.

Cartographer: Bois St. Lys.

Size: 39.5 x 59 cm.

Colored.

This map was acquired with the purchase of a large collection of maps about St. Domingue and other Gulf of Mexico areas from Henry Schwenk Tanner on 14 December 1835.

(725: [1800]: S27rpg Large)

32 (47) 1800

Plan de la ville et des environs de Jacmel, dans l'isle de Saint Domingue assiégée par le Général en chef Toussaint l'Overture et defendue par le G[énér]al André Rigaud revolté en 1800. Avec la position des troupes qui la bloquent et des batteries qui ont été dressées contre elle. Dessiné pour Georges Bois St. Lys. 1800.

Scale: 1.9 cm. = 200 toises.

Cartographer: Georges Bois St. Lys.

Size: 55 x 79.5 cm.

Colored.

This map was acquired with the purchase of a large collection of maps about St. Domingue and other Gulf of Mexico areas from Henry Schwenk Tanner on 14 December 1835.

(725: 1800: Sa27jdt Large)

32 (48) 1800

Plan de la ville & de port de Samana dans l'isle de St. Domingue. Fait par A. P. Advenier de Breuilly. 1800.

Scale: 1.7 cm. = 100 toises.

Cartographer: A. P. Advenier de Breuilly.

Size: 41.1 x 60.6 cm.

Colored.

This map was acquired with the purchase of a large collection of maps about St. Domingue and other Gulf of Mexico areas from Henry Schwenk Tanner on 14 December 1835.

(725: 1800: Ad11psd Large)

32 (49) 1800

Plan de l'entrée de la baye de St. Yago dans l'isle de Cuba. Fait et réduit par G. J. Bois de St. Lys. 1800.

Scale: 2.3 cm. = 50 toises.

Cartographer: G. J. Bois St. Lys.

Size: 49.8 x 63 cm.

Colored.

This map was acquired with the purchase of a large collection of maps about St. Domingue and other Gulf of Mexico areas from Henry Schwenk Tanner on 14 December 1835.

(723: 1800: S27bsy Large)

32 (50) [1800]

Plan du bourg de La Croix des Bouquets dans l'isle de St. Domingue. [1800].

Scale: 1.8 cm. = 200 toises.

Cartographer: G. J. Bois St. Lys.

Size: 41 x 87 cm.

Detail: Le Fort la Constitution servant de citadelle à la Croix de Bouquets.

Colored.

This map was acquired with the purchase of a large collection of maps about St. Domingue and other Gulf of Mexico areas from Henry Schwenk Tanner on 14 December 1835.

(725: [1800]: Sa27cdb Large)

32 (51) 1800

Plan du Port Français dans la partie du nord de Saint Domingue. Réduit et fait par G. J. Bois St. Lys pour Mr. E. Stevens. 1800.

Scale: 3.4 cm. = 50 toises.

Cartographer: G. J. Bois St. Lys.

Size of paper: 53.3 x 64.9 cm.

Colored.

This map was acquired with the purchase of a large collection of maps about St. Domingue and other Gulf of Mexico areas from Henry Schwenk Tanner on 14 December 1835.

(725: 1800: Sa25fra Large)

32 (52) 1800

Plan d'une partie de la Plaine du Cap Français et de la côte, depuis la baye de l'Acul, jusques et après le bourg de la Petite Anse, dans l'isle de St. Domingue. Fait par Georges Bois St. Lys. 1800.

Scale: 2.5 cm. = 400 toises.

Cartographer: Georges Bois St. Lys.

Size of paper: 57.3 x 87 cm.

Colored.

This map was acquired with the purchase of a large collection of maps about St. Domingue and other Gulf of Mexico areas from Henry Schwenk Tanner on 14 December 1835.

(725: 1800: Sa27cf Large)

32 (53) 1800

Plan d'une partie de la plaine du fond de l'isle à Vache et de la côte, depuis de Grande Baye du Mesle, jusqu'à la pointe de l'Abacou, dans l'isle de St. Domingue. Fait par A. P. Advenier de Breuilly. 1800.

Scale: Une lieue marine de 2833 toises.

Cartographer: A. P. Advenier de Breuilly.

Size: 53.4 x 87.9 cm.

Colored.

This map was acquired with the purchase of a large collection of maps about St. Domingue and other Gulf of Mexico areas from Henry Schwenk Tanner on 14 December 1835.

(725: 1800: Ad11pLv Large)

32 (54) [ca. 1800]

Survey of the lands adjoining Monticello in Albemarle county, belonging to Nicholas Meriwether and Abraham Lewis, and later to Meriwether Lewis. By T. Jefferson. [ca. 1800]. Photograph.

Cartographer: Thomas Jefferson.

Size: 39.1 x 29 cm.

Provenance: From original manuscript in the Library of Congress.

(654: [ca.1800]: J354acv Small)

32 (55) 1801

Carte réduite des débouquemens de St. Domingue desinée d'après celle levée en 1785, par Mr. le Cte. de Chastenet Puysegur . . . par Georges Bois St. Lys. 1801.

Cartographers: Le comte de Chastenet Puysegur and Georges Bois St. Lys.

Size of paper: 58 x 97.7 cm.

Colored.

This map was acquired with the purchase of a large collection of maps about St. Domingue and other Gulf of Mexico areas from Henry Schwenk Tanner on 14 December 1835.

(725: 1801: Sa27cpb Large)

32 (56) 1801

Plan de la baye des Gonaives dans l'isle de St. Domingue. Par Georges Bois St. Lys. 1801.

Scale: 1.8 cm. = 30 toises.

Cartographer: Georges Bois St. Lys.

Size: 56.3 x 69.8 cm.

Colored.

This map was acquired with the purchase of a large collection of maps about St. Domingue and other Gulf of Mexico areas from Henry Schwenk Tanner on 14 December 1835.

(725: 1801: Sa27bg Large)

32 (57) 1801

Plan de la baye du Fond la Grange dans l'isle de St. Domingue. By Georges Bois St. Lys. 1801.

Scale: 3.6 cm. = 100 toises.

Cartographer: Georges Bois St. Lys.

Size: 46.1 x 62.4 cm.

Colored.

This map was acquired with the purchase of a large collection of maps about St. Domingue and other Gulf of Mexico areas from Henry Schwenk Tanner on 14 December 1835.

(725: 1801: Sa27fLg Large)

32 (58) 1801

Plan de la baye du Moustique dans la partie nord de l'isle de St. Domingue. By G. J. Bois St. Lys. 1801.

Scale: 1.9 cm. = 50 toises.

Cartographer: G. J. Bois St. Lys.

Size: 43.7 x 62.5 cm.

Colored.

This map was acquired with the purchase of a large collection of maps about St. Domingue and other Gulf of Mexico areas from Henry Schwenk Tanner on 14 December 1835.

(725: 1801: Sa27mou Large)

32 (59) 1801

Plan de la baye du Port à l'Ecu dans l'isle de St. Domingue. By G. Bois St. Lys. 1801.

Scale: 2.6 cm. = 50 toises.

Cartographer: G. Bois St. Lys.

Size: 50.9 x 61.8 cm.

Colored.

This map was acquired with the purchase of a large collection of maps about St. Domingue and other Gulf of Mexico areas from Henry Schwenk Tanner on 14 December 1835.

(725: 1801: Sa27pbe Large)

32 (60) 1801

Plan de la ville de Jacmel dans l'isle de St. Domingue. By G. J. Bois St. Lys. 1801.

Scale: 2.8 cm. = 50 toises.

Cartographer: G. J. Bois St. Lys.

Size: 49.8 x 64.9 cm.

Colored.

This map was acquired with the purchase of a large collection of maps about St. Domingue and other Gulf of Mexico areas from Henry Schwenk Tanner on 14 December 1835.

(725: 1801: Sa27uj Large)

32 (61) 1801

Plan de la ville et de la baye de St. Marc dans l'isle de St. Domingue. By G. J. Bois St. Lys. 1801.

Scale: 3.2 cm. = 300 toises.

Cartographer: G. J. Bois St. Lys.

Size: 49.5 x 70.7 cm.

Colored.

This map was acquired with the purchase of a large collection of maps about St. Domingue and other Gulf of Mexico areas from Henry Schwenk Tanner on 14 December 1835.

(725: 1801: Sa27vsm Large)

32 (62) 1801

Plan de l'anse à Chouchou dans la partie nord de l'isle de St. Domingue. By G. J. Bois St. Lys. 1801.

Scale: 1.7 cm. = 50 toises.

Cartographer: G. Bois St. Lys.

Size of paper: 43.5 x 62 cm.

Colored.

This map was acquired with the purchase of a large collection of maps about St. Domingue and other Gulf of Mexico areas from Henry Schwenk Tanner on 14 December 1835.

(725: 1801: Sa27std Large)

32 (63) 1801

Plan de Saint Jean de Puerto Rico. Réduit sur celui de Dn. Thos. Lopez par Georges Joseph Bois St. Lys. 1801.

Scale: 1.7 cm. = 100 toises.

Cartographers: George Joseph Bois St. Lys. and Don Thomas Lopez.

Size: 51.2 x 71.3 cm.

Colored.

This map was acquired with the purchase of a large collection of maps about St. Domingue and other Gulf of Mexico areas from Henry Schwenk Tanner on 14 December 1835.

(728: 1801: Sa27jpr Large)

32 (64) 1801

Plan du mouillage et rade de la Basse Terre dans l'isle de la Tortue. By Georges Bois St. Lys. 1801.

Scale: 2.3 cm. = 50 toises.

Cartographer: G. Bois St. Lys.

Size of paper: 49.5 x 62.1 cm.

Colored.

This map was acquired with the purchase of a large collection of maps about St. Domingue and other Gulf of Mexico areas from Henry Schwenk Tanner on 14 December 1835.

(725: 1801: Sa27pm Large)

32 (65) 1801

Special map, representing Man of War Shoal, its distance and bearings from the island of St. Martins, conformable to the discovery of Captain Foulke, Esqr., commanding the British frigate, the Proselite, in the month of May 1801. By Samuel Fahlberg. 1801.

Cartographer: Samuel Fahlberg.

Size: 20.9 x 27.3 cm.

Provenance: Presented by S. Fahlberg, 16 April 1802.

Nicholas Collin delivered this map to the Society on 16 April 1802. Collin wrote that Fahlberg "presentes by me a special map of a lately discovered dangerous shoal . . . as an appendage to his map of it [St. Bartholomew], sometimes ago received by the Society; and expresses a solicitious desire that it may be published for the benefit of the many American vessels that trade to that part of the West Indies: believeing this execution the more necessary, as the very frigate in which the discovery was made, though then commanded by another captain, ran aground on that shoal." The members desired Secretary John Vaughan to write and see "if any account can be published" of this "Rock lately discovered south of S. Martins."

(733.1: 1800: F143ssm Small)

32 (66) 1802

A sketch of the Muscle Shoals of the Tennessee River. Laid down from a Scale of two Computed Miles to the Inch. The Width of the River being doubled. By J. W[ilkinson]. 1802.

Cartographer: James Wilkinson.

Size: 58.6 x 45 cm.

Provenance: Presented by Samuel Brown, April 1802.

Colored.

Annotated: "N.B. The red dots mark the Canoe Track by which I descended. The black, note the deepest channel."

John Vaughan, secretary of the American Philosophical Society, was asked by the Members to get "some further account of it for publication" from the donor. Brown replied on 10 June 1802 that the map "of the Muscle Shoals which you did me the favor to present to the Society, was taken under the direction of Gen[era]l Wilkinson who transmitted it to a friend together with a short description which unluckily was attached to a private confidential letter, from which his correspondent would not even suffer me to make an extract. Gen[era]l Wilkinson will no doubt supply the deficiency on his arrival at Philadelphia."

(663: [1802]: M975sms Small)

32 (67) 1807

[Pennsylvania and neighboring states]. By Frederick Pursh. 1807.

Cartographer: Frederick Pursh.

Size: 36 x 50 cm.

Provenance: Presented by Asa Gray, 3 April 1868.

Colored.

The endorsement on the back reads: "Found by me in a parcel of Plants collected by F. Pursh w[hi]ch made part of ye Lambert Herbarium, & was bought by me in London at the Lambert sale in 1842. Edward Tuckerman."

In 1807 Pursh kept a "Journal of a botanical excursion in the Northeastern parts of Pennsylvania & in the state of New York." The manuscript was found among the papers of his patron, Benjamin Smith Barton, in 1817. It has been published twice: first in Philadelphia in 1869, and then in 1923 for the Onondaga Historical Association, Syracuse, New York, with notes by William M. Beauchamp.

On 3 April 1868 Thomas P. James spoke of various properties of Pursh. He told how the map was drawn and used in the botanical expedition, and he spoke of the history of the map. Asa Gray, the famous botanist, owned it, and wished to present it to the American Philosophical Society and reunite it with the journal. "The Journal and map are now, after a separation of sixty years, united."

(640.2: [1807]: P376ns Small)

32 (68) 1814

Map of the country nine miles west of the city of Philadelphia, and between Darby Creek & Young's ford on the River Schuylkill. Surveyed by order of Gen. Jonathan Williams . . . of the sub-committee of defence. By William Strickland. Philadelphia: 28 September 1814.

Scale: 8 inches = 1 mile.

Cartographers: William Strickland, Robert Brooke, and William Kneass.

Provenance: Presented by Jonathan Williams, 3 February 1815.

Size: 214 x 180 cm.

(649.762: 1814: S87dcyf Large)

32 (69) [ca. 1815]

Plan hidrografico de la laguna de Maracaybo. [ca. 1815].

Scale: 25 Millas Colombianos de 60 al grado.

Provenance: Presented by John Vaughan, 17 July 1815.

Size: 61.6 x 40.6 cm.

(753: [ca.1815]: M335Lam Large)

32 (70) 1821

Plano del puerto de Guaimas, situado en la costa de Sonora [Mexico] . . . levantado de orden del gobierno en 1819 . . . p[a]r defensa del puerto y puntos adjacentes à la costa . . . By Pedro Celestino Negrete. 1821.

Scale: Escale de une milla maritima.

Cartographer: Pedro Celestino Negrete.

Size: 41.9 x 45.2 cm.

Colored.

The bay contains soundings.

(705.3: 1821: N316msg Small)

32 (71) 1823

[Plano] de [Orizava] y Xalapa, en la parte que media des la sierra a la costa . . . By Don Diego Garcias. 1823.

Scale: Escala de neuve legs. del reyno de n.e. de 5000 vars. cadavna.

Cartographer: Diego Garcias.

Size: 35.9 x 53 cm.

Annotated: "Copiado por Santiago Wilkinson en Mexico."

Projected route from Vera Cruz with distances to Orizaba, Xalapa, etc.

(703.4: 1823: G163xrp Small)

32 (72) 1830

Chart of the northwest coast of the island of Cuba, between the meridians of Havana and the Cape St. Antonio. By Juan J. Martinez. 1830.

Cartographer: Juan J. Martinez.

Provenance: Presented by Juan J. Martinez, 7 December 1832.

Size: 47.7 x 60.9 cm.

(723: 1830: M365cha Large)

32 (73) [ca. 1830]

Chesapeake & Delaware canal. [ca. 1830]. 3 pieces.

Scale: 1.3 cm. = 1,000 yards; 1.9 cm. = 1 mile.

Cartographer: Henry S. Tanner.

Size: 19.6 x 37.8 cm.

Contains topographical view and vertical section.

(651.4: [ca.1830]: T158cdp Small)

32 (74) 1831

Lac Supérieur & autres lieux ou sont les missions des pères de la compagnie de Iesus comprises sous le nom d'Ovtaovacs; Mont Ste. Marie. 1831.

Scale: 50 lieues.

Size: 54 x 42.5 cm.

Annotated: "Facsimile d'une carte attachée à un des volumes des lettres edifiantes pour le Canada dont les Jesuites publierent plus de 40 volumes de 1611 à 1678." *(Relations des Jesuites en la Nouvelle France.* Paris: Cramoisy: 1632-1673.)

A note reads: "rough fac smilie-this and the Scutcheon above are Well drawned and Engraven."

(626.2: 1831: Su77cd Small)

32 (75) 1831

Situate in the counties of Schuylkill and Northampton showing a route for a rail road commencing at the five locks at the Schuylkill canal below Pottsville and ending at the mouth of Lizzard Creek where the same falls into the Lehigh River, being in length 22 miles and 102 perches, intended to constitute a part of the general rail road now in contemplation between Philadelphia and Pottsville. Surveyed at the request of William Audenried, Esqr. during the month of June 1831. By John Dreher. 1831.

Cartographer: John Dreher.

Size of paper: 46.4 x 187.4 cm.

Colored.

(649: 1831: D812sur Large)

32 (76) No item.

32 (77) [ca. 1835]

[Susquehanna, Schuylkill and little Schuylkill rivers, in Pennsylvania]. [ca. 1835].

Scale: 0.5 inch = 1 mile.

Cartographer: Christian Brobst.

Size of paper: 48.8 x 77.7 cm.

Annotated: "A copy of a Christian Brobst traycing [and] Cattawissa," and, "W[illia]m J. Duane has the original."

(649: [ca.1835]: B781msc Large)

32 (78) [ca. 1852]

[Middle Atlantic states. ca. 1852].

Scale: 1.5 cm. = 60 miles.

Size: 14.4 x 23.5 cm.

Colored.

The route of the Pennsylvania Central Railroad is shown from Philadelphia to Pittsburgh.

(640.2: [ca.1852]: At61mas Small)

32 (79) 1853-1855

Charts pertaining to the Second U.S. Grinnell Expedition in search of Sir John Franklin, under the command of Elisha Kent Kane. 1853-1855.

1. [Melville Bay and surrounding area]. 1853-1855.

Cartographer: Elisha Kent Kane.

Size: 60.4 x 73 cm.

2. [Unidentified]. 1853-1855.

Cartographer: Elisha Kent Kane.

Size: 44.8 x 63.4 cm.

3. [Base of Sylva Mountain, observatory and brig]. 1853-1855. Draft.

Cartographer: Elisha Kent Kane.

Size: 36 x 40.5 cm.

4. Sylva Mountain, position of observatory and brigg. 1853-1855.

Cartographer: Elisha Kent Kane.

Size: 18.3 x 20.8 cm.

There is also a printed copy of Number 4.

A member of the American Philosophical Society, Kane spoke on 16 November 1855 of various geographical features of the Arctic regions he had seen and displayed his map of the area. He spoke of the progress of the expedition and the difficulties encountered. On 7 December 1855 he exhibited two charts showing his corrections and the new features described on them. These may be the charts Kane displayed and spoke about. See: American Philosophical Society *Proceedings*, vol. 7 (1885): pp. 154, 162.

(172.3:1853-55: G895exp Large)

32 (80) [ca. 1854]

Wythe County, Virginia. Profile, showing geological strata from Big Walker Mountain to Peak Mountain. By J. P. Lesley. [ca. 1854].

Cartographer: J. Peter Lesley.

Size: ca. 172.5 x 308 cm.

Colored.

(654: [ca.1854]: L565geo Large)

32 (81) 1856

Cumberland & Frobisher Sts. from a drawing by an Eskimo in 1856 given to Capt. Salter of the Clara of Peterhead. [By Franz Boas]. 1856.

Size of paper: 51 x 43.8 cm.

See the description of the Boas map in No. 6, above.

(617: 1856: C911cfs Small)

32 (82) 1859

Map of the military reservation [on the Columbia River] at Fort Vancouver, W[ashington] T[erritory]. Surveyed under the direction of Capt. Geo. Thom . . . by Lts. J. B. Wheeler and J. Dixon . . . By order of Brig. Gen. W. S. Harney, 1859. Photographs. 8 pieces total.

Scale: 10 inches = 1 mile.

Cartographers: George Thom, J. B. Wheeler, J. Dixon, and W. S. Harney.

Size: ca. 34 x 25.8 cm.

Originals in General Land Office Records, Abandoned Military Reservation Series, Fort Vancouver, Washington. Box 100, National Archives, Washington, D.C.

Includes photograph of so-called "Covington Map." (Original in Hudson's Bay Company Archives, London, 1846.)

Includes also photographs from other sources.

(697: 1859: Un38fvc Large)

32 (83) 1865

Profile showing the comparative levels of the bituminous coal fields and oil district of West Virginia, Ohio and western Pennsylvania. By P. W. Sheafer. March 1865. On linen.

Scales: 5 miles per inch horizontal, and 500 feet per inch vertical.

Cartographer: P. W. Sheafer.

Size of linen: 31 x 112 cm.

(649: 1865: S37wvop Large)

32 (84) [ca. 1869]

Section of the lake, river and canal navigation from Lake Superior to the Gulf of the St. Lawrence. [ca. 1869].

Scales: Horizontal: 1 inch = 60 miles; Vertical: 1 inch = 400 feet.

Cartographer: T. E. Blackwell.

Size: 23 x 85.4 cm.

Colored.

With this is a water color painting of "Sketch from the Mountain of Montreal shewing part of the plain extending south east & southwards to the Green Mountain Range in eastern townships of Canada and Vermont & the Adirondack Mountain regions of New York. To illustrate Mr. Blackwell's paper on the basin of the St. Lawrence." See: American Philosophical Society *Transactions,* Series 2, vol. 13 (1869).

(670: [ca.1869]: B561Lsg Large)

32 (85) 1874

Chart of Niantilic harbor by William A. Mintzer, U.S. Navy, from observations by Lieut. Wilkins, U.S. Navy, navigating officer of the U.S. St. Tigress on *Polaris* search. 1874.

Scale: ½ inch = 1 mile.

Cartographers: W. A. Mintzer and Lt. Wilkins.

Size of paper: 25.9 x 20.6 cm.

Niantilic Harbor is on Baffin Bay, Cumberland Sound. The *Polaris* had disappeared, and the *Tigress* and another ship were sent to locate it. The readings for this map are: North lat. 64° 56′ - West long. 66° 21′. Also in folder is a map showing "Padloaping. Nettiling Fjord and Nettiling."

(178: 1874: M667nia Small)

32 (86) [ca. 1885]

[United States coast bordering the Atlantic Ocean and the Gulf of Mexico]. [ca. 1885]. 8 maps in 13 pieces.

Size: ca. 96 x 60 cm.

Colored.

(635: [ca.1885]: Un38atL Large)

32 (87) 1896-1897

Geological map of the northern part of the Indian territory. By N. F. Drake. 1896-1897.

Scale: 1 inch = 4 miles.

Cartographer: N. F. Drake.

Size: 110 x 89.9 cm.

See: N. F. Drake, "A geologial reconnaissance of the coal fields of the Indian territory." American Philosophical Society *Proceedings*, vol. 36 (1897): pp. 326-419.

(667: 1897: D782ind Large)

32 (88) 1897

[Arctic regions]. By Fridtjof Nansen. Enlarged by James B. Sword, Philadelphia. 1897.

Cartographers: James B. Sword and Fridtjof Nansen.

Size: 156.5 x 160.3 cm.

Colored.

This is an enlargement of a portion of a map. See: Fridtjof Nansen, *Farthest north* (New York: Harper: 1897), vol. 1.

(170: 1897: N166arm Large)

32 (89) 1932

Philadelphia as William Penn knew it, 1684. Commemoration of the 250th anniversary of the first arrival, October 24, 1682, of William Penn in America, 1682-1932. Prepared for the program committee [of the 250th anniversary], Albert Cook Myers, chairman. Philadelphia: 1932.

Scale: 1¾ inches = 600 feet.

Cartographers: William Wilson Pollard and Albert Cook Myers.

Size: 42.5 x 26.6 cm.

Provenance: Presented by Albert Cook Myers, 28 March 1935.

(649.962: 1932: P766wpp Small)

32 (90) 1948

Plan of ground and site of the residence of Benjamin Franklin. Federal Project "C," Public Law 795, 80th Congress. Approved June 28, 1948. Plate II. 1948.

Scale: 1 inch = 20 feet.

Cartographer: Fred J. Gorman.

Size: 39 x 30.7 cm.

Colored.

(649.962: 1948: G656rbf Large)

32 (91) n.d.

Albemarle sound, North Carolina. n.d.

Size of paper: 38.6 x 48.7 cm.

Colored.

Roanoke Island is shown, as is Edenton, Hallifax (sic), and the Virginia line.

(656: n.d.: AL21asm Small)

32 (92) n.d.

[Delaware Bay and River]. By Joshua Fisher? n.d.

Cartographer: Joshua Fisher?

Size: 41.5 x 23 cm.

This appears to have been adapted directly from the map published in London by Laurie & Whittle in 1794.

(640.2: n.d.: F533bdr Small)

32 (93) n.d.

A chart of the world, upon Mercator's projection, with the new discoveries. [Also, the eastern and western hemispheres]. n.d. 3 pieces.

Sizes: 39 x 45.6 cm; and 36 x 36 cm.

Colored.

These are printed maps with heavy manuscript markings indicating air (?) currents and observations drawn and crudely colored. Antarctica is also drawn on each map.

The maps were printed in London by "Jas Wyld," and "J. Wyld & Son, Charing Cross East 1836." (One map has no publication date on it.)

(100: n.d.: C382wmd Small)

32 (94) n.d.

[Geological map of part of Switzerland?]. n.d.

Size: 12.5 x 18.9 cm.

Colored.

(237: n.d.: Swgeo Small)

32 (95) n.d.

[Central Pennsylvania, showing Danville, Catawissa, and other cities]. n.d. In pencil.

Size: 32.9 x 47.6 cm.

This is on the reverse of C. G. Childs. Plan of Keim's addition to Port Clinton.

(649.964: n.d.: C432kpc Small)

32 (96) n.d.

[Sections of Old Philadelphia during the 17th, 18th & 19th centuries]. Philadelphia: n.d. 6 sheets. Photostats.

Size: each sheet ca. 45.2 x 55.5 cm.

Photostats of manuscript maps at the Historical Society of Pennsylvania.

(649.962: n.d.: P376hist Small)

32 (97) n.d.

[Shwemyo district, central Burma. Lat. 20° 3′, long. 96° 14"]. n.d. In Burmese.

Size: 47 x 60 cm.

Colored.

Four lines in Burmese in the lower left-hand corner are the key to the colors used in the map. Note on back in Burmese identifies location as Shwemyo district. See letters from E. W. Burlingame and Cecil Hobbs to the American Philosophical Society, June 1914 and 27 January 1961.

(457: n.d.: B922shw Small)

32 (98) n.d.

Tabula novae Franciae. 1660. Copy. n.d.

Cartographer: François Du Creux [Creuxius]?

Size: 34.6 x 44.3 cm.

Colored.

Insert: Chorographia regionis Huronum. Hodie desertae.

Copied from François Du Creux, *Historiae Canadensis seu novae Franciae libri decem ad annum usque Christi 1656* (Paris: 1664). This was printed in the Champlain Society *Publications*, vols. 30-31, translated by Percy J. Robinson (Toronto: 1951).

A note by P. J. Robinson states that this map deserves study and is of special value on account of the inset map of the Huron country, the improved outline of the lakes, especially Lake Ontario, and Georgian Bay, the marking of the trails to the Hudson Bay and of the new route discovered by the Groseilliers and Radisson from Lake Nipigon.

(611: n.d.: C162tnf Small)

33. MISCELLANEOUS MANUSCRIPTS COLLECTION

This collection is precisely what the title indicates: a miscellany. It consists of individual manuscripts which do not fit into any established collection.

33 (1) 1743

[Survey of lots at Sassafras and Fourth Streets, Philadelphia]. By Nicholas Scull. 16 August 1743.

Cartographer: Nicholas Scull.

Size: 19 x 31.8 cm.

Scull made this survey by order of the Orphans Court, Philadelphia, dividing the "three Messuages or Tenements and Lots of Land once the property of Thomas Bristol Case, deceased." The division was made to settle the claims of Case's five children.

33 (2) 1743

[Survey of the property of Samuel Carpenter]. 9 March 1743.

Size: 45.7 x 61 cm.

Survey of property on Chestnut Street between Sixth and Seventh Streets in Philadelphia. The lots are identified and the buildings are marked.

33 (3) 1745

Survey of land. By John Harris. 17 July 1745.

Cartographer: John Harris.

Size: 19 x 20.3 cm.

Draught of 405 as. & allowance situate in Antrim Town[shi]p Lanc[aste]r County, formerly granted to Jeremiah and Joseph Harris (sons of Jn. Harris) by two Warrts. from y[e] [?] each dated y[e] 4th 7ber 1738, who not complying with the Conditions of the sd. Warrts. nor either of them the same became void. And by Wart of y[e] 3d 7ber 1742 the same Land was surveyed for y[e] sd. Jno. Harris And returned for his use into the Survy's Office the 17th July A.D. 1745.

33 (4) [ca. 1755]

[Kittanning, near Shippensburgh, Pennsylvania]. By John Armstrong. [ca. 1755]. 2 pieces.

Cartographer: John Armstrong.

Size of paper: 31.8 x 19.7 cm.; and, 33.6 x 42.5 cm.

Proposal for attack on Kittanning, an Indian town, by Armstrong, based upon a report by "John Baker Soldier at Fort Shierley, who last Winter made his escape from the Indians at the Kittanning." The attack was made in 1756; Kittanning was taken and burned.

33 (5) 1765-1766

[Survey of lands for Michael Snyder, J. M. Moyer, and Jacob Shallus]. By John Scull. 30 April 1765; 30 January and 4 December 1766.

Cartographer: John Scull.

Size: 21.6 x 34.3 cm.

Annotated: "A.B. two Pieces surveyed to Snyder (who is now contented) in pursuance of his Warrent dated 30th April 1765 for 50.00.

"C. Sur[veyed] to Moyer in pursuance of his Application No. 1318, dated 30 Jan[uar]y 1766, altho' not the very Spot mentioned in his Description;-

"Note. Mr. Miller wrote me a Letter, in which he mentions the Opinion of Mr. Tilghman, concerning the above Lands & that the Secretary desir'd him to write(?) to me that it was his Opinion, which allows Moyer his g[ ]ns, & Snyder the Land in two Pieces adj[oinin]g his old [tenet?].

"D. surveyed for Jacob Shallus, in pursuance of his Application No. 2132, 4 Decr. 1766 for 50.00, being the Remainder of the Vacancy."

33 (6) 1785

[Little and Big Beaver Creeks]. By Benjamin Smith Barton. 1 October 1785. 12 pieces.

Cartographer: Benjamin Smith Barton.

Size: 12.1 x 19.1 cm.

Barton, while practicing surveying, kept a notebook labelled "No. 3. Geographycal Notes October 1st 1785. Waters of Little Beaver Creek. No. 4. Waters of Big Beaver." There are twelve pages of sketches of maps of the drainage of these creeks in this volume, but there are few notes which would help identify them, so they are not listed separately here.

33 (7) 1803

[Richmond and Norfolk Canals]. By Benjamin Henry Latrobe. 18 December 1803. 2 pieces.

Cartographer: Benjamin Henry Latrobe.

Size: 19.1 x 22.9 cm.

Annotated: "Sketch of canals executed or proposed near Norfolk - the latter by B. H. Latrobe." These maps have been reproduced and described in Darwin H. Stapleton, ed., *The engineering drawings of Benjamin Henry Latrobe* (New Haven: Yale University Press: 1980).

33 (8) 1820

[Survey of land in the County of Lincoln on the Ohio]. By George Rogers Clark. 20 January 1820.

Surveyor: George Rogers Clark.

Size: 30.5 x 27.3 cm.

Annotated: "Surveyed for land for Brig. Gen. George Rogers Clark 37,000 Acres of Land by virtue of 25 Treasury Warrants." The land was near the Tennessee and Ohio Rivers.

33 (9) 1947

[Site of Fort Allen, Pennsylvania]. By John A. McConomy. 24 November 1947.

Cartographer: John A. McConomy.

Size: 22.9 x 20.3 cm.

Identifies Old Fort Allen, built by Benjamin Franklin, at Gnadenhuetten in 1756. McConomy says his identification is the "probable site" on Franklin and Allen Streets, near the Lehigh River in the "n.w. corner Weissport Borough." This map accompanies a sketch of the fort and an article on the fort by Rev. John A. McConomy.

33 (10) n.d.

Copper mine, on the land of John Shannon . . . of Montgomery County, Pennsylvania. n.d.

Scale: 50 fathoms = 17 cm.

Size: 39.4 x 31.8 cm.

This map shows eight shafts which were sunk and five are described as being on the vein, e.g., "No. 2 A Pit sunk 12 feet deep the vein 4 feet wide some five stones of Copper ore a great prospect." Also, the "road leading from Perkiomen to the White Horse tavern" is shown.

(Misc. Mss. Collection)

34. MUHLENBERG, GOTTHILF HEINRICH ERNST. Botanical notebook.

Muhlenberg was a Lutheran clergyman whose avocation was botany. He was well known on both sides of the Atlantic Ocean for his botanical work. He lived in Lancaster, Pennsylvania, and studied the flora there, keeping several volumes of notes. The maps in this notebook, part of a large collection of Muhlenberg's writings on botany and natural history, seem to be plots for gardens he maintained.

34 (1) n.d.

[Plots for gardens?]. By G. H. E. Muhlenberg. n.d. 4 pieces.

Cartographer: G. H. E. Muhlenberg.

Size: 17.8 x 10.8 cm.

(580: M89bot)

35. MURPHY, ROBERT CUSHMAN. Papers.

Murphy was the world's foremost authority on sea birds; he was curator and later chairman of ornithology at the American Museum of Natural History. The maps that follow appear in his journals and travel diaries.

35 (1) 1919

Chincha Islands. By R. C. Murphy. 14 October 1919.

Cartographer: R. C. Murphy.

Size: 14 x 21.6 cm.

Colored.

(B: M957, vol. 12)

35 (2) 1919

South Island, Chincha. By R. C. Murphy. 14 October 1919.

Cartographer: R. C. Murphy.

Size: 14 x 21.6 cm.

Colored.

(B: M957, vol. 12)

35 (3) 1919

Bay of Pisco and waters about Paracas Peninsula. By R. C. Murphy. 4 November 1919.

Cartographer: R. C. Murphy.

Size: 14 x 21.6 cm.

(B: M957, vol. 12)

35 (4) 1919

Santa Rosita Island and Isla Viegas. By R. C. Murphy. 16 November 1919.

Cartographer: R. C. Murphy.

Size: 14 x 21.6 cm.

(B: M957, vol. 12)

35 (5) 1919-1920
Peruvian coastline. By R. C. Murphy. 1919-1920.
Cartographer: R. C. Murphy.
Size: 5.1 x 18.4 cm.
(B: M957, vol. 12)

35 (6) 1919
San Gallan Island. R. C. Murphy. 1919.
Cartographer: R. C. Murphy.
Size: 14 x 21.6 cm.
Colored.
(B: M957, vol. 12)

35 (7) 1920
Chincha Islands, Middle Island. By R. C. Murphy. 1 January 1920.
Cartographer: R. C. Murphy.
Size: 14 x 21.6 cm.
Colored.
(B: M957, vol. 12)

35 (8) 1920
Islas de Guanape. By R. C. Murphy. 1 January 1920.
Cartographer: R. C. Murphy.
Size: 14 x 21.6 cm.
Colored.
(B: M957, vol. 12)

35 (9) 1920
Isla de Lobos de Afuera. By R. C. Murphy. 1920.
Cartographer: R. C. Murphy.
Size: 14 x 21.6 cm.
(B: M957, vol. 12)

35 (10) 1920
Islas de Macabi. By R. C. Murphy. 1920.
Cartographer: R. C. Murphy.
Size: 14 x 21.6 cm.
(B: M957, vol. 12)

35 (11) 1920
Islas de Lobos de Tierra. By R. C. Murphy. 1920.
Cartographer: R. C. Murphy.

Size: 14 x 21.6 cm.
Colored.

(B: M957, vol. 12)

35 (12) 1937
[Choco, Columbia]. By R. C. Murphy. 1937.
Cartographer: R. C. Murphy.
Size: 16.5 x 17.1 cm.

(B: M957, vol. 29)

35 (13) 1937
Map of Panama and Colombia Indian tribes. By R. C. Murphy. 1937.
Cartographer: R. C. Murphy.
Size: 12.7 x 15.2 cm.

(B: M957, vol. 38)

35 (14) 1940
Hollins Island, Long Island. By R. C. Murphy. 24 August 1940. 2 pieces.
Cartographer: R. C. Murphy.
Size: 8.3 x 15.2 cm.

(B: M957, vol. 1)

35 (15) 1941
Malpelo Island. By R. C. Murphy. 30 March 1941. 2 pieces.
Cartographer: R. C. Murphy.
Size: 28 x 21.6 cm.

(B: M957, vol. 15)

35 (16) 1948
[R. C. Murphy's New Zealand travel route]. By Murray McCaskill. 8 March 1948.
Cartographer: Murray McCaskill.
Size: 17.1 x 19.1 cm.

(B: M957, vol. 16)

35 (17) 1949
Mauna Loa, Hawaii; and Island of Hawaii. By R. C. Murphy. 30 January 1949.
Cartographer: R. C. Murphy.
Size: 21.6 x 28 cm.

(B: M957, vol. 17)

35 (18) 1951

Northeastern end of Bermuda. By R. C. Murphy. 1951.

Cartographer: R. C. Murphy.

Size: 16.5 x 25.4 cm.

(B: M957, vol. 19)

35 (19) 1953

Coastal valley north of Lima, Peru. By R. C. Murphy. 12 December 1953.

Cartographer: R. C. Murphy.

Size: 10.2 x 17.8 cm.

(B: M957, vol. 13)

35 (20) 1955

Riverhead tract, Long Island. By R. C. Murphy. 16 September 1955.

Cartographer: R. C. Murphy.

Size: 18.4 x 21 cm.

(B: M957, vol. 6)

36. NEWMAN, JOHN. A short account of the situation, soil, production &c. of the State of Tennessee; 1797.

Newman was a North Carolina physician who sent this map and a paper on Tennessee to the American Philosophical Society.

36 (1) 1797

Plan of settlements near the Natches [Trace, Tennessee], particularly Mr. William Dunbars. By. John Newman. 1797.

Cartographer: John Newman?

Size: 23.5 x 37.5 cm.

Annotated: "Rec'd from C. Ross, Augt 1803."

(917.68: N46)

37. PARSONS, ELSIE CLEWS. Papers.

Parsons received the Ph.D. from Barnard College in 1899. During her career as an anthropologist, sociologist, and folklorist, she became fascinated with the culture of the Pueblo peoples and visited, studied and wrote extensively about them. She served as the president of the American Anthropological Association and published a number of books, including: *American Indian life* (1922), *Hopi and Zuni ceremonialism* (1933), and *Pueblo Indian religion* (1939); she also edited *Pueblo Indian journal* (1925) and Alexander M. Stephen's *Hopi journal* (1936).

There are 42 maps in this collection, chiefly of Picuris Pueblo and of Hopi materials. Some of the maps are diagrams of dwellings and dance areas. Most of the maps have no title and were made in her notebooks.

The initial accession of Parsons material came to the Library in 1949; more recent accessions, not yet catalogued, may contain additional maps.

(572: P25.1)

38. PEALE, CHARLES WILLSON. Papers.

Peale is known as "the artist of the American Revolution." He not only painted during that period, but he also was a soldier in the Revolution. He was best known for his museum in Philadelphia, which contained natural history artifacts as well as a gallery of portraits of the major American figures of the Revolution.

38 (1) [ca. 1776]

[New York, showing British and American troop positions]. By Charles Willson Peale. [ca. 1776].

Cartographer: Charles Willson Peale.

Size: 33.7 x 41.6 cm.

(B: P31.50)

38 (2) 1817

[Ground plan of City Hall, New York City]. By Charles Willson Peale. 7 June 1817.

Cartographer: Charles Willson Peale.

Size: 12.7 x 10.2 cm.

(B: P31, vol. 22)

39. PIKE, ZEBULON MONTGOMERY. Collection of biographical materials.

Pike was an American explorer and army officer. In 1805 he led an exploring party to search for the source of the Mississippi River; he later explored the headwaters of the Arkansas and Red Rivers, and the Spanish settlements as far as the site of Pueblo, Colorado. He was a brigadier general during the War of 1812 and was killed while leading his troops during the successful assault on York [Toronto], Canada.

This collection consists of materials collected by W. Eugene Hollon for his biography of Pike, *The lost pathfinder* (1949).

39 (1) 1946

Map of Fort Bellefontaine. By Edwin Mills. 1946.

Cartographer: Edwin Mills.

Size: 21.6 x 35.4 cm.

Fort Bellefontaine was the first United States Army post west of the Mississippi River. It was established by General James Wilkinson in 1804.

(B: P63)

40. RUSSELL, RICHARD JOEL. Papers.

Russell was one of the most eminent geographers and geologists of the United States. He studied the river deltas of America (chiefly the Mississippi River), Spain, France, and Turkey. He also studied the coastal morphology of the Gulf of Mexico; stream patterns; arid areas of the United States; and he worked on a climatic map of California.

The approximately 150 maps in this collection are in various notebooks which give a general indication of the area and sometimes a date. The maps generally are quite small and were sketched on the site of the topographical or geological study then underway.

(B: R91)

41. SANDERSON, IVAN TERENCE. Papers.

Sanderson was a naturalist who travelled extensively, collected animals, and published a number of books. Two of the following maps were made by Winifred Duncan for the publication *Your Mexico.*

41 (1) n.d.

Map D: Mexico N.W. [Area from Mexico City to Guanajuato to Manzanilla]. By Winifred Duncan. n.d.

Cartographer: Winifred Duncan.

Size: 21.6 x 28 cm.

41 (2) n.d.

Mazanilla. By Winifred Duncan. n.d.

Cartographer: Winifred Duncan.

Size: 21.6 x 28 cm.

41 (3) n.d.

[Patos Island, between Trinidad and Venezuela]. By Ivan T. Sanderson. n.d.

Cartographer: Ivan T. Sanderson.

Size: 20.3 x 25.4 cm.

(B: Sa3)

42. SELLERS, GEORGE ESCOL. Papers.

A descendant of Charles Willson Peale and the paper-maker Nathan Sellers, George Escol Sellers (1808-1899) was a mechanic and inventor who spent much of his life in the paper-making trade. Born and raised in Philadelphia, he also lived in Upper Darby Township, Pa., Cincinnati, southern Illinois, and Chattanooga.

42 (1) n.d.

[Coalfields along Ohio River near Caseyville (Kentucky?)] n.d.

Size of paper: 67 x 35.2 cm.

(B: P31, 30-P)

43. SELLERS FAMILY PAPERS

The members of the Sellers family, related to the Peale family through the marriage of Sophonisba Peale to Coleman Sellers, were important manufacturers of wire, iron goods, and paper in the Philadelphia area during the eighteenth and nineteenth centuries.

43 (1) 1788

[150 acres of land off Darby Creek, Tinecum Township, Chester County, Pennsylvania]. By John Sellers. 21 April 1788.

Cartographer: John Sellers.
Size: 41 x 33 cm.

43 (2) [ca. 1804]

[Lands of N. and D. Sellers and George Sellers, showing the grist mill, West Chester Road, Marshall Road, Garrett Road, etc.; Philadelphia and Delaware Counties, Pennsylvania]. [ca. 1804].

Size: 30 x 39 cm.
Colored.

43 (3) 1822

Draft of land of N[athan] Sellers in 3 fields of 7 acres each taking only as near the lines as can be cultivated with the plow. 29 May 1822.

Size of paper: 20.3 x 33 cm.

43 (4) 1822

Rough draft of George Sellers's land. 235¾ acres and 8 p., Delaware County, Pennsylvania. December 1822.

Size: 39 x 39.5 cm.

43 (5) n.d.

Roads in the west side of Schuylkill [River, near Darby, Springfield, West Chester, Joseph Gibbons tavern, etc]. n.d.

Size: 77.5 x 42.5 cm.

43 (6) n.d.

[John Sellers's property located on Cobbs Creek and West Chester Pike (Red Lion Road); Delaware County, Pennsylvania]. By John Sellers. n.d.

Cartographer: John Sellers.
Size: 36 x 30.5 cm.

(B: P31, 50)

44. SEYBERT, ADAM. Commonplace book.

Seybert was a Philadelphia scientist, physician, and Pennsylvania congressman. He was well known as a chemist and a mineralogist.

44 (1) n.d.

Quebec and its environs with the operation of the siege made by order of Admiral Saunders. By Adam Seybert. n.d. Copy.

Cartographer: Adam Seybert.
Size: 19.1 x 15.9 cm.
Colored.

Legend: "Battle of the Plains of Abraham 13th Septr. 1759. A boom laid across the mouth of the River St. Charles. Battle of Montmorency July 13th 1795. P.L. Point Levy. Encampment of the British Troops 30th June 1759 under . . . "

This map is pasted in a commonplace book on p. 167 amid Seybert's comments on navies and armies.

(B: Se95)

45. SHIPPEN FAMILY COLLECTION

The Shippen family was one of the most industrious and illustrious of Philadelphia in the eighteenth and nineteenth centuries. They were active in government, the military, and in their various businesses: land acquisition and speculation, housebuilding, etc. The activities of Joseph Shippen, Jr., while with the army under Col. Burd during the French and Indian war, are well documented in this collection.

45 (1) 1757

A draught of the west branch of Susquehanna & part of the Ohio River. Fort Augusta, 26 January 1757. 2 pieces.

Cartographer: Joseph Shippen, Jr.

Size of paper: 31.1 x 48.3 cm.

45 (2) 1758

Draught of part of the River Susquehanna by Joseph Shippen. January 1758.

Cartographer: Joseph Shippen, Jr.

Size of paper: 38.7 x 47 cm.

45 (3) 1759

Rough draught of the Monongahela River from Fort Burd to the confluence of Muddy & Cheat Rivers, taken by J. Shippen, Jr., November 1759. 2 pieces.

Cartographer: Joseph Shippen, Jr.

Size of paper: 64.8 x 21 cm.

45 (4) [ca. 1763]

[Plan of the English fort at Pittsburgh. ca. 1763].

Cartographer: Joseph Shippen, Jr.

Size of paper: 30.8 x 38.7 cm.

45 (5) 1769

Draught of the west branch of Susquehanna, taken from Capt. Patterson. 1 April 1769.

Cartographer: Col. Patterson.

Size of paper: 21 x 30.5 cm.

(B: Sh62)

46. SHIPPEN FAMILY. Real property surveys.

The following is a listing of maps of various pieces of property in the Southwark section of the City of Philadelphia which once belonged to members of the Shippen family. The maps were taken, in part, from the Philadelphia City records: other maps are up-to-the-minute surveys done when the property was sold or subdivided. Basically, the maps are of the early part of the nineteenth century and many are undated. The names of the purchasers are sometimes listed.

46 (1) 1741

Plan of subdivision recorded with release of Edward Shippen to Joseph Shippen, Philadelphia, Second Street to Delaware River. 1 August 1741.

Size: 27.5 x 36.5 cm.

46 (2) 1744

[Partition of land belonging to Duche, Knowles, Parham and Cox]. 16 August 1744.

Size: 30.5 x 48 cm.

46 (3) 1750

[Partition of land of Arkton(?) and wife, and Samuel Wheeler]. 3 September 1750.

Size: 30.5 x 48.5 cm.

46 (4) 1756

Partition of the Shippen estate on South S[treet]. 11 October 1756.

Size: 33 x 49.5 cm.

46 (5) 1762

[Lot of Dr. William Shippen on east side of Shippen's Lane]. By William Scull. 18 August 1762.

Cartographer: William Scull.

Size: 20.5 x 33.5 cm.

46 (6) 1769

[Land on the Delaware River fronting League Island]. 22 July 1769.

Size: 30 x 40 cm.

46 (7) 1773

Mary Garrigue's estate [on Queen Street]. 1773.

Size: 35.4 x 52.5 cm.

46 (8) 1774

Partition of N. Pryors estate. 11 March 1774.

Size: 20.4 x 34.2 cm.

46 (9) 1788

Plan of 64 acres . . . called Solitude, the property of Samuel and Israel Morris. 1788.

Size: 38.5 x 84 cm.

46 (10) 1792

Road to be opened from Grays Ferry. 3 November 1792.

Size: 38.4 x 31.3 cm.

46 (11) 1794

[Survey of the property of John Duche]. 14 November 1794.

Size: 22.4 x 19.4 cm.

46 (12) 1795

[Lands of Bankson estate, May Road, Turner's Lane, and Moyamensing Road]. 5 November 1795.

Size: 30.5 x 48.5 cm.

46 (13) 1799

Hardie's property, Pass[yunk] Road and Oak Street. 1799.

Size: 25.6 x 21.3 cm.

Colored.

46 (14) 1802

[Lot covering square in Philadelphia between Fourth and Fifth Streets and Shippen Street]. 4 August 1802.

Size: 40 x 34 cm.

46 (15) 1815

Plots on Shippens Lane: Plots on Catharine Street. 6 October 1815.

Size: 31 x 47 cm.

On the reverse is: Drafts of plots. n.d.

46 (16) [ca. 1816]

[Lots on south side of Cedar Street with names of Edward, William, and Joseph Shippen as owners]. By D.(?) Groves. [ca. 1816].

Cartographer: D.(?) Groves.

Size: 25 x 38.5 cm.

Colored.

46 (17) 1816

The lots on the south side of Cedar Street [with names of Edward, William, and Joseph Shippen as owners]. By D.(?) Groves. 1816.

Cartographer: D.(?) Groves.

Size: 19 x 53 cm.

Colored.

46 (18) 1825

Survey of "late Dr. Shippen's estate." 28 April 1825.

Size: 20 x 24.8 cm.

46 (19) 1831

[Tracts fronting on the Delaware River. By D. Coombs]. 17-21 May 1831.

Cartographer: D. Coombs.

Size: 42.3 x 36.3 cm.

In addition, there are forty undated surveys pertaining to the same section of Philadelphia. The names of the owners, purchasers, estates, etc. are: Anthony Duche, Edward S. Burd, John Flower, William G. Flower, John Graeme, — Hulbert, John Inglis, Lawrence McCall, John McGoffin, — Marriott, — Mifflin, Francis Moore, A. Noble, and — Penrose.

(649.962: Various dates: Sh6m Small)

47. SMITH, ERWIN FRINK. Papers.

Smith (1854-1927) was a plant pathologist with the U.S. Department of Agriculture.

47 (1) 1877

Map of my trip in Ronald [Michigan] . . . By Erwin Frink Smith. 16-20 August 1877.

Cartographer: Erwin Frink Smith.

Size: 19.7 x 70 cm.

Colored.

(B: Sm53, Box 1)

48. SPECK, FRANK GOULDSMITH. Papers.

Speck was an eminent ethnologist whose special interest was the Indians of the Eastern Woodlands. A student of Franz Boas, Speck spent his entire career at the University of Pennsylvania. Many of the maps in this collection were annotated by Speck with ethnographic data on hunting territories, tribal boundaries, etc.

48 (1) n.d.

Chart showing distribution of Montagnais-Naskapi bands of the lower St. Lawrence River. N.d.

Cartographer: F. G. Speck.

Size: 49.5 x 61.3 cm.

48 (2) n.d.

Distribution of the Catawba and related Siouan tribes, and certain related life forms in the southeast functioning in Catawba culture. N.d.

Cartographer: F. G. Speck.

Size of paper: 48.3 x 28 cm.

48 (3) n.d.

Family hunting territories of the Penobscot. N.d.

Cartographer: F. G. Speck.

Size: 24.1 x 34 cm.

Colored.

48 (4) n.d.

[Hudson Bay area. Montagnais-Naskapi territories]. N.d.

Cartographer: F. G. Speck.

Size: 21.3 x 28 cm.

48 (5) n.d.

Index to published maps of Indian hunting territories [Montagnais-Naskapi]. N.d.

Cartographer: F. G. Speck.

Size: 88.9 x 90.5 cm.

Colored.

48 (6) n.d.

Malecite hunting grounds. N.d.

Cartographer: F. G. Speck.

Size: 43.5 x 45.4 cm.

On linen.

48 (7) n.d.

[James Bay area]. N.d. 2 pieces.

Cartographer: F. G. Speck.

Size: 21.3 x 28 cm.

48 (8) n.d.

Map of Attislopi(?), Labrador. Montagnais-Naskapi. N.d.

Cartographer: F. G. Speck.

Size: 41.3 x 69.2 cm.

On linen.

48 (9) n.d.

Map of Maine showing family hunting territories of the Penobscot. N.d.

Cartographer: F. G. Speck.

Size: 41.6 x 52.4 cm.

Colored.

48 (10) n.d.

Map of river flowing into Lake St. John, Labrador. N.d.

Cartographer: F. G. Speck.

Size: 21.6 x 19 cm.

48 (11) n.d.

Map showing approximate distribution of Eskimo . . . in the Labrador peninsula . . . and on the coasts in the 17th and 18th centuries . . ., and approximate recorded distribution of Montagnais-Naskapi groups. N.d.

Cartographer: F. G. Speck.

Size: 22.5 x 24.1 cm.

48 (12) n.d.

Map showing approximate location, since about 1850, of local groups or bands of Montagnais-Naskapi and Eskimo. N.d.

Cartographer: F. G. Speck.

Size: 22.5 x 24.8 cm.

48 (13) n.d.

Map showing direction of journeys of Hero-Transformer. N.d.

Cartographer: F. G. Speck.

Size: 26.4 x 34 cm.

48 (14) n.d.

Map showing the family hunting territories of the Mistassini Indians, the Montagnais of Lake St. John and the St. Lawrence and the tribal territories of some of the neighboring bands of Cree. N.d.

Cartographer: F. G. Speck.

Size: 57.2 x 58 cm.

Colored.

48 (15) n.d.

Montagnais-Naskapi bands in Labrador near Ungava Bay. N.d.

Cartographer: F. G. Speck.

Size: 42.5 x 42.5 cm.

48 (16) n.d.

Rappahannock territory. N.d.

Cartographer: F. G. Speck.

Size: 21.6 x 23.2 cm.

(572.97: Sp3)

49. U.S. WORK PROJECTS ADMINISTRATION. Reports.

From 1941 through 1948 the Work Projects Administration funded the excavations of American Indian tumuli in Pennsylvania and southern New York. Sketches were made and maps were drawn of the localities and excavations. Twenty-nine sites were excavated and written up for the American Philosophical Society by Edward Snow Carpenter, as described in Carpenter's report, "The ancient mounds of Pennsylvania" (913.748:C223).

There are ca. 50 maps in this collection.

(913.748: Un3)

50. VARLEY, CROMWELL J. Journal of astronomical observations.

Varley was a brother of the more noted astronomer, John Varley.

50 (1) 1845

Mars as he was once seen this month [surrounded by Delphinus, Antinous, Andromeda and Hercules]. 23 September 1845.

Cartographer: Cromwell J. Varley.

Size of paper: 17.8 x 11.4 cm.

50 (2) 1846

[Nebulosity]. 31 March 1846.

Cartographer: Cromwell J. Varley.

Size of paper: 17.8 x 11.4 cm.

50 (3) n.d.

[Saturn and constellations]. n.d.

Cartographer: Cromwell J. Varley.

Size of paper: 17.8 x 11.4 cm.

(522.1942: V42)

51. VAUGHAN, BENJAMIN. Papers.

Vaughan was the son of Samuel Vaughan, the first Benefactor mentioned in the American Philosophical Society's minutes, and brother of John Vaughan, the first active Librarian of the Society. A friend of Benjamin Franklin, he helped edit Franklin's autobiography while visiting him in France. He moved to Maine, where he was active in the development of the state and in philanthropic works.

51 (1) 1797

Draft of the town of Winthrop, Maine, by Jno. Jones, surveyor. 19 May 1797.

Cartographer: John Jones.

Size of paper: 32.7 x 24.8 cm.

Colored.

Provenance: Presented by Mary Vaughan Marvin, 1950.

(B: V46p)

52. WARNER, JOHN. Papers.

This amateur mathematician and naturalist (d. 1873) resided in Pottstown, Pennsylvania. He published on engineering subjects and organic morphology. He lived abroad from 1862 to 1868, visiting England, France, Italy, the Levant, Egypt, Greece, Constantinople, the Balkans, and Germany. Warner's pocket diaries from these travels include 23 maps.

(B: W243d)

PART II

Printed Maps

SPACE Nos. 53-61

53. 1772

A chart of the zodiacal stars, used in finding the longitude at sea by the moon. And their hourly positions to the circumpolar stars, both above and below the north and south poles. By S. Dunn. London: 1772 [4 July 1772].

Cartographer: S. Dunn.

Size of paper: 38.1 x 90.1 cm.

Colored.

Provenance: Presented by Samuel Dunn, 8 October 1772.

(523.29: 1772 Large)

54. [ca. 1832]

[Map of the moon accompanied by an explanation of the figures and letters]. N.p.: [ca.1832].

Cartographers: J. J. Littrow and Albert Richard.

Size: 40 cm. diameter.

Provenance: Presented by C. Nagy, 15 May 1833.

Enclosed in original wrapper marked: DERMOND.

(022: [ca.1832]: L722mar Large)

55. 1835

Sketch of the solar system for the use of schools. Printed at the Lith. Press of His Majesty the King of Oude. Lucknow: 1835.

Size: 75.2 x 73.9 cm.

Provenance: Presented by J. P. Engles.

(523.2: Sk2s Large)

56. 1837

General-Karte der sichtbaren Seite der Mondoberflaeche zugleich als übersichts-Blatt zur grössern Mondkarte von Wilh. Beer und Joh. Heinr. Mädler. Berlin: Simon Schropp et Comp.: 1837.

Cartographers: Wilh. Beer and Joh. Heinr. Mädler.

Engraver: E. Leidenfrost.

Size of paper: 59 x 48.3 cm.

Provenance: Presented by A. D. Bache, 6 December 1839.

(022: 1837: B397gks Small)

57. [ca. 1884]

Celestial charts made at the Litchfield Observatory of Hamilton College, Clinton, N.Y., by C. H. F. Peters. N.p.: [ca. 1884]. 20 charts.

Size of paper: 58 x 40 cm.

Original wrappers.

(525.5: L71c Large)

58. [ca. 1895]

Lick Observatory atlas of the moon. N.p.: [ca. 1895]. 20 plates.

Size of paper: 50.3 x 40.5 cm.

Published by the gift of W.W. Law, from original negatives.

There is also one photograph entitled: Mond-Atlas von L. Weinek.

(022: [ca. 1895]: L627atm Small)

59. 1911

Egyptian chart of the heavens. Constellations of the northern hemisphere. By Lee H. McCoy. Pueblo, Colo.: Franklin Press Co.: 1911 [24 July 1911].

Cartographer: Lee H. McCoy.

Size of paper: 43.2 x 41 cm.

Provenance: Presented by Lee H. McCoy, 3 August 1911.

McCoy wrote that this chart of the ancient Egyptian constellations contained "but about one half" of the Egyptian constellations he had worked out and that he would like to hear from the Society about it.

(523.89: M13e Small)

60. n.d.

Copernican system. N.p.: n.d.

Size: 19.2 x 20.3 cm.

(521.5: C79s Small)

61. n.d.

[Lunar orbiter material. Missions 1-5]. U.S. National Aeronautics and Space Administration. N.d.

Provenance: Presented by Donald Menzel, October 1971.

(523.39: Un3L)

THE WORLD Nos. 62-87

62. 1508

Universalior cogniti orbis tabula ex recentibus confecta observationibus. Published for Cl. Ptolomaei Geographia. Rome: 1508.

Size: 40.6 x 54.1 cm.

Facsimile. Contains facsimile autograph of John Farden.

(100: 1508: P956rom Large)

63. 1700

Nova & accuratissima totius terrarum orbis tabula nautica variationum magneticarum index juxta observationes anno 1700. Habitas constructa per Edm. Halley. 1700. London: Vincent Brookes, Day & Son, 1870.

Cartographer: Edmund Halley.

Size: 48 x 114.6 cm. [The size of the copy in the British Museum is 48 x 20½ in.]

Provenance: Presented by Greenwich Observatory, 6 October 1871.

Contains autograph of G. B. Airy.

Reproduced by photolithography, September 1870, with the permission of the Principal Librarian of the British Museum, from the copy (presumed to be of the original edition) preserved in the Library of the Museum.

(100: 1700: H143ter Large)

64. 1759

Mappe-monde divisée en ses quatre parties. N.p.: 1759.

Size: 15.6 x 30 cm.

(100: 1759: M325gbe Small)

65. 1765

Carte des parties principales du globe terrestre pour servir à l'histoire des deux premiers siècles depuis la création du monde. Dediée à Monseig[neu]r le Comte de Saint Florentin, ministre & secrétaire d'état, Commandeur des Ordres du Roy. Par Monsieur Luneau de Boisjermain. Paris: 1765.

Cartographer: H. Gravelot.

Engraver: Jac. Charpentier.

Size: 59.8 x 82.3 cm.

Inset: Supplement à la carte du Paradis Terrestre.

"Premier feuille."

(100: 1765: L965ter Large)

66. 1785

Chart of the world, according to Mercator's projection, shewing the latest discoveries of Capt. Cook. London: C. Dilly and G. Robinson: 1785. 15 June 1785.

Size: 37.2 x 48.6 cm.

Colored.

Provenance: Presented by Mathew Carey, 18 October 1805.

Engraved for Guthrie's new system of geography.

Contains manuscript notations for use in Carey's edition. The *Transactions* (Vol. 6) state that Carey presented the "Materials from which Guthrie's Geography were compiled." This is one of those maps. See: William Guthrie, *The general atlas for Carey's edition of his geography improved.* Philadelphia: Carey: 1795 [1 May 1795] (912.G98c).

(100: 1785: D582mer Large)

67. 1785

A map of the world from the best authorities. London: C. Dilly & G. Robinson: 1785.

Size of plate: 30 x 54 cm.

Colored.

Provenance: Presented by Mathew Carey, 18 October 1805.

Engraved for William Guthrie. *Atlas to his system of geography* . . . London: Dilly: 1785.

Shows tracks of James Cook's voyages.

The *Transactions* (Vol. 6) state that Carey presented the "Materials from which Guthrie's Geography were compiled." This is one of those maps. See: William Guthrie, *The general atlas for Carey's edition of his geography improved.* Philadelphia: Carey: 1795 [1 May 1795].

(100: 1785: D582wor Small)

68. 1786

[Map of the world in two hemispheres]. Western, New World Hemisphere; and, Eastern, Old World Hemisphere. London: William Faden: 1786 [28 October 1786].

Cartographer: William Faden.

Engraver: William Faden.

Size of plate: 36.7 x 73 cm.

Colored.

Provenance: Purchased from the library of William Priestman, 15 July 1831.

(120: 1786: F121weh Large)

69. 1787

A new general chart of the world, exhibiting the whole of the discoveries made by the late Capt. James Cook, F.R.S. with the track of the ships under his command. Also those of Capt. Phipps (now Lord Mulgrave) in his expedition to the North Pole. London: Wm. Faden: 1787 [1 January 1787].

Cartographer: William Faden.

Size: 41.7 x 57 cm.

Colored.

Provenance: Purchased from the library of William Priestman, 15 July 1831.

(100: 1787: F121ccv Large)

70. 1788

A map of the world, drawn & engraved from d'Anville's two sheet map, with improvements for I. Harrison No. 115, Newgate Street. [London]: I. Harrison: 1788. 1 December 1788.

Engraver: Neele.

Size of paper: 47.6 x 78 cm.

Colored.

In two hemispheres.

(100: 1788: H243amw Large)

71. [1789]

[Two designs for projecting the map of the world. 1789].

Engraver: Eliza Colles.

Size of plate: 18.8 x 27.1 cm.

In: Christoper Colles. *The geographical ledger and systematized atlas.*

(Pam., vol. 349, no. 4)

72. [ca. 1790]

Carte magnétique des deux hemisphères. N.p.: [ca. 1790].

Size: 48.2 x 93.3 cm.

"No. 8."

(100: [ca. 1790]: C242cmh Large)

73. [ca. 1790]

A chart of the world, according to Mercators projection; shewing the latest discoveries of Capt. Cook. London: [ca. 1790].

Engravers: Woodman & Mutlow.

Size: 36.4 x 47.2 cm.

Provenance: Presented by Mathew Carey, 18 October 1805.

Showing the tracks of the *Endeavour* and the *Resolution.* Engraved for Chambers's edition of Guthrie's new system of geography.

The *Transactions* (Vol. 6) state that Carey presented the "Materials from which Guthrie's Geography were compiled." This is one of those maps. See: William Guthrie, *The general atlas for Carey's edition of his geography improved.* Philadelphia: Carey: 1795 [1 May 1795].

(100: [ca. 1790]: W859ccv Large)

74. 1790

To George Washington president of the United States of America this magnetic atlas or variation chart is humbly inscribed by John Churchman. Philadelphia: James Johnson: 1790. First edition. Trimmed.

Cartographer: John Churchman.

Size of paper: 60.6 x 62.4 cm.

Colored.

Provenance: Presented by the author to the Society, 17 September 1790, with a description. Another copy presented by Johann R. Valltravers, 19 May 1797.

In: Pam. 538.7:C47e. John Churchman, *An explanation of the magnetic atlas or variation chart.* Philadelphia: James Johnson: 1790. Front.

Churchman was an American who was devoted to his own theories of the variations of the magnetic needle and other things. He had presented his scheme to the Society in 1787 and was given mild encouragement. He founded his belief on the

> hypothesis of two bodies (besides the moon) revolving round the earth, in small circles parallel to the equator; one near the north pole, and the other was near the south pole; and the needle, being wholly governed by the attraction of these magnetic satellites, will, in whatever part of the world, always rest in the plane of the circle, passing through them and the given place. [*The American Museum,* Sept. 1789: p. 218].

He persevered in his work and at the 17 September 1790 meeting his "Navigation Chart, or Magnetick Atlas" was presented. It disappeared, but another copy was given by Johann Rodolph Valltravers on 19 May 1797. The Valltravers copy is the one now at the Society.

Wheat: 6

(100: 1790: C492mag Large)

74a. [1792]

Dialing. A "Universal Dial" incorporating a "terrestrial globe," indicates Africa, Europe, Asia, and America on the globe. [1792].

Scale: Diameter of the sphere = 2.8 cm.

Engraver: Scot.

Size: 27 x 20.6 cm.

Provenance: Presented by Thomas Dobson, 1798.

In: Thomas Dobson, publisher, *Encyclopaedia,* vol. 5, opposite p. 792.

Wheat: 15

(032: En2)

74b. [1792]

Geography. Map of the world, comprehending the latest discoveries. Captain Cook's last voyage is shown with his Alaskan discoveries. Zaara or The Desert appears for the Sahara in Africa. [1792].

Scale: Diameter of sphere, 17.4 cm.

Size: 19.6 x 35.5 cm.

Provenance: Presented by Thomas Dobson, 1798.

In: Thomas Dobson, publisher, *Encyclopaedia,* vol. 7, following page 662.

Wheat: 16

(032: En2)

74c. [1792]

Geography. A map of the world in three sections, describing the polar regions to the tropics in which are traced the tracts of Lord Musgrave and Captain Cook towards the North and South Poles and the torrid zone or tropical regions with the new discoveries in the South Sea. [North and south hemispheres and a belt map of the tropical regions]. [1792].

Engraver: W. Barker.

Size of plate: 23.5 x 43.5 cm.

Provenance: Presented by Thomas Dobson, 1798.

In: Thomas Dobson, publisher, *Encyclopaedia,* vol. 7, following p. 662.

Wheat: 17

(032: En2)

74d. [1792]

Geography. [With six diagrams, four of which show portions of the earth's surface. All the continents are shown on one or more of the hemispheres. California is shown as an island]. [1792].

Size of plate: 23.7 x 17.5 cm.

Provenance: Presented by Thomas Dobson, 1798.

In: Thomas Dobson, publisher, *Encyclopaedia,* vol. 7, opposite p. 650.

Wheat: 14

(032: En2)

75. 1793

Die Obere oder Nördliche Halbkugel der Erde auf den Horizont von Berlin stereographisch entworfen von T. E. Bode Astronom der Königl. Pr. Acad. d. Wissensch. 1793, [and] Die Untere oder Südliche Halbkugel der Erde . . . [Berlin]: 1793. 2 maps.

Sizes: 45.7 x 43.5 cm. and 43.5 x 43.3 cm.

(100: 1793: B631ger Large)

76. [ca. 1793]

The world from the best authorities. [Boston]: Thomas & Andrews: [ca. 1796].

Size of plate: 20.4 x 38.2 cm.

Engraver: Amos Doolittle.

Engraved for Morse's geography. [*American Universal Geography*, 1, front.]

Wheat: 19

(100: [ca. 1793]: D722bos Small)

77. 1794

A correct chart of the terraqueous globe, on which are described lines shewing the variation of the magnetic needle in the most frequented seas; originally composed in the year 1700 by the celebrated Dr. Edmund Halley; renewed by William Mountaine and James Dodson, F R S according to observations made about the year 1756. London: Laurie & Whittle: 1794 [12 May 1794].

Cartographers: Edmund Halley, William Mountaine, and James Dodson.

Size: 52.1 x 122 cm. *Size of paper:* 55.3 x 147.9 cm.

On either side are: An account of the variation chart, and, Remarks on the variation lines &c.

(100: 1794: M865haL Large)

78. [1794]

A general chart of the globe, shewing the course of the Gulph Stream, and various tracks to and from the East Indies, China, Europe &c. [This is a chart on Mercator's projection with eight different lines lettered from "A" to "H" showing Truxtun's courses between various points. 1794].

Scale: 2.5 cm. = ca. 690 mi.

Cartographer: Thomas Truxtun.

Size: 45.5 x 89.4 cm.

Provenance: Presented by Thomas Truxtun, 17 February 1797.

In: Thomas Truxtun, *Remarks, instructions, and examples relating to the latitude and longitude . . .*, frontispiece.

Wheat: 22

(527: T77)

78a. [1798]

General chart on Mercator's projection. [1798].

Scale: 2.5 cm. = ca. 35° latitude.

Engraver: Rollinson.

Size: 17.9 x 24.9 cm.

See: John Payne, *A new and complete system of universal geography,* vol. 1, opposite p. v.

Wheat: 39

(910: P29)

78b. [1798]

The world from the best authorities [1798].

Engraver: Rollinson.

Size of page: 21 x 38 cm.

In: John Payne. *A new and complete system of universal geography,* vol. 1, frontispiece.

Wheat: 40

(910: P29)

79. 1817

Chart of the variation of the magnetic needle, for all the known seas comprehended within sixty degrees of latitude north and south; with a new and accurate delineation of the magnetic meridians, accompanied with suitable remarks & illustrations. By Thomas Yeates. [London]: Black, Barbury & Allen: 1817 [22 August 1817].

Cartographer and engraver: J. Walker.

Size: 53 x 122.8 cm. *Size of paper:* 60.6 x 154.4 cm.

Provenance: Presented by John Garnett, prior to 1818.

Contains also: Remarks historical and explanatory; Deductions and experiments made by Capt. M. Flinders relating to the variation of the magnetic needle; and Example of the aberration of the needle caused by a change in the ship's head.

(100: 1817: Y39varc Large)

80. 1817

The world on Mercator's projection; revised and improved to 1818 by John Melish. Philadelphia: John Melish & Saml. Harrison: 1817 [16 October 1817].

Cartographer: John Melish.

Engravers: S. Harrison and G. Murray.

Size: 91 x 123 cm.

Colored.

Provenance: Presented by John Melish, 5 November 1819.

Contains statistical table of the several countries shown on the map. Contains also tracks of various explorers: James Cook; Vancouver; Furneaux; La Perouse; Clerke; Capt. Gore (*Resolution*); Commander James from Bombay; Capt. Phipps, Lord Musgrave; and of certain famous ships.

(100: 1817: M485wmp Large)

81. 1831

A new and authentic map of the world embracing all the recent discoveries, and exhibiting particularly the nautical researches of the most distinguished circumnavigators from the latest & best authorities with numerous corrections & additions, by H. S. Tanner, 1831. Philadelphia: Tanner: 1831.

Cartographer: H. S. Tanner.

Engraver: E. B. Dawson.

Size: 89.8 x 168.5 cm.

Colored.

Provenance: Presented by Henry Schenck Tanner, 15 April 1831.

Contains also: comparative lengths of the principal canals in the world, projected on a scale of 34 miles to an inch; statistics for North America, South America, West Indies; Summary of Africa, Asia, Oceania, and Europe statistics; heights of mountains; British possessions in Africa, America, Asia, Oceania; possessions of European powers in Asia, Oceania, Africa, America; vertical sections of North America, projected on uniform scales.

(100: 1831: T158wnr Large)

82. 1842

The world. Philadelphia: Engraved and published by W. Williams: 1842.

Engraver: William Williams.

Size: 51.6 x 63.9 cm.

Colored.

Insets: District of Columbia; Sandwich Islands; Map of Palestine and the Holy Land; and Map of the American Colony of Liberia.

Contains tracks of ships of Cook, Biscoe, Furneaux, Vancouver, La Perouse; Clarke; Gore, James Waddell, Columbus, etc.

(100: 1842: W679eah Large)

83. 1855

Colton's map of the world on Mercator's projection. New York: J. H. Colton & Co.: 1855.

Size: 41 x 62.4 cm.

Colored.

Contains isothermal lines of frigid or cold, temperate, tropical and equatorial zones; and, tracks of famous explorers and ships: Cook, La Perouse, Vancouver, Furneaux, *U.S.S. Vincennes,* etc.

(100: 1855: C672cwm Large)

84. 1888

Maps showing the location of the diplomatic and consular offices of the United States of America. 1 March 1888. Washington: Government Printing Office: 1888. 8 maps.

Cartographers: A. C. Roberts and G. Noetzel.

Lithographer: N. Peters, Photo-Lithographer, Washington, D.C.

Size: 44.7 x 54.7 cm.

Contains index to the maps.

(100: 1888: Un38dip Large)

85. [ca. 1908]

Principal transportation routes of the world. U.S. Department of Commerce and Labor, Bureau of Statistics. N.p.: [ca. 1908].

Size: 57.6 x 136.3 cm. *Size of paper:* 81.3 x 145 cm.

Colored.

At bottom are distances from New York, New Orleans, San Francisco, and Port Townsend to the principal ports of the world and the principal cities of the United States.

(100: [ca.1908]: P932trw Large)

86. [ca.1943]

[Map of the world]. Chicago: Geographical Publishing Co.: [ca. 1943].

Size: 75.5 x 66 cm.

Colored.

Contents include scenes from foreign lands and pictures of U.S. presidents to F. D. Roosevelt.

(100: [ca.1943]: G293abc Roller)

87. n.d.

Terrestrial magnetism. By Colonel Edward Sabine. Physical atlas. Plate 23. London: William Blackwood & Sons: n.d.

Size of paper: 53 x 66.4 cm.

Colored.

Provenance: Presented by Maj. Gen. Sabine.

Contains seven maps of the world.

Accompanied by a broadside which describes the plate.

(100: n.d.: Salltem Large)

EASTERN HEMISPHERE Nos. 87a-87b

87a. [1796]

[Eastern hemisphere. A topographical map showing Europe, Africa and Asia. 1796].

Scale: 13.7 cm. diameter.

Size: 14.6 x 15.9 cm.

In: Constantin F. C. Volney, *The Ruins,* opposite p. 29.

Wheat: 29

(904: V88r)

87b. [1792]

Geography. [Three spheres, one of the heavens; one depicting an imaginary island; and, one showing Europe and Africa. 1792].

Engraver: Smither.

Sizes of spheres: Diameters of 5.7 cm.; 3.4 cm.; and 5.7 cm.

Provenance: Presented by Thomas Dobson, 1792.

In: Thomas Dobson, publisher, *Encyclopaedia,* vol. 7, following p. 660.
Wheat: 13

(032: En2)

WESTERN HEMISPHERE Nos. 88-91

88. 1786

A new map of the whole continent of America, divided into north and south and West Indies. Wherein are exactly described the United States of North America as well as the several European possessions according to the preliminaries of peace signed at Versailles Jan. 20, 1783. Compiled from Mr. D'Anville's maps of that continent, with the additions of the Spanish discoveries in 1775 to the north of California & corrected in the several parts belonging to Great Britain, from the original materials of Governor Pownall, M.P. London: Robert Sayer: 1786 [15 August 1786]. 2 sheets.

Scale: 60½ English mi. = 1 degree.

Cartographers: D'Anville and Governor Pownall.

Size: 51.7 x 119.5 cm.

Colored.

Inset: The supplement to North America containing the countries adjoining to Baffins and Hudsons Bays.

Provenance: Purchased from the library of William Priestman, 15 July 1831.

(120: 1786: D196nac Large)

89. [ca. 1787]

A new and correct map of [North and South] America, with the West India Islands. London: Robert Sayer & Co.: [ca. 1787]. 2 copies.

Size: 55 x 47.2 cm. *Size of paper:* 61 x 103.1 cm.

Colored.

Insets: Port Royal Harbor, Boston Harbor, Georgia, the North Pole, and how fish are cured and dried in Newfoundland. Also, on each side, are scenes of the American Indian, chiefly of Virginian, Mexican, and Peruvian natives.

(122: [ca. 1787]: Sa97amw Large)

90. 1794

A new map of the whole continent of America, divided into North and South and West Indies. Wherein are exactly described the United States of North American as well as the several European possessions according to the preliminaries of peace signed at Versailles Jan. 20, 1793. Compiled from Mr. d'Anville's maps of that continent, with the addition of the Spanish discoveries in 1775 to the north of California & corrected in the several parts belonging to Great Britain, from the original materials of Governor Pownall. London: Laurie & Whittle: 1794 [12 May 1794]. 2 copies.

Scale: 69½ English mi. = 1 degree.

Cartographers: d'Anville and Gov. Pownall.

Size: 103 x 119.7 cm.

Colored.

Insets: Countries adjoining to Baffins and Hudsons Bays; and Possessions of various countries in North and South America.

(122: 1794: L365con Large)

91. 1796

America nach der zweyten Ausgabe von Arrowsmiths Weltcharte und dessen Globular Projection, nach den Berichten der Jesuiten und anderer Reisebeschreiber, und nach Raynals, Gatterers, Angaben entworfen von C. Mannert. Herausgegeben mit Kaiser. Allergl. Privil. Nürnberg: Adam Gottlieb Schreiber u. Weigel: 1796. [North and South America].

Cartographers: C. Mannert and Arrowsmith.

Size: 52.6 x 60.8 cm.

Colored.

(120: 1796: M315awc Large)

THE ARCTIC Nos. 92-100

92. [ca. 1790]

[Circular map of the arctic regions]. N.p.: [ca. 1790].

Size: 24 cm. diameter.

Endorsed: Late eighteenth-century hand: North Pole.

Provenance: Presented by Mathew Carey, 18 October 1805.

Contains: Mr. Hearn's "Rout" in 1771 from Fort Churchill.

Published for William Guthrie, *Atlas to his system of geography.* The *Transactions* (vol. 6) state that Carey presented the "Materials from which Guthrie's Geography were compiled." This is one of those maps. See: William Guthrie, *The general atlas for Carey's edition of his geography improved.* Philadelphia: Carey: 1795 [1 May 1795].

(171: [ca. 1790]: Ar21oce Small)

93. 1855

Arctic Sea. Baffin Bay. Sheet 1. 1853. London: Hydrographic Office of the Admiralty. Corrected to 1855. [London]. 1855.

Engravers: J. & C. Walker.

Size: 63.3 x 48.5 cm.

Insets: Omenak Fiord; and Disko Bay.

There are ten engravings of capes, inlets, islands, etc.

(171: 1855: Ar27bab Large)

94. 1855

Chart exhibiting the discoveries of the Second American Grinnell Expedition in search of Sir John Franklin. Unrevised from the original material and projected on the spot. E. K. Kane. Deposited 15 November 1855.

Scale: 2.8 cm. = 10 mi.

Cartographer: E. K. Kane.

Lithographer: Julius Bien.

Size: 89 x 56.2 cm.

Provenance: Presented by E. K. Kane.

(172.3: 1855: G895exp Large)

95. 1856

Arctic Sea. Melville Sound. Sheet II. 1856. [London]: 1856 [26 May 1856].

Engravers: J. & C. Walker.

Size: 63.7 x 49.7 cm.

(171: 1856: Ar27mvs Large)

96. 1865

Chart showing the discoveries tracks and surveys of the Arctic Exploring Expedition of 1860 and 1861. I. I. Hayes, M.D., commanding. Newly projected from revised materials discussed for the Smithsonian Institution by Charles A. Schott. Washington: 1865.

Scale: 1:1,200,000.

Cartographer: Charles A. Schott.

Engraver: H. S. Barnard and A. Petersen.

Size: 57.8 x 36.2 cm.

(171: 1865: Sch67aee Small)

97. 1897

The Arctic regions comprising the most recent explorations of Robert E. Peary, Fridtjov Nansen, and F. Jackson, by Prof. Angelo Heilprin. Philadelphia: Geographical Society of Philadelphia: 1897.

Cartographers: Angelo Heilprin and J. W. Ross.

Size of page: 45 x 49.5 cm.

(170: 1897: H343pnj Small)

98. [ca. 1900]

Arctic Regions [showing bank ice, cold currents & warm currents]. N.p.: [ca. 1900].

Size of paper: 26.9 x 23.1 cm.

Colored.

(171: [ca. 1900]: Ar21npc Small)

99. 1903

Polar regions Baffin Bay to Lincoln Sea, showing the recent discoveries and routes of exploration of Civil Engineer Robert E. Peary U.S.N. together with the work of earlier explorers. Based on the chart of the Hydrographic Office. Washington: 1903.

Cartographer: F. B. Greene.

Size: 48.2 x 44.2 cm.

Colored.

To accompany *Bulletin,* vol. IV, no. 1, Geographical Society of Philadelphia.

"The name United States Coast, designating the most northernly land-mass, was suggested by Commander Perry in letter, Feb. 22, 1904."

(176: 1903: P316bbL Small)

100. 1924

Physical map of the Arctic. Translated and revised by the American Geographical Society of New York from map in Andrée's *Handatlas*, 8th edition, 1924. Printed in Germany. Copyright, 1929, by the American Geographical Society of New York. New York: 1929.

Size of page: 46 x 58.6 cm.

Scale of main map: 1:20,000,000.

Colored.

Insets: Novaya Zemlya, Jan Mayen Island, Bear Island, Ice Fiord (Spitsbergen), Spitsbergen and Bear Island, Franz Josef Land, Eastern Greenland, Bering Strait, Northernmost Greenland, Southern Greenland, and Smith Sound to Robeson Channel.

(170: 1924: Am31phy Small)

THE ANTARCTIC Nos. 101-104

101. [ca. 1822]

[Three charts of the Antarctic (South Shetland Islands) comprising chart of Edward Bransfield, press mark "Ae 1 S92," Georges Bay, press mark "Ae 1 S90," and chart of New South Britain, discovered by Capt. Smith in the brig, "Williams," the 19th. February, 1819]. Great Britain. Hydrographic Department. [London: ca. 1822].

Cartographers: Edward Bransfield and Capt. Smith.

Size: Various.

Provenance: Presented by J. A. Edgell, August 1941.

See: *Geographical Journal*, XCIV, 315 (Oct. 1939): map C, and *Geographical Review*, XXXI, 491 (July 1941).

Photostats of the original documents in the Hydrographic Office, Great Britain.

(183.54: [ca. 1822]: G813ant Roller)

102. 1929

Bathymetric map of the Antarctic (southern Atlantic, Indian, and Pacific Oceans) compiled by the American Geographical Society of New York. New York: A. Hoen & Co.: 1929.

Scale: 1:20,000,000.

Size: 55.3 x 60 cm.

*Colored,*with a scale of submarine relief.

(180: 1929: Am31bath Large)

103. 1929

Map of the Antarctic, compiled by the American Geographical Society of New York. Wilkins-Hearst Antarctic expedition 1928-1929. New York: 1929.

Scale: 1:12,500,000.

Size: 88 x 128 cm.

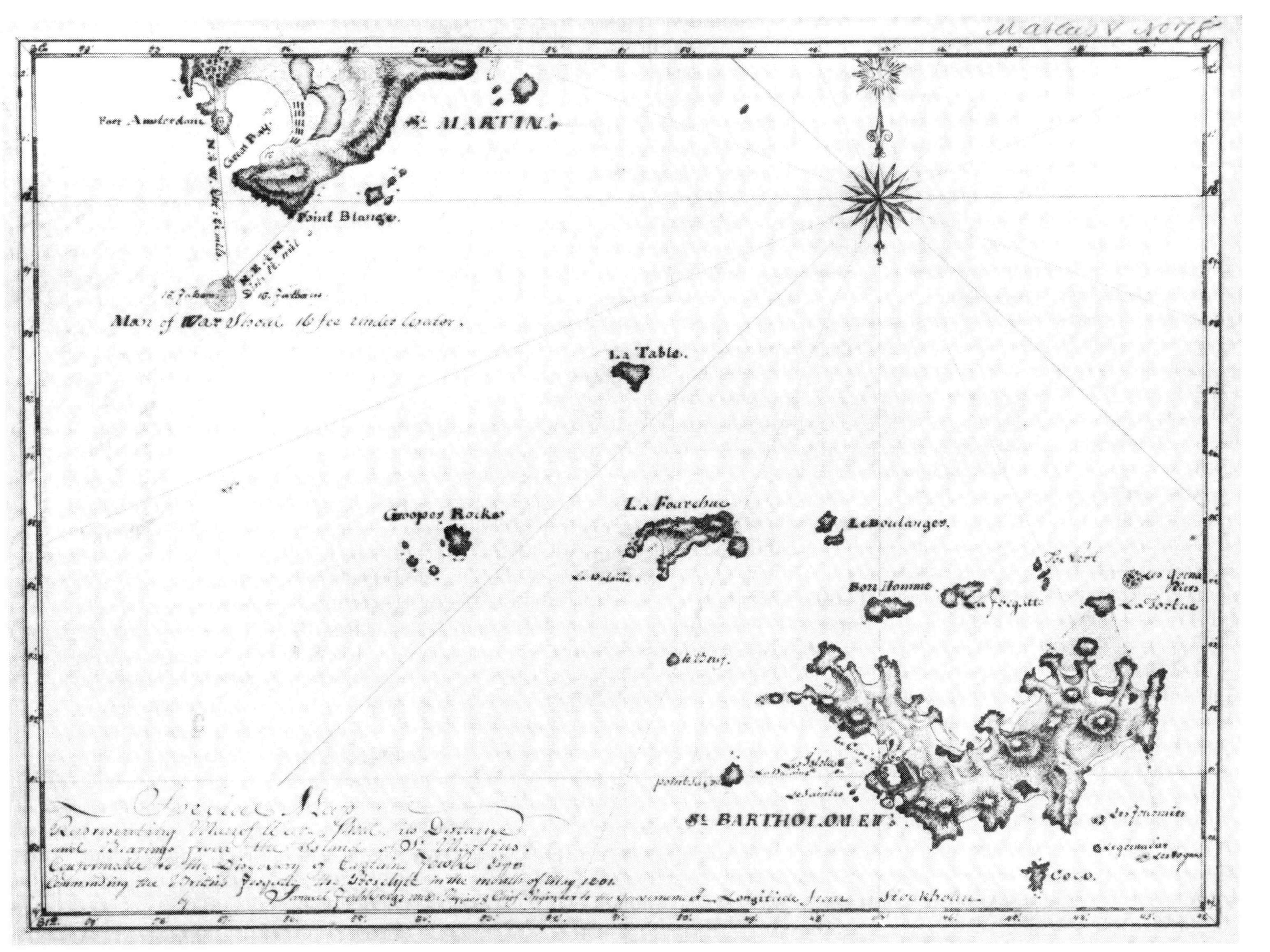

Samuel Fahlberg. Special map, representing Man of War Shoal . . . 1801.
[No. 32 (65)]

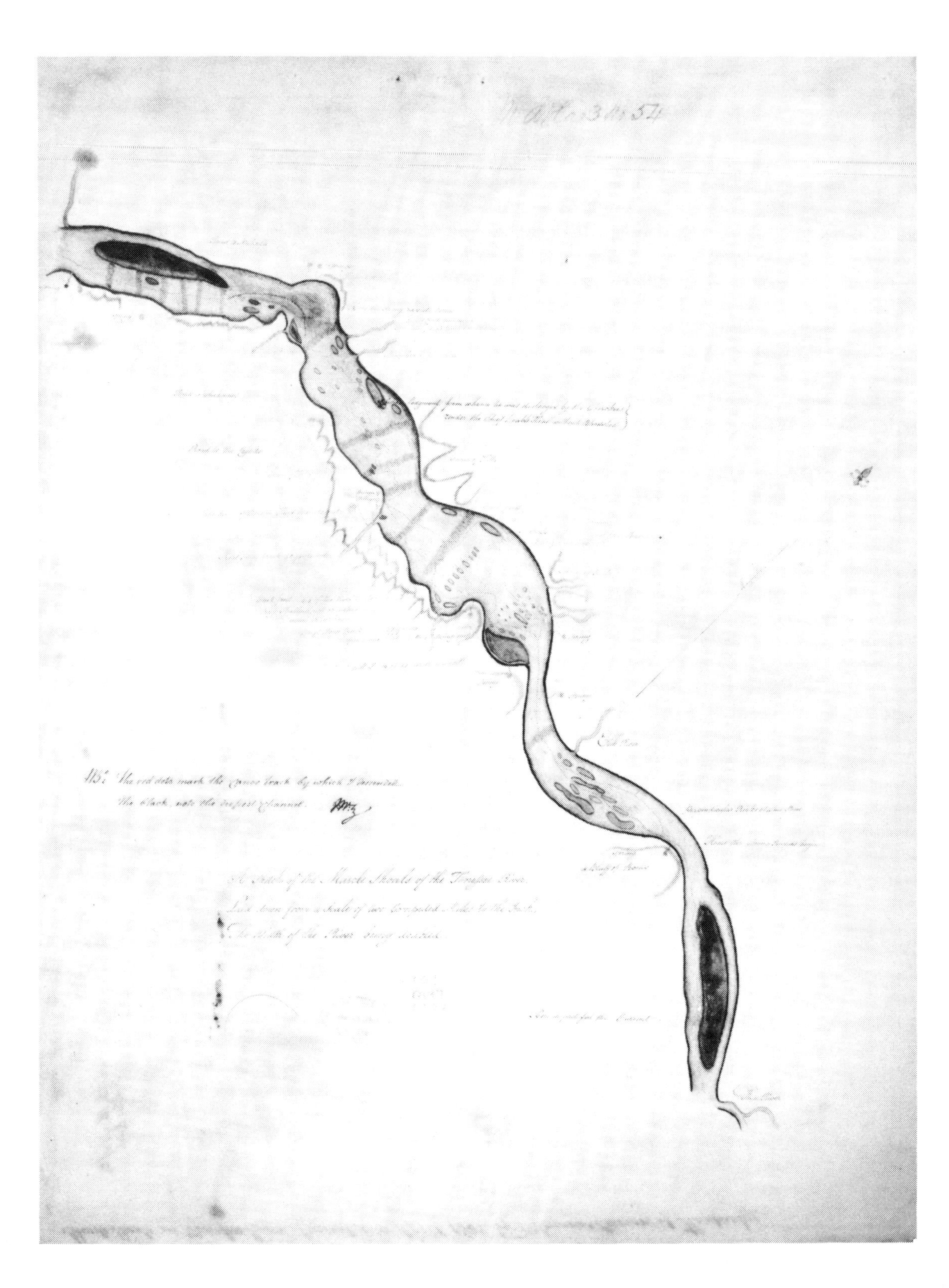

James Wilkinson. A sketch of the Muscle Shoals of the Tenessee [sic] River. 1802. [No. 32 (66)]

Colored.

Insets: Map of the Antarctic archipelago, showing route of Sir Hubert Wilkins's flight, 20 December 1928. Also, photographs of fiords, islands, etc.

Provenance: Presented by Isaiah Bowman, 1929.

(180: 1929: Am31exp Roller)

104. 1943

Antarctica, compiled from all available sources to 1943, including the results of all American explorations from the United States Exploring Expedition, 1839-1840, to the United States Antarctic Service, 1940-1941. Washington: Hydrographic Office: 1943.

Scale (latitudinal): 1:11,250,000.

Size: 78 x 122 cm.

Colored.

Azimuthal equidistant projection.

(180: 1943: Un36hyd Roller)

EUROPE Nos. 105-118

105. 1759

L'Europe divisée en ses grands états 1759. N.p.: 1759.

Size: 15.9 x 19.5 cm.

(200: 1759: Eu72for Small)

106. [ca. 1780]

A new map of Europe. With its empires, kingdoms, republicks, etc. justly described and graduated to the true longitude and latitude of the latest and best discoveries. And to render this map more generally useful and entertaining a summary view of the several kingdoms and dominions of Europe is engraved on the sides in geographical and political tables. By Richard Bennett Engraver.

Cartographer and engraver: Richard Bennett.

Size: 58.8 x 74.8 cm. Size of paper: 66.5 x 106.5 cm.

Colored.

Contains historical data in two columns at each side of the map.

(200: [ca. 1780]: B431ekr Large)

107. 1785

An accurate map of Europe from the best authorities. London: C. Dilly & G. Robinson: 1785 [1 January 1785].

Scale: 69 British statute mi. = 1 degree.

Size: 35.7 x 37.4 cm.

Colored.

Inset: Spitsbergen, East and West Greenland, and Iceland.

Provenance: Presented by Mathew Carey, 18 October 1805.

Engraved for William Guthrie, *Atlas to his system of geography* . . . (London: Dilly: 1785).

Contains 38 contemporary manuscript notes relating to validity of map.

The Transactions (vol. 6) state that Carey presented the "Materials from which Guthrie's Geography were compiled." This is one of those maps. See: William Guthrie, *The general atlas for Carey's edition of his geography improved.* Philadelphia: Carey: 1795 [1 May 1795].

(200: 1785: D582eur Small)

108. 1787

Europe divided into its empires, kingdoms, states, republics, &ca. By Thomas Kitchin. With many additions and improvements from the latest surveys and observations. London: Robert Sayer: 1787 [1 January 1787].

Cartographer: Thomas Kitchin.

Size: 51.3 x 105 cm. *Size of paper:* 54.4 x 132.7 cm.

Colored.

Provenance: Purchased from the library of William Priestman, 15 July 1831.

On each side is a printed column of tables with an explanation.

(200: 1787: K655eks Large)

109. [1792]

Europe. [1792].

Engraver: Creed.

Size: 15 x 19.1 cm.

Provenance: Presented by Thomas Dobson, 1798.

In: Thomas Dobson, publisher, *Encyclopaedia,* vol. 7, following p. 40.

Wheat: 749

(032: En2)

110. [1808]

L'Europe [géographique, politique et statistique] en 1808. N.p.: [1808].

Size: 37.1 x 44 cm. Size of paper: 53.1 x 69.6 cm.

Colored.

Contains notes on three sides concerning the geographical, political, and statistical remarks for Europe.

(200: [1808]: Eu12egps Large)

111. [ca. 1812]

Carte générale du théâtre de la guerre. Plate ler. N.p.: [ca. 1812].

Scale: 25 French common leagues = 1 degree.

Size: 35.4 x 48.3 cm.

(204: [ca. 1812]: Eu72nap Small)

112. [ca. 1815]

Nouvelle carte de l'Europe dans son état actuel après la traité de Vienne et la réunion de la Hollande à l'Empire Français divisé en 130 départemens. Comprenant en outre une partie de l'Asie, le golfe Persique, le Bassin de la Méditerranée, et tout la côte d'Afrique. Nouvellement dessinée, revue corrigée et considerablement augmentée par M. Herisson Ingénieur-Géographe, d'après les cartes les plus nouvelles et le plus estimées, nationales et étrangères, les changemens politiques, les observations de savants, &c. &c. Paris: Desray: [ca. 1815].

Scale: 25 French common leagues = 1 degree.

Cartographer: Herisson.

Engraver: M. Glot.

Size: 94.8 x 124.7 cm.

Colored.

(200: [ca. 1815]: H423tvh Large)

113. 1829

Tableau théorique de la succession et de la disposition la plus général en Europe, des terrains et roches qui composent l'ecorce de la terre. Un exposition graphique de tableau des terrains, publiée en 1829 par M. Alexandre Brongniart. Paris: Levrault: 1829.

Cartographers: Constant Viguier and Alexandre Brongniart.

Engraver: J. Collon.

Lithograph: Knecht & Roissy.

Size: 51 x 65.4 cm.

Colored.

(200: 1829: B781eur Large)

114. [ca. 1860]

Geological map of Europe executed under the direction of Joseph Prestwich, F.R.S. by William Topley, F.G.S., and J.G. Goodchild, F.G.S. from the latest surveys. Edinburgh: W. and A. K. Johnston: [ca. 1860].

Scale: 69 English statute mi. = 1 degree.

Cartographers: William Topley and J. G. Goodchild.

Engravers: W. and A. K. Johnston.

Size: 44.3 x 57.1 cm.

Colored.

Printed for Prestwich's *Geology*, vol. 2, Clarendon Press.

See: Sir Joseph Prestwich, *Geology chemical, physical, & statigraphical.* Oxford: Clarendon Press: 1886-1888. 2 vols.

(200: [ca. 1860]: P926geo Small)

115. 1895-1913

Carte géologique internationale de l'Europe executée sous la direction de Beyrich et Hauchecorne. Berlin: Reimer: 1895-1913. 47 maps, t.p., and page identifying the colors used. Berliner Lithog. Institut.

Scale: 1:1,500,000.

Cartographers: Beyrich et Hauchecorne.

Lithographer: Berliner Lithog. Institut.

Size: 48.7 x 57 cm.

Colored.

Original wrappers.

One hundred sixty German marks were sent 7 May 1896 to the Geographische Verlagshandlung Dietrich Reimer for a copy of this international geological map of Europe [Archives. P. Frazer to E. Vohsen, 7 May 1896]. A receipt for the money was mailed 11 May 1896 [Archives. D. Reimer to APS, 11 May 1896]. The Society had acquired a typewriter by this date and Persifor Frazer wrote J. Sergeant Price 7 May 1896 that the typewriter was useful and was now paid for. He spoke of the purchase of German marks for this map in this letter [Archives].

(200: 1895-1913: B461geo Large)

116. [1905]

Eclipse totale de soleil des 29 et 30 Août 1905 lieux des points d'où l'on peut en observer les phases. Carte dressée sous la direction des Bureau des longitudes. Paris: Gauthier-Villars: 1905.

Size: 93.5 x 101.8 cm.

Colored.

(200: [1905]: Ec56ets Book map)

117. 1921

Europe and Asia Minor (with inserts of Africa and Oceana). Compiled by the U.S. Army Military Intelligence Division. Based upon various data, including maps by Edward Stanford, John Bartholomew and Co., and U.S. Hydrographic Office. Map No. 62. 1921. [Washington]: U.S. Geological Survey: 1921. Map in 2 sheets.

Scale: 1:3,200,000.

Cartographers: U.S. Army Military Intelligence Division, Edward Stanford, John Bartholomew and Co., and U.S. Hydrographic Office.

Size: 145.7 x 167.5 cm.

The boundaries shown upon this map should not be regarded as having political significance, or as involving recognition of new governments, or of new boundaries, or of transfers of territory, except as the United States Government has already made such recognition in a formal and official manner.

(200: 1921: Un38eam Large)

118. n.d.

[Medieval map of Europe, the Mediterranean and North Africa]. Published by A. Lesoüef. N.p.:n.d.

Size: 74.3 x 103 cm.

Provenance: Presented by A. Lesoüef.

Contains stamp: Institution Ethnographique.

(100: n.d.: M465emn Large)

ENGLAND AND WALES Nos. 119-143

119. [ca. 1665]

Plan of the late fire in Cornhill [Street, London]. N.p.: M. Payne: n.d.

Size: 11.3 x 18 cm.

Provenance: Presented by James B. Nolan, 1958.

Contains list of tenants burnt out by the fire. This is from a publication, pp. 148-149.

(212: [1665]: C812ngb Small)

120. 1760

A new map of the counties of Cumberland and Westmoreland, divided into their respective wards from the best surveys & intelligence. Illustrated with historical extracts relative to natural history, produce, trade and manufacture. Shewing also the rectories & vicarages, with various other improvements. London: T. Bowles, Robt. Sayer and John Bowles: 1760.

Scale: 69½ British statute mi. = 1 degree.

Size: 68.2 x 52.1 cm.

Colored.

(212: 1760: Sa97cew Large)

121. [ca. 1762]

An accurate map of Northampton shire. Divided into its hundreds, and laid down from the best authorities, assisted by the most approved modern maps, with various improvements. Illustrated with historical extracts relative to its natural produce, trade, manufacture, & present state of its principal towns. By Eman[uel] Bowen geographer to His Majesty. [London]: Robt. Wilkinson & Robt. Sayer: [ca. 1762].

Scale: 69 British statute mi. = 1 degree.

Size: 70 x 53 cm.

Colored.

(214: [ca. 1762]: B671nsh Large)

122. 1770

A new and correct map of England and Wales laid down according to ye accurate improvements of Senex, Moll & other modern geographers; exhibiting all ye cities, borough and market towns: also all ye direct & principal cross roads, w[i]th all the post towns as they are at present regulated by the Postmaster Gener[a]l together with ye computed distances between town & town. 1770. [London]: Robt. Sayre [Sayer]: 1770.

Scale: 60 English mi. = 1 degree.

Cartographers: Senex and Moll.

Engraver: Em[anue]l Bowen.

Size: 57.8 x 52.5 cm.

Colored.

Contains on each side coats of arms of all the cities and shire-towns of England and Wales; and an alphabetical table of the cities, borough and market towns.

(211: 1770: B67lewb Large)

123. [ca. 1770]

A new map of England & Wales with the roads and distances between town and town, taken from all the surveys hitherto published, describing the cities, boroughs & market towns in each county, and their distances from London. London: Robt. Sayer: [ca. 1770].

Size: 55.4 x 50 cm.

Colored.

Contains on each side views of London, Bristol, Hull, Newcastle, Liverpool, Yarmouth, Portsmouth, and Plymouth.

(211: [ca. 1770]: Sa97gbew Large)

124. [ca. 1774]

The post roads through England and Wales, by T. Jefferys. London: Jefferys & Faden: [ca. 1774].

Scale: 69½ British mi. = 1 degree.

Cartographer: T. Jefferys.

Size: 57 x 48.5 cm.

Colored.

Provenance: Purchased from the library of William Priestman, 15 July 1831.

(211: [ca. 1774]: J377prd Large)

125. 1774

The traveller's guide or a new and accurate map of England & Wales, containing all the cities, market towns, rivers, bridges, &c. also the great & small roads, with the exact number of miles in figures from town to town. London: R. Marshall: 1774.

Size: 59.1 x 51.6 cm.

Colored.

Contains perspective views on each side of: London, Bristol, Hull, Newcastle, Chatham, Liverpool, Yarmouth, Portsmouth, Plymouth, and Sheerness.

(211: 1774: M355tgw Large)

126. [ca. 1778]

The environs, or countries twenty miles round London, drawn from accurate surveys, by Thomas Kitchin Geographer. [London: ca. 1778].

Scale: 69 British statute mi. = 1 degree.

Cartographer: Thomas Kitchin.

Size: 47.5 x 54.5 cm.

Colored.

Endorsed with note by B. S. Barton: "Contains *no* papers relating to my work . . . to any other work, or object in which I am engaged. B.S.B."

(211.949: [ca. 1778]: K644ec Small)

127. 1788

A new map of England & Wales, with all the great and cross roads and arms of the cities and principal towns in the kingdom. [London]: Robert Sayer: 1788 [10 April 1788].

Size: 57.2 x 51.8 cm.

Colored.

On each side are the coats of arms of all the cities and shire towns of England, and an alphabetical table of the cities, borough and market towns.

(211: 1788: Sa97cra Large)

128. [ca. 1790]

An accurate map of England and Wales, with the principal direct and post roads, from the best authorities. N.p.: [ca. 1790].

Engravers: Woodman & Mutlow.

Size: 34.5 x 32.4 cm.

Provenance: Presented by Mathew Carey, 18 October 1805.

Engraved for Chambers's edition of Guthrie's new system of geography.

The *Transactions* (vol. 6) state that Carey presented the "Materials from which Guthrie's Geography were compiled." This is one of those maps. See: William Guthrie, *The general atlas for Carey's edition of his geography improved.* Philadelphia: Carey: 1795 [1 May 1795].

(211: [ca. 1790]: W859eng Small)

129. 1790

Caesars camp at Hollwood in the county of Kent, about 14 miles southeast from London, occupies very high ground, commanding extensive prospects on the northeast, and south sides: the intrenchments have been cut with great labour in a hard gravelly soil, and therefore those on the west and north sides remain to this day in good preservation. Plate X. Surveyed and drawn by Thos. Milne, 11th Oct. 1790. London: Society of Antiquaries: 1806 [11 April 1806].

Scale: 1 in. = 5 chains.

Cartographer: Thos. Milne.

Engraver: James Basire.

Size: 54.4 x 40.3 cm.

Shows William Pitt's Hollwood house.

(216: 1790: B291cch Large)

130. [1792]

England. [1792].

Scale: 2.5 cm. = 75 mi.

Engraver: Scot.

Size: 16.3 x 18.7 cm.

Provenance: Presented by Thomas Dobson. 1798.

In: Thomas Dobson, publisher, *Encyclopaedia,* vol. 6, following p. 582.

Wheat: 748

(032: En2)

131. [1793]

[The Royal Exchange, London. 1793].

Scale: 2.5 cm. = ca. 60 ft.
Size: 8.4 x 10 cm.
Provenance: Presented by Thomas Dobson, 1798.
In: Thomas Dobson, publisher, *Encyclopaedia,* vol. 10, following p. 253.
Wheat: 757

(032: En2)

132. 1794

London docks. Plan of the River Thames with the proposed docks and cut. London: 1794.
Size: 27.8 x 50.6 cm.
Colored.
Provenance: Presented by William Vaughan.
See: William Vaughan, *Plan of the London dock* . . ., 1794.

(216: 1794: L845doc Small)

133. 1794

New map of England and Wales. Describing the post and cross roads, with the distances, from place to place. London: R. Wilkinson: 1794 [1 January 1794].
Size: 56.7 x 47.2 cm.
Colored.

(211: 1794: W659pcd Large)

134. [ca. 1800]

[Diagram showing plan for two buildings on ground surrounded by low land, level land and deep way, possibly for a volume on archaeology]. Plate first. Page 31. [London: ca. 1800].
Scale: 30 perches = 1 in.
Engraver: W. Darton.
Size: 18.5 x 28 cm.

(211: [ca. 1800]: D242dia Small)

135. [ca. 1800]

Map of the country seven miles round Taunton. R. Fenner: [ca. 1800].
Scale: 7.5 cm. = 2 mi.
Size: 55.4 x 48.1 cm.
Inset: Map of Taunton.
Engraved for the history of Taunton.

(211.981: [ca. 1800]: F353ste Large)

136. [ca. 1800]

A plan of the Leicester shire and Northampton shire intended Union Canal; shewing its connection with the Grand Junction Canal at Northampton, and with the river Trent by means of the Leicester navigation: by [...] channels it will communicate with most of the present inland navigations in the kingdom. [London: ca. 1800].

Scale: 12 mi. = 1 in.

Size: 43.5 x 41.5 cm.

(214: [ca. 1800]: L535Lnc Large)

137. 1801

Fores's traveller's companion through England and Wales, exhibiting the turnpike and cross roads, with the great rivers and all the navigable canals; also the market and borough towns, with the distances marked from one town to another, and the exact admeasurement from the principal places to the metropolis. London: Fores: 1801.

Scale: 69½ British mi. = 1 degree.

Engraver: S. J. Neele.

Size: 61.9 x 53.5 cm.

Colored.

(211: 1801: F743tce Small)

138. 1815

A delineation of the strata of England and Wales, with part of Scotland: exhibiting the colleries and mines, the marches and fen lands originally overflowed by the sea, and the varieties of soil according to the variations in the sub-strata, illustrated by the most descriptive names by W. Smith . . . Aug. 15, 1815. London: J. Cary: 1815.

Scale: 69½ British statute mi. = 1 degree.

Cartographer: William Smith.

Size: 257 x 179.8 cm.

Colored.

Endorsed: Smith's delineation of the Strata of England and Wales 1815.

Contains a sketch of the succession of strata and their relative altitudes.

(210.2: 1815: Sm58ew Large)

139. 1830

Map of the inland navigation, canals & railroads with the situations of the various mineral productions throughout Great Britain, from actual surveys projected on the basis of the trigonometrical survey made by order of the Honourable the Board of Ordnance, by J. Walker, land and mineral surveyor, Wakefield, accompanied by a book of reference, compiled by Joseph Priestly Esqre. of the Aire & Calder Navigation. Wakefield: Nicholls: 1830 [1 January 1830].

Scale: 8 cm. = 19 mi.

Cartographer: J. Walker.

Engravers: Franks & Johnson, Wakefield.

Size: 187.7 x 155.6 cm.

Inset: Scotland.

Provenance: Presented by the Board of Ordnance of Great Britain, 2 October 1840.

Endorsed by John Vaughan: Priestley & Walker Survey of G.B.

(210: 1830: W149cnm Large)

140. [ca. 1834]

England. Philadelphia: Tanner: [ca. 1834].

Scale: 69½ British statute mi. = 1 degree.

Cartographer: H. S. Tanner.

Engraver: D. Haines.

Size: 27.7 x 22.6 cm.

Colored.

From Tanner's *Universal Atlas.*

(210.2: [ca. 1834]: T158eng Small)

141. 1838

Railway map of England & Wales. 1838. London: W. A. Robertson: 1838 [1 August 1838].

Scale: 20 mi. = 1 in.

Cartographer and Lithographer: J. R. Jobbins.

Size: 50.9 x 41 cm.

Colored.

Provenance: Presented and endorsed by William Vaughan, 4 January 1839.

Indicates railways for which acts have been obtained and proposed railways.

(211: 1838: J57rLw Small)

142. 1839

A physical and geological map of England & Wales. London: Geological Society: 1819 and 1839. 2nd ed., 1 November 1839.

Scale: 6 mi. = 1 in.

Cartographer: George Bellas Greenough.

Size: 188.5 x 158 cm.

Colored.

Provenance: Presented by the Geological Society of London, 21 August 1840.

Endorsed: Geological map of England and Wales by G. B. Greenough.

(211: 1839: G823pge Large)

143. [ca. 1840]

London. Published under the superintendence of the Society for the Diffusion of Useful Knowledge. [London: ca. 1840].

Size: 35.7 x 63.8 cm.

Colored.

Inset: A view of the Tower from London Bridge.

(211.949: [ca. 1840]: S17kLgb Small)

IRELAND Nos. 144-147

144. [ca. 1712]

[Ireland]. N.p.: [ca. 1712].

Scale: 69 British measured mi. = 1 degree.

Cartographer and engraver: I. Senex.

Size: 96.7 x 67 cm.

Colored.

Inset: A map of the British Isles.

Mutilated.

(211: [ca. 1712]: Se57msLb Large)

145. 1714

A new map of Ireland divided into its provinces, counties and baronies, wherein are distinguished the bishopricks, borroughs, barracks, bogs, passes, bridges &c. with the principal roads, and the common reputed miles. According to the newest and most exact observations. By Herman Moll geographer. [London]: I. Bowles *et al.*: 1714.

Scale: 60 mi. of Great Britain = 1 degree.

Cartographer: Herman Moll.

Size: 102.5 x 61.7 cm.

Colored.

Insets: maps of the British Isles; Isle of St. Patrick's purgatory; Waterford; Gyants Cawsway; Gallway; harbour of Kinsale; Limrick; Dublin; and, Cork. Also, a catalogue of ye towns & places where barracks are erected . . .

This is a map of Ireland showing post roads travelled by Benjamin Franklin in 1771.

(211: 1714: M735bfi Large)

146. 1786

A new map of Ireland divided into its provinces, counties &c. By Thomas Kitchin. [London]: Robert Sayer: 1786 [1 August 1786].

Scale: 2 cm. = 10 British statute mi.

Cartographer: Thos. Kitchin.

Size: 56.3 x 63.5 cm.

Colored.

Provenance: Purchased from the library of William Priestman, 15 July 1831.

(223: 1786: K655pro Large)

147. [1793]

Ireland. [1793].

Scale: 2.5 cm. = 50 British statute mi.

Engraver: S. Allardice.

Size: 16.7 x 16.7 cm.

Provenance: Presented by Thomas Dobson, 1798.

In: Thomas Dobson, publisher, *Encyclopaedia,* vol. 9, opposite p. 344.
Wheat: 755

(032: En2)

SCOTLAND Nos. 148-158

148. 1714

The north part of Great Britain called Scotland. With considerable improvements and many remarks not extant in any map. According to the newest and exact observations. By Herman Moll geographer 1714. [London]: 1714.

Scale: 60 English mi. = 1 degree.

Cartographer: Herman Moll.

Size: 60 x 101 cm.

Colored.

Inset: Shetland & Orkney Islands. On each side are views of Aberdeen; Edinburg[h] castle; Sterling castle; Dunotyr castle in Merns; the Bass; Channery town in Ross; Edinburg[h]; Glasgow; St. Andrews; Sterling; and Montrose.

(221: 1714: M735npb Extra-oversize)

149. [ca. 1760]

A new and exact mapp (sic) of Scotland or North-Britain, described by N. Sanson geographer to the French King translated into English at the expence of Robert Sayer of London mapseller. London: Robert Sayer: [ca. 1760].

Scale: 9 mi. = 1 in.

Cartographers: N. Sanson and Robert Sayer.

Engraver: Sutton Nicholls.

Size: 85.5 x 58 cm.

Colored.

Insets: Stirling castle and Edinburg[h] castle.

(221: [ca. 1760]: Sa57nob Large)

150. 1778

North Britain; or Scotland, divided into its counties. Corrected from the best surveys & astronomical observations by Thos. Kitchin. London: W. Faden: 1778 [1 December 1778].

Scale: 2 cm. = 10 mi.

Cartographer: Thos. Kitchin.

Size: 66.2 x 53.3 cm.

Colored.

Provenance: Purchased from the library of William Priestman, 15 July 1831.

(221: 1778: K655cou Large)

151. [ca. 1778]

The north part of Great Britain, called Scotland. With considerable improvements and many remarks not extant in any map according to the newest and exact observations. By Herman Moll geographer. London: R. Marshall: [ca. 1778].

Scale: 60 British mi. = 1 degree.

Cartographer: Herman Moll.

Size: 59.5 x 63.5 cm.

Colored.

Trimmed on both sides.

(221: [ca. 1778]: M735maL Large)

152. 1782

New map of Scotland or North Britain, wherein all the post and other public roads are correctly delineated; with adjacent parts of England and Ireland, compiled from the best surveys regulated by the latest observations, by Marcus Armstrong geogr. [London]: Sayer and Bennett: 1782 [15 April 1782].

Scale: 69½ British measured mi. = 1 degree.

Cartographer: Marcus Armstrong.

Size: 124.5 x 106.1 cm.

Colored.

Insets: Zetland or Shetland Islands; and distance of the principal stages on the great roads leading from Edinburgh.

(221: 1782: Ar61sbi Large)

153. [ca. 1790]

Scotland, with the principal roads, from the best authorities. N.p.: [ca. 1790].

Scale: 2.6 cm. = 25 British statute mi.

Size: 35.4 x 32.8 cm.

Inset: Shetland Isles.

Provenance: Presented by Mathew Carey, 18 October 1805.

The *Transactions* (vol. 6) state that Carey presented the "Materials from which Guthrie's Geography were compiled." This is one of those maps. See: William Guthrie, *The general atlas for Carey's edition of his geography improved.* Philadelphia: Carey: 1795 [1 May 1795]. (912:G98c).

Engraved for Chambers's edition of Guthrie's new system of geography.

(221: [ca. 1790]: Sco37std Small)

154. [1796]

Scotland. [1796].

Scale: 2.5 cm. = 50 British statute mi.

Engravers: Scot & Allardice.

Size: 17.4 x 19.1 cm.

Inset: Shetland Isles.

Provenance: Presented by Thomas Dobson, 1798.

In: Thomas Dobson, publisher, *Encyclopaedia,* vol. 16, opposite p. 722.

Wheat: 787

(032:En2)

155. [ca. 1800]

Map of the soil of Sutherland. N.p.: [ca. 1800].

Scale: 4 mi. = 1 in.

Engraver: Neele.

Size: 32.9 x 39.9 cm.

Colored.

(221.2: [ca. 1800]: N296sth Small)

156. 1803

Plan of certain farms on the river Thurso, the property of Sir John Sinclair, bart. M.P. with the new divisions thereof into regular fields; together with the subdivisions of that part of an extensive common allotted to Sir John Sinclair in right of those farms, amounting to 1706 Scotch or 2134 English acres; intended partly to be let in small lots on improving leases to new settlers, and partly to the neighboring farmers: drawn up by Captain John Henderson, anno 1803. Plate A. N.p.: 1803.

Scale: 42 chains = 2.6 cm.

Cartographer: John Henderson.

Engraver: S. J. Neele.

Size: 18.7 x 29.1 cm.

See: *An account of the improvements carried on by Sir John Sinclair on his country estate in Scotland.* London: McMillan: 1812.

(221.2: 1803: H383thu Small)

157. 1912

The Hebrides. International map of Europe. North 0 29. Compiled from the ordnance surveys of Great Britain. Southampton: 1912.

Scale: 1 in. = 15.78 mi.

Size: 46.4 x 39.3 cm.

Colored.

(221.3: 1912: G813heb Large)

158. 1912

Scotland-The Highlands. International map of Europe. North 0 30. Compiled from the ordnance surveys of Great Britain. Southampton: 1912.

Scale: 1 in. = 15.78 mi.

Size: 46.5 x 39.3 cm.

Colored.

(221.5: 1912: G813scd Large)

SWEDEN Nos. 159-163

159. 1788

A new map of the northern states containing the kingdoms of Sweden, Denmark, and Norway, with the western part of Russia, Livonia, Courland &ca. London: Robert Sayer: 1788 [10 October 1788].

Scale: 69½ British mi. = 1 degree.

Size: 46.8 x 64.4 cm.

Colored.

Provenance: Purchased from the library of William Priestman, 15 July 1831.

(225: 1788: N422sdn Large)

160. [ca. 1790]

Sweden; Denmark; Norway and Finland. From the best authorities. N.p.: [ca. 1790].

Scale: 100 mi. = 1 in.

Engravers: Woodman & Mutlow.

Size: 32.8 x 36.1 cm.

Inset: West Greenland and Iceland.

Provenance: Presented by Mathew Carey, 18 October 1805.

The *Transactions* (vol. 6) state that Carey presented the "Materials from which Guthrie's Geography were compiled." This is one of those maps. See: William Guthrie, *The general atlas for Carey's edition of his geography improved.* Philadelphia: Carey: 1795 [1 May 1795].

Engraved for Chambers's edition of Guthrie's new system of geography.

(225: [ca. 1790]: W859sca Small)

161. 1861-1873

[Geological atlas of Sweden]. Pa. offentlig bekostnad utförd under ledning af A. Erdmann. Stockholm: Inrättn: 1861-1873. 53 maps.

Scale: 1:50,000.

Cartographer: A. Erdmann.

Size: 46 x 61 cm.

Colored.

Provenance: Presented by Sweden. Geologiska Undersökning, 21 April 1865-1 October 1875.

Contains index map.

See: Sweden. Geologiska undersökning. *Nagra ord till upplysning om bladet* . . . Stockholm: 1861-1873.

(227: 1861-1873: Sw37sto Large)

162. 1863

Ofversigtskarta öfver Södra Delen af Sverige. Sammandragen för de Geologiska Undersökningarnes Behof År 1863. Inrättn: 1866.

Scale: 1:1,000,000.

Size: 70.3 x 51.9 cm.

Colored.

(227: 1863: Er22gla Large)

163. 1870

Geologisk öfversigtskarta öfver Bergarterna på östra dal. Sveriges Geologiska Undersökning. Gen. Stabs. Lithogr. Inrättn.: 1870.

Scale: 1:200,000.

Size: 38.8 x 22.5 cm.

Colored.

Contains profiles of certain elevations.

(227.6: 1870: Sw37geo Small)

NORWAY Nos. 164-165

164. [ca. 1880]

Le royaume de Norvége. [Paris: ca. 1880].

Scale: 1:2,400,000.

Size: 67.4 x 50.3.

Inset: Norway divided into sousprefectures.

Color chart showing elevations.

(226: [ca. 1880]: N756nor Small)

165. 1940

Bathymetrical map in seven sheets of the Norwegian coastal waters and adjoining seas. Norske Videnskaps-Akademi i Oslo. Oslo: Jacob Dybwad: 1940.

Scales: Various.

Size: Various.

Colored.

Kartbilag til Olaf Holtedahl: "The submarine relief off the Norwegian coast" utgitt av det Norske Videnskaps-Akademi i Oslo, 1940. Text in Norwegian and English.

(226: 1940: N756sea Large)

DENMARK Nos. 166-185

166. [ca. 1750]

Grund Tegning afden Kongelige Residenz stad Kiøbenhavn. N.p.: [ca. 1750].

Engraver: Fridrich.

Size: 19.3 x 30.1 cm.

Provenance: Presented through Peder Pedersen.

Has identification of 51 places of interest.

(228.923: [ca. 1750]: F913dem Small)

167. 1768

Den Nord–Østlige Fierdedeel af Siaeland under det Kongl. Viidenskabernes Societets . . . tegnet af Caspar Wessel. [Copenhagen]: 1771.

Scale: 2 Danish mi. = 24,000 ells.

Cartographer: Caspar Wessel.

Engraver: Martin 1771.

Size: 56.5 x 66.5 cm.

Provenance: Presented through Peder Pedersen.

(228.5: 1768: W519fsc Large)

168. 1770

Den Syd Østilge Fierdedeel af Siaelland under det Kongl. Viidenskabernes Societets . . . tegnet af C. Wessel. [Copenhagen]: 1770.

Scale: 2 Danish mi. = 24,000 ells.

Cartographer: C. Wessel.

Engraver: Martin.

Size: 57 x 66.8 cm.

Provenance: Presented through Peder Pedersen.

(228.5: 1770: W519dmc Large)

169. 1772

Den Sydvestlige Fierdedeel af Siaelland under det Kongl. Viidenskabernes Societets . . . tegnet af C. Wessel. [Copenhagen]: 1772.

Scale: 2 Danish mi. = 24,000 ells.

Cartographer: C. Wessel.

Engraver: Guiter.

Size: 56.6 x 66.7 cm.

Provenance: Presented through Peder Pedersen.

(228.5: 1772: W519sfs Large)

170. 1776

Kort over Møen Falster og Laaland under det Kongl. Viidenskabernes Societets . . . tegnet af H. Skanke. [Copenhagen]: 1776.

Scale: 2 Danish mi. = 24,000 ells.

Cartographer: H. Skanke.

Engraver: Guiter.

Size: 57 x 87 cm.

Provenance: Presented through Peder Pedersen.

(228.5: 1776: Skal7me Large)

171. 1777

Kort over Siaelland og Möen med tilgraendsende Kyster af Skanne Falster Laaland Langeland Thorsinge Fyen Samsöe og Jylland under det Kongl. Viidenskabernes Societets . . . tegnet af C. Wessel og H. Skanke. [Copenhagen]: 1777.

Scale: 1 Danish mi. = 12,000 ells.

Cartographer: C. Wessel and H. Skanke.

Engraver: Guiter.

Size: 61.1 x 67.1 cm.

Provenance: Presented by P. Pedersen, 1 May 1818.

(228.5: 1777: W519msL Large)

172. 1780

Kort over den nordlige Deel af Fyen med tilgraendsende Kyster af Jylland og Schleswig under det Kongl. Videnskabernes Societets . . . tegnet af C. Wessel. [Copenhagen]: 1780.

Scale: 1½ Danish mi. = 18,000 ells.

Cartographer: C. Wessel.

Engraver: Guiter.

Size: 48.5 x 85.3 cm.

Provenance: Presented through Peder Pedersen.

(228.4: 1780: W519fjs Large)

173. 1783

Kort over den Sydlige Deel af Fyen tilligemed det tilgraendsende Stykke af Hertugdommet Schleswig saavelsom øerne Langeland, Taasinge, Aerøe, Als. og mellemliggendesmaae Øer under det Kongelige Viidenskabernes Societets . . . tegnet af H. Skanke. [Copenhagen]: 1783.

Scale: 2 Danish mi. = 24,000 Danish ells.

Cartographer: H. Skanke.

Engraver: Guiter.

Size: 57 x 85.2 cm.

Provenance: Presented through Peder Pedersen.

(228.4: 1783: Skal7sL Large)

174. 1789

Kort over Haureballegards, Stiernholms, Aakier og Skanderborg samt Stykker af Silkeborg og Kildinghuus Amter under det Kongl. Viidenskabernes Societets . . . tegnet af O. Warberg. No. 1. [Copenhagen]: 1789.

Scale: 2 Danish mi. = 24,000 Danish ells.

Cartographer: O. Warberg.

Engraver: N. Angelo: 1789.

Size: 48.4 x 84.5 cm.

Provenance: Presented through Peder Pedersen.

(228.3: 1789: W199hss Large)

175. 1790

A map of the kingdom of Denmark, with the duchy of Holstein. By William Faden. 1790. London: William Faden: 1790 [2 August 1790].

Scale: 69½ British statute mi. = 1 degree.

Cartographer: William Faden.

Size: 69.5 x 51.3 cm.

Colored.

Provenance: Purchased from the library of William Priestman, 15 July 1831.

Contains manuscript note: Wm. Priestman, Philadelphia, 266 Arch St. 1800.

(228: 1790: F121noL Large)

176. 1791

Kort over Dronningborg og Kaløe Amter samt stykker af Aalborghuus, Halds, Mariager, Silkeborg, Haureballegaards og Skanderborg Amter under det Kongelige Viideskabernes Societets . . . tegnet af O. Warberg. No. 2. [Copenhagen]: 1791.

Scale: 2 Danish mi. = 24,000 Danish ells.

Cartographer: O. Warberg.

Engraver: N. Angelo: 1791.

Size: 51 x 85 cm.

Provenance: Presented through Peder Pedersen.

(228.3: 1791: W199dka Large)

177. [1792]

Denmark, Norway, Sweden, and Finland. [1792].

Scale: 2.5 cm. = 160 British mi.

Engraver: J. Smither.

Size: 17.6 x 21.4 cm.

Provenance: Presented by Thomas Dobson, 1798.

Inset: [Northern tip of Norway and East Greenland].

In: Thomas Dobson, publisher, *Encyclopaedia,* vol. 5, following p. 752.

Wheat: 747

(032: En2)

178. 1793

Kort over en deel af Aalborghuus oc Seiglstrup Amter samt af Vendsyssel under det Kongl. Viidenskabernes Societets . . . tegnet af P. Harboe. [Copenhagen]: 1793.

Scale: 1½ Danish mi. = 18,000 Danish ells.

Cartographer: P. Harboe.

Engraver: N. Angelo: 1793.

Size: 51.5 x 84.5 cm.

Provenance: Presented through Peder Pedersen.

(228.3: 1793: H213asv Large)

179. 1795

Kort over Aastrup og Børglum Amter under det Kongl. Viidentskabernes Societets . . . tegnet af P. Harboe. [Copenhagen]: 1793.

Scale: 1½ Danish mi. = 18,000 Danish ells.

Cartographer: P. Harboe.

Engraver: G. N. Angelo: 1795.

Size: 51.6 x 85.5 cm.

Provenance: Presented through Peder Pedersen.

(228.3: 1795: H213ajd Large)

180. [1797]

Kort over Ørum amt samt stykker af Aalborghuus og Vestervig amter og Vensyssel under det Kongl. Viidensk. Societets . . . tegnet af P. Harboe. Copenhagen: [1797].

Scale: 2 Danish mi. = 18,000 Danish ells.

Cartographer: P. Harboe.

Engraver: N. Angelo: 1797.

Size: 51.5 x 85.4 cm.

Provenance: Presented through Peder Pedersen.

(228.3: 1795: H213avv Large)

181. 1800

Kort over Skivehuus, Bøvling og Lundenaes Amter samt stykker af de tilstödende amter, under det Kongelige Viidenskabernes Societets . . . tegnet af P. Harboe. Copenhagen: 1800.

Scale: 2 Danish mi. = 24,000 Danish ells.

Cartographer: P. Harboe.

Engraver: N. Angelo.

Size: 52 x 85.5 cm.

Provenance: Presented through Peder Pedersen.

(228.3: 1800: H213ksb Large)

182. 1802

Carte des états danois par Lapie. Paris & Strasbourg: Treuttel et Würtz: 1802.

Scale: 15 Danish mi. = 1 in.

Cartographer: Pierre Lapie.

Engraver: Blondeau.

Size: 50.7 x 59.6 cm.

Insets: Islande; Isles Faeroe; Groenland; Afrique Guinée; Bengal; Isles Nicobar; Isles des Antilles; Mysore; and Carnate.

See: J. P. G. Catteau-Calleville, *Tableau des états danois, envisages sous les rapports du mécanisme social.* Paris: 1802.

(225: 1802: L315dan Book map)

183. 1803

Kort over Lundenaes amt samt stykker af Bövling, Koldinghuss, Riberhuus og Silkeborg Amter under det Kongl. Viidenskabernes Societets . . . tegnet af P. Harboe. Copenhagen: 1803.

Scale: 2 Danish mi. = 24,000 Danish ells.

Cartographer: P. Harboe.

Engraver: G. N. Angelo 1806.

Size: 48.5 x 85 cm.

Provenance: Presented through Peder Pedersen.

(228.3: 1803: H213Lbk Large)

184. 1804

Kort over en Deel af Koldinghuus og Riberhuus Amter udi Iylland samt af Haderslewhuus amt udi Hertugdømmet Schleswig under det Kongelige Viidenskabernes Societets . . . tegnet af T. Bugge og F. Wilster. Copenhagen: 1804.

Scale: 1½ Danish mi. = 18,000 Danish ells.

Cartographers: T. Bugge and F. Wilster.

Engraver: I. Sonne.

Size: 48.3 x 85.3 cm.

Provenance: Presented through Peder Pedersen.

(228.3: 1804: B861hhs Large)

185. 1805

Kort over Bornholm under det Kongel. Viidenskabernes Societets . . . tegnet af T. Bugge og F. Wilster 1805. [Copenhagen]: 1805.

Scale: 1 Danish mi. = 12,000 ells.

Cartographers: T. Bugge and F. Wilster.

Engraver: G. N. Angelo.

Size: 72.7 x 61 cm.

Provenance: Presented by P. Pedersen, 1 May 1818.

(228.7: 1805: B861sdb Large)

THE LOW COUNTRIES Nos. 186-193

186. 1709

The VII United Provinces. Corrected from the observations communicated to the Royal Society at London, and ye Royal Academy at Paris. Humbly dedicated to Elihu Yale Esq. 1709.

Scale: 2 cm. = 2 common English mi.

Cartographers: Ion. Senex and Ion. Maxwell, geographers to the Queen.

Engraver: John Senex.

Size: 63.8 x 94.1 cm.

Colored.

(232: 1709: Se57up Large)

187. 1714

A new and correct map of the ten Spanish provinces according to the new observations communicated to the Royal Society at London & the Royal Academy at Paris by John Senex and John Maxwell. [London]: 1714.

Scale: 60 English mi. = 1 degree.

Cartographer: John Maxwell.

Engraver: John Senex.

Size: 62.8 x 94.7 cm.

Inset: Fortified place in the style of Vauban with several military engines.

Colored.

(232: 1714: S57tsp Large)

188. 1784

A new map of Zealand; with the rivers Schelde, part of Holland, Flanders & Brabant; shewing the situation of the present dispute between the Emperor and the Dutch. London: Jno. Stockdale: 1784 [29 November 1784].

Scale: One in. = 4 mi.

Size: 41.2 x 52.5 cm.

Colored.

Provenance: Purchased from the library of William Priestman, 15 July 1831.

(228.5: 1784: N423zca Large)

189. 1789

A map of the Austrian possessions in the Netherlands or Low Countries, with the principalities of Liege and Stavelo, &ca. Reduced from the trigonometrical survey made by order of H.R.H. the prince Charles of Lorraine. London: William Faden: 1789 [1 January 1789].

Scale: 3 in. = 25 English statute mi.

Size: 53 x 92 cm.

Colored.

Provenance: Purchased from the library of William Priestman, 15 July 1831.

(232: 1789: M328aup Large)

190. 1789

A map of the frontiers of the emperor and the Dutch in Flanders and Brabant, including the greatest part of these two countries, with the isles of Zealand, the mouths of the Scheldt, &ca. From the survey made under the direction of the comte de Ferraris. London: Wm. Faden: 1789 [1 January 1789].

Scale: 1 in. = 3 English statute mi.

Cartographer: Comte de Ferraris.

Size: 60 x 91.5 cm.

Colored.

Provenance: Purchased from the library of William Priestman, 15 July 1831.

(232: 1789: M322fbr Large)

191. 1789

A map of the Seven United Provinces, with the land of Drent, and the generality lands. By W. Faden. London: 1789 [1 January 1789].

Scale: 2 cm. = 5 British statute mi.

Cartographer: W. Faden.

Size: 71.5 x 53.5 cm.

Colored.

Provenance: Purchased from the library of William Priestman, 15 July 1831.

(232: 1789: F121hoL Large)

192. [ca. 1790]

The Austrian French and Dutch Netherlands, from the best authorities. N.p.: [ca. 1790].

Engravers: Woodman & Mutlow.

Size: 28.7 x 35.5 cm.

Provenance: Presented by Mathew Carey, 18 October 1805.

Engraved for Chambers's edition of Guthrie's new system of geography.

The *Transactions* (vol. 6) state that Carey presented the "Materials from which Guthrie's Geography were compiled." This is one of those maps. See: William Guthrie, *The general atlas for Carey's edition of his geography improved.* Philadelphia: Carey: 1795 [1 May 1795].

(230: [ca. 1790]: W859afd Small)

193. 1794

A new map of The Netherlands or Low Countries, with the south part of the provinces of Holland, Utrecht & Gelders and the whole of Zeeland. London: Laurie & Whittle: 1794 [12 May 1794].

Scale: 4 mi. = 1 in.

Size: 95.4 x 117.7 cm.

Colored.

Inset: British siege of a city in the 17th-18th centuries.

(230: 1794: N386hgz Large)

THE NETHERLANDS Nos. 194-213

194. [ca. 1750]

De Stad Leyden. N.p.: [ca. 1750].

Size: 23.7 x 35.8 cm.

Contains profile of the city with identification of certain buildings.

(232: [ca. 1750]: N386Ley Small)

195. [ca. 1790]

Seven United Provinces of Holland, Groningen, Gelders, Friesland, Overyssel, Utrecht and Zealand. N.p.: [ca. 1790].

Scale: 69½ British mi. = 1 degree.

Engraver: Richard Woodman, Sr.

Size: 29.2 x 34.2 cm.

Provenance: Presented by Mathew Carey, 18 October 1805.

The *Transactions* (vol. 6) state that Carey presented the "Materials from which Guthrie's Geography were compiled." This is one of those maps. See: William Guthrie, *The general atlas for Carey's edition of his geography improved.* Philadelphia: Carey: 1795 [1 May 1795].
(232: [ca. 1790]: W859nea Small)

196. [1792]
Fortification[s. Three illustrations, for Hunningen 1792].
Scale: 2.5 cm. = 160 toises.
Size of plate: 22.1 x 17.2 cm.
Provenance: Presented by Thomas Dobson, 1798.
In: Thomas Dobson, publisher, *Encyclopaedia,* vol. 7, following p. 368.
Wheat: 750
(032: En2)

197. [ca. 1865]
Bargerveen. 12. Ministerie van Oorlog. Topographisch Bureau. N.p.: [ca. 1865].
Scale: 1:200,000.
Size: 28 x 33.1 cm.
Colored.
Provenance: Presented by the Minister of the Interior of the Netherlands, 19 May 1865.
(232: [ca. 1865]: N386bar Small)

198. [ca. 1865]
Betuwe. 19. Ministerie van Oorlog. Topographisch Bureau. N.p.: [ca. 1865].
Scale: 1:200,000.
Size: 28 x 43 cm.
Colored.
Provenance: Presented by the Minister of the Interior of the Netherlands.
(232: [ca. 1865]: N386bet Small)

199. [ca. 1865]
Biesbosch. 18. Ministerie van Oorlog. Topographisch Bureau. N.p.: [ca. 1865].
Scale: 1:200,000.
Size: 28.1 x 43 cm.
Colored.
Provenance: Presented by the Minister of the Interior of the Netherlands, 19 May 1865.
(232: [ca. 1865]: N386bie Small)

200. [ca. 1865]
Hunsingoo. 4. Ministerie van Oorlog. Topographisch Bureau. N.p.: [ca. 1865].
Scale: 1:200,000.
Size: 28 x 33 cm.

Colored.
Provenance: Presented by the Minister of the Interior of the Netherlands.
(232: [ca. 1865]: N386hun Small)

201. [ca. 1865]
Kennemerland. 10. Ministerie van Oorlog. Topographisch Bureau. N.p.: [ca. 1865].
Scale: 1:200,000.
Size: 28 x 43 cm.
Colored.
Provenance: Presented by the Minister of the Interior of the Netherlands, 19 May 1865.
(232: [ca. 1865]: N386ken Small)

202. [ca. 1865]
Munsterland. 20. Ministerie van Oorlog. Topographisch Bureau. N.p.: [ca. 1865].
Scale: 1:200,000.
Size: 28 x 33 cm.
Colored.
Provenance: Presented by the Minister of the Interior of the Netherlands.
(232: [ca. 1865]: N386mun Small)

203. [ca. 1865]
Oostergoo. 7. Ministerie van Oorlog. Topographisch Bureau. N.p.: [ca. 1865].
Scale: 1:200,000.
Size: 28 x 43 cm.
Colored.
Provenance: Presented by the Minister of the Interior of the Netherlands, 19 May 1865.
(232: [ca. 1865]: N386oos Small)

204. [ca. 1865]
Peel. 23. Ministerie van Oorlog. Topographisch Bureau. N.p.: [ca. 1865].
Scale: 1:200,000.
Size: 28 x 43 cm.
Colored.
Provenance: Presented by the Minister of the Interior of the Netherlands, 19 May 1865.
(232: [ca. 1865]: N386pee Small)

205. [ca. 1865]
Schouwen. 17. Ministerie van Oorlog. Topographisch Bureau. N.p.: [ca. 1865].
Scale: 1:200,000.
Size: 28 x 33 cm.
Colored.
Provenance: Presented by the Minister of the Interior of the Netherlands.

(232: [ca. 1865]: N386sch Small)

206. [ca. 1865]
Texel. 6. Ministerie van Oorlog. Topographisch Bureau. N.p.: [ca. 1865].
Scale: 1:200,000.
Size: 28 x 43 cm.
Colored.
Provenance: Presented by the Minister of the Interior of the Netherlands.
(232: [ca. 1865]: N386tex Small)

207. [ca. 1865]
Twenthe. 16. Ministerie van Oorlog. Topographisch Bureau. N.p.: [ca. 1865].
Scale: 1:200,000.
Size: 28 x 33.1 cm.
Colored.
Provenance: Presented by the Minister of the Interior of the Netherlands.
(232: [ca. 1865]: N386twe Small)

208. [ca. 1865]
Veluwe. 15. Ministerie van Oorlog. Topographisch Bureau. N.p.: [ca. 1865].
Scale: 1:200,000.
Size: 28 x 43 cm.
Colored.
Provenance: Presented by the Minister of the Interior of the Netherlands.
(232: [ca. 1865]: N386veL Small)

209. [ca. 1865]
Wadden. 3. Ministerie van Oorlog. Topographisch Bureau. N.p.: [ca. 1865].
Scale: 1:200,000.
Size: 28 x 43 cm.
Colored.
Provenance: Presented by the Minister of the Interior of the Netherlands.

210. [ca. 1865]
Walcheren. 21. Ministerie van Oorlog. Topographisch Bureau. N.p.: [ca. 1865].
Scale: 1:200,000.
Size: 28 x 33 cm.
Colored.
Provenance: Presented by the Minister of the Interior of the Netherlands, 21 December 1866.
(232: [ca. 1865]: N386waL Small)

211. [ca. 1865]

Westerwolde. 8. Ministerie van Oorlog. Topographisch Bureau. N.p.: [ca. 1865].

Scale: 1:200,000.

Size: 28 x 33.2 cm.

Colored.

Provenance: Presented by the Minister of the Interior of the Netherlands, 21 December 1866.

(232: [ca. 1865]: N386wes Small)

212. [ca. 1865]

Zuiderzee. 11. Ministerie van Oorlog. Topographisch Bureau. N.p.: [ca. 1865].

Scale: 1:200,000.

Size: 28 x 43 cm.

Colored.

Provenance: Presented by the Minister of the Interior of the Netherlands, 21 December 1866.

(232: [ca. 1865]: N386zui Small)

213. n.d.

Plan très exact de la fameuse ville marchande d'Amsterdam. Gravé et mis au jour par Henry de Leth. Amsterdam: Cövens & Mortier en J. Cövens, Junior: n.d.

Scale: 1500 Rhinelandic ft.

Cartographer and engraver: Henry de Leth.

Size: 58 x 88.7 cm.

Inset: Carte van Amstelland.

Provenance: Presented by William Short, 5 November 1824.

(232.912: n.d.: L565pda Large)

BELGIUM Nos. 214-216

214. [ca. 1710]

A map of the county of Flanders. By William de l'Isle. London: Senex and Maxwell: [ca. 1710].

Scale: 1 in. = 3½ mi.

Cartographer: William de l'Isle.

Size: 48.5 x 60.9 cm.

Colored.

(234.3: [ca. 1710]: Is14cfb Large)

215. [ca. 1710]

A new map of the provinces of Hainault, Namur and Cambray. By I. Maxwell and John Senex, Geographers to the Queen. [London]: 1710.

Scale: 20 British and French leagues = 1 degree.

Cartographers: John Senex and I. Maxwell.

Size: 46.9 x 63.8 cm.

Colored.

(234.6: 1710: M455hmc Large)

216. [ca. 1900]

Plan-guide d'Anvers indiquant toutes les lignes des tramways omnibus et canots à vapeur. [Antwerp]: Dressé et publié par Martin Ghys: [ca. 1900].

Scale: 1:12,500.

Cartographer: Martin Ghys.

Size: 41.3 x 56 cm.

Colored.

Inset: Environs d'Anvers.

(234.912: [ca. 1900]: G353anv Small)

LUXEMBOURG No. 217

217. [ca. 1877]

Carte géologique du Grand-Duché de Luxembourg par N. Wies . . . et P. M. Siegen . . . publiée par les soins de la section des sciences naturelles de l'Institut Royal Grand-Ducal . . . Paris: Lemercier et Cie: 1877. 8 sheets.

Scale: 1:40,000.

Cartographers: N. Wies and P. M. Siegen.

Engraver: Georges Erhard Schielde.

Size: ca. 67 x 54 cm.

Colored.

See: N. Wies, *Guide de la carte géologique du Grand-Duché de Luxembourg.* Pierre Bruck: 1877.

(235: 1877: W639geo Large)

SWITZERLAND Nos. 218-244

218. [ca. 1750]

Ditionis pagi Scaphusiani, qui confoederatae Helvetiorum reipublicae pars est descriptio geographica nova studio adornata à Matth. Seuttero. [ca. 1750].

Scale: 6.5 cm. = 3000 Schritt; or 1 Stund wegs.

Cartographer: Matthew Seutter.

Engraver: Tob. Conr. Lotter.

Size: 49.1 x 57.5 cm.

Colored.

Provenance: Presented by Ferdinand Rudolph Hassler, 20 March 1807.

(237.772: [ca. 1750]: S87sch Large)

219. [ca. 1750]

Mappa geographica illustris Helvetiorum Reipublicae Bernensis, cum adjacentibus pagorum et dynastiarum confiniis accurata delineata à Matth. Seuttero.

Scale: 3 cm. = 6 German mi.

Cartographer: Matthew Seutter.

Engraver: T. C. Lotter.

Size: 48.8 x 57 cm.

Colored.

Provenance: Presented by Ferdinand Rudolph Hassler, 20 March 1807.

(237: [ca. 1750]: S87hrb Large)

220. 1763

Canton Lucern sive illustris Helvetiorum Respublica Lucernensis cum subditis suis et confinibus de nuo correcta a Gabr. Walser. N.p.: Homann: 1763.

Scale: 1½ in. = 1 Stund.

Cartographer: Gabriel Walser.

Size: 47.8 x 57.7 cm.

Colored.

Inset: Die Herrschafft Griessenberg im Thurgeu, dem Canton Lucern.

Provenance: Presented by Ferdinand Rudolph Hassler, 20 March 1807.

(237.749: 1763: W149cLh Large)

221. 1766

La Turgovie avec le Lac de Constance, et des Pays Circonvoisins. Nova landgraviatus Turgoviae chorographica tabula, ubi Scaphusiensis, Abbatisque Cellanae respublicae specialiter designatae proponuntur: nec non abbatia Sancti Galli, Episcopatus Constantinus, Toggenburgensisque comitatus, cum omnibus insertis, et contiguis comitatibus Dominüs, et Confinibus; omnia ex probatissimis subsidüs accuratissime et juxta limites delineata per Ioh. Ant. Rizzi Zannoni. Nürnberg: Homann: 1766.

Cartographer: Giovanni Antonio Rizzi Zannoni.

Size: 42 x 55.5 cm.

Colored.

Provenance: Presented by Ferdinand Rudolph Hassler, 20 March 1807.

Damaged.

(237: 1766: Z29cons Small)

222. 1767

Canton Freiburg sive pagus Helvetiae Friburgensis cum confinibus recenter delineatus per Gabrielem Walserum. Nürnberg: Homann: 1767.

Scale: 6 cm. = 2 Stunden.

Cartographer: Gabriel Walser.

Engraver: Seb. Dorn.

Size: 47.5 x 57.5 cm.

Colored.

Provenance: Presented by Ferdinand Rudolph Hassler, 20 March 1807.
(237.733: 1767: W149sfc Large)

223. 1767

Canton Schweiz sive pagus Helvetiae Suitensis cum confinibus recenter delineatus per Gabrielem Walserum. Nürnberg: Homann: 1767.

Scale: 1½ in. = 2 Stund.

Cartographer: Gabriel Walser.

Size: 47.4 x 57.2 cm.

Colored.

Inset: Prospect des Hoch-Fürstischen Stiffts Einsidlen.

Provenance: Presented by Ferdinand Rudolph Hassler, 20 March 1807.
(237.733: 1767: W149csh Large)

224. 1767

Canton Unterwalden sive pagus Helvetiae subsylvanus cum adjacentibus terrarum tractibus recenter delineatus per Gabrielem Walserem. N.p.: Homann: 1767.

Scale: 2 cm. = ½ Stund.

Cartographer: Gabriel Walser.

Size: 47.6 x 52.7 cm.

Colored.

Provenance: Presented by Ferdinand Rudolph Hassler, 20 March 1807.
(237.786: 1767: W149cuh Large)

225. 1768

Canton Glarus sive pagus Helvetiae Glaronensis cum satrapia Werdenberg recenter delineatus per Gabrielem Walserum. Nürnberg: Homann: 1768.

Scale: 1½ in. = 2 Stund.

Cartographer: Gabriel Walser.

Size: 45.2 x 55.5 cm.

Colored.

Inset: Die Grafschaft Werdenberg.

Provenance: Presented by Ferdinand Rudolph Hassler, 20 March 1807.
(237.735: 1768: W149hgw Large)

226. [ca. 1768]

Pagus Helvetiae Uriensis cum subditis suis in Valle Lepontina accuratissima delineatio cura et sumptibus Matthei Seutteri. [ca. 1768].

Scale: 2 cm. = ½ Stund.

Cartographer: Matthew Seutter.

Size: 57.7 x 49.3 cm.

Colored.

Inset: Vallis Lepontina.

Provenance: Presented by Ferdinand Rudolph Hassler, 20 March 1807.
(237: [ca. 1768]: S87huL Large)

227. 1768

Vallesia superior, ac inferior, Wallis le Valais, geographice repraesentata, cura et studio Gabrielis Walseri. Nürnberg: Homann: 1768.

Scale: 2 cm. = 1 Stund.

Cartographer: Gabriel Walser.

Size: 47 x 57 cm.

Colored.

Insets: Four alpine scenes.

Provenance: Presented by Ferdinand Rudolph Hassler, 20 March 1807.
(237: 1768: W149vsv Large)

228. 1768

Die zwey erste Zugewandte orte der Eidgenosschaft nemlich: der fürstliche Abt von St. Gallen und die Stadt St. Gallen samt dem Toggenburg und denen angränzenden Orten Appenzell, Thurgeu und Rheinthal, neu gezeichnet, von Gabriel Walser. Nürnberg: Homann: 1768.

Scale: 2.7 cm. = 1 Stund Milliare Germanicum.

Cartographer: Gabriel Walser.

Size: 47.5 x 57.3 cm.

Colored.

Provenance: Presented by Ferdinand Rudolph Hassler, 20 March 1807.
(237: 1768: W149sgg Large)

229. 1776

Canton Solothurn sive pagus Helvetiae Solodurensis cum confinibus recenter delineatus per Gabrielem Walserum. Nürnberg: Homann: 1766.

Scale: 6.8 cm. = 2 Stunden.

Cartographer: Gabriel Walser.

Size: 47.5 x 57.5 cm.

Colored.

Provenance: Presented by Ferdinand Rudolph Hassler, 20 March 1807.
(237.776: 1766: W149hcs Large)

230. 1777

Plan de la ville de Genève corrigé sur les lieux, en 1777. Dedié aux magnifiques et très honorés Seigneurs, Sindic et Conseil de la Ville et République de Genève par . . . C. B. Glot. Geneva: Hardy: 1777.

Scale: 6 cm. = 200 toises.

Cartographer: C. B. Glot.

Size: 50.4 x 65 cm.

Insets: Plan de Genève ancienne; Plan de Genève en 1715; and Carte du Bailliage de Gex en France . . .

Provenance: Presented by William Short, 5 November 1824.

(237.934: 1777: G513pvg Large)

231. 1783

Geognostische Karte des Sanct Gotthard. Aufgenommen in den Jahren 1864, 1865, 1866 u. 1871 durch Dr. Karl von Fritsch. Winterthur: Wurster, Randegger: 1873.

Scale: 1:50,000.

Cartographer: Karl von Fritsch.

Size: 49.9 x 72.2 cm.

Colored.

(237: 1873: F913ksg Large)

232. 1784

Die Grafschafft Toggenburg von Ion: Iacob Büler. Augsburg: Michael Probst: 1784.

Scale: 4.05 cm. = 10,000 schu.

Cartographer: John Jakob Büler.

Size: 56.9 x 45.4 cm.

Colored.

Provenance: Presented by Ferdinand Rudolph Hassler, 20 March 1807.

(237: 1784: B851dgt Large)

233. [ca. 1790]

Switzerland according to the best authorities. N.p.: [ca. 1790].

Scale: 21.2 x 28.5 cm.

Provenance: Presented by Mathew Carey, 18 October 1805.

Engraved for Chambers's edition of Guthrie's new system of geography.

The *Transactions* (vol. 6) state that Carey presented the "Materials from which Guthrie's Geography were compiled." This is one of those maps. See: William Guthrie, *The general atlas for Carey's edition of his geography improved.* Philadelphia: Carey: 1795 [1 May 1795].

(237: [ca. 1790]: Sw57cha Small)

234. [ca. 1799]

Carte en perspective du nord au midi, d'après le plan en relief et les mésures du Gen. Pfyffer. Zug: [ca. 1799]. 2 copies.

Scale: League of 2400 toises.

Cartographer: F. X. Pfyffer von Altishofen.

Engraver: Jos. Clausner.

Size: 78.7 x 64.5 cm.

Colored.

Provenance: Presented by Ferdinand Rudolph Hassler, 20 March 1807, and one purchased with the Benjamin Vaughan papers, 1950.

(237: [ca. 1799]: C512cpm Large)

235. [ca. 1800]

Die Eisgebyrge des Schweizerlandes, mit allen dabey vorkommenden Mineralien. In ihrem zusammenhang vorgestelt von G. S. Gruner. Paris: A. Zingg: [ca. 1800].

Cartographer: G. S. Gruner.

Size: 38.5 x 81 cm.

Provenance: Presented by Ferdinand Rudolph Hassler, 18 July 1808.

(237: [ca. 1800]: G923esm Large)

236. 1842–1861

Carte géologique de la Suisse publiée aux frais de la confédération, par la Commission Géologique de la Société Helvétique des Sciences Naturelles . . . par E. Renevier et H. Schardt. Winterthur-Neuchâtel: Wurster, Randegger: 1842–1861. Nos. 2–3, 6–11, 15–16, 20, 22, 24.

Scale: 1:100,000.

Cartographers: Eugène Renevier and Hans Schardt.

Size: 48 x 69.5 cm.

See pamphlet accompanying maps: *Notice explicative de la feuille xvi* (2nd ed.). 1899.

(237: 1842–1861: R297geo Large)

237. [ca. 1861–1866]

Geologische Karte der Umgebung von Brugg (Aargau) von C. Moesch. Neuenburg: H. Furrer, litho.: [ca. 1861–1866].

Scale: 1:25,000.

Cartographer: C. Moesch.

Lithographer: H. Furrer.

Size: 40.6 x 29 cm.

Colored.

Contains also profiles of certain elevations.

(237: [ca. 1861–1866]: M715geo Small)

238. 1862

Karte vom Canton Basel entworfen von Andreas Kündig im Verlag von C. Detloff in Basel. Geologisch aufgenommen u. dargestellt von Dr. Albr. Müller in Basel 1862. Basel: C. Detloff: 1862.

Scale: 1:50,000.

Cartographers: Andreas Kündig, C. Detloff, and A. Müller.

Size: 68.7 x 80.1 cm.

Colored.

See: Schweizerische Naturforschende Gesellschaft, Geologische Commission, *Beiträge zur geologischen Karte der Schweiz.* 1862.

(237.715: 1862: M915kcb Large)

239. 1871

Carte géologique du Canton de Genève par Alphonse Favre. Publiée sous les auspices de la classe d'agriculture et la Société des Arts, d'après la carte levée par ordre de gouvernement dans le années 1837 et 1838, revue en 1871. Winterthur: Wurster, Randegger & Cie: 1878. 4 sheets.

Scale: 1:25,000.

Cartographers: Alphonse Favre and J. J. Goll.

Engraver: Bressanini.

Size: 49.5 x 65.5 cm.

Colored.

See: Alphonse Favre, *Description géologique du Canton de Genève.* Société des arts de Genève. *Bulletin de la classe d'agriculture.* Geneva: 1879.

(237.734: 1871: F273cgg Large)

240. 1873

Geolog: Profile zur Sentis-Karte v. Arnold Escher v. d. Linth. Plates 1–2. Winterthur: Wurster, Randegger: 1873. 2 plates.

Cartographer: Arnold Escher von der Linth.

Size: 79.5 x 58.2 cm. each.

Colored.

(237.735: 1873: Es12sen Large)

241. 1873

Geologische Karte des Sentis aufgenommen von Arnold Escher von der Linth in den Jahren 1837–1872. Herausgegeben von der Geologischen Commission der Schweizerischen Naturforschenden Gesellschaft. Winterthur: Wurster, Randegger: 1873.

Scale: 1:25,000.

Cartographer: Arnold Escher von der Linth.

Size: 73.2 x 109.1 cm.

Colored.

(237.735: 1873: Es12gks Large)

242. 1875

Carte géologique de la partie sud des alpes Vaudoises et des portions limitrophes du Valais comprenant les Massifs des Diablerets, Muveran d[en]t de Morcles, etc. par E. Renever . . . Publiée par la Commission Géologique Fédérale. Pl.1. Winterthur: Wurster, Randegger & Cie: 1875.

Scale: 1:50,000.

Cartographer: Eugène Renevier.

Size: 49.3 x 46.5 cm.

Colored.

On side are lists of conventional signs and colors.

(237: 1875: R297cgv Large)

243. 1884

Carte du phénomène erratique et des anciens glaciers du versant nord des alpes Suisses et de la chaîne du Mont-Blanc. Par Alphonse Favre. Publiée par la Commission Géologique de la Société Helvétique des Sciences Naturelles. Winterthur: Wurster, Randegger & Cie: 1884. 4 sheets.

Scale: 1:250,000.

Cartographer: Alphonse Favre.

Size: ca. 51 x 73 cm. each.

Colored.

(237: 1884: F273psa Large)

244. n.d.

Nouvelle carte de la Suisse dans laquelle sont exactement distinguées les Treize Cantons, leurs alliés, et leurs swets, dressée sur les mémoires les plus corrects, et assujettie aux observations astronomique. London: W. Faden: n.d.

Scale: 69½ English mi. = 1 degree.

Size: 62.3 x 82.5 cm.

Colored.

Insets: Vue de la vallée de Chamouny . . . ; and Section des principals montagnes.

Provenance: Purchased from the library of William Priestman, 15 July 1831.

(237: n.d.: N856swi Large)

FRANCE Nos. 245-275

245. [ca. 1644]

Le pais de Brie. Guiljelmus Blaeu excudit. [Amsterdam: ca. 1644].

Scale: 8.1 cm. = 4 common Gallic mi.

Cartographer: W. J. Blaeu.

Size: 39.1 x 49.8 cm.

(243.2: [ca. 1644]: B561fpb Small)

246. [ca. 1708–1712]

[Map of France]. [London: Jno Senex & Jno Maxwell: ca. 1708–1712].

Scale: 60 British mi. = 1 degree.

Cartographer: John Maxwell.

Engraver: John Senex.

Size: 63.7 x 91.5 cm.

Colored.

Mutilated: title is torn off.

(240: [ca. 1708–1712]: S57mis Large)

247. 1713

Map of the provostship and vicounty of Paris. Drawn from a great number of particular memoirs made according to the observations of the Royal Academy of Sciences by G. Delisle. London: Senex & Maxwell: 1713.

Scale: 5 cm. = 2 French leagues.

Cartographer: Guillaume de L'Isle.

Size: 49.2 x 62.9 cm.

Colored.

(240.761: 1713: D372pvp Large)

248. 1713

The province of Artois and the country adjacent. By William de l'Isle, geographer. London: Senex & Maxwell: 1713.

Scale: 1 in. = 3½ British computed mi.

Cartographer: Guillaume de l'Isle.

Engraver: John Senex.

Size: 48 x 59.8 cm.

Colored.

(242.7: 1713: D372paa Large)

249. 1744

La Champagne divisée par elections suivant les dernieres observations. Paris: Le Rouge: 1744.

Scale: 1 cm. = 1 short league.

Size: Ca. 57.5 x 48.5 cm.

Provenance: Purchased from the library of William Priestman, 15 July 1831.

Contains manuscript note: Wm. Priestman, Philadelphia, 266 Arch St. 1800.

(243.2: 1744: C354dob Large)

250. 1772

The post roads of France, from the map of Jaillot, published by the order of the Duke de Choiseul, post-master general of that kingdom. London: Robt. Sayer: 1772 [2 March 1772].

Scale: 2 cm. = 10 mi.

Cartographer: Jaillot.

Size: 46.6 x 57.7 cm.

Colored.

Provenance: Purchased from the library of William Priestman, 15 July 1831.

(240: 1772: J211prd Large)

251. [ca. 1773]

The kingdom of France distinguished according to the extent of all its provinces and conquests. In Spain, Italy, Germany, Flanders, Artoise, Haynault, Namur, and Luxembourg. Provinces of the Low Countries. By N. Sanson, geographer to the French King. [London]: Robert Sayer: [ca. 1773].

Scale: 1 in. = 30 mi.

Cartographer: N. Sanson.

Size: 54.8 x 86 cm.

Colored.

(240: [ca. 1773]: Sa57kfp Large)

252. 1774

Carte du canal royal de la province de Languedoc. [Paris]: J. B. Dutertre: 1774. 4 sheets: 1. Toulouse jusqu'à Renneville. 49.6 x 227 cm.; 2. Renneville jusqu'à Trebes. 50 x 385.7 cm.; 3. Trebes jusqu'à Capestang. 51.4 x 333 cm.; 4. Capestang jusqu'à l'Etang de Thau. 49.5 x 261 cm.

Scale: De deux mille toises à 5 lignes de pied de roy pour 100 toises.

Engraver: Nicolas Chalamandrier.

Sizes: See above.

See: Louis Lacoste, *Précis historique du canal du Languedoc ou des deux mers.* Paris: 1810.

(247.7: 1774: C352cpL Large)

253. 1779

Tableau général des postes, l'année 1779. N.p.: n.d.

Size: 45.9 x 41.5 cm.

Slightly mutilated.

Acquired with the Franklin-Bache purchase in 1936. This purchase greatly enhanced the Society's collection of Benjamin Franklin manuscripts.

(240: 1779: P847adp Small)

254. 1785

France, divided into its military governments, with the Netherlands, from the best authorities. London: C. Dilly & G. Robinson: 1785.

Scale: 25 French leagues = 1 degree.

Size: 33.3 x 37.7 cm.

Colored.

Provenance: Presented by Mathew Carey, 18 October 1805.

Engraved for Guthrie's new system of geography.

The *Transactions* (vol. 6) state that Carey presented the "Materials from which Guthrie's Geography were compiled." This is one of those items. See: William Guthrie, *The general atlas for Carey's edition of his geography improved.*

(240: 1785: D582fra Small)

255. 1786

Plan historique de la ville de Lyon et de ses quartiers, assujetti à ses accroissements embellissments, et projets. Paris: Moithey: 1786.

Scale: 150 toises.

Size: 44.2 x 35.5 cm.

Provenance: Presented by William Short, 5 November 1824.

(248.949: 1786: M725fra Small)

256. 1787

Carte très utile aux voyageurs qui designe les différentes routes pour aller de Paris à Rome, et à Naples et aux villes considérables d'Italie, qui l'on peut voir en faisant ce voyage dressée par le S.I.B. Nolin Géographe à S.A.R. Monsieur à Paris 1700. Revué, corrigée, et augmentée en 1787. [Paris]: 1787.

Scale: 25 common leagues.

Cartographer: I. B. Nolin.

Size: 24.6 x 33.6 cm.

Provenance: Presented and endorsed by William Short, 5 November 1824.

(248: 1787: N716par Small)

257. 1789

France, divisée en départements &c. avec les postes. London: R. Wilkinson: 1789.

Scale: 1 in. = 25 English mi.

Size: 62.4 x 65 cm.

Insets: Isle de Corse.

Colored.

(240: 1789: W659fdd Large)

258. 1790

Carte de France suivant sa nouvelle division, en LXXXIII départements. Dressée sur la même échelle que la carte de provinces de Guil. Delisle, par Dezauche. Paris: Dezauche: 1790.

Scale: 2.2 cm. = 10 French common leagues.

Cartographers: L'Isle and Dezauche.

Engraver: P. F. Tardieu.

Size: 49 x 61.7 cm.

Colored.

Provenance: Purchased from the library of William Priestman, 15 July 1831.

(240: 1790: D196pro Large)

259. [1792]

France. [1792].

Scale: 60 to a degree [2.5 cm. = 95 mi.].

Size: 16.3 x 20.2 cm.

Provenance: Presented by Thomas Dobson, 1798.

In: Thomas Dobson, publisher, *Encyclopaedia,* vol. 7, following p. 446.

Wheat: 751

(032: En2)

260. 1798

France divided into circles and departments. Philadelphia: J. Stewart & Co.: 1798 [17 February 1798].

Scale: 1⅛ in. = 60 British statute mi.

Size: 31.4 x 35.5 cm.

See: John Gifford [John Richards Green], *The history of France.* 1798, vol. 4, frontispiece.

(944: G36h)

261. 1798

Plan of part of a line of circumvallation [of Arras in 1654]. 1798.

Scale: 2.5 cm. = 55 fathoms.

Engraver: A. Lauson [Lawson].

Size: 3 x 17 cm.

Inset: Profile of same.

Provenance: Presented by Thomas Dobson, 1798.

In: Thomas Dobson, publisher, *Encyclopaedia,* vol. 18, following p. 768.

Wheat: 813

(032: En2)

262. 1799

A new and accurate map of France divided into departments, with the Netherlands &c. 1799.

Scale: 2.5 cm. = 75 mi.

Engraver: A. Anderson.

Size: 21.5 x 22.3 cm.

Inset: Plan of Toulon &c.

In: John Payne, *A new and complete system of universal geography,* vol. 3, opposite p. 607.

Wheat: 839

(910: P29)

263. 1800

Atlas topographique en XVI feuilles des environs de Paris, à la distance d'environ 8 myriamètres, ou 18 lièves . . . par Dom G. Coutans . . . revu, corrigé & considerablement augmenté, d'après nombre de cartes precieuses & plans particuliers . . . par Charles Picquet, géograph. graveur. Paris: C. Picquet: 1800. 16 folding maps in 2 volumes.

Scale: 3.1 cm. = 2 km.

Cartographers: G. Coutans and Charles Picquet.

Engraver: Charles Picquet.

Size: 38 x 61.5 cm.

Engraved title page contains index map: Tableau d'assemblage.

Received 1 March 1805.

(243: 1800: C83a Book map)

264. 1800

Plan routier de la ville et faubourg de Paris divisé en 12 mairies, 1800. [Paris]: Jean: 1800.

Scale: 3.7 cm. = 200 toises.

Size: 55.3 x 84.8 cm.

Colored.

Provenance: Legacy of and endorsed by Thomas Peters Smith, 1802.

(240.961: 1800: J344pfp Large)

265. 1801

Nouveau plan routier de la ville et faubourgs de Paris avec ses principaux édifices et nouvelles barriers. Par M. Pichon, Ingénieur Géographe. Paris: Esnault: 1801 An 9.

Scale: 13.7 cm. = 500 toises.

Cartographer: Pichon.

Engravers: C. B. Glot and E. Voysard.

Size: 99.8 x 142 cm.

Colored.

Insets: Some of the splendid views of Paris and environs.

(240.961: 1801: P586nrp Large)

266. 1813

Carte de l'empire Français et des états limitrophes, dessinée par Hérisson, Ingénieur Géographe. Revue, corrigée et augmentée par H. Brué. Paris: Desray et J. Goujon: 1813.

Scale: 25 French common leagues = 1 degree.

Cartographers: Eustache Hérisson and A. H. Brué.

Size: 125 x 193 cm.

(240: 1813: H423cef Large)

267. [ca. 1820]

Carte pour l'intelligence de la route d'Annibal, depuis le passage du Rhône jusqu' à Turin. N.p.: [ca. 1820].

Scale: 2283 toises = 1 league.

Engraver: Ambroise Tardieu.

Size: 16.8 x 19.3 cm.

(248: [ca. 1820]: T178fra Small)

268. 1822

Carte de la France divisée en 86 départemens et en 27 cours royales. Paris: Selves Fils, Lithographe: 1821.

Scale: 25 au dégré.

Lithographer: Selves fils.

Size: 7 x 21.4 cm.

Colored.

Provenance: Presented by Comte Jean D. Lanjuinais, August 1826 (?).

(240: [1822]: L225fra)

268a. 1826

Plano de la Bahia de S[a]n Juan de Luz levantado en 1826 por los Ingenieros Hidrografos de la Marina Francesa. Publicado en la Dirección de Hidrografia. Madrid: 1850.

Scale: 2.5 cm. = .02 nautical mi.

Engravers: J. Noguera and F. Bregante.

Size: 58 x 41.5 cm.

Provenance: Presented by Madrid, Spain, Dirección de Hidrografia, 19 October 1852.
(246.4: 1826: Sp27sjdL Large)

269. [ca. 1833]

France. Philadelphia: H. S. Tanner: [ca. 1833].

Scale: 69½ British statute mi. = 1 degree.

Cartographer: H. S. Tanner.

Engraver: J. Knight.

Size: 22.7 x 27.6 cm.

Colored.

Inset: Corsica.

See: H. S. Tanner, *Universal atlas.* Philadelphia: Tanner: 1833.
(240: [ca. 1833]: T158fra Small)

270. 1843

Vues & coupes du Cap de la Hève. Profit du Cap. Le Havre: Lemercier: 1843.

Cartographer and lithographer: Charles A. Lesueur.

Size: 45.3 x 62.5 cm.

Provenance: Presented by Charles A. Lesueur, 14 February 1845.
(242.1: 1843: L565vch Large)

271. [ca. 1875]

Carte géologique de la France. Ministère des Travaux Publiques. [Paris]: Erhard fres.: [ca. 1875]. Planches 11–21.

Scale: 1:10,000.

Cartographers: Elie de Beaumont and Dufrénoy.

Engraver: L. Wuhrer.

Sizes: Various sizes.

Colored.

Provenance: Le Bureau des Travaux Publiques de France presented on 21 October 1864 six sheets of "La Carte Géologique de la France." Dufrénoy and Elie de Beaumont prepared the map from 1841 through 1848. [Donation book].

In separate folder: "Les assises crétaciques et tertiaries du nord de la France." Fasc. III (Region de Béthune).
(248: [ca. 1875]: F853geo Large)

272. 1875

Monographie géologique des anciens glaciers et du terrain érratique de la partie moyenne du bassin du Rhône par A. Falsan & E. Chantre. Lyon: Pitrat aîné: 1875. 6 maps in original wrappers.

Cartographers: A. Falsan and E. Chantre.

Size: 54.2 x 85.

Colored.
(248: 1875: F183gLa Small)

273. [1887]

Plan de la ville de Paris. Période révolutionnaire [1790–1794]. Ce plan a été exécuté conformément a la décision prise par le conseil municipal de Paris dans sa séance du 30 Décembre 1887. [Paris: 1887].

Scale: 1 cm. = 1,000 m.

Cartographers: Lucien Faucou, Laporte, Bibert, and Bizard.

Engravers: Wuhrer and Sulpis.

Size: 79.2 x 99.5 cm.

Colored.

(240.961: [1887]: F273ppr Large)

274. [1887]

Plan de la ville de Paris en 1789. Ce plan a été exécuté conformément à la décision prise par le conseil municipal de Paris dans sa séance du 30 Décembre 1887. [Paris: 1887].

Scale: 1 cm. = 1,000 m.

Cartographers: Lucien Faucou, Laporte, Bibert, and Bizard.

Engravers: Wuhrer and Sulpis.

Size: 79.6 x 99.6 cm.

Colored.

(240.961: [1887]: F273dvp Large)

275. [ca. 1915]

Bartholomew's map of north-eastern France, Belgium and the Rhine. The Edinburgh Geographical Institution. Edinburgh: John Bartholomew, & Co.: [ca. 1915].

Scale: 1:1,000,000 or 1 in. = 16 English mi.

Size: 54.3 x 74.1 cm.

Colored.

(243.1: [ca. 1915]: B231fbr Book map)

CENTRAL EUROPE Nos. 276-287

276. [ca. 1830]

[Lon]gitudinal section of the first 38⅔ English miles of the railway for joining the [Mol]dau and the Danube . . . with the Elbe, and the Black Sea with the northern ocean, by means of the continent, executed under the superintendence of [the Ch]evalier Francis Anthony Gerstner. The [ac]t of the Emperor of Austria was . . . 1824, the work begun 28, July 1825 and the 38⅔ english miles were completely finished at the end of the year 1828. N.p.: [ca. 1830]. 5 sheets.

Size of paper: each ca. 59.5 x 72 cm.

Insets: Transverse section of an excavation; A horse takes 3 waggons, each loaded with 2 or 2½ tons and goes with his load on a level railway 20 English miles in one day; and Transverse section of an embankment, shewing the dry-walling in the middle.

Provenance: Presented by the Rev. Mr. Barry.

Mutilated.

(281: [ca. 1830]: L862mda Large)

277. 1832

Strassen-Karte des Koenigreiches Ungarn. Vienna: 1832.

Scale: 1 Wiener Zoll = 6,000 Wiener Klafter.

Lithographer: A. Floder, Wien.

Size: 128 x 189.3 cm.

Colored.

Provenance: Presented by Károly Nagy, 20 August 1833.

(293: 1832: F653sck Large)

278. 1849

Geognostische Karte der Umgebungen von Krems und vom Manhardsberge. Im Flächenraume von 34 Quadrat Meilen von Joh. Czjzek, aufgenommen in Monaten August, September, October 1849. N.p.: 1849.

Scale: 1 Meile = 4 Zoll.

Cartographer: Joh. Czjzek.

Size: 69.1 x 88.8 cm.

Colored.

Contains geological profiles, etc.

(282.2: 1849: C992gkk Large)

279. 1862

Carte géologique des parties de la Savoie, du Piédmont et de la Suisse voisines du Mont-Blanc. Par Alphonse Favre. Winterthur: J. Wurster et Cie.: 1862.

Scale: 1:150,000.

Cartographer: Alphonse Favre.

Size: 81.8 x 62.4 cm.

Colored.

Contains: Explication des signes et des couleurs, and Hauteurs principales au dessus de la mer.

(204.3: 1862: F273gmb Large)

280. [ca. 1865]

Plan der K. K. Chaussée und der im Bau begriffenen Eisenbahn von Budweis bis zum Scheidungspunkte bei Leopoldschlag nebst ihren nächsten Umgebungen. Vienna: Laurenz Herr and H. F. Müller: [ca. 1865].

Scale: Prifils, 1 Zoll = 1200 Klafter.

Cartographer: Casimir von V. Pilarski.

Lithographer: E. K. Frühwirth.

Size: 50.6 x 70.8 cm.

Provenance: Presented by the Rev. Mr. Barry.

(283: [ca. 1865]: P696rbL Large)

281. [ca. 1873]

[Geological profile maps of the Alps]. Winterthur: Wurster, Randegger: [ca. 1873]. 4 maps.

Sizes: 25.7 x 51.6; 25.6 x 43.5; 25 x 50.4; 24.5 x 73.2 cm.

Colored.

(204.3: [ca. 1873]: AL61geo Large)

282. 1914

Map of Central Europe prepared in the War College Division, General Staff. Washington: Norris Peters Co.: 1914.

Scale: 4.7 cm. = 100 km.

Size: 124.3 x 112.5 cm.

Shows population, army, and reserves for Triple Alliance and Triple Entente; also Vessels building or authorized.

(204: 1914: Un38wgs Large)

283. 1914

Map showing Franco-German frontier. Prepared in the War College Division, General Staff. August 1914.

Scale: 2 cm. = 10 mi.

Size: 64.6 x 77.5 cm.

Includes parts of Switzerland, Belgium, and all of Luxembourg.

(204: 1914: Un38fgf Large)

284. 1915

Strategic map of Central Europe showing the international frontiers. Prepared in the War College Division, General Staff War Department. Washington: Eckert Lithographing Co.: 1915.

Scale: 1:2,280,960.

Cartographers: Arch. B. Williams & George F. Bontz, draftsmen.

Size: 145.1 x 178.5 cm.

Colored.

(204: 1915: Un38sme Large)

285. 1939

Central Europe and the Mediterranean as of September 1, 1939. Compiled and drawn in the cartographic section of the National Geographic Society for the National Geographic Magazine. Gilbert Grosvenor, editor. Baltimore: A. Hoen & Company: 1939.

Scale: 1:5,000,000.

Size: 64.4 x 89.4 cm.

Colored.

Provenance: Presented by Alban W. Hoopes, February 1940.

(200: 1939: G293cem Large)

286. n.d.

. . . Map of the present seat of war, between the Russians, Poles, and Turks . . . By Andrew Drury. N.p.: n.d. Second ed. 2 sheets.

Cartographer: Andrew Drury.

Engravers: J. Caldwell and T. W. Tueur, Jr.

Size: 84.3 x 141 cm.

Colored.

Provenance: Purchased from the library of William Priestman, 15 July 1831.

(205: n.d.: D842eew Large)

287. n.d.

Le royaume de Boheme divisée en ses douze cercles. Carte réduite sur celle de 25 feuilles par Müller. Amsterdam: Cövens et Mortier: n.d.

Scale: 2.5 cm. = 4 Bohemian leagues.

Cartographer: Müller.

Size: 47.1 x 54.8 cm.

Colored.

Provenance: Purchased from the library of William Priestman, 15 July 1831.

(283: n.d.: M889sdc Large)

GERMANY Nos. 288-306

288. [ca. 1716]

[Map of Germany]. N.p.: [Senex and Maxwell: 1716].

Scale: 1 in. = 20 mi.

Cartographer: John Maxwell.

Engraver: John Senex.

Size: 65.5 x 96.6 cm.

Colored.

Double line denotes Dr. Brown's travels from Norwich through Germany.

Mutilated.

(260: [ca. 1716]: S57tto Large)

289. [ca. 1755]

A map of the German empire divided into its circles to which is added the Kingdom of Prussia, the whole laid down from the most accurate surveys and chiefly from the map of Marshal de Schmettau, lately published at Berlin by the Royal Academy of Sciences by L. de la Rochette. London: Carrington Bowles & Robert Wilkinson: [ca. 1755].

Scale: 1 in. = 18 mi.

Cartographer: Samuel von Schmettau.

Size: 101.7 x 119.3 cm.

Colored.

Mutilated.

(260: [ca. 1775]: R587pms Large)

290. 1773

Basis novae chartae Palatinae . . . By Christian Mayer. N.p.: January 1773. 2 copies.

Scale: Pariser toises, oder klaffter zu 6 schube.

Cartographer: Christian Mayer.

Engraver: E. Vevelst.

Size: 21.5 x 32 cm.

Provenance: Presented by Christian Mayer, 15 December 1773.

The Society thanked Mayer "for his ingenious performances" and wished to have "the continuation of his valuable correspondence." In return, the members promised that they "will esteem themselves happy in communicating anything in their power to assist him in the Prosecution of his Philosophical Researches." [Minutes: Archives].

(273: 1773: M455pal Small)

291. 1784

A new map of the German empire, and the neighbouring states with their principal post roads originally published by the Royal Academy of Berlin. Engraved with several additions and improvements. William Faden. London: William Faden: 1784 [9 November 1784].

Scale: 69½ English mi. = 1 degree.

Engraver: William Faden.

Size: 63 x 73.5 cm.

Colored.

Provenance: Purchased from the library of William Priestman, 15 July 1831.

(260: 1784: F121gem Large)

292. 1789

Karte von Deutschland in XVI. Blatt nach des H.O.C. Büsching Erdbeschreibung u. den besten Hülfsmitteln entworffen von D.F. Sotzmann. Berlin: Carl Jäck: 1792.

Scale: Maasstab von 15 geographische Meilen [= 9.8 cm.].

Cartographers: D. F. Sotzmann and J. W. Meil.

Engraver: Carl Jäck.

Size: 97 x 113 cm.

Colored.

(260: 1789: So87kvd Large)

293. 1789

A new map of the king of Great Britain's dominions in Germany, or the electorate of Brunswick-Luneburg and its dependencies. By Thomas Jefferys. With various improvements and emendations. London: W. Faden: 1789 [1 June 1789].

Scale: 69½ British mi. = 1 degree.

Cartographer: Thomas Jefferys.

Size: 59 x 52.3 cm.

Colored..

Provenance: Purchased from the library of William Priestman, 15 July 1831.

(260: 1789: J377ebl Large)

294. 1791

Carte générale de l'empire d'Allemagne, par Mr. Chauchard, Captaine d'Infanterie et Ingénieure Militaire. Paris: Dezauche: [1791]. 9 leaves.

Scale: 15 German common leagues = 1 degree.

Cartographer: M. Chauchard.

Engraver: Bellanger.

Size: 177 x 212.5 cm.

Colored.

Provenance: Presented by Peter Stephen du Ponceau, 1 November 1839.

Accompanied by: Carte réduite de la carte générale d'Allemagne, pour servir à rassembler les neuf Feuilles dont cette carte est composée, par M. Chauchard [61.2 x 72.2 cm.]; Carte d'une partie des Pays-Bas, pour servir de supplément à la carte de l'empire d'Allemagne, par Mr. Chauchard (incomplete) [39.4 x 39.6 cm.]; and Tableau d'assemblage de sa carte d'Allemagne en neuf-feuilles grand aigle par M. Chauchard [24.4 x 35.4 cm.].

(260: 1791: C392cga Large)

295. [1792]

Germany. [1792].

Scale: 2.5 cm. = 110 mi.

Engraver: W. Barker.

Size: 17.3 x 21.1 cm.

Provenance: Presented by Thomas Dobson, 1798.

In: Thomas Dobson, publisher, *Encyclopaedia,* vol. 7, opposite p. 696.

Wheat: 752

(032: En2)

296. [1798]

War. Attack of fortified places [Philippsburg: 1798].

Scale: 2.6 cm. = 140 ft.

Engraver: T. Clarke.

Size: 16 x 11.1 cm. (Plate: 23.9 x 18.3 cm.).

Provenance: Presented by Thomas Dobson, 1798.

In: Thomas Dobson, publisher, *Encyclopaedia,* vol. 18, opposite p. 808.

Wheat: 811

(032: En2)

297. [1798].

War. Attack of fortified places. Plan of the attacks of Landau in 1713. [1798].

Scale: 2.7 cm. = 200 fathoms.

Engraver: T. Clarke.

Size: 9.8 x 17 cm. *Size of Plate:* 23.7 x 18.5 cm.
Provenance: Presented by Thomas Dobson, 1798.
In: Thomas Dobson, publisher, *Encyclopaedia,* vol. 18, following p. 784.
Wheat: 810

(032: En2)

298. [1798]

War. Plan of the circumvallation and attacks of Philippsburg in 1734. [1798].
Scale: 2.5 cm. = 550 fathoms.
Engraver: A. Lauson [Lawson].
Size: 9.9 x 17.3 cm. Plate: 24.2 x 18.8 cm.
Insets: Plan of part of the circumvallation of Philippsburg in 1734; also Profile of same.
Provenance: Presented by Thomas Dobson, 1798.
In: Thomas Dobson, publisher, *Encyclopaedia,* vol. 18, following p. 768.
Wheat: 812

(032: En2)

299. [1798]

War. [Two maps of the environs of Soest, Westphalia. 1798].
Engraver: Lawson.
Size of plate: 22.3 x 16.4 cm.
Provenance: Presented by Thomas Dobson, 1798.
In: Thomas Dobson, publisher, *Encyclopaedia,* vol. 18, following p. 752.
Wheat: 809

(032: En2)

300. 1805

Kort over Tønder og Lugumcloster Amter, samt deele af Haderslebhuus Apenrade Flensborg og Bredsted Amter udi Hertug dømmet Schleswig, under Det Kongelige Viderskabernes Societets Direction . . . Reduceret og tegnet af T. Bugge og F. Wilster . . . [Copenhagen]: 1805.
Scale: 2 Danish mi. = 24,000 Danish ells.
Cartographers: T. Bugge and F. Wilster.
Engraver: G. N. Angelo.
Size: 56.5 x 85 cm.
Provenance: Presented through P. Pedersen, 1 May 1818.

(261.1: 1805: B851tLh Large)

301. [1808]

Des contrées transrhenanes on de l'Allemagne politique en 1808. [Paris: Didot l'Ainé: 1808].
Scale: 50 lieues en deux dégrés.
Size: 32.6 x 40.7 cm.

Colored.

Surrounded on three sides with data pertaining to the countries depicted and to some of Napoleon's victories.

(260: [1808]: G313ctr Large)

302. [ca. 1862]

Karte der Regenhöhen Deutschlands gezeichnet von Theodor Schade in Görlitz. Görlitz: [ca. 1860].

Cartographer: Theodor Schade.

Lithographer: Oscar Diessler.

Size: 56.5 x 50 cm.

Colored.

For article containing the information on which this map is based, see: Goerlitz, in *Naturforschende Gesellschaft* . . . , Moellendorff, G. von, ed., "Die Regenverhältnisse Deutschlands . . . " Goerlitz: 1862.

(262.6: [ca. 1862]: Schl7rd Large)

303. 1866

Geologische Übersichskarte der Rheinprovinz und der Provinz Westfalen im Auftrage des Königl. Ministers für Handel, Gewerbe und öffentliche Arbeiten Herrn Graf von Itzenplitz bearbeitet von Dr. H. von Dechen. Berlin: Simon Schropp'sche Hof-Landkartenhandlung: 1866.

Scale: 1:500,000.

Cartographer: H. von Dechen.

Lithographer: Berliner Lith. Institut.

Size: 78.4 x 55.6 cm.

Colored.

(261.7: 1866: D362grw Large)

304. 1871

Geognostische Karte der Umgegend von Hainichen im Königreich Sachsen von Carl Naumann. Leipzig: Engelmann: 1871.

Scale: 3 cm. = 1/8 mi.

Cartographer: Carl Naumann.

Lithographer: J. G. Bach.

Size: 37.3 x 58.9 cm.

Colored.

See: C. F. Naumann, "Erläuterungen zu der geognostischen Karte . . . ." Leipzig: Engelmann: 1871.

(262.7: 1871: N226gks Large)

305. n.d.

Geologische Karte von Deutschland. F. A. Brockhaus Geogr.-artist. Anstalt Leipzig: n.d.

Cartographer: F. A. Brockhaus.

Size: 20.2 x 28.2 cm.

Colored.

(260: n.d.: B781deu Small)

306. n.d.

Germany divided into circles; with the kingdom of Prussia, from the best authorities. N.p.: n.d.

Size: 30 x 36.6 cm.

Engraved for Chambers's edition of Guthrie's new system of geography.

(260: n.d.: G313ger Small)

POLAND Nos. 307-311

307. 1787

A new map of the kingdom of Poland with its dismembered provinces. London: Robert Sayer: 1787 [1 January 1787].

Scale: 69½ British mi. = 1 degree.

Size: 47.4 x 64.5 cm.

Colored.

Provenance: Purchased from the library of William Priestman, 15 July 1831.

(310: 1787: N422pro Large)

308. [ca. 1790]

Poland with the possessions of Russia, Prussia & Austria from the best authorities. Woodman & Mutlow: [ca. 1790].

Scale: 20 Polish mi. = 1 degree.

Engraver: Mutlow.

Size: 32.2 x 36.3 cm.

Provenance: Presented by Mathew Carey, 18 October 1805.

The *Transactions* (vol. 6) state that Carey presented the "Materials from which Guthrie's Geography were compiled." This is one of those maps. See: William Guthrie, *The general atlas for Carey's edition of his geography improved.* Philadelphia: Carey: 1795 [1 May 1795].

(310: [ca. 1790]: W859poL Small)

309. [1796]

Poland, Lithuania and Prussia. [1796].

Scale: 2.5 cm. = 130 British statute mi.

Engraver: Vallance.

Size: 20.3 x 16.9 cm.

Provenance: Presented by Thomas Dobson, 1798.

In: Thomas Dobson, publisher, *Encyclopaedia,* vol. 15, opposite p. 272.

Wheat: 785

(032: En2)

310. 1809

Carte hydrographique de Pologne présentée au feu roi Stanislas Auguste par Mr. le Colonel de Perthes, géographe de sa majesté . . . Cette carte a été réduite d'après la grande carte topographique inédite de Pologne, assujettie aux observations du Père Rostan astronome distingué, et rédigée par Mr. de Perthes, sur les plans detaillés des cantons levés dans le cours de 21 ans sous les ordres et par les soins du feu roi. Paris: Komarzewsky: 1809.

Scale: 25 lieues communes de France = 1 degré.

Cartographer: de Perthes.

Engraver: E. Collin.

Size: 64.2 x 80 cm.

Provenance: Presented by Kamarzewski, 2 February 1810.

(310: 1809: P41chp Large)

311. [ca. 1900]

Carte des travaux astronomiques, géodésiques et topographiques, exécutés depuis la fondation du Corps Topographique, jusqu'à 1890, dressé par le Général-Major Koversky. [Ca. 1900].

Scale: 1:8,400,000.

Cartographer: E. Koversky.

Size: 48.9 x 42.7 cm.

Colored.

(321: [ca. 1900]: K844rft Small)

USSR (RUSSIA) Nos. 312-320

312. 1788

[The Russian Empire]. The European part . . . from the maps published by the Imperial Academy of St. Petersburg on the new provinces on the Black Sea: and, the Asiatic part . . . from maps published by the Imperial Academy of St. Petersburg with the new discoveries of Captn. Cook &c. London: Robert Sayer: 1788 [1 May 1788].

Scale: 60½ British mi. = 1 degree.

Size: 47.1 x 127.5 cm.

Colored.

Provenance: Purchased from the library of William Priestman, 15 July 1831.

(321: 1788: R92lccv Large)

313. [ca. 1795]

Ducatuum Estoniae et Livoniae tabula cum cursu fluvii Dwinae. No. 3. [St. Petersburg: ca. 1795].

Scale: 7.1 cm. = 10 Russian mi.

Size: 49.8 x 57.6 cm.

(331: [ca. 1795]: Es82deL Large)

314. [ca. 1795]

Territorium Mesenense et Pustoserense cum adiacentibus insulis et territoriis. No. 6. [ca. 1795].

Size: 48.8 x 54.3 m.

(322: [ca. 1795]: R917tmp Large)

315. [1796]

Russia or Moscovy. [1796].

Scale: 2.5 cm. = 500 Russian versts.

Engravers: Scot & Allardice.

Size: 16.8 x 21 cm.

Provenance: Presented by Thomas Dobson, 1798.

In: Thomas Dobson, publisher, *Encyclopaedia,* vol. 16, opposite p. 554.

Wheat: 786

(032: En2)

316. 1821

Geographical and statistical map of Russia. Map of the Russian empire. Intended for the elucidation of Lavoisne's historical atlass. Improved to 1821. Philadelphia: Palmer: 1821.

Size: 26.5 x 39.2 cm. *Size of paper:* 47 x 52.6 cm.

Colored.

Provenance: Presented by Mr. Ivanoff, consul general of Russia, 1 June 1821.

Third Philadelphia edition, 1821—printed on Gilpin's machine paper by T.H. Palmer for M. Carey & Son, from the London edition of 1817, with numerous corrections and additions.

Surrounded on three sides by printed information on Russia.

(320: 1821: R917Lha Large)

317. [1842]

Karte vom Ural Gebirge gegründet auf die astronomischen Beobachtungen von Wischnewsky, Schubert, A.v. Humboldt, Ad. Erman, und auf handschriftliche Specialkarten. Als Beilage zu G. Rose's Mineralogisch-Geognostischem Bericht von Humboldt's Sibirischer Reise. [1842].

Scale: 1:2,000,000.

Size: 55.2 x 74.6 cm.

Colored.

Inset: Bjelaja Gora mit den Umgebungen.

(421: [1842]: R72ahu Book map)

318. 18421885

[Maps of Russia, northern Asia, and islands in adjacent oceans]. Gidrograficheskii departament, Russia. St. Petersburg: 1842–1885. 43 maps.

Engravers: Various.

Sizes: Various.

(320: 1842–1885: R91m Large)

319. 1892

Carte geólogique de la Russie d'Europe éditée par la Comité Géologique. Par A. Karpinsky, S. Nikitin, Th. Tschernyschev, N. Sokolov, A. Mikhalsky, J. Mouschketov, A. Sorokin, S. Simonowitch, A. Konchin, N. St. Barbot-de-Marny, V. Ramsay, J. Sederholm, E. Federov, A. Gourov, and P. Armaschevsky. St. Petersburg: 1893. 6 pieces.

Scale: 1:520,000.

Cartographers: A. Karpinsky, S. Nikitin, Th. Tschernyschev, N. Sokolov, A. Mikhalsky, J. Mouschketov, A. Sorokin, S. Simowitch, A. Konchin, N. St. Barbot-de-Marny, V. Ramsay, J. Sederholm, E. Federov, A. Gourov, and P. Armaschevsky.

Size: 56 x 66.7 cm. each sheet.

Colored.

Provenance: Presented by S. Nikitin and A. Karpinsky, 1 September 1893.

Accompanied by a pamphlet of text.

(321: 1892: R927geo Large)

320. n.d.

Tiefenkarte des Goktscha Sees. St. Petersburg: n.d.

Scale: 1:102,000.

Cartographer: A. A. Iljin.

Size of paper: 58.4 x 70.2 cm.

Colored.

(327.5: n.d.: IL44azr Large)

FINLAND Nos. 321-322

321. 1897

Distribution des dépôts quaternaires en Finlande. Par J. J. Sederholm. Helsinki: F. Tilgmann: 1897.

Scale: 1:2,000,000.

Cartographer: J. J. Sederholm.

Engraver: F. Tilgmann.

Size: 57.4 x 42 cm.

Colored.

(335: 1897: Se27fin Small)

322. 1897

Geologische Übersichtskarte von Finland, und den angrenzenden Landteilen von J. J. Sederholm. Helsinki: F. Tilgmann: 1897.

Scale: 1:2,500,000.

Cartographer: J. J. Sederholm.

Engraver: F. Tilgmann.

Size: 52.5 x 37.5 cm.

Colored.

(335: 1897: Se27geo Small)

MEDITERRANEAN AREA Nos. 323-325

323. [1799]

A correct map of the Mediterranean Sea with the countries adjacent. [1799].

Size: 18.4 x 40.9 cm.

In: John Payne, *A new and complete system of universal geography,* vol. 2, opposite p. 346.

Wheat: 860

(910: P29)

324. 1829

Map of the countries found in sacred classic and ecclesiastical writers in their ancient and modern names carefully compiled from the best authorities and drawn by C. Wiltberger Junr. Philadelphia: S. Tiller: 1829.

Scale: 1 in. = 60 mi.

Cartographer: C. Wiltberger.

Engravers: John & William W. Warr, Philadelphia.

Size: 59 x 138.5 cm.

Colored.

Inset: Palestine or the Holy Land.

See: C. Wiltberger, Jr., "Vocabulary of all the places contained in the map . . ."

(207: 1829: W719sbL Large)

325. n.d.

[Map of the Mediterranean world]. Western part of the Mediterranean Sea with the coasts of Spain, France, Italy, Barbary. Eastern part of the Mediterranean Sea with the coasts of Turkey, Anatolia, Syria, Egypt and Barbary. [London: W. Faden: n.d.].

Scale: 69½ British statute mi. = 1 degree.

Size: 51.2 x 106 cm. *Size of plate:* 53.4 x 126.5 cm.

Colored.

Provenance: Purchased from the library of William Priestman, 15 July 1831.

On each side are some statistics, etc., of the countries depicted.

(207: n.d.: W524sfi Large)

SPAIN AND PORTUGAL Nos. 326-340

326. 1773

Plan of the River Viguo, and Port de Rande. Situated in the north part of Spain, in the kingdom of Gallice . . . by Joseph Smith Speer. London: Speer: 1773 [14 June 1773].

Scale: 5 marine mi. 60 to a degree.

Cartographer: Speer.

Engraver: Hall.

Size: 33 x 47 cm.

Provenance: Presented by the Direccion de Hidrografia, Madrid, 19 October 1852.

(341: 1773: Sp27vig Small)

327. 1787

The kingdoms of Spain and Portugal, divided into their great provinces. London: Robert Sayer: 1787 [2 March 1787].

Scale: 69½ British mi. = 1 degree.

Size: 47 x 64 cm.

Colored.

Provenance: Purchased from the library of William Priestman, 15 July 1831.

(340: 1787: K589pro Large)

328. 1788

Geo-hydrographic survey of the isle of Madeira with the Dezertas and Porto Santo Islands geometrically taken in the year 1788 by William Johnston. London: W. Faden: 1791 [1 January 1791].

Scale: 69½ Enlgish statute mi. = 1 degree.

Cartographers: William Johnston and Andrew Skinner.

Size: 58.4 x 114.5 cm.

Insets: View of the city of Funchal; plan of the road of Funchal; plan of the town of Funchal by Captain Skinner, 1775; and two views of the island of Madeira.

(349: 1788: J624ghm Large)

329. [ca. 1789]

Ilha de Leão com asilhetas e fortes adjacentes ciudade e porto de Cadix com scus baixos &c. D. Vicente Tofino. [Lisbon?: ca. 1789].

Scale: 69½ British statute mi. = 1 degree.

Cartographer: Vincente Tofino.

Engraver: Queiroz.

Size: 42.3 x 33.3 cm.

(346: [ca. 1789]: T578Leo Small)

330. 1794

Mapa de las carreras de postas de España . . . por D[o]n Bernardo Espinalt y Garcia, administrador principal de los correos, y postas de la ciudad y reyno de Valencia. N.p.: 1794.

Scale: 8.7 cm. = 35 common Spanish leagues.

Cartographer: Bernardo Espinalt y Garcia.

Engraver: Pasqual Cucó.

Size: 39.7 x 56.7 cm.

Upper left-hand corner torn off.

(341: 1794: Es62spm Large)

331. [1797]

Spain and Portugal. [1797].

Scale: 2.5 cm. = 120 British statute mi.

Engraver: Vallance.
Size: 16 x 21.3 cm.
Provenance: Presented by Thomas Dobson, 1798.
In: Thomas Dobson, publisher, *Encyclopaedia,* vol. 17, opposite p. 619.
Wheat: 796

(032: En2)

332. 1810

The bay, harbour, and environs of Cadiz: reduced from the survey of Tofino. London: Robert Laurie & James Whittle: 1810 [24 May 1810].
Scale: 4.9 cm. = 3 nautical or geographic mi.
Cartographer: Tofino.
Size: 21.1 x 28.1 cm.
Colored.

(341.921: 1810: Sp17cad Small)

333. [ca. 1827]

Mappa geographica do reino de Portugal. [Paris: ca. 1827]. 4 sheets.
Scale: Legoas Portugzas de 18 ao graá.
Cartographer: J. P. Cardozo Casado Giraldes.
Engraver: E. Collin.
Size: 47.3 x 24 cm. *Size of paper:* ca. 59.5 x 73.7 cm.
Colored.
Provenance: Presented by Cardozo Casado Giraldes, 21 September 1827.
Contains historical and geographical statistics of Portugal by Cardozo Casado Giraldes.

(348: [ca. 1827]: C262mgp Large)

334. [ca. 1833]

Spain and Portugal. Philadelphia: H. S. Tanner: [ca. 1833].
Scale: 69½ British statute mi. = 1 degree.
Size: 22.8 x 28 cm.
Colored.
Provenance: Presented by H. S. Tanner.
See: H. S. Tanner's *Universal atlas.* Philadelphia: Tanner: 1833.

(341: [ca. 1833]: T158spp Small)

335. 1850

Plano de la ria de Vigo levantado de orden superior por el Capitan de Fragata de la Armada D. Antonio Doral y publicado en la Dirección de Hidrografia. Madrid: 1850.
Scale: 8.3 cm. = 3 nautical mi.
Cartographer: F. Bregante.
Engraver: J. Noguera and F. Bregante.

Size: 45 x 60.9 cm.

Provenance: Presented by the director of the Dirección de Hidrografia in Madrid through the hands of Don Severiano Moreledo of Cadiz, 7 January 1853.

(341: 1850: Sp27prv Large)

336. 1851

Carta esférica de la isla de Fuerteventura en las Canarias, levantada en 1835 por el Teniente Arlett, R.N. de la M.R. Inglesa; publicada en la Dirección de Hidrografia. Madrid: 1851.

Cartographer: Arlett.

Engravers: J. Noguera and F. Bregante.

Size: 67.2 x 47.2 cm.

Inset: Enlarged area of the s.w. of Fuenteventura.

Provenance: Presented by the director of the Dirección de Hidrografia in Madrid through the hands of Don Severiano Moreledo of Cadiz, 7 January 1853.

(346.7: 1851: S27fLp Large)

337. 1852

Carta esférica de la isla de Lanzarote en las Canarias, levantada en 1835 por el Teniente Arlett, R.N. de la M.R. Inglesa: publicada en la Dirección de Hidrografia. Madrid: 1852.

Cartographer: Arlett.

Engravers: J. Noguera and F. Bregante.

Size: 60.8 x 45.8 cm.

Inset: Enlargement of northern tip of Lanzarote.

Provenance: Presented by the director of the Dirección de Hidrografia in Madrid through the hands of Don Severiano Moreledo of Cadiz, 7 January 1853.

(346.7: 1852: Sa27LiL Large)

338. 1852

Carta esférica de la isla de Tenerife en las Canarias, levantada por el Capitan A. T. E. Vidal en 1838, publicada en la Dirección de Hidrografia. Madrid: 1852.

Cartographer: A. T. E. Vidal.

Engravers: J. Noguera and F. Bregante.

Size: 47.6 x 62.3 cm.

Provenance: Presented by the director of the Dirección de Hidrografia in Madrid through the hands of Don Severiano Moreledo of Cadiz, 7 January 1853.

(346.7: 1852: Sp27ctc Large)

339. 1899

Carta geologica de Portugal por J. F. N. Delgado e Paul Choffat. Direcçao trabalhos Geologicos. Paris: L. Wuhrer: 1899.

Scale: 1:500,000.

Cartographer: J. F. N. Delgado and Paul Choffat.

Size: 122.3 x 77.6 cm.

Colored.

(348: 1899: P836cgp Large)

340. 1906

Carta hypsometrica de Portugal. Comissão do Serviço Geologico. Paris: L. Wuhrer: 1906. 2 sheets.

Scale: 1:500,000.

Cartographer: P. Guedes.

Engraver: L. Wuhrer.

Size: 122 x 76.9 cm.

Colored.

See: Portugal, Direccão dos trabalhos geologicos, *Communicações*, vol. 7.

(348: 1906: P836chy Large)

ITALY Nos. 341-358

341. [ca. 1753]

Urbis Romae veteris ac modernae accurata delineato edita a Ioh. Bapt. Homanno. Nürnberg: [ca. 1753].

Scale: 3 cm. = ⅔ Italian mi.

Cartographer: Johann Baptist Homann.

Size: 48.8 x 58.9 cm.

(350.968: [ca. 1753]: H753urv Large)

342. [ca. 1770]

Pianta topografica della città di Napoli in Campagna Felice. N.p.: V. A. P. Nicolaus Carletti: [ca. 1770].

Cartographer: Nicolaus Carletti.

Engraver: Philip Morghen, 1770.

Size: 27.6 x 37.9 cm.

Provenance: Presented by William Short, 5 November 1824.

(363.956: [ca. 1770]: C212nap Small)

343. 1773

La topografia di Roma digio battanoli dalla Maggiore in questaminor tavola dal medesimoridota. N.p.: 1773.

Scale: 4.8 cm. = 2500 palmi.

Engraver: Benedetti.

Size of paper: 46.7 x 68.5 cm.

Provenance: Presented by William Short, 5 November 1824.

(350.968: 1773: B431trm Large)

344. [ca. 1780]

Isola dell' Elba. Tab. 1. N.p.: [ca. 1780].

Scale: 4.5 cm. = 4 Florentine mi.

Size: 16.7 x 18.2 cm.

(350: [ca. 1780]: El12nap Small)

345. 1783

Carte général des États du roi de Sardaigne réduite d' après la grande carte pour trouver facilement ce qu'on a envie de chercher dans celle-la. [London]: W. Faden: 1783 [24 February 1783].

Scale: 69½ British statute mi. = 1 degree.

Size: 51.7 x 54.2 cm.

Colored.

Inset: Morceau détaché qui représente l'accroissement des États de sa Majesté le roi de Sardaigne, depuis l'an 1680.

Provenance: Purchased from the library of William Priestman, 15 July 1831.

(368: 1783: C248ets)

346. 1784

Pianta della città di Firenze nelle sue vere misure; colla descrizione dei luoghi più notabili di ciascun quartiere. N.p.: 1784.

Cartographer and engraver: Giuseppe Poggiali.

Size: 26.6 x 36 cm.

Provenance: Presented by William Short, 5 November 1824.

(350.932: 1784: P756fir Small)

347. 1786

Italy, divided into its several dominions, and the islands of Sicily, Sardinia and Corsica. From Mr. D'Anville, by the late Thomas Jefferys . . . to which has been added the post roads, &ca. London: R. Sayer: 1786 [1 May 1786].

Scale: 69½ English mi. = 1 degree.

Cartographers: D'Anville and Thomas Jefferys.

Size: 57.6 x 50.2 cm.

Colored.

Provenance: Purchased from the library of William Priestman, 15 July 1831.

(350: 1786: D196ssc Large)

348. [1787]

Nuova pianta dell' inclita città di Venezia recolata l'anno 1787. Ka. Lodovico Ughi. Lodovico Lurlanetto: [1787].

Scale: 2.5 cm. = 100 Veneti passi.

Cartographer: Ka. Lodovico Ughi.

Size: 49.5 x 68 cm.

Provenance: Presented by William Short, 5 November 1824.

Beneath the map is the key.

(350.988: [1787]: Ug18npv Large)

349. [1788]

Nuova planta della reale città di Torino. Con i nomi di ciascheduna isola con la tavola per ritrovare le chiese, è luoghi più co' numeri segnati, e tavola alfabetica per ritrovare i luoghi più principali della città. 1778. Giuseppe Rameletti: [1788].

Cartographer: Giuseppe Rameletti.

Size: 59.2 x 45.8 cm.

Provenance: Presented by William Short, 5 November 1824.

Along left side of map are the keys to the map.

(350.984: [1788]: R147npt Large)

350. [ca. 1790]

Italy and Sardinia, from the best authorities. [London]: Woodman & Mutlow: [ca. 1790].

Scale: 65½ British statute mi. = 1 degree.

Engraver: Mutlow.

Size: 33.3 x 36.4 cm.

Provenance: Presented by Mathew Carey, 18 October 1805.

Engraved for Chambers's edition of Guthrie's new system of geography.

The *Transactions* (vol. 6) state that Carey presented the "Materials from which Guthrie's Geography were compiled." This is one of those maps. See: William Guthrie, *The general atlas for Carey's edition of his geography improved.* Philadelphia: Carey: 1790 [1 May 1790].

(350: [ca. 1790]: W859sa Small)

351. 1791

Carte de la partie septentrionale de l'Italie par M. Chauchard. Paris: Dezauche: 1791.

Scale: 25 French common leagues = 1 degree.

Cartographer: Chauchard.

Size: 89.5 x 160.2.

Colored.

Provenance: Presented by Peter S. du Ponceau, 1 November 1839.

(350.2: 1791: C392sLi Large)

352. [1793]

Italy. [1793].

Scale: 2.5 cm. = 102 British statute mi.

Engraver: S. Allardice.

Size: 17.2 x 20.2 cm.

Provenance: Presented by Thomas Dobson, 1798.

In: Thomas Dobson, publisher, *Encyclopaedia,* vol. 9, opposite p. 390.

Wheat: 756

(032: En2)

353. 1793

Topografia dell' agro Napoletano con le sue adjacenze. Delineata dal Ro Geografo G. A. Rizzi Zannoni. N.p.: 1793.

Cartographer: G. A. Rizzi Zannoni.

Engraver: Gius. Guerra.

Size: 57.3 x 87.7 cm.

(363.956: 1793: Z29tnm Large)

354. [1796]

Island of Ponza [map, with two views. 1796].

Scale: 2.5 cm. = 1¼ mi.

Engraver: Thackara.

Size: 4.3 x 11.3 cm.

Provenance: Presented by Thomas Dobson, 1798.

In: Thomas Dobson, publisher, *Encyclopaedia,* vol. 15, opposite p. 372.

Wheat: 784

(032: En2)

355. [ca. 1798]

Italia. Cioè tutte le grandi e picciole sovranità repubbliche d'Italia, divise nelli nuovi loro confini, comprese le strade pubbliche, le stazioni di posta e loro distanze. Disegnata secondo le più recenti notizie da Ignazio Heymann. Trieste: author: [ca. 1798].

Scale: 3.6 cm. = 20 Italian common miles.

Cartographer: Ignazio Heymann.

Engraver: Felice Zuliani.

Size: 106.5 x 101.5 cm.

Colored.

Inset: Malta

(350: [ca. 1798]: H513sri Large)

356. [ca. 1800]

Piano della città di Milano, e suo castello. N.p.: [ca. 1800].

Scale: 9 cm. = 200 trabuchi di Milano.

Cartographer and engraver: G. C. Bianchi.

Size: 60.2 x 59.8 cm.

Provenance: Presented by William Short, 5 November 1824.

(353.953: [ca. 1800]: B471pcm Large)

357. 1814

Carte de l'isle d'Elbe, dressée d'après les opérations trigonométriques extraites du traité de géodésie de Mr. Puissant. Par Ch[ar]les Picquet. Paris: 1814.

Scale: 25 French common leagues = 1 degree.

Cartographer: Charles Picquet.

Engraver: Alexandre Blondeau.

Size: 37.8 x 50.7 cm.

Inset: Map of Porto-Ferrajo.

(350: 1814: P586tri Small)

358. 1834

Carta di cabotaggio della costa del regno delle Due Sicilie bagnata dall' Adriatico dal fiume Tronto al capo Sta. Maria di Leuca . . . Reale Officio Topographico. Napoli: 1834. 14 maps.

Scale: 1:1,400,000.

Size: 38.5 x 57 cm.

Provenance: Presented by the King of the Two Sicilies through Chevalier Morelli, 1 April 1836.

(350.6: 1834: It14sic Small)

TURKEY Nos. 359-362

359. 1755

Turquie européenne par le Sr. Robert. [Paris]: 1755.

Scale: 60 millespas géometriques = 1 degree.

Cartographer: Robert.

Engraver: Groux.

Size: 47.2 x ca. 44 cm.

Colored.

Provenance: Purchased from the library of William Priestman, 15 July 1831.

(390: 1755: R546teu Large)

360. 1772

Carte réduite de la Mer Noire, dressée par le service des vaisseaux du roy. Par ordre de M. de Boynes . . . Par S. Bellin. N.p.: 1772.

Scale: 20 French and English nautical leagues = 1 degree.

Cartographer: Bellin.

Size: 53.8 x 82.6 cm.

Colored.

Provenance: Purchased from the library of William Priestman, 15 July 1831.

(388.1: 1772: B419bbe Large)

361. [ca. 1790]

Turkey, in Europe and Hungary; from the best authorities. [London]: Woodman & Mutlow: [ca. 1790].

Scale: 13⅓ Hungarian mi. = 1 degree.

Engraver: Mutlow.

Size: 33 x 36.6 cm.

Provenance: Presented by Mathew Carey, 18 October 1805.

Engraved for Chambers's edition of Guthrie's new system of geography.

The *Transactions* (vol. 6) state that Carey presented the "Materials from which Guthrie's Geography were compiled." This is one of those maps. See: William Guthrie, *The general atlas for Carey's edition of his geography improved.* Philadelphia: Carey: 1795 [1 May 1795].

(390: [ca. 1790]: W859tuh Small)

362. 1912

Istambul [Constantinople]. International Map of Europe North K 35. Compiled from Great Britain War Office, Geographical Section, General Staff. [London]: 1912.

Scale: 1:1,000,000.

Size: 45.4 x 53 cm.

Colored.

(390.942: 1912: G813con Large)

ASIA Nos. 363-370

363. 1787

Asia and its islands according to d'Anville; divided into empires, kingdoms, states, regions, &ca. with the European possessions and settlements in the East Indies and an exact delineation of all the discoveries made in the eastern parts by the English under Captn. Cook. London: Robert Sayer: 1787 [6 January 1787]. 2 sheets.

Size: 102.2 x 242.5 cm.

Colored.

Provenance: Purchased from the library of William Priestman, 15 July 1831.

(400: 1787: D196ccv Large)

364. [ca. 1787]

New and correct map of Asia laid down from the latest and best observations. [London]: R. Sayer: [ca. 1787].

Engraver: W. Roads.

Size: 56 x 48 cm.

Colored.

(400: [ca. 1787]: R537rsL Large)

365. [ca. 1788]

A new and correct map of Asia, according to Sre d'Anville & other modern geographers. London: Robert Sayer: [ca. 1788].

Scale: 69 British mi. = 1 degree.

Size: 58.3 x 101.7 cm.

Colored.

(400: [ca. 1788]: Sa97nca Large)

366. [1790]

Asia. [1790].

Size: 16.5 x 21.4 cm.

Provenance: Presented by Thomas Dobson, 1798.

In: Thomas Dobson, publisher, *Encyclopaedia,* vol. 2, opposite p. 394.

Wheat: 863

(032: En2)

367. [ca. 1790]

Asia, according to the best authorities. [London]: Woodman & Mutlow: [ca. 1790].

Engraver: Mutlow.

Size: 32.6 x 35.8 cm.

Provenance: Presented by Mathew Carey, 18 October 1805.

Engraved for Chambers's edition of Guthrie's new system of geography.

The *Transactions* state that Carey presented the "Materials from which Guthrie's Geography were compiled." This is one of those maps. See: William Guthrie, *The general atlas for Carey's edition of his geography improved.* Philadelphia: Carey: 1795 [1 May 1795].

(400: [ca. 1790]: W859aba Small)

368. [ca. 1793]

Asia. Boston: Thomas & Andrews: [ca. 1793].

Engraver: Amos Doolittle.

Size: 16.3 x 21 cm.

Engraved for Morse's *American universal geography,* vol. 1, opposite p. 384

Wheat: 873

(400: [ca. 1793]: D722asa Small)

369. [1798]

Asia from the latest authorities. [1798].

Engraver: Rollinson.

Size: 18.7 x 22.1 cm.

In: John Payne, *A new and complete system of universal geography,* vol. 1, opposite p. 3.

Wheat: 903

(910: P29)

370. 1799

Asia and its islands according to d'Anville; divided into empires, kingdoms, states, regions &ca, with the European possessions and settlements in the East Indies and an exact delineation of all the discoveries made in the eastern parts by the English under capts. Cook, Vancouver & Peyrouse. London: Laurie & Whittle: 1799 [2 February 1799].

Size: 102.7 x 120 cm.

Colored.

(400: 1799: As41adc Large)

NEAR EAST Nos. 371-377

371. 1823

Carte comprenant le pays de Nedjd ou Arabie Centrale, l'Égypte et partie des autres régions occupées en 1820, par les troupes de Mohammed-Aly, pour servir à l'intelligence de l'histoire de l'Égypte sous le gouvernement de Mohammed-Aly. Par M. E. J[omard] D. L. Paris: 1823.

*Scale:*1:4,000,000.

Cartographer: Edmé François Jomard.

Size: 42.5 x 54.3 cm.

Inset: Environs d'el Derréyeh.

(415.4: 1823: J69ara Small)

372. 1835

Plan of Jerusalem by F. Catherwood, Architect, July, 1835. London: author: 1835.

Scale: 2½ in. = 1,500 feet.

Cartographer: F. Catherwood.

Engraver: S. Bellin.

Size: 26.8 x 21.2 cm.

See "F. Catherwood architect," (in New-York Historical Society *Quarterly*, vol. 30, no. 1.)

(414.944: 1835: C282jer Small)

373. [ca. 1850]

Map of Mr. Southgate's route through Turkey and Persia: chiefly from original observations. [New York: ca. 1850].

Scale: 69.1 English mi. = 1 degree.

Engravers: Sherman & Smith, New York.

Size: 32 x 82.3 cm.

(411: [ca. 1850]: So87tu Large)

374. 1855

Vestiges of Assyria [in three sheets] . . . Ninevah . . . Nimrúd and Selamiyeh . . . [and] River Tigris & the upper Zab . . . constructed from a trigonometrical survey in the spring of 1852 at the command of the government of India by Felix Jones . . . [and] J. M. Hyslop. [Baghdad]: John Walker: 1855. 3 maps.

Scale: 1760 yd. = 1 English statute mi.

Cartographers: Felix Jones and J.M. Hyslop.

Engravers: J. & C. Walker.

Size: 122.8 x 69 cm.; 123 x 69 cm.; 123 x 69 cm.; 123.2 x 69.1 cm.

Colored.

Provenance: Presented by the Royal Asiatic Society of Calcutta, 1 February 1856.

(417: 1855: J714nnt Large)

375. [1863]

A map of Turkestan to illustrate the travels of Arminius Vambéry in 1863. New York: Harper: [1863].

Scale: 1 in. = 40 English statute mi.

Size: 36.7 x 64. cm.

Colored.

(421: [1863]: T848avt Large)

376. [ca. 1939]

Distribution of tribes in Iraq. Map A. N.p.: [ca. 1939].

Scale: 1:2,000,000.

Cartographer: Richard A. Martin.

Size: 41 x 52.1 cm.

To accompany: "The anthropology of Iraq," by Henry Field. Field Museum of Natural History. Anthropological series, vol. 30.

(417: [ca. 1939]: M365tri Small)

377. [ca.1939]

Distribution of tribes in western Iran. Map B. N.p.: [ca. 1939].

Scale: 1:2,000,000.

Cartographer: Peter Gerhard.

Size: 52 x 41.2 cm.

To accompany: "Contributions to the anthropology of Iran," by Henry Field. Field Museum of Natural History Anthropological series, vol. 29, 1939.

(418:[ca. 1939]: G322tri Small)

FAR EAST Nos. 378-380

378. [ca. 1795]

Ostium fluvii Amur cum parte australiori terrae Kamtschatkae variisque in oceano sitis insulis inter quas pars eminet Iaponiae. N.p.: [ca. 1795].

Scale: 3 cm. = 100 russticorum.

Size: 51.2 x 56.8 cm.

(425.2: [1795]: Am11ofa Large)

379. 1904

Map of Korea and Manchuria prepared by the Second Division, General Staff (Military Information Division) War Department. Washington: Norris-Peters: 1904.

Scale: 1:2,534,400.

Size: 99.8 x 85.7 cm.

Colored.

Inset: Index map of East Asia.

(405: 1904: Un38kmc Large)

380. n.d.

Large print map of the Far East. Produced under the direction of Alexander Gross, F.R.G.S. New York: n.d.

Scale: 2 in. = 300 mi.

Cartographer: Alexander Gross.

Size: 97.2 x 134.5 cM.

Colored.

Insets: Indian Ocean; Korea; Korea and Saishu; the Pacific Ocean basin; Formosa.

Provenance: Presented by Mrs. A. I. Hallowell, November 1983.

(405: n.d.: G922fea Large)

JAPAN Nos. 381-386

381. [ca. 1895]

Korea or Cho-sen of the Japanese. Compiled from the map published by the Japanese War Ministry, Tôkiô, in 1875, and from the latest coast surveys. Names transliterated by Dr. Ernest Satow. Washington: Norris Peters Co.: [ca. 1895].

Scale: 1:1,700,000.

Cartographer: Charles H. Ourand.

Size: 53.1 x 28 cm.

Colored.

Inset: Quelpart I[slan]d Ché-Ju of the Koreans.

(433: [ca. 1895]: K844tok Small)

382. 1899

Topographical map of the Japanese Empire. Imperial Geological Survey of Japan. 1899. Tokyo: G. Yoshii: 1899. 13 sheets.

Scale: 1:1,000,000.

Cartographers: T. Togawa and K. Suzuki.

Engraver: N. Matsudaira.

Size: Various sizes.

Colored.

See: Geological Survey of Japan, *Outlines of the geology of Japan.* 1902.

(430: 1899: J274top Small)

383. [ca. 1902]

General map [of Japan] showing the distribution of volcanoes. Imperial Geological Survey of Japan. [Tokyo: ca. 1902]. 2 sheets.

Scale: 1:5,000,000.

Size: 33.1 x 50.6 cm.

See: Geological Survey of Japan, *Outlines of the geology of Japan.* 1902.

(430: [ca. 1902]: J274vol Small)

384. 1902

Geological map of the Japanese Empire . . . compiled by the Imperial Geological Survey of Japan. Tokyo: G. Yoshii: 1902. 12 sheets.

Scale: 1:1,000,000.

Cartographers: T. Togawa, E. Tamura, and K. Suziki.

Engraver: N. Matsudaira.

Size of paper: ca. 41 x 57.5 cm. each.

Colored.

See: Geological Survey of Japan, *Outlines of the geology of Japan.* 1902.

(430: 1902: J274geo Small)

385. [ca. 1902]

Hypsometrical and bathymetrical chart [of Japan]. Imperial Geological Survey of Japan. [Tokyo: ca. 1902]. 2 sheets.

Scale: 1:5,000,000.

Size: 33.2 x 50.6 cm., and, 37 x 50.6 cm.

Colored.

See: Geological Survey of Japan, *Outlines of the geology of Japan.* 1902.

(430: [ca. 1902]: J274voL Small)

386. [ca. 1902]

Political division of the Japanese Empire. Imperial Geological Survey of Japan. Tokyo: [ca. 1902].

Scale: 1:10,000,000.

Size: 38 x 50.5 cm.

Colored.

See: Geological Survey of Japan, *Outlines of the geology of Japan.*

(430: [ca. 1902]: J274voL Small)

CHINA Nos. 386a-398

386a. [1797]

Carte de la Chine dressée pour servir au voyage de l'ambassade de la Compagnie des Indes Hollandaises vers l'empéreur de la Chine, dans les années 1794 & 1795 [1797].

Scale: 2.5 cm. = 30 mi.

Engraver: Vallance.

Size: 105.9 x 53 cm.

Provenance: Presented by John Vaughan, 15 March 1839.

In: Andreas Everard van Braam Houckgeest, *Voyage de l'Ambassade,* vol. 1, opposite p. XLIX.

Wheat: 892

(915. l: B72)

387. [1798]

Plan de la ville de Macao à la Chine possédée par les Portugais. 1795 [1798].

Scale: [2.5 cm. = ca. 4,500 ft.].

Engraver: Vallance.

Size: 17.3 x 36 cm.

Provenance: Presented by John Vaughan, 15 March 1839.

In: Andreas Everard van Braam Houckgeest, *Voyage de l'Ambassade,* vol. 2, opposite p. 219.

Wheat: 896

(915.1: B72)

388. [ca. 1820]

[Chinese maps of Canton and Hong Kong, China. Ca. 1820].

Size of paper: 57.2 x 34.4 cm. Canton: 21.8 x 28.2 cm., and Hong Kong: 19.5 x 37.5 cm.

Colored.

Provenance: Presented by Joshua L. Oakford, 8 November 1820.

(443: [ca. 1820]: C162kwa Small)

389. [ca. 1900]

Bartholomew's special map of China, Japan, and Korea. The Edinburgh Geographical Institute. Edinburgh: John Bartholomew & co.: [ca. 1900].

Scale: 1 in. = 11¾ mi.

Size: 52.3 x 85.7 cm.

Colored.

Insets: Asia, Hong Kong, and Canton areas; Korea to Viet Nam coastal regions; Tokio Bay; Peking; Peking to Ta-Ku area showing railroads and proposed railroads.

(405: [ca. 1900]: B231cjk Large)

390. 1904

Map of Kwantung peninsula prepared in the Second Division, General Staff, War Department, July, 1904. Washington: Andrew B. Graham: 1904.

Scale: 2 cm. = 1 mi.

Size: 56.2 x 82 cm.

Colored.

(441: 1904: Un38kwp Large)

391. [ca. 1910]

A commercial map of China and its dependencies, showing the areas owned or leased by foreign powers. Issued by the China League. London: Edward Stanford: [ca. 1910].

Scale: 120 mi. = 1 in.

Size: 55.3 x 78.4 cm.

Colored.

(440: [ca. 1910]: C342ccf Large)

392. [ca. 1919]

[China]. N.p.: n.d., ca. [1919].

Size: 69.5 x 86.8 cm.

Colored.

Provenance: Presented by the Rockefeller Institute and James Thomas Flexner, 1964.

(441: [ca. 1919]: C442chi Large)

393. 1922

General map of China and adjacent regions, showing treaty ports and railways. Prepared for the Department of State by the Topographic Branch, U.S. Geological Survey, 1921. Revised Jan. 1922. Lambert conformal conic projection. [Washington]: 1922.

Scale: 1:10,000,000.

Size: 50.1 x 63.7 cm.

Colored.

(441: 1922: Un38ctp Large)

394. 1924

China. Showing mission stations except the Roman Catholic. Compiled by S. W. Boggs and R. Beach. New York: Missionary Education Movement of the U.S. and Canada: 1924.

Scale: 1:3,250,000.

Cartographers: S. W. Boggs and R. Beach.

Size: 91.2 x 116.3 cm.

Colored.

Inset: China provinces, special administrative districts, and outer territories.

Provenance: Presented by the Rockefeller Institute and James Thomas Flexner, 1964.

Contains an index of mission stations.

(441: 1924: C442ms Large)

395. 1926

General geological map of China. V.K. Ting and W.H. Wong, directors of the Geological Survey of China. Compiled in 1924. N.p.: 1926. Sheet N J-19.

Scale: 1:1,000,000.

Cartographers: J. F. Na, C.C. Wang, Frank A. Herald, F. G. Clapp, M.L. Fuller, E. L. Estabrook, C. Li, Bailey Willis, V. K. Ting, and E. T. Nyström.

Size: 47 x 57 cm.

Colored.

(441: 1926: C442gmc Large)

396. n.d.

[Peking, China]. N.p.: n. d.

Size of paper: 95 x 54.6 cm.

Colored.

Provenance: Presented by Joshua L. Oakford, 8 November 1820.

(444.962: n.d.: P366cmp Large)

397. n.d.

[China]. N.p.: n.d.

Size of paper: 120.5 x 73.4 cm.

Colored.

Insets: Two, of groups of islands.

Provenance: Presented by Joshua L. Oakford, 8 November 1820.

(441: n.d.: C342cmc Large)

398. n.d.

[China]. N.p.: n.d.

Size of paper: 115 x 55 cm.

Provenance: Presented by F. J. Eckhard, 20 March 1807.

(441: n.d.: C342mce Large)

INDIA Nos. 399-401

399. 1786

A map of Bengal, Bahar, Oude & Allahabad with part of Agra and Delhi. Exhibiting the course of the Ganges from Hurdwar to the sea. By James Rennell. London: Wm. Faden: 1786 [1 January 1786].

Scale: 2 in. = 40 British mi.

Cartographer: James Rennell.

Size: 70.8 x ca. 106.5 cm.

Colored.

Provenance: Purchased from the library of William Priestman, 15 July 1831.

(455: 1786: R295bba Large)

400. [ca. 1790]

An accurate map of Hindostan or India, from the best authorities, N.p.: Woodman & Mutlow: [ca. 1790].

Scale: 69½ British mi. = 1 degree.

Engraver: John Mutlow.

Size: 39.5 x 41.6 cm.

Provenance: Presented by Mathew Carey, 18 October 1805.

The *Transactions* (vol. 6) state that Carey presented the "Materials from which Guthrie's Geography were compiled." This is one of those maps. See: William Guthrie, *The general atlas for Carey's edition of his geography improved.* Philadelphia: Carey: 1790 [1 May 1790].

(450: [ca. 1790]: W859hin Small)

401. 1791

The southern countries of India from Madras to Cape Comorin, describing Colonels Fullarton and Humberston during the campaigns of 1782, 1783, & 1784. Surveyed by Col. Kelly, Capt. Wersebe, and others. London: Wm. Faden: 1791 [8 February 1791].

Scale: 69½ British statute mi. = 1 degree.

Cartographers: Col. Kelly and Capt. Wersebe.

Size: 87.3 x ca. 105 cm.

Colored.

Provenance: Purchased from the library of William Priestman, 15 July 1831.

(456: 1791: K288mcc Large)

INDOCHINA Nos. 402-403

402. 1939

Carte archéologique de la région d'Ankor. D'après les travaux de l'École Française d'Extrême-Orient de l'Aviation Militaire et du Service Géographique. Héliogravé et imprimé par le Service Géographique de l'Indochine: 1939.

Scale: 1:40,000.

Size: 56.7 x 87.1 cm.

Colored.

(461:1939: In244 Large)

403. 1949

Carte ethnolinguistique. Carte de l'Indochine. Dressé sous la direction de l'École Française d'Extrême-Orient. Dessiné et publié par le Service Géographique de l'Indochine: 1949.

Scale: 1:2,000,000.

Size: 84.1 x 60.2 cm.

Colored.

Has key to languages.

(460: 1949: In243 Large)

INDONESIA Nos. 404-407

404. 1768

The East Indies, with the roads. By Thomas Jefferys. [London]: 1768. 2 sheets.

Cartographer: Thomas Jefferys.

Size: 108 x 274 cm.

Colored.

Provenance: Purchased from the library of William Priestman, 15 July 1831.

(470: 1768: J377rds Large)

405. [1793]

East Indies. [1793].

Scale: 60 mi. = 1 degree.

Engraver: S. Alardice.

Size: 17.2 x 27.5 cm.

Provenance: Presented by Thomas Dobson, 1798.
In: Thomas Dobson, publisher, *Encyclopaedia,* vol. 9, opposite p. 218.
Wheat: 871

(032: En2)

406. 1797

Chart of the new discoveries east of New Holland and New Guinea. 1797.
Engraver: Callender.
Size: 16.7 x 24.4 cm.
In: Jedidiah Morse, *The American gazetteer,* opposite "Terra Austral."
Wheat: 911

(917.3: M83am)

407. [1798]

East Indies from the best authorities. [1798].
Scale: 2.5 cm. = 600 American statute mi.
Engraver: Rollinson.
Size: 18.6 x 26.4 cm.
In: John Payne, *A new and complete system of universal geography,* vol. 1, opposite p. 312.
Wheat: 913

(910: P29)

THE PHILIPPINE ISLANDS Nos. 408-420

408. 1850

Plano de la bahia de Nin y puerto de Mandao, situado en la costa ocidental de la Ysla de Masbate; levantado en 1828 por . . . D. Federico Vargas y publicado en la Dirección de Hidrografia. Madrid: 1850.
Scale: Scale of 1 nautical mi. divided in 1/10s.
Cartographers: Federico Vargas and F. Bregante.
Engraver: J. Noguera.
Size: 29.8 x 45.7 cm.
Provenance: Presented by the director of the Dirección de Hidrografia in Madrid through the hands of Don Severiano Moreledo, of Cadiz, 7 January 1853.
On same sheet is: Plano de puerto de Batan. . .

(482: 1850: Sp27ppb Large)

409. 1850

Plano del puerto de Batan situado en la costa W. de Panay, levantado en 1842 por . . . Comisión Hidrografica. Madrid: 1850.
Scale: Scale of 1 nautical mi. divided in 1/10s.
Engraver: Noguera.
Size: 34.2 x 45.7 cm.

Provenance: Presented by the director of the Dirección de Hidrografia in Madrid through the hands of Don Severiano Moreledo, of Cadiz, 7 January 1853.

On same sheet is: Plano de la bahia de Nin . . .

(482: 1850: Sp27ppb Large)

410. 1850

Plano del puerto de Busainga, situado en la parte n. e. de la isla de Burias; levantado en 1841 por . . . la Comisión Hidrografica. Madrid: 1850.

Scale: Scale of ½ nautical mi. divided in 1/10s.

Cartographer: F. Bregante.

Engravers: J. Noguera and F. Bregante.

Size: 32.2 x 46 cm.

Provenance: Presented by the director of the Dirección de Hidrografia in Madrid through the hands of Don Severiano Moreledo, of Cadiz, 7 January 1853.

On the same sheet is: Plano del puerto de Laguimanor . . .

(482: 1850: Sp27ppL Large)

411. 1850

Plano del puerto de Laguimanor situado en la costa s. de la isla de Luzon, levantado en 1841 por el comandante . . . de la Comisión Hidrográfica, en el Archipiélago Filipino. Madrid: 1850.

Scale: Scale of ½ nautical mi. divided in 1/10s.

Cartographer: F. Bregante.

Engravers: J. Noguera and F. Bregante.

Size: 32 x 46 cm.

Provenance: Presented by the director of the Dirección de Hidrografia in Madrid through the hands of Don Severiano Moreledo, of Cadiz, 7 January 1853.

On same sheet is: Plano del puerto de Busainga . . .

(482: 1850: Sp27ppL Large)

412. 1850

Plano del puerto de Zebú, levantado por el comandante y oficiales de la Comisión Hidrográfica el el archipiélago Filipino . . . Madrid: 1850.

Scale: 11.5 cm. = 2 nautical mi.

Cartographer: F. Bregante.

Size: 46 x 63.9 cm.

Provenance: Presented by the director of the Dirección de Hidrografia in Madrid through the hands of Don Severiano Moreledo, of Cadiz, 7 January 1853.

(482: 1850: Sp27ppzf Large)

413. [1890]

Isla de Panay bosquejo geológico por D. Enrique Abella y Casariego. Manila: Carmelo y Bauermann: [1890]. Map in 2 pieces.

Scale: 1:200,000.

Cartographer: Enrique Abella y Casariego.

Size: 99.7 x 76.4 cm.

Colored.

See Enrique Abella y Casariego, *Descripción física, geológica y mineria en la isla de Panay.* Manila: 1920.

(484: [1890]: Abllpan Large)

414. 1890-1905

[Maps of the Philippine Islands belonging to the *Bulletins* of the Division of Geology and Mines, Philippine Islands] 3-5. Manila: 1890-1903. 11 maps.

Scale: various.

Size: various.

Colored.

See Philippine Islands. Mining Bureau. Annual report.

(480: 1890-1905: P536mgd Large)

415. 1898

[Map of the Philippine Islands, chiefly from Spanish surveys]. Washington: Hydrographic Office: 1898.

Lithographer: Julius Bien & Co.

Size: 260 x 126 cm.

Colored.

This is a composite of several maps.

(484: 1898: Un36ppa Large)

416. 1900

Carta general (en dos hojas) del Archipiélago Filipino levantada principalmente por la Comisión Hidrografia. [Por] Claudio Montero y Gay . . . 1870 . . . 1875 . . . Issued by the War Department, Adjutant General's Office, Military Information Division, 1900. Second ed. [Washington]: 1900.

Cartographers: Claudio Montero y Gay and J. Noguera.

Engraver: J. Noguera.

Size: 127 x 96 cm.

Colored.

(480: 1900: Sp27cga Large)

417. 1900

Map of the Viscayan Group Philippine Islands including Panay, Negros, Samar, Leyte, Cebu, Mosbate, Bohol and adjacent islands. Prepared in the War Department, Adjutant General's Office. Military Information Division. 1900. [Washington]: Julius Bien & Co.: 1900.

Scale: 1 cm. = 5 km.

Cartographer: C. H. Stone.

Engravers: C. H. Ourand and E. M. Eastwood.

Size: 83.2 x 96.6 cm.

Colored.

(484: 1900: Un38vgp Large)

418. 1902

Map of the Philippines prepared in the Bureau of Insular Affairs, War Department. Washington: Andrew B. Graham: 1902.

Scale: 1 in. = 18 statute mi.

Lithographer: Andrew B. Graham.

Size: 166.3 x 106.3 cm.

Colored.

Inset: Manila.

(484: 1902: Un38bia Large)

419. 1903

Map of the Philippine Islands compiled from the latest official data in the Military Information Division, Adjutant General's Office, War Department, Washington, U.S.A. 1903. [Washington]: 1903. 4 pieces.

Scale: 2 cm. = 10 mi.

Size: 239.3 x 156 cm.

Colored.

Inset: Manila.

(484: 1903: Un38mdp Large)

420. [1912]

Geologic reconnaissance map of Mindanao. U.S. Bureau of Science, Division of Mines. Based on U.S. Coast and Geodetic chart. [1912].

Cartographer: Warren D. Smith, geologist.

Engraver: A. Moskaira.

Size: 54 x 59 cm.

Colored.

(486: [1912]: P536rec Small)

AFRICA Nos. 421-426

421. 1754

A new and correct map of Africa, done from the latest and best observations. London: R. Sayer: 1754.

Size: 54.7 x 46.8 cm.

Colored.

Inset: A prospect of the Cape of Good Hope.

(500: 1754: Af81ncm Large)

422. 1772

Africa, according to Mr. D'Anville with several additions and improvements, with a particular chart of the Gold Coast, wherein are distinguished all the European forts and factories, the whole illustrated with a summary description relative to the trade & natural produce, manners & customs of that part of the world. London: Robert Sayer: 1772 [1 August 1772].

Scale: 7.25 cm. = 400 English mi.

Size: 103 x 122.5 cm.

Colored.

Insets: A chart of the Gold Coast and the Azores.

(500: 1772: Af81dag Large)

423. 1787

Africa with all its states, kingdoms, republics, regions, islands &ca. Improved and inlarged from d'Anville's map; to which have been added a particular chart of the Gold Coast, wherein are distinguished all the European forts and factories. By S. Boulton. And also a summary description relative to the trade and natural produce, manners and customs of the African continent and islands. London: Robert Sayer: 1787 [6 January 1787]. 2 sheets.

Scale: 5½ in. = 300 English mi.

Cartographers: D'Anville and S. Boulton.

Size: 102.6 x ca. 245 cm.

Colored.

Inset: A particular chart of the Gold Coast.

Provenance: Purchased from the library of William Priestman, 15 July 1831.

(500: 1787: D196skr Large)

424. [ca.1790]

Africa according to the best authorities. [London]: Woodman & Mutlow: [ca. 1790].

Scale: 60 British statute mi. = 1 degree.

Size: 33.7 x 36.3 cm.

Provenance: Presented by Mathew Carey, 18 October 1805.

The *Transactions* (vol. 6) state that Carey presented the "Materials from which Guthrie's Geography were compiled." This is one of those maps. See: William Guthrie, *The general atlas for Carey's edition of his geography improved.* Philadelphia: Carey: 1795 [1 May 1795].

Engraved for Chambers's edition of Guthrie's new system of geography.

(500: [ca. 1790]: W859cgg Small)

425. [1790]

Africa from the best authorities. [1790].

Engraver: R. Scot.

Size: 17.7 x 21.5 cm.

Provenance: Presented by Thomas Dobson, 1798.

In: Thomas Dobson, publisher, *Encyclopaedia,* vol. 1, following p. 228.

Wheat: 840

(032: En2)

426. [1799]

Africa from the best authorities. [1799].

Engraver: Rollinson.

Size: 18.2 x 21.8 cm.

In: John Payne, *A new and complete system of universal geography,* vol. 2, frontispiece.

Wheat: 859

(910: P29)

EGYPT Nos. 427-433

427. 1820

Carte de l'oasis de Thèbes, comprenant le vallon d'El Khargeh et celui d'El Dakel ainsi que les routes qui y conduisent, à partir d'Esné, de Girgeh et de Syout. Dressée d'après les itinéraires de M. M. Cailliaud et Drovetti par M. Jomard. [Paris]: 1820.

Scale: 25 leagues = 1 degree.

Cartographer: E. F. Jomard.

Engraver: Chocarne.

Size: 25 x 35.7 cm.

Indicates routes of Cailliaud and Drovetti. To accompany Cailliaud, *Voyage à l'Ouest.*

(524: 1820: J694tkd Small)

428. 1820

Carte itinéraire du désert situé entre le Nil et la Mer Rouge comprenant la montagne de Zabarah, les mines d'eméraude, la mine de soufre et les vestiges de l'ancienne route du commerce entre l'Egypte et l'Inde. Dressée d'après l'itinéraire de Mr. Cailliaud . . . par Mr. Jomard. N.p.: 1820. Plate 1.

Scale: 25 leagues = 1 degree.

Cartographer: E. F. Jomard.

Engraver: Blondeau.

Size: 29.2 x 36.3 cm.

See: Cailliaud, *Voyage à l'Est.*

(521: 1820: J694nmr Small)

429. 1820

Pente du Nil comparée à celle du plusieurs autres fleuves. N.p.: [ca. 1820].

Scale: 1:32,000,000.

Cartographer: E. F. Jomard.

Size of paper: 26.6 x 41.8 cm.

To accompany Jomard's *Mémoire sur la communication du Niger et du Nil.*

(521: [ca. 1820]: J694nmr Small)

430. [ca. 1820]

Tableau d'assemblage pour la carte topographique de l'Égypte, en 47 feuilles et pour la carte géographique en 3 feuilles. Dépôt Général de la Guerre. Paris: [ca. 1820].

Scale: 1 mm. = 50 m.

Size: 49.5 x 33.6 cm.

(521: [ca. 1820]: F843tae Small)

431. 1821

Carte de l'Égypte et d'une partie des contrées adjacentes, dressée pour l'usage des Collèges; d'après les observations de voyageurs Français, et les découvertes les plus récentes. Par M. E[dme] J[omard] D.L. Paris: Selves fils: 1821.

Scale: 25 common leagues = 1 degree.

Cartographer: E. Jomard.

Size: 42.6 x 30.3 cm.

(521: 1821: J694epc Small)

432. [ca. 1829]

Le Kaire. Plan particulier de la ville. Vol. 1, Pl. 26. N.p.: [ca. 1829].

Scale: 6 cm. = 300 m.

Cartographers: Simonel, Jomard, Bertre, Lecesne, et dirigé par Mr. Jacotin.

Engravers: Vicq and Dandeleux.

Size: 60.4 x 94.5 cm.

Provenance: Presented by Jomard, 17 July 1829.

(522.921: [ca. 1829]: V668cai Large)

433. 1910

Geological map of Egypt. Survey Department Egypt. N.p.: 1910.

Scale: 1:2,000,000.

Size: 59 x 57.8 cm.

Colored.

(521: 1910: Eg92sdg Large)

WEST AFRICA Nos. 434-442

434. 1768

The western coast of Africa; from Cape Blanco to Cape Virga, exhibiting Senegambia proper. By T. Jefferys. London: Robert Sayer: 1768 [12 August 1768].

Scale: 20 British leagues = 1 degree.

Cartographer: T. Jefferys.

Size: 70.7 x 52.2 cm.

Colored.

Provenance: Purchased from the library of William Priestman, 15 July 1831.

Contains "Mr. Moore's account of the English settlements on the River Gambia in 1730."

(503: 1768: J377cbv Large)

435. 1775

Guinée entre Serre-Lione et le passage à la ligne. Par Sr. d'Anville. [Paris]: 1775.

Scale: 25 common French leagues = 1 degree.

Cartographer: D'Anville.

Engraver: Guill. de la Haye.

Size: 30.5 x ca. 67.5 cm.

Colored.

Inset: Côte d'Or.

Provenance: Purchased from the library of William Priestman, 15 July 1831.

(549: 1775: D196sLL Large)

436. [1794]

A map of Barbary comprehending Morocco, Fez, Algiers, Tunis and Tripoli. [1794].

Engraver: J. T. Scott.

Size: 12.5 x 26 cm.

Provenance: Presented by John Vaughan, June 1815.

In: [Mathew Carey], *A short account of Algiers . . .*, frontispiece.

Wheat: 845

(Pam., vol. 146, no. 1)

437. 1795

Plan of Sierra Leone and the parts adjacent. N.p.: 1795.

Scale: 2.5 cm. = 1 mi. 4½ furlongs.

Engraver: Vallance.

Size: 26 x 24.5 cm.

Provenance: One copy presented by Estate of Richard Gimbel, 1974.

In: Sierra Leone Company, *Substance of the report delivered by the court of directors*, frontispiece.

Wheat: 848

(Pam., vol. 153, no. 6)

438. 1822

Esquisse d'une carte itinéraire pour les voyages de mr. Frédéric Cailliaud de Nantes et du plusieurs autres voyageurs. Par M. J[omard]. Paris: 1822.

Cartographer: E. Jomard.

Size: 36.5 x 32 cm.

Indicates routes of Cailliaud, Brown, and Hornemann.

(521: 1822: C122mve Small)

439. 1844

A chart of the western coast of Africa, extending from Sierra Leone and the Isle of Los, to the Cape of Good Hope. With enlarged plans of the principal islands, harbours, &

roadsteads; constructed chiefly from the surveys of Captain W. F. W. Owen, R.N. by J. W. Norie, hydrographer. 1834, additions &c. 1844. London: J. W. Norie: 1844.

Scale: Various.

Cartographers: J. W. Norie and W. F. W. Owen.

Size: 76.4 x 197.5 cm.

Insets: 27, of islands, harbors, and roadsteads.

(503: 1844: N776sLL Large)

440. 1845

Map of Liberia. Compiled, from data on file in the office of the American Colonization Society, under the direction of the Revd. W. McLain, Secy. By R. Coyle. Baltimore: E. Weber & Co.: 1845.

Scale: 1 in. = 12 mi.

Cartographers: R. Coyle and J. Ashman.

Size: 44.5 x 62 cm.

Inset: Vicinity of Monrovia. Surveyed by J. Ashmun, 1825.

(535: 1845: C832acs Large)

441. 1850

Carte esferica de una parte de la costa ocidental de Africa en el golfo de Guinea, que comprende desde Cabo S[a]n Pablo hasta el de Formosa levantado en 1846 pour el capitan de la M.R.B. H.M. Denham Commandante de Avon, publicada en la Dirección de Hidrografia. Madrid: 1850.

Cartographers: F. Bregante and H. M. Denham.

Engravers: J. Noguera and F. Bregante.

Size: 60.3 x 93.3 cm.

Provenance: Presented by the director of the Dirección de Hidrografia in Madrid through the hands of Don Severiano Moreledo, of Cadiz, 7 January 1853.

(500.3: 1850: Sp27gga Large)

442. 1868

Carte de l'Etbaye ou pays habité par les arabes Bicharieh, comprenant les contrées des mines d'or connues des anciens sous le nom d'Ollaki. Par Mr. Linant de Bellefonds. Paris: Lemercier: 1868.

Scale: 1:250,000.

Cartographer: Linant de Bellefonds.

Size: 55 x 60.4 cm.

See: Maurice Adolphe Linant de Bellefonds, *L'Etbaye: pays inhabité par les Arabes Bicharieh.* Paris: A. Bertrand: 1868.

(525: 1868: L635etb Large)

CENTRAL AFRICA Nos. 443-446

443. 1884

Croquis de l'Afrique équatorial contenant les derniers renseignements recuellis par les agents de l'Association Internationale du Congo. Brussels: Institut National de Géographie: 1884.

Scale: 1:4,000,000.

Size: 43.1 x 92.4 cm.

Colored.

Inset: Africa.

(555: 1884: B411gnb Large)

444. 1886

Croquis de l'Afrique centrale mis au courant des dernières explorations. Par. A. J. Wauters. Brussels: Institut National de Géographie: 1886. Seventh ed.

Scale: 1:10,000,000.

Cartographer: A. J. Wauters.

Engraver: A. Verwest.

Size: 25.3 x 37.5 cm.

Colored.

Insets: Africa and Belgium.

(504: 1886: W369cfx Small)

445. 1889

Carte de l'État Indépendant du Congo dressée par A. J. Wauters. Brussels: Institut National de Géographie: 1889.

Scale: 1:7,000,000.

Cartographer: A. J. Wauters.

Size: 31.1 x 33.1 cm.

Colored.

Inset: Belgium.

Endorsed: This copy is valuable. Handle with care. No other obtainable.

(555: 1889: W369cdb Small)

446. 1911

Comité Special de Katanga. Carte des réserves minières. J. Malvaux: 1911.

Scale: 1:2,000,000.

Size: 55.6 x 48 cm.

Colored.

(555: 1911: K154crk Large)

SOUTH AFRICA Nos. 447-448

447. 1899

Map of the seat of war in Africa prepared in the War Department, Adjutant General's office, Military Information Division. Washington: Norris Peters Co.: 1899.

Scale: 2.7 cm. = 10 mi.

Size: 105.9 x 85.7 cm.

Colored.

Inset: South Africa (24° to 34° latitude).

(500: [1899]: Un38swa Large)

448. 1911

Kenhardt. International Map. South H 34. [Compiled by] Geographical Section, General Staff. [London]: 1911.

Scale: 1 in. = 15.78 mi.

Engravers: W. and A. K. Johnston, Ltd.

Size: 46.2 x 58.7 cm.

Colored.

(571: 1911: G813ken Large)

NORTH AMERICA Nos. 449-470

449. 1733

Map of the British empire in America with the French and Spanish settlements adjacent thereto. London: 1733.

Cartographers: Henry Popple and C. Lempriere.

Engraver: B. Baron.

Size: 232.5 x 229.6 cm.

Colored.

Inset: New York harbor.

This huge map of the North American British Empire was given to the Society by Florens Rivinus on 21 March 1834. It was one of the great maps which hung in Independence Hall in July 1776. Rivinus offered it for sale to the Society and a committee was appointed on 5 May 1826 to report on its purchase. The committee thought the asking price was too high, so he presented it eight years later.

During 1975-76 the map was deacidified and restored and was a major exhibition piece in the united 1976 Bicentennial Exhibition of the Library Company of Philadelphia, Historical Society of Pennsylvania, and the American Philosophical Society.

For photograph see: 635: 1733: P816bea.p. Small.

(635: 1733: P816bea Extra-oversize)

450. ca. 1744

A map of North America with the European settlements & whatever else is remarkable in ye West Indies, from the latest and best observations. [London: ca. 1744].

Cartographer and engraver: R. W. Seale.

Size: 37.5 x 47 cm.

See: Paul Rapin de Thoyas, *The history of England.* Continued by N. Tindal. London: 1744: vol. 3.

(600: [ca. 1744]: Se17nae Small)

451. 1755

North America from the French of Mr. D'Anville improved with the back settlements of Virginia and course of Ohio. Illustrated with geographical and historical remarks. [London]: T. Jefferys: 1755 [May 1755].

Scale: 69½ English mil. = 1 degree.

Cartographers: D'Anville and T. Jefferys.

Size: 46 x 51.1 cm.

Colored.

Provenance: Presented by C. J. Ingersoll, 7 April 1843.

Ingersoll wrote 14 April 1837 that he had "four antient maps . . . some of which I believe are rare and valuable." He added that he was too busy, even to attend meetings, but "absence does not diminish at all the great respect I entertain" for the Society.

[This is the only map I can determine that came from Ingersoll which is "rare and valuable"]: MDS.

(601: 1755: An91nav Large)

452. [1757]

A map of the British and French dominions in North America with the roads, distances, limits, and extent of the settlements. Jno. Mitchell. [London: 1757] 2nd ed.

Scale: 69½ English mi. = 1 degree.

Cartographer: John Mitchell.

Engraver: Thomas Kitchin, Clerkenwell Green.

Size: 135 x 194.6 cm.

Inset: A new map of Hudson's Bay and Labrador.

(602: [1757]: M695bfd Large)

453. 1774

A map of the British Empire in North America; by Samuel Dunn. London: Robert Sayer: 1774 [10 January 1774].

Scale: 69½ English mi. = 1 degree.

Cartographer: Samuel Dunn.

Size: 47 x 30.4 cm.

Colored.

Former Secretary of the Treasury, Albert Gallatin, wrote 7 November 1828 requesting the loan of particular maps for use in settling the Northeast Boundary Dispute with Great Britain. Secretary John Vaughan acquired some for him from Samuel Hazard and the "City Library," as well as some from the Society's library. Gallatin wrote that "in every case where we have but one engraved copy of a Map, we are bound to deliver a manuscript copy to the British Minister here on 1st of January next." He stated that the map he wanted

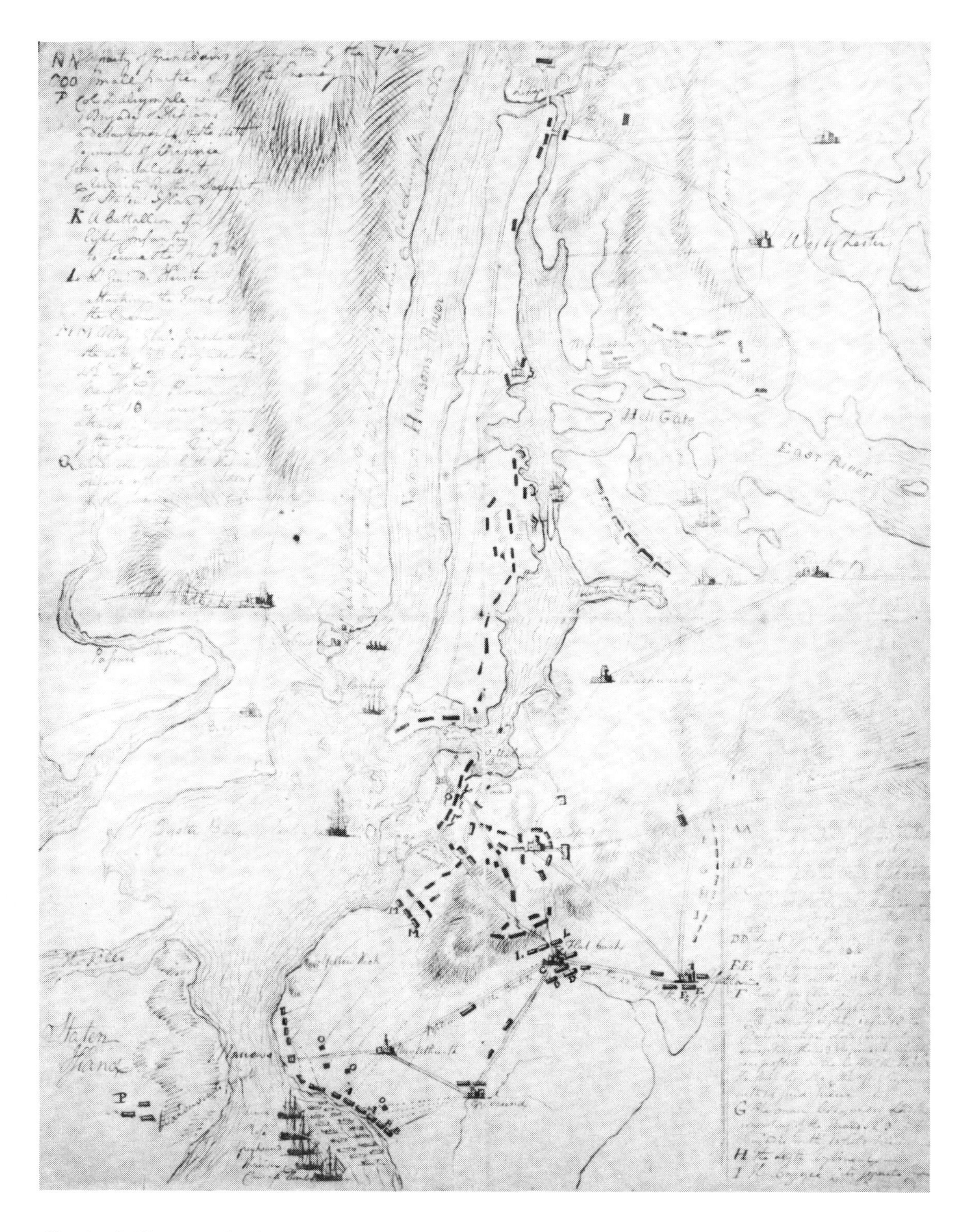

Charles Willson Peale. [New York, showing British and American troop positions.] ca. 1776. [No. 38 (1)]

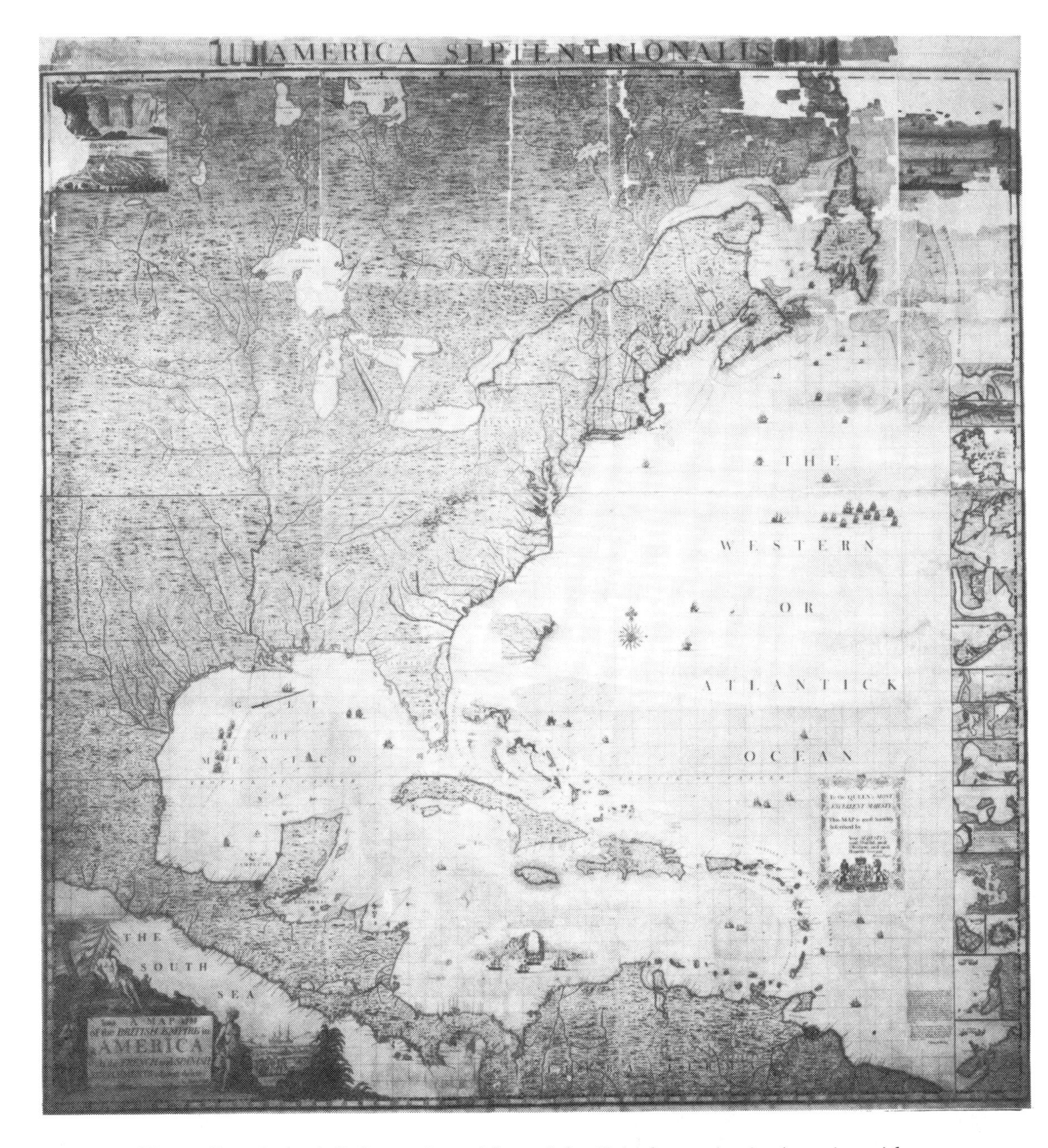

Henry Popple and C. Lempriere. Map of the British empire in America with the French and Spanish settlements adjacent thereto. 1733. [No. 449]

more than any other, is in truth the same map, but *bearing on the face of it* the date of 1776 to 1782 (or 1777 I believe) and *not* having the word "United States" on it or in it. The title is in substance as follows vizt:

A new and correct map of North America with the West Indian Islands, divided according to the last treaty of peace concluded at Paris 10th February 1763, corrected from the original materials of Govr. Pownal M.P. Sayer & Bennet 1777. But it may vary a little in the said title & been any date from 1775 to 1782 [Archives].

Vaughan shipped the maps 16 November, and Gallatin retained two which belonged to the Society [22 November 1828, Archives]. As "U.S. Agent for the N.E. boundary," he signed for these two on 26 December: "Sayer & Bennett's U.States-Feby 1783" and "Dunn's British Empire 1774." They were to be "laid before the Arbiter" and they would be "returned, replaced, or, in case of unavoidable accident, compensated for by the United States" [Archives, Receipt]. He returned the unused maps 2 January 1829 and sent his "thanks and those of the Government for your zealous and useful services on this important occasion" [Archives].

These maps were not returned by 1843 so the members instructed Secretary George Ord on 15 December to recover them. Ord wrote immediately to Gallatin who replied 19 December, explaining what happened to them:

Sayer & Bennet's United States, delivered to the British Minister at Washington, as part of the evidence to be laid before the Arbiter and which, in conformity with the Convention, the U. States were bound to communicate to the British Government on the 31st Dec[emb]er 1828. – cannot be recovered. *Dunn's British Empire,* laid before the Arbiter. All the maps laid before the King of Holland were bound together at Washington. That Atlas was, I understand, brought back from Holland by Mr. Treble and must be in the Department of State [Archives].

On 18 May 1852, Acting Secretary of State William Hunter returned the Sayer and Bennett map. It had been found "this morning" and Hunter hastened "to restore it to your Society" [Archives].

The Dunn map was returned 15 December 1852 by Secretary of State Edward Everett. He wrote member C. B. Trego that the map was returned by direction of the President "with the thanks of this Department for its use" [Archives].

For the Sayer and Bennett map, see: No. 550.

(601: 1774: D912bna Large)

454. 1774

North America, as divided amongst the European Powers. By Samuel Dunn. London: Robert Sayer: 1774 [10 January 1774].

Cartographer: Samuel Dunn.

Size: 30.5 x 44.5 cm.

Colored.

The American Military Pocket Atlas (Pam. 912.7:Am3). Maps removed from binding 5 November 1962.

(600: 1774: D912naep Small)

455. [1776?]

Carte nouvelle de l'Amérique angloise contenant tout ce que les anglois possédent sur le continent de l'Amérique Septentrionale savior le Canada, la nouvelle Écosse ou Acadie, les treize provinces unies qui sont les quatre colonies de la Nouvelle Angleterre 1. New Hampshire, 2. Massachusetsbaye, 3. Rhode-Island & 4. Connecticut, 5. la Nouvelle York, 6.

Nouvelle Jersey, 7. Pensilvanie, 8. les comtés de Newcastle, Kent et Sussex sur la Delaware, 9. Mariland, 10. Virginie, 11. la Caroline Septentrionale, 12. la Caroline Méridionale et 13. Géorgie avec la Floride. Augsburg: [1776?].

Scale: 1:20 leagues.

Cartographer and engraver: Matthieu Albrecht Lotter.

Size: 60 x 49.1 cm.

Colored.

(602: [1776?]: L915asc Large)

456. 1787

North America and the West Indies, with the opposite coasts of Europe and Africa. Chart, containing the coasts of California, New Albion, and Russian discoveries to the north; with the peninsula of Kamtschatka, in Asia, opposite thereto; and islands, dispersed over the Pacific Ocean, to the north of the line. London: Robert Sayer: 1787 [10 November 1787].

Size of plate: 45.3 x 113 cm.

Colored.

Shows routes of early navigators.

(130: 1787: Sa97naw Small)

457. [1790]

A general map of North America from the best authorities. [1790].

Engraver: Scot.

Size: 17.8 x 22.6 cm.

Provenance: Presented by Thomas Dobson, 1798.

In: Thomas Dobson, publisher, *Encyclopaedia,* vol. 1, opposite p. 538.

Wheat: 51

(032: En2)

458. 1797

A new map of North America showing all the new discoveries.

Engraver: Hill.

Size: 19.3 x 12.5 cm.

In: Jedidiah Morse, *The American gazetteer,* frontispiece. Incomplete.

Wheat: 60

(917.3: M83am)

459. 1799

A map of North America from the latest authorities. 1799.

Size: 18.3 x 21.7 cm.

In: John Payne, *A new and complete system of universal geography,* vol. 4, opposite p. 11.

Wheat: 65

(910: P29)

460. 1806

Charte von Nord America nach astronomischen Bestimmungen und den neuesten Charten von Dalrymple, Arrowshmit, Edwards, u.a.m. neu entworfen von F. L. Güssefeld. Nürnberg: Homann: 1806.

Scale: 69 12/100 English mi. = 1 Grad.

Cartographer: F. L. Güssefeld.

Engraver: I. Rausch.

Size: 45.4 x 54.1.

Colored.

(600: 1806: R197cna Small)

461. 1819

Map of North America exhibiting the boundaries arranged agreeably to the late British and Spanish treaties. Philadelphia: Tanner, Vallance, Kearny & Co.: 1819.

Engraver: Tanner.

Size: 45.3 x 51 cm.

Colored.

(130: 1819: T158bst Small)

462. 1868

North America. A working map for illustrating, by coloration, the geographical distribution of life. Prepared for the Boston Society of Natural History by William C. Cleveland. Authorities Johnston's Royal Atlas, & recent discoveries . . . Isothermal data by Prof. Arnold Guyot. [Boston]: American Photo-Lithographic Co.—Osborne's Process: 1868.

Scale: 4.1 cm. = 200 English mi.

Cartographers: Samuel H. Scudder and William C. Cleveland.

Lithographer: American Photo-Lithographic Company.

Size: 56.6 x 44 cm.

Provenance: Presented by William C. Cleveland, 15 January 1869.

(601: 1868: C592gna Large)

463. 1912

Map of North America, reduced from map on scale of 1:5,000,000. Compiled by the U.S. Geological Survey, George Otis Smith, director. [Washington]: U.S. Geological Survey: 1912.

Cartographer: U.S. Geological Survey.

Scale: 1:10,000,000.

Engraver: U.S. Geological Survey.

Size: 95.4 x 72.9 cm.

Colored.

Insets: Windward Islands and Aleutian Islands.

(120: 1912: Un38int Large)

464. 1941

Map of North American Indian languages. Compiled and drawn by C. F. and E. W. Voegelin. New York: J. J. Augustin, Inc.: 1941.

Cartographers: C. F. and E. W. Voegelin.

Size: 84 x 90.5 cm.

Colored.

See: American Ethnological Society, *Publications*, No. 20.

(130: 1941: V857nai Large)

465. 1945

Glacial map of North America published by the Geological Society of America. First edition. N.p.: 1945. 2 pieces.

Scale: 1:4,555,000.

Cartographers: Richard F. Flint, Wm. C. Alden, E. T. Apfel, H. S. Bostock, S. R. Capps, J. W. Goldthwait, L. M. Gould, George F. Kay, M. M. Leighton, Frank Leverett, Paul MacClintock, D. A. Nichols, G. W. H. Norman, F. T. Thwaites, George W. White, and G. A. Young.

Size: 133.4 x 200.8 cm.

Colored.

Insets: Glacial map of the north polar hemisphere; Index map; Principal loess distribution in central United States.

(130: 1945: F644gLm Large)

466. 1946

Geologic map of North America . . . Under grants from the Geological Society of America, the American Philosophical Society, the American Association of Petroleum Geologists. Bailey Willis, grantee. Washington: Geological Society of America: 1946. 2 sections.

Scale: 1:5,000,000.

Cartographers: Bailey Willis.

Size: 96 x 134 cm.

Photograph.

(130: 1946: St67nha Roller)

467. 1961

[New chart of the English empire in North America]. Engraven and printed by Fra. Dewing, Boston, New England, 1717. Reproduced in collotype by the Meriden Gravure Company: 1961. 4 pieces.

Scale: 4 cm. = 100 mi.

Cartographer: Cyprian Southack.

Engraver: Francis Dewing.

Size: Each piece: 35.5 x 41 cm.

Facsimile.

Contains four columns of place names.

Original in John Carter Brown Library, Providence, Rhode Island, 1942.

Notes of Southack map accompany the facsimile.

Wheat: 44

(635: 1961: So87cyp Small)

468. [1966]

Map of North American Indian languages. Compiled and drawn by C. F. and F. W. Voegelin. Copyright 1966 by the American Ethnological Society. Prepared and printed by Rand McNally & Company, U.S.A. Revised edition. N.p.: [1966].

Scale: 1 inch = 100 mi.

Cartographers: C. F. and F. M. Voegelin.

Size: 149.4 x 120 cm.

Provenance: Presented by C. F. Voegelin, May 1967.

(130: [1966]: V857nai Extra-oversize)

469. 1974

North America at the time of the Revolution. A collection of eighteenth century maps with introductory notes by Louis De Vorsey, Jr. Part II. Ashford, Kent: Harry Margery, Lympne Castle, Kent: 1974. 22 sheets: 5 maps (title page, 2 p. of introduction, and 19 map sheets).

Sizes: various.

Facsimile reproductions.

Contents: John Mitchell. A map of the British colonies in North America. London: 1755 (8 sheets); Joshua Fry and Peter Jefferson. A map of the most inhabited part of Virginia. London: 1755 (4 sheets); C. J. Sauthier. A topographical map of Hudson River. London: 1776 (1 sheet); Samuel Holland. The provinces of New York and New Jersey with part of Pensilvania. London: 1776 (2 sheets); and Thomas Jefferys. A map of the most inhabited part of New England. London: 1774 (4 sheets).

(601: 1974: N81d Large)

470. n.d.

Europe and North America on Mercator's projection. N.p.: n.d. Map in 6 pieces.

Scale: 5 cm. = 110 statute mi.

Size: 90 x 58 cm. each.

Inset: Europe and North America.

(610: n.d.: Eu72noa Large)

CANADA Nos. 471-482

471. [1882]

Map of the Dominion of Canada. Geologically colored from surveys made by the Geological Corps, 1842 to 1882. Alfred R. C. Selwyn, director. Geological and Natural History Survey of Canada. Montreal: Burland: [1882].

Scale: 1:2,851,200.

Cartographer: Alfred R. C. Selwyn.

Size: 65.2 x 199.8 cm.

Colored.

(610: [1882]: C162gns Large)

472. 1900

Map of the Dominion of Canada. Map illustrating water power of Canada. To accompany presidential address. Royal Society, Canada, May 1899. James White, geographer. N.p: 1900. Second ed.

Scale: 1:6,336,000.

Cartographer: James White.

Size: 43.5 x 92.9 cm.

Colored.

(610: 1900: C162cwp Large)

473. 1900

Relief map of Canada and the United States. George M. Dawson, director. Orographic features of the Dominion of Canada compiled from various sources. Data for the United States from the map published by the U.S. Geological Survey. Geological Survey of Canada. Toronto: Toronto Lith. Co.: 1900.

Scale: 1 in. = 350 mi.

Cartographer: George M. Dawson.

Size: 51 x 44.5 cm.

Colored.

(610: 1900: C162srm Large)

474. [1904]

Relief map of the Dominion of Canada. James White, geographer. Department of the Interior of Canada. N.p. [1904].

Scale: 1:6,336,000.

Cartographer: James White.

Size: 43.6 x 92.9 cm.

Colored.

(610: [1904]: C162rmd Large)

See: Department of the Interior, Canada, *Dictionary of altitudes* . . . 1903.

475. [1908]

Map of the Dominion of Canada. 1908. James White, chief geographer. Department of the Interior of Canada. N.p.: [1908].

Scale: 1:6,366,000.

Cartographer: James White.

Size: 49.5 x 88.6 cm.

Colored.

(610: [1908]: C162mdc Large)

476. 1909

Railway map of the Dominion of Canada. Department of the Interior of Canada. James White, geographer. N.p.: 1909. 8 sheets.

Scale: 1:2,217,600.

Cartographer: James White.

Size of each sheet: Ca. 67 x 65 cm.

Colored.

(610: 1909: C162dom Extra-oversize)

477. [1920]

Railway map of the Dominion of Canada. Department of the Interior of Canada. 1920. J. E. Chalifour, chief geographer. N.p.: [1920]. 8 sheets.

Scale: 1:2,217,600.

Cartographer: J. E. Chalifour.

Size of each sheet: Ca. 78 x 65 cm.

Colored.

(610: [1920]: C162dom Extra-oversize)

478. 1925

Canada. J. E. Chalifour, chief geographer. Department of the Interior of Canada. Natural Resources Intelligence Service. N.p.: 1925.

Scale: 1:3,801,600.

Cartographer: J. E. Chalifour.

Size: 99.2 x 151.6 cm.

Colored.

Inset: Canada [northern North America].

(610: 1925: C162nrs Large)

479. [1927]

Map of the Dominion of Canada. Department of the Interior of Canada. Natural Resources Intelligence Service. J. E. Chalifour, chief geographer. 1927. N.p.: [1927].

Scale: 1:6,336,000.

Cartographer: J. E. Chalifour.

Size: 49.8 x 89.4 cm.

Colored.

Contains railroad and steamship routes, with legends.

(610: [1927]: C162nrs Small)

480. 1937

Dominion of Canada. Department of Mines and Resources of Canada. J. Wardle, director. Ottawa: 1937.

Scale: 1:6,336,000.

Cartographer: J. Wardle

Size: 61 x 87.6 cm.

Colored.

Provenance: Presented by Mrs. A. I. Hallowell, November 1983.

(610: 1937: W211dmr Large)

481. 1950

Tectonic map of Canada. Lambert conformal conic projection—standard parallels 47° 30′ and 65° 30′. Prepared by the Geological Association of Canada with the support of the Geological Society of America. Washington: Williams & Heintz, lith.: 1950.

Scale: 1:3,801,600.

Lithographer: Williams & Heintz.

Size: 101.5 x 145.3 cm.

(610: 1950: C162tec Roller)

482. n.d.

Canada. Index to maps of the old geographic series. Department of Mines and Resources of Canada. J. M. Wardle, director. N.d.

Scales: 1:250,000 and 1:500,000.

Cartographer: J. M. Wardle.

Size: 22.2 x 46.9 cm.

Colored.

Inset: British Columbia and Alberta.

Provenance: Presented by Mrs. A. I. Hallowell, November 1983.

(610: n.d.: C162dmr Small)

EASTERN CANADA Nos. 483-491

483. [1814]

A new correct map of the seat of war in lower Canada protracted from Holland's large map compiled from actual survey made by order of the provincial government laid down with many late additions and improvements by Amos Lay. Philadelphia: A. Lay & J. Webster: [1814].

Scale: 1 in. = 7 mi.

Cartographers: Holland and Amos Lay.

Engraver: H . S. Tanner.

Size: 54.9 x 84.9 cm.

Colored.

Provenance: Presented by Amos Lay, 20 May 1814.

(612: [1814]: L455swc Large)

484. [1825]

A map of the province of Upper Canada and the adjacent territories in North America compiled by James G. Chewitt shewing the districts, counties and townships in which are situated the lands purchased from the crown by the Canada Company. Incorporated 1825. N.p.: [1825].

Cartographer: James G. Chewett.

Engraver: I. S. Cox.

Size: 98.6 x 179.1 cm.

Colored.

Provenance: Presented by John Vaughan, 7 March 1834.

(612: [1825]: C412puc Large)

485. 1884

Maps to illustrate reports by Loring Woard Bailey, George Frederick Matthew, Robert Walpole Ellis and Albert Peter Low. Geological Survey of Canada. 1882-1883. Montreal: Canada Bank Note Co.: 1884. 10 maps.

Scale: 1 in. = 4 mi.

Cartographers: Wallace Broad, William McInnes, Robert Wheelock Ells, Albert Peter Low, and N. J. Giroux.

Colored.

Province of New Brunswick. 32.1 x 47.7 cm.; Province of Quebec. 32 x 47.4 cm.; Province of Quebec. 32.1 x 47.2 cm.; Part of New Brunswick and Prince Edward Island. 31.9 x 47.1 cm.; Prince Edward Island. 32 x 47.2 cm.; Prince Edward Island. 16.7 x 47.1 cm.; Province of Quebec. 32 x 47.2 cm.; Province of Quebec. 32 x 47 cm.; Province of Quebec. 32.1 x 47.3 cm.; Province of Quebec. 32 x 47.3 cm.

(610: 1884: C162rep Small)

486. [ca. 1888]

The Eskimo tribes of north-east America. By Franz Boas. [New York]: Julius Bien & Co. Lith.: [ca. 1888].

Cartographer: Franz Boas.

Lithographer: Julius Bien & Co.

Size: 19⅜ in. x 17³⁄₁₆ in.

Colored.

Provenance: Presented by Mrs. Helene Boas Yampolsky, 1961-1962.

See: U.S. Bureau of American Ethnology. Franz Boas, "The Central Eskimo." *Sixth Annual Report.*

(610: [ca. 1888]: B638etr Small)

487. [ca. 1888]

[Map showing in detail the geographical divisions of territory occupied by the Eskimo tribes of northeastern America]. By Franz Boas. [New York]: Julius Bien & Co.: [ca. 1888].

Cartographer: Franz Boas.

Lithographer: Julius Bien & Co.

Size: 20⅞ in. x 18¾ in.

Colored.

Provenance: Presented by Mrs. Helene Boas Yampolsky, 1961-1962.

See: U.S. Bureau of American Ethnology. Franz Boas, "The Central Eskimo." *Sixth Annual Report.*

1. Oqo and Akudnirn; 2. Frobisher Bay; 3. Eclipse Sound and Lyon inlet; 4. Repulse Bay and Lyon inlet; 5. Boothia Isthmus and King William Land.

(610: [ca. 1888]: B638est Small)

488. 1901

Index maps for annual report. Geological Survey of Canada. Part A, vol. XIV. Robert Bell . . . acting director.

Cartographers: C. O. Senécal and Robert Bell.

Colored.

1. New Brunswick, Prince Edward Island & part of Quebec. 27 x 40.5 cm.; 2. Nova Scotia. 28.4 x 38.2 cm.; 3. Northwestern Ontario. 17.7 x 30.6 cm.; 4. Southern British Columbia. 22.2 x 40.5 cm.

(610: 1901: C162sLs Small)

489. 1905

Geological map of the northeastern part of the Dominion of Canada, to illustrate the report on the cruise of D. G. S. Neptune to Hudson Bay and the Arctic Islands, by A. P. Low. Geological Survey of Canada. 1905.

Scale: 1 in. = 50 mi.

Cartographers: A. P. Low, C. F. King, and P. Frèreault.

Size: 92 x 68.9 cm.

Colored.

Inset: Fullerton harbor.

Provenance: Presented by Mrs. A. I. Hallowell, November 1983.

(610: 1905: L955 Large)

490. 1930

Highways map of eastern Canada and adjacent states. [Toronto]: Imperial Oil Ltd.: 1930.

Scale: 1 in. = 40 mi.

Cartographer: A. R. Thomson.

Lithographer: Rolph Clark Stone, Ltd.

Size: 57.3 x 88.8 cm.

Colored.

Insets: Trans-Canada highway (proposed); Canada; District of Montreal, Quebec; Rideau Lakes, Ontario; Muskoga and Lake of Bays, Ontario; Kawartha Lakes, Ontario; and 40 smaller places.

Provenance: Presented by Mrs. A. I. Hallowell, November 1983.

(610.5: 1930: T388eca Large)

491. [ca. 1939]

[Turner valley], Geological Survey of Canada. N.p.: [ca. 1939]. 9 maps.

Sizes: Various.

(610: [ca. 1939]: T857val Large)

NOVA SCOTIA AND NEW BRUNSWICK, CANADA Nos. 492-496

492. [ca. 1758]

A new and accurate map of the Province of Nova Scotia, in North America; from the latest observations. N.p.: [ca. 1758].

Size: 28.2 x 33.4 cm.

(611.5: [ca. 1758]: N856nsa Small)

493. [ca. 1790]

[Map of Pasamaquody Bay and the St. Croix River, including Deer, Campo Bello and Grand Manan Islands, New Brunswick, Canada]. N.p.: [ca. 1790].

Scale: 1 in. = 3/4 mi.

Size: 150.5 x 119 cm.

(611.3: [ca. 1790]: N426bru Large)

494. 1841

A new pocket map of the peninsula of Nova Scotia. Intended as a topographical guide, also to illustrate its geological structure, according to the observations of Messrs. Jackson & Alger. Compiled from the most recent surveys. 1841.

Scale: 1½ cm. = 10 mi.

Cartographers: Jackson and Alger.

Size: 23.6 x 42.6 cm.

Colored.

Provenance: Presented by Jackson and Alger, 15 October 1841.

(611.5: 1841: J134nov Book map)

495. 1895

New Brunswick and Nova Scotia. Atlas of the world. Maritime provinces, p.77. Rand, McNally & Co.: 1895.

Scale: 1 in. = 37 mi.

Size: 23.5 x 31.8 cm.

Colored.

Provenance: Presented by Mrs. A. I. Hallowell, November 1983.

On same sheet is: Manitoba, *Atlas of the world* . . . 1895.

(611: 1895: N428nns Small)

496. 1917

Whiteburn gold district Queens County Nova Scotia. Department of Mines of Canada. Geological Survey. William McInnes, directing geologist. Publication No. 1690. Economic geology. Structural plan. N.p.: 1917.

Scale: 1:4,800.

Cartographers: C.-O. Senécal, A. S. Jost, and William McInnes.

Size: 50.4 x 62 cm.

Colored.

(611.5: 1917: C162wgq Small)

QUEBEC, CANADA Nos. 497-503

497. [1775]

A plan of Quebec, metropolis of Canada in North America. [1775]
Scale: [2.5 cm. = ca. 300 yds.].
Engraver: Aitken.
Size: 11 x 17.7 cm.
In: *Pennsylvania Magazine,* vol. 1 (1775), opposite p. 563.
Wheat: 90

(050: P383)

498. [1775]

Plan of the town & fortifications of Montreal or Ville Marie in Canada. [1775].
Scale: 2.5 cm. = 90 French toises.
Engraver: Aitkin [sic].
Size: 24.4 x 16.6 cm.
In: (Philadelphia) *Pennsylvania Magazine,* vol. 1 (1775), opposite p. 517.
Wheat: 91

(050: P383)

499. 1777

Map of the inhabited part of Canada, from the French surveys; with the frontiers of New York and New England; from the large survey by Claude Joseph Sauthier. To Major General John Burgoyne. . . . Wm. Faden. London: W. Faden: 1777.
Scale: 69½ British statute mi. = 1 degree.
Cartographer: Claude Joseph Sauthier.
Engraver: William Faden.
Size: 57.5 x 87.4 cm.
Colored.
Provenance: From the George Ord purchase, 21 June 1839. Librarian George Ord visited France several times and was authorized to purchase books for the Library. Occasionally he purchased either manuscripts or maps as well.
See: W. Faden, North American Atlas. London: W. Faden: 1777. No. 4.

(612: 1777: S97fsc Large)

500. 1813

Map of Upper Canada, describing all the new settlements, townships, etc. with the countries adjacent, from Quebec to Lake Huron. Compiled at the request of His Excellency Major General John G. Simcoe. . . . By David William Smyth. With additions and corrections from Holland's three sheet map. New York: Prior & Dunning: 1813.

Cartographers: Sir David William Smyth and Holland.

Engraver: James D. Stout.

Size: 54 x 84 cm.

Colored.

(612: 1813: Sm97uc Large)

501. 1813

Plan of Quebec and adjacent country showing the principal encampments & works of the British & French armies during the siege by General Wolfe in 1759. Reduced from the mss. map of Captain J. G. Glegg, by John Melish. The different parts of this map were surveyed as follows. The east coast of the Falls of Montmorency, the point of Orleans & the south side of the River St. Lawrence, by Capt. Digby, Engineer in Ordinary. The coasts of Beauport, from the River St. Charles to the Falls of Montmorency, by Capt. Holland, of the Royal Americans, Assis[tan]t Engineer. The ground between the River St. Lawrence & the River St. Charles, by Lieut. Desbarres, of the Royal Americans, Assis[tan]t Engineer. Philadelphia: Melish: 1813.

Scale: 1 in. = 1000 yds.

Cartographers: J. B. Glegg, John Melish, Capt. Digby, Samuel Jan Holland, and J. F. W. Desbarres.

Engraver: H. S. Tanner.

Size: 23.2 x 50.8 cm. *Size of paper:* 44.8 x 57 cm.

Colored.

Inset: View of Quebec from Point Levi.

In: John Melish, *A military and topographical atlas of the United States; including the British possessions & Florida.* Philadelphia: 1813.

(612.966: 1813: M485que Large)

502. 1920

Quebec. Montreal and Quebec sheet. Standard topographical map. Department of the Interior of Canada. Sheet 11. 1920.

Scale: 1:500,000.

Cartographer: J. E. Chalifour, geographer.

Size: 69.2 x 49.7 cm.

Colored.

Provenance: Presented by Mrs. A. I. Hallowell, November 1983.

Contains manuscript notes and markings pertaining to the Indians.

(612: 1920: C357qmo Large)

503. 1920

Quebec & Ontario. Gatineau Sheet. Department of the Interior of Canada. Standard topographical map. Sheet 10 and part of 33. 1920. Edition of 1920.

Scale: 1:500,000.

Cartographer: J. E. Chalifour, chief geographer.

Size: 77.3 x 50.2 cm.

Colored.

Provenance: Presented by Mrs. A. I. Hallowell, November 1983.

Contains manuscript material.

(612: 1920: C162qon Large)

ONTARIO, CANADA Nos. 504-514

504. 1854

Sketch of the navigation through East Neebish rapids River St. Mary from surveys by Captain E. P. Scammon . . . W. H. Hearding, and C. S. Cole . . . under the direction of Capt. J. N. Macomb, 1853. U.S. War Department. Survey of the lakes. Washington: 1854.

Scale: 1:15,000.

Cartographers: F. Herbst, E. P. Scammon, W. H. Hearding, C. S. Cole, and J. N. Macomb.

Engraver: W. H. Dougal.

Size: 48.1 x 36 cm.

Provenance: Presented by G. G. Meade, 19 July 1861.

(613: 1854: Un38smc Small)

505. [ca. 1858]

Chart No. 1 of River Ste. Marie from Point Iroquis to East Neebish. Survey of the Northern and North Western lakes. Surveyed under the direction of Captain J. N. Macomb U.S. Army with additions to date under the orders of Captain G. G. Meade. 1857 [ca. 1858].

Scale: 8 cm. = 1 mi.

Cartographers: J. N. Macomb, G. G. Meade, and J. U. Mueller.

Engraver: W. H. Dougal.

Size: 61 x 80.3 cm.

Inset: Continuation from Round Id. Point to Point Iroquois.

Provenance: Presented and endorsed by George G. Meade, 4 April 1861.

(613: [ca. 1858]: Un38smm Extra-oversize)

506. [1858]

Chart no. 2 of River Ste. Marie including the part from the entrance of Mud Lake to the East Neebish. U.S. War Department. Bureau of Topographical Engineers. Survey of the Northern and North Western lakes. Surveyed under the direction of Capt. J. N. Macomb in 1853 and 1854. Reduction for engraving under the direction of Capt. G. G. Meade, in 1858. [Washington: 1858].

Scale: 1:40,000.

Cartographers: J. U. Müller, J. N. Macomb, and G. G. Meade.

Engraver: W. H. Dougal.

Size: 61.8 x 62.3 cm.

Provenance: Presented by G. G. Meade, 19 July 1861.

(613: [1858]: Un38mLn Large)

507. 1911

Mineral district northern Ontario. R. E. Young, chief geographer. Department of the Interior of Canada. N.p.: 1911. 2 sheets.

Scale: 1:250,000.

Cartographer: R. E. Young.

Size: 81.4 x 124.3 cm.

Colored.

(613: 1911: C162mdo Large)

508. 1915

Map showing branches of chartered banks in Ontario & Quebec. Prepared in the railway lands branch under the direction of F. C. C. Lynch . . . Department of the Interior of Canada. N.p.: 1915. 2nd ed.

Scale: 1 in. = 25 mi.

Cartographer: F. C. C. Lynch.

Size: 50.6 x 81.2 cm.

Colored.

Insets: Northern Ontario; Gaspé Peninsula.

With this is a chart showing branches of chartered banks and post office savings banks in Ontario and Quebec.

(612: 1915: C162bcb Small)

509. 1919

Mattagami sheet. Ontario. Portions of Algoma, Sudbury and Timiskaming districts. Department of the Interior of Canada. J. E. Chalifour, chief geographer. Sheet 31. Standard topographical map. N.p.: 1919.

Scale: 1:500,000.

Cartographer: J. E. Chalifour.

Size: 69.2 x 47.4 cm.

Colored.

(613: 1919: C162oms Small)

510. 1928

Ontario. Department of the Interior of Canada. 1928.

Scale: 1:2,217,600.

Cartographers: F. C. C. Lynch and J. E. Chalifour.

Size: 84.6 x 95 cm.

Colored.

Provenance: Presented by Mrs. A. I. Hallowell, November 1983.

(613: 1928: C162dic Large)

511. 1943

Ontario. Index to map sheets of the national topographic series. Department of Mines and Resources of Canada. Ottawa: 1943.

Scale: 1 in. = 1 and 2 mi.

Size: 47.3 x 55.2 cm.

Colored.

Provenance: Presented by Mrs. A. I. Hallowell, November 1983.

On same sheet as: Ontario, Index to map sheets . . . 1943.

(613: 1943: C162dmr Large)

512. 1943

Ontario. Index to map sheets of the national topographic series. Department of Mines and Resources of Canada. Ottawa: 1943.

Scale: 1 in. = 4 and 8 mi.

Size: 47.3 x 55.2 cm.

Colored.

Provenance: Presented by Mrs. A. I. Hallowell, November 1983.

Contains manuscript notations.

On same sheet is: Ontario. Index to map sheets . . . 1943.

(613: 1943: C162dmr2 Large)

513. n.d.

Ontario. Index to map sheets of the national topographic series. Department of the Interior of Canada. Topographical Survey of Canada. N.d.

Scale: 1 in. = 50 mi.

Size: 52.1 x 44.1 cm.

Colored.

Inset: Eastern Ontario.

Provenance: Presented by Mrs. A. I. Hallowell, November 1983.

On reverse is map of Canada. Contains manuscript notations on the American Indian.

(613: n.d.: C162ims Large)

514. n.d.

Ontario. Sheet No. 52 N. Department of the Interior of Canada. Topographical Survey of Canada. N.p.: n.d.

Size: 45.9 x 56.2 cm.

Colored.

Provenance: Presented by Mrs. A. I. Hallowell, November 1983.

Contains manuscript notations. Trimmed and incomplete.

(613: n.d.: C162tos Small)

MANITOBA, CANADA Nos. 515-526

515. 1895

Manitoba, *Atlas of the world.* P. 78. Rand McNally & Co.: 1895.

Scale: 1 in. = 24 mi.
Size: 24 x 32.3 cm.
Colored.
Provenance: Presented by Mrs. A. I. Hallowell, November 1983.
On same sheet as: New Brunswick and Nova Scotia, *Atlas of the world* . . . 1895.
(614.1: 1895: M315ppr Small)

516. [1904]
Railways in Manitoba, Assiniboia, Alberta and Saskatchewan. 1904. James White, geographer. Department of the Interior of Canada. N.p.: [1904].
Scale: 1:2,217,600.
Cartographer: James White.
Size: 40.2 x 81.6 cm.
Colored.
(610.3: [1904]: C162maa Large)

517. [1908]
Index showing the townships in Manitoba, Saskatchewan, Alberta and British Columbia. Plans of which have been printed up to the 1st July 1908. Department of the Interior of Canada. Topographical Survey Branch. Ottawa: [1908].
Scale: 1:2,217,600.
Size: 49.7 x 77.4 cm.
Colored.
Inset: S.W. British Columbia.
(614: [1908]: C162msa Large)

518. 1909
Map of Manitoba, Saskatchewan and Alberta. James White, geographer. Department of the Interior of Canada. N.p.: 1909. 4 sheets.
Scale: 1:792,000.
Cartographer: James White.
Size: 96.2 x 225.7 cm.
Colored.
(610.3: 1909: C162ska Large)

519. [1909]
Map showing elevators in Manitoba, Saskatchewan and Alberta. James White, chief geographer. Department of the Interior of Canada. N.p.: [1909]. Edition of April 20th. 1909.
Scale: 1 in. = 25 mi.
Cartographer: James White.
Size: 40.7 x 87.3 cm.
Colored.
Capacity of elevators in Canada printed on the borders.
(610.3: [1909]: C162sec Large)

520. 1927

Grand Rapids, Manitoba. (Provisional edition.) Sheet no. 63-G. Department of the Interior of Canada. Topographical Survey of Canada. Ottawa: 1927.

Scale: 1:253,440.

Size of paper: 59.9 x 76.1 cm.

Colored.

Insets: Key map; declination of the compass needle, July 1927.

Provenance: Presented by Mrs. A. I. Hallowell, November 1983.

(614.1: 1927: C162grm Large)

521. 1929

Map of Manitoba. Department of the Interior of Canada. N.p.: 1929. 2 sheets.

Scale: 1:792,000.

Cartographer: J. E. Chalifour.

Size of each: 74 x 86.4 cm.

Colored.

Provenance: Presented by Mrs. A. I. Hallowell, November 1983.

(614.1: 1929: C162dic Large)

522. [1929]

Norway House, Manitoba. (Provisional edition.) Sheet No. 63 H. Department of the Interior of Canada. Topographical Survey of Canada. Ottawa: [1929].

Scale: 1:253,440.

Size of paper: 60.8 x 76 cm.

Colored.

Inset: Index map of Manitoba; and the declination of the compass needle, January 1929.

Provenance: Presented by Mrs. A. I. Hallowell, November 1983.

(614.1: [1929]: C162nhm Large)

523. 1929

Oxford House, Manitoba. Department of the Interior of Canada. Topographical Survey of Canada. Sheet No. 53-I. Ottawa: 1929.

Scale: 1:253,440.

Size of paper: 60.8 x 75.8 cm.

Colored.

Insets: Areas not mapped shaded green; index map for Manitoba; the declination of the compass needle, July 1929.

Provenance: Presented by Mrs. A. I. Hallowell, November 1983.

(614.1: 1929: C162man Large)

524. [1930]

Selkirk, Manitoba. Sheet No. 62 I. Department of the Interior of Canada. Topographical Survey of Canada. Ottawa: [1930].

Scale: 1:253,440.

Size of paper: 61 x 76.3 cm.

Colored.

Insets: Index map of Manitoba; the declination of the compass needle, January 1930.

Provenance: Presented by Mrs. A. I. Hallowell, November 1983.

(614.1:[1930]: C162sLk Large)

525. n.d.

[Lake Winnipeg and territory to the east]. N.p.: n.d.

Size of paper: 89.5 x 121 cm.

Colored.

Provenance: Presented by Mrs. A. I. Hallowell, November 1983.

Consists of four maps pasted together. Contains manuscript notes pertaining to the American Indian.

(614.1: n.d.: W73letr Large)

526. n.d.

Manitoba. Southern sheet. Canada. Compiled and engraved at the Chief Geographer's Office. N.p.: n.d.

Scale: 1:792,000.

Size: 71 x 86.6 cm.

Colored.

Provenance: Presented by Mrs. A. I. Hallowell, November 1983.

Manuscript markings and notes on the Indian.

(614.1: n.d.: C162cgo Large)

SASKATCHEWAN, CANADA Nos. 527-530

527. 1928

Saskatchewan and Manitoba. Index to map sheets of the national topographic series. Department of the Interior of Canada. Ottawa: 1928.

Scale: 1 in. = 50 mi.

Size: 44 x 52.3 cm.

Colored.

Provenance: Presented by Mrs. A. I. Hallowell, November 1983.

Contains manuscript notations. On the reverse is a map of Canada.

(614: 1928: C162skm Large)

528. 1928

Saskatchewan and Manitoba. Index to map sheets of the national topographic series. Department of the Interior of Canada. Topographical Survey of Canada. Ottawa: 1928.

Scale: 1 in. = 50 mi.

Size: 43.9 x 52.3 cm.

Colored.

Provenance: Presented by Mrs. A. I. Hallowell, November 1983.

Contains manuscript notes.

(614: 1928: C162skm2 Large)

529. 1929

Saskatchewan. Department of the Interior of Canada. N.p.: 1929. 2 sheets.

Scale: 1:792,000.

Cartographer: J. E. Chalifour.

Size of each: 81 x 83.5 cm.

Colored.

Provenance: Presented by Mrs. A. I. Hallowell, November 1983.

Contains manuscript notations pertaining to the American Indian.

(614.4: 1929: C162skw Large)

530. n.d.

Saskatchewan. Department of the Interior of Canada. James White, geographer. N.p.: n.d.

Cartographer: James White.

Size: ca. 106.5 x 55.5 cm.

Colored.

(614.4: n.d.: W582sas Extra-oversize)

ALBERTA, CANADA Nos. 531-536

531. 1897

Map shewing the country between the lower portions of the Peace and Athabasca Rivers north of Lesser Slave Lake. Geological Survey of Canada. George M. Dawson, director. Nos. 1-3. N.p.: 1897.

Scale: 1 in. = 10 statute mi.

Cartographers: C. O. Senécal, G. M. Dawson and R. G. McConnell.

Lithographer: H. B. Baine.

Sizes: 42.1 x 64.8 cm.; 66.5 x 42 cm.; 42.1 x 64.8 cm.

Colored.

(614.7: 1897: C162dpa Small)

532. 1902

Topographical map of the Rocky Mountains. Banff sheet. Department of the Interior of Canada. James White, geographer. N.p.: 1902.

Scale: 1 in. = 2 mi.

Cartographers: James White and W. J. Graham.

Size: 34.7 x 47.4 cm.

Colored.

(614.7: 1902: C162rmb Small)

533. 1902

Topographical map of the Rocky Mountains. Lake Louise sheet. Department of the Interior of Canada. James White, geographer. N.p.: 1902.

Cartographer: James White.

Size: 43.5 x 50.4 cm.

Colored.

(614.7: 1902: C162rmL Large)

534. 1908

Northern Alberta. Peace River district. James White, geographer. Department of the Interior of Canada. N.p: 1908. Preliminary edition.

Scale: 1:800,000.

Cartographer: James White.

Size: 94.9 x 70.3 cm.

Colored.

(614.7: 1908: C162prd Large)

535. 1930

Province of Alberta. Department of the Interior of Canada. N.p.: 1930. 2 sheets.

Scale: 1:792,000.

Cartographer: J. E. Chalifour.

Size of each: ca. 82 x 89.9 cm.

Colored.

Provenance: Presented by Mrs. A. I. Hallowell, November 1983.

Contains manuscript notations.

(614.7: 1930: C162aLb Large)

536. n.d.

Map of Alberta. Department of the Interior of Canada. James White, chief geographer. N.p.: n.d. 2 sheets.

Cartographer: James White.

Size: 83.5 x 91 cm.

Colored.

Inset: Peace River block.

(614.7: n.d.: W582aLb Extra-oversize)

BRITISH COLUMBIA, CANADA Nos. 537-544

537. [1886]

Maps of the principal auriferous creeks of the Cariboo mining district. Surveyed and drawn by Amos Bowman assisted by James McEvoy. Department of the Interior of Canada. Geological and Natural History Survey. 1885-1886. N.p.: Mortimer & Co.: [1886]. 9 sheets.

Cartographers: Amos Bowman and James McEvoy.

Lithographer: Mortimer & Co.

Colored.

1. Williams Creek, 82.1 x 57.5 cm.; 2. Lightning Creek, 56.6 x 71.4 cm.; 3. Little Snowshoe and Keithley Creeks, 69.2 x 32.5 cm.; 4. Sugar Creek, Hardscrabble, Slough Creek, Nelson Creek, Willow River and Dragon Creek, 67.3 x 43.2 cm.; 5. Cunningham Creek, 58.4 x 32.cm., 6. Antler Creek, 59.5 x 41.7 cm.; Grouse Creek, 38.4 x 28.7 cm.; 8. Harvey Creek, 30.2 x 22.1 cm.; 9. Section along Harvey Creek, Size of paper: 34.4 x 52.3 cm.

To illustrate the report of Amos Bowman. See: *Note to accompany maps . . .* by George M. Dawson, April 1895. Geological and Natural History survey of Canada. *Annual report.* 1886. Part R.

(615:[1886]: C162min Large)

538. 1888

Map of a portion of the southern interior of British Columbia embodying the explorations made in 1877 by G. M. Dawson & in 1882-1884 by Amos Bowman. Department of the Interior of Canada. Geological and Natural History Survey. Montreal: Burland Lith. Co.: 1888. Preliminary edition not geol. colored, Dec. 1887.

Scale: 1:506,880.

Cartographers: G. M. Dawson, Amos Bowman, and Israel Marion.

Size: 59.7 x 51.9 cm.

(615: 1888: C162gnh Large)

539. 1903

Map of Yukon district with the adjacent northern part of British Columbia. 1903. James White, geographer. Department of the Interior of Canada. N.p.: 1903.

Scale: 1:750,000.

Cartographer: James White.

Size: 100.2 x 70.9 cm.

Colored.

(615.8: 1903: C162ydc Large)

540. 1904

British Columbia railway belt. Department of the Interior of Canada. 1904. James White, geographer. [Ottawa: 1904].

Scale: 1:500,000.

Cartographer: James White.

Size: 66.5 x 103.5 cm.

Colored.

(615: 1904: C162bcr Large)

541. 1908

British Columbia railway belt. James White, geographer. Special edition prepared under the direction of R. E. Young. Corrected to February 1st 1908. Preliminary edition. Department of the Interior of Canada. [Ottawa]: 1908. 2 sheets.

Scale: 1:500,000.

Cartographers: James White and R. E. Young.

Size: 66.2 x 136 cm.

Colored.

Inset: Vancouver Island.

(615: 1908: C162bcr Large)

542. 1908

Topographical sheet. Special map of Rossland British Columbia by W. H. Boyd. Surveyed 1905-1906. No. 1001. Department of Mines of Canada. Geological Survey Branch. N.p.: 1908.

Scale: 1:4,800.

Cartographers: W. H. Boyd, A. Dickison, and C.-O. Senécal.

Size: 44 x 52.5 cm.

Colored.

(615: 1908: C162ros Large)

543. 1925

British Columbia & Alberta. Calgary sheet. Standard topographical map. Department of the Interior of Canada. Natural Resources Intelligence Service. Sheet 20. N.p.: 1925.

Scale: 1 in. = 7.89 mi.

Cartographer: J. E. Chalifour.

Size: 69 x 46 cm.

Colored.

(615: 1925: C162bcc Small)

544. 1926

British Columbia & Alberta Jasper sheet. Standard topographical map. Sheet 45. Department of the Interior of Canada. Natural Resources Intelligence Service. J. E. Chalifour, chief geographer. N.p.: 1926.

Scale: 1 in. = 7.89 mi.

Cartographer: J. E. Chalifour.

Size: 68.8 x 44.5 cm.

Colored.

(615: 1926: C162bca Small)

NORTHERN CANADA Nos. 545-549

545. 1874

Map showing the townships surveyed in the province of Manitoba and North-West Territory in the Dominion of Canada. Department of the Interior of Canada. Dominion Lands Branch. Published by order of . . . David Laird . . . 1874. J. Johnston, chief draughtsman. New York: H. H. Lloyd & Co.: 1874.

Scale: 1 in. = 6 mi.

Cartographer: J. Johnston.

Engraver: H. H. Lloyd.

Size: 64.3 x 95.5 cm.

(617:1874: C162man Large)

546. [1904]

Explorations in northern Canada and adjacent portions of Greenland and Alaska. 1904. James White, chief geographer. Department of the Interior of Canada. [Ottawa: 1904].

Cartographer: James White.

Size: 64.5 x 90.6 cm.

Colored.

(610.2: [1904]: C162cga Large)

547. 1905

Yukon Territory Kluane, White, and Alsek Rivers from surveys by International Boundary Commission 1893-1895, J. J. McArthur, 1900, A. D. Talbot, 1899, and J. B. Tyrrell, 1898. Department of the Interior of Canada. James White, geographer. N.p.: 1905.

Scale: 1:400,000.

Cartographer: James White, J. J. McArthur, A. D. Talbot, and J. B. Tyrrell.

Size: 88.7 x 57.8 cm.

Colored.

(615.8: 1905: C162kwa Large)

548. 1906

Map of part of southwestern coast of Hudson Bay from York factory to Fort Severn. Geological Survey of Canada. From micrometer & compass survey by Owen O'Sullivan. 1905. No. 915. Robert Bell, acting director. N.p.: 1906.

Scale: 1 in. = 16 mi.

Cartographers: P. Frèreault, Owen O'Sullivan, and Robert Bell.

Size: 19.1 x 36.4 cm.

Colored.

See: Geological Survey of Canada, *Summary report.* 1905.

(617.6: 1906: C162hby Small)

549. [1910]

Rocky Mountains between lat. 51° and 53° 10′. R. E. Young, chief geographer. Department of the Interior of Canada. N.p.: [1910]. 2 sheets.

Scale: 1:253,440.

Cartographer: R. E. Young.

Size: 99 x 87.1 cm.

Colored.

(610.3: [1910]: C162rmL Large)

UNITED STATES OF AMERICA Nos. 550-623

550. 1783

The United States of America with the British possessions of Canada, Nova Scotia, & of Newfoundland, divided with the French; and the Spanish territories of Louisiana and Florida according to the preliminary articles of peace signed at Versailles the 20th of Jan[uar]y 1783. London: R. Sayer & J. Bennett: 1783 [9 February 1783].

Scale: 69½ English mi. = 1 degree.

Cartographer: T. Jeffery.

Size: 46.1 x 51.3 cm.

Colored.

Originally in T. Jefferys, *The American atlas.* London: Sayer and Bennett: 1782.

Loaned to the U.S. Government in 1828 for use during the Northeast Boundary dispute with Great Britain. For story, see No. 453.

(635: 1783: Un38cns Large)

551. 1784

Carte des États-Unis de l'Amérique suivant le traité de paix de 1783. Dédiée et présentée à, S. Excellence Mr. Benjamin Franklin ministre plenipotentiare des États-Unis de l'Amérique près la Cour de France, anc. présid. de la conventiõ de Pensilvanie et de la société philosophique de Philadelphia, &c. &c. Par son très humble et très obeissant serviteur Lattré. 1784. Paris: Lattré: 1784.

Scale: 70 English mi. = 1 degree.

Cartographer: Lattré.

Size: 54 x 75.6 cm. *Size of paper:* 56.9 x 101.7 cm.

Colored.

Inset: Supplément à la Floride.

On the right and left sides of the map are: Principaux événemens militaires entre les Américains et les Anglois.

The Society also owns a reproduction of this map taken from the copy in the Newberry Library.

(635: 1784: L355eua Large)

552. 1785

The United States of North America. With the British and Spanish territories according to the Treaty, of 1784. [London]: Wm. Faden: 1785.

Scale: 69½ English mi. = 1 degree.

Engraver: William Faden.

Size: 52.6 x 62.8 cm.

Colored.

Provenance: Purchased from the library of William Priestman, 15 July 1831.

(630: 1785: F121bst Large)

553. [1791]

The United States of America laid down from the best authorities agreeable to the peace of 1783. Certificate of correctness signed by Osgood Carleton attached to title. Boston: I. Norman: [1791].

Scale: 69½ mi. = 1 degree.

Cartographer: Osgood Carleton.

Size: 81.8 x 111.5 cm.

Colored.

Inset: A map of the lakes and rivers between Lake Superior and the North Sea.

This map was assembled from two different issues from the same plate.

Wheat: 119

(635: [1791]: Un38msn Large)

554. 1795

A map of the United States: compiled chiefly from the state maps, and other authentic information, by Sam'l Lewis. 1795. Philadelphia: M. Carey: 1795.

Cartographer: Samuel Lewis.

Engraver: W. Harrison.

Size: 61.7 x 89 cm.

Inset: Florida and the Gulf coast past New Orleans.

See: Mathew Carey, *General atlas.* Philadelphia: M. Carey: 1796, no. 24.

(635: 1795: L585usa Large)

555. [1796]

A map of the United States, exhibiting the post-roads, the situations, connections & distances of the post-offices stage roads, counties, ports of entry and delivery for foreign vessels, and the principal rivers. By Abraham Bradley, Junr.

Scale: 1 cm. = 15 mi.

Cartographer: Abraham Bradley, Jr.

Engraver: William Harrison, Jr.

Size of paper: 89 x 94.4 cm.

Inset: Progress of the mail on the main line.

Wheat: 128

(635: [1796]: B721usp Large)

556. [1796]

The United States of America. [1796].

Scale: 2.5 cm. = 150 mi.

Engraver: W. Barker.

Size: 24.9 x 32.5 cm.

Provenance: Presented by Suzanne Wister Eastwick and Joseph L. Eastwick, July 1975.

See: Mathew Carey, *Carey's American pocket atlas,* following p. 4.

Wheat: 131

(917.3: C25c)

557. 1799

Map of the United States, Canada &c. intended to illustrate the travels of the Duke de la Rochefoucauld Liancourt. London: R. Phillips: 1799 [1 July 1799].

Scale: 69½ English mi. = 1 degree.

Engravers: Smith and Jones and T. Foot.

Size: 40.5 x 42.3 cm.

Photostat.

On map in ink: "Rapport sur les Archives de France, relative à l'histoire du Canada, par J. Edmond Roy - Canada - Public Archives, pub. no. 6-1911."

See: La Rochefoucauld Liancourt, F. A. F., duc de, *Travels through the United States.* London: R. Phillips: 1799.

(630: 1799: Sm67drL Large)

558. [1799]

The United States of America. [1799].

Scale: 2.5 cm. = 150 mi.

Size: 24.7 x 31.1 cm.

In: John Payne, *A new and complete system of universal geography,* vol. 4, opposite p. 50.

Wheat: 142

(910: P29)

559. 1804

Map of the United States, exhibiting the post-roads, the situations, connexions and distances of the post-offices stage roads, counties & principal rivers. By Abraham Bradley, Junr. District of Columbia: 1804.

Scale: 1 in. = 30 mi.

Cartographer: Abraham Bradley, Jr.

Size: 95 x 132.2 cm.

Colored.

Inset: Map of North America.

This map was first published in 1796. See map No. 555 above.

(635: 1804: B721usb Large)

560. [1805]

A map of part of the continent of North America from longitude [blank] w. of Washington City to the Pacific and between lattitude 35 and 52 north compiled from the information of the best informed travellers through that quarter of the Globe in which, the

Missouri Jefferson Lewis and the upper part of the Clarks river and the lower part of the Columbia River is [partially?] corrected by celestial observations from the junction of the Missouri and Mississippi to the enterance of the Columbia into the pacific ocean in Longitude 47° 57′ west of Washington City and 124° 57" west of Greenwich and in Lattitude 46° 19′ 11" north. Showing Lewis and Clarks rout over the Rocky Mountains in 1805 on their rout to the pacific from the United States. By William Clark. [Meriden Gravure: ca. 1950]. 4 sheets.

Scale: 1 in. = 50 mi.

Cartographer: William Clark.

Size: 147.6 x 262.7 cm.

Colored.

Facsimile of the original manuscript map in the William Robertson Coe collection of Western Americana, Yale University Library, 1950.

(630: [1805]: C552Lcr Large)

561. 1806

New map of the United States of America including part of Louisiana drawn from the latest authorities. Revised and corrected by Osgood Carleton, Esqr. Boston: John Sullivan: 1806.

Cartographer: Osgood Carleton.

Size: 121 x 137.6 cm.

Colored.

Contains tables of the ports of entry in the United States, and situation and extent of each state.

Mutilated.

(630: 1806: Un38maL Large)

562. 1809

Die Vereinigten Staaten von Nord-America, nach den sichersten Bestimmungen, neuesten Nachrichten und Charten, in der Alber'schen Projection entworfen, von C. G. Reichard. Nürnberg: Homanns: 1809.

Scale: 69 English mi. = 1 degree.

Cartographer: C. G. Reichard.

Engraver: C. Trummer.

Size: 58.9 x 73.8 cm.

Colored.

(630: 1809: R277vsa Book map)

563. 1811

[Map of the United States]. Philadelphia: T. L. Plowman: 1811 [copy 2].

Size: 50.1 x 64.9 cm.

Two nearly identical copies. Copy 2 marked "2nd Edition. Containing the Seat of War &c."

(630: 1811: P726mus Large)

564. 1812

The United States of America confirmed by treaty 1783. London: R. Wilkinson: 1812 [10 February, 1812].

Scale: 1 in. = 150 British statute mi.

Engraver: B. Smith.

Size: 22.2 x 28.3 cm.

Colored.

Shows the state of Franklinia.

(630: 1812: Sm57uso Small)

565. 1813

Map of the United States of America. Philadelphia: Melish: 1813 [June 1813].

Scale: 1 in. = 100 mi.

Cartographer: John Melish.

Engraver: H. S. Tanner.

Size: 36.8 x 46.2 cm.

In: [Melish], *Documents relative to the negotiations for peace between the United States and Great Britain.* Philadelphia: Palmer: 1814.

(Pam., vol. 131, no. 1)

566. [1814]

A map of Lewis and Clark's track, across the western portion of North America from the Mississippi to the Pacific Ocean. [Philadelphia: 1814].

Cartographer: Samuel Lewis and William Clark.

Size: 30.3 x 70.3 cm.

Impression from original copper plate in the possession of the American Philosophical Society. It was first printed in 1814 and this pulling from the plate in 1979 is the first supplemental pulling ever done.

For original pulling, see: *History of the expedition under the command of Lewis and Clark . . .* Philadelphia: Bradford and Inskeep; New York: Abrm. H. Inskeep, J. Maxwell: 1814.

(630: [1814]: C55mir Large)

567. [ca. 1815]

A map historical and biographical chart. Of the United States by David Ramsay. [Philadelphia: Carey: ca. 1815].

Cartographer: David Ramsay.

Size: 31.5 x 35.3 cm. *Size of paper:* 48.2 x 73.8 cm.

Colored.

Contains map and chart with four columns of text.

See: Robert L. Brunhouse, "David Ramsey, selections from his writings." American Philosophical Society. *Transactions,* n.s., vol. 55.

(635: [ca. 1815]: R147hbc Large)

568. 1815

A new and correct map of the United States of North America; exhibiting the counties, towns, roads &c. in each state. Carefully compiled from surveys and the most authentic documents, by Samuel Lewis. Philadelphia: Emmor Kimber: 1815.

Scale: 69½ American mi. = 1 degree.

Cartographer: Samuel Lewis.

Engravers: W. and S. Harrison.

Size: 171.5 x 186.5 cm.

Colored.

Inset: United States boundary, crossing the Lake of the Woods.

Provenance: Presented by Emmor Kimber, 3 November 1815.

(630: 1815: L585ncm Large)

569. 1816

Map of the United States with the contiguous British & Spanish possessions. Compiled from the latest and best authorities by John Melish. Philadelphia: Melish: 1816 [6 June 1816].

Scale: 1 in. = 50 mi.

Cartographer: John Melish.

Engravers: Vallance and H. S. Tanner.

Size: 89 x 144 cm.

Colored.

Inset: West Indies.

Provenance: Presented by John Melish, 2 December 1816.

(630: 1816: M485mbs Large)

570. [ca. 1818]

Map of the United States of America. Designed to illustrate the geological memoir of William Maclure, Esqr. Entered as the Act directs, and published by John Melish, Philadelphia [ca. 1818].

Scale: 1 in. = 90 mi.

Cartographer: William Maclure.

Size: 33 x 44.3 cm.

Colored.

Published in the *Transactions* of the American Philosophical Society, vol. 1, n.s., to illustrate William Maclure's geological observations, as read to the Society, 16 May 1817.

Map shows proposed canals, boundary lines, and roads and distances.

(630: [ca. 1818]: Un38geo Small)

571. 1819

United States of America compiled from the latest and best authorities, by John Melish 1819. Philadelphia: Melish: 1819 [10 July 1819].

Scale: 1 in. = 120 mi.

Cartographer: John Melish.

Engraver: Samuel Harrison.

Size: 40.3 x 49 cm.

Colored.

Provenance: Presented by John Melish prior to 1826.

(630: 1819: M485un Small)

572. 1820

Map of the United States with the contiguous British and Spanish possessions compiled from the latest and best authorities by John Melish. Philadelphia: Melish: 1820 [16 June 1820].

Scale: 1 in. = 50 mi.

Cartographer: John Melish.

Engravers: J. Vallance and H. S. Tanner.

Size: 107.8 x 142.3 cm.

Colored.

Inset: West Indies.

(630: 1820: M485bps Large)

573. [1825]

A map of the United States of America. Philadelphia: H. S. Tanner: [1825].

Scale: 1 in. = 125 mi.

Cartographer: H. S. Tanner.

Size: 39 x 65.2 cm.

Colored.

Inset: Profile or vertical section of the Erie Canal, exhibiting its various locks, levels . . .

(630: [1825]: T158pkt Book map)

574. 1825

Map of the United States of North America compiled from the latest and most authentic information by David H. Vance. Philadelphia: Anthony Finley: 1825 [12 July 1825].

Cartographer: David H. Vance.

Engraver: J. H. Young.

Size: 131 x 155.3 cm.

Colored.

Insets: Map of North America, including all the recent geographical discoveries to 1825; comparative elevations of the principal mountains and hills in the United States.

Provenance: Presented by David H. Vance, 21 October 1825.

(630: 1825: V278usa Large)

575. 1827

Lay's map of the United States. Compiled from the latest and best authorities and actual surveys by Amos Lay. New York: J. M. Bowen: 1827.

Cartographer: O. H. Throop and Amos Lay.

Engravers: O. H. and D. S. Throop and William Chapin.

Size: 132.5 x 153.5 cm.

Colored.

Inset: Florida.

Contains statistical table of the United States.

(620: 1827: L455mus Large)

576. 1829

United States of America. Philadelphia: H. S. Tanner: 1829.

Cartographer: H. S. Tanner.

Size: 117.7 x 151.3 cm.

Insets: South part of Florida, and, Oregon and Mandan district. Individual insets of Washington, Baltimore, Philadelphia, New York, Boston, Charleston and New Orleans: environs of Albany, Boston, New York, Philadelphia and Trenton, Baltimore and Washington, and Savannah. Also, profiles of the following canals: Erie, Ohio, Morris (N. J.), Union, Chesapeake and Ohio. Also, profile of Schuylkill navigation, Penna., and the Columbia railroad.

Provenance: Presented by Henry Schenck Tanner, 18 September 1829.

(630: 1829: T158moe Large)

577. 1835

Mitchell's reference & distance map of the United States by J. H. Young. Published by S. Augustus Mitchell. Philadelphia: Mitchell & Hinman: 1835.

Scale: 69.2 mi. = 1 degree.

Cartographer: J. H. Young.

Engravers: J. H. Young, F. Dankworth, E. Yeager, and E. F. Woodward.

Size: 32.6 x 170.7 cm.

Colored.

Insets: General map of the United States with the contiguous British and Mexican possessions; vicinity of Boston; vicinity of New York; vicinity of Philadelphia; north part of Maine; south part of Florida; vicinity of Albany; vicinity of Baltimore and Washington; vicinity of Cincinnati; vicinity of Charleston; vicinity of New Orleans; vicinity of the Falls of Niagara; and vicinity of Rochester, N.Y.

Provenance: Presented by James H. Young, 15 May 1835. Publisher Mitchell presented a copy, ca. 1837.

(630: 1835: Y79rdm Large)

578. [1849]

Skeleton map showing the rail roads completed and in progress in the United States, and those projected through the public lands and their connection with the principal harbours on the lakes and on the seaboard. Printed by order of the Senate of the United States, the 9th Sess., 31st Congress, 1848-49. [Washington]: C. B. Graham, lith.: [1849].

Scale: 4.7 cm. = 100 mi.

Lithographer: C. B. Graham.

Size: 116.9 x 93.4 cm.

Colored.

This was accompanied with a report from the Hon. Sidney Breeze, for granting the land to the state of Illinois to aid in the completion of her railroads.

(630: [1849]: Un38srr Book map)

579. [ca. 1850]

United States. London, Edinburgh & Dublin: J. & F. Tallis: [ca. 1850].

Scale: 1 in. = 250 mi.

Cartographer: John Rapkin.

Engravers: J. Rogers and J. Marchant.

Size: 24.3 x 32.2 cm.

Colored.

Insets: Portraits of Washington and Franklin; the Capitol, Washington; Penn's treaty with the Indians; Buffalo hunt; Washington's monument.

(630: [ca. 1850]: R187wfp Small)

580. 1855

Bien & Sterner's new rail road map of the United States and the Canadas compiled from the latest surveys and locations under the direction of Zerah Colburn editor of the Rail Road Advocate. New York: A. Ranney: 1855.

Scale: 3½ in. = 200 mi.

Cartographer: Zerah Colburn.

Lithographers: Bien & Sterner.

Size: 67.9 x 79.5 cm.

Colored.

In: Zerah Colburn, *Railroad economy, the revised and extended report on European and American railways.* New York: Wynkoop: [1855].

(630: 1855: R177usc Book map)

581. 1856

Equal magnetic declination for the year 1850. U.S. Coast Survey. A. D. Bache, supdt. N.p.: 1856.

Scale: 1:20,000,000.

Cartographer: A. D. Bache.

Size: 16.6 x 25.7 cm.

Provenance: Presented by A. D. Bache, 7 March 1857.

(630: 1856: B121mde Small)

582. 1856

Lines of equal magnetic dip and horizontal intensity for the year 1850. U.S. Coast Survey. A. D. Bache, supdt. N.p.: 1856.

Scale: 1:20,000,000.

Cartographer: A. D. Bache.

Size: 17 x 25.6 cm.

Provenance: Presented by A. D. Bache, 7 March 1857.

(630: 1856: B121emd Small)

583. 1858

J. Sage & Son's new & reliable rail road map comprising all the railroads of the United States and Canadas with their stations and distances, compiled from the most accurate statistics. Buffalo: J. Sage & Sons: 1858.

Scale: 1 in. = 33⅓ mi.

Size: 91.2 x 110.8 cm.

Inset: Eastern Massachusetts, & part of N. Hampshire, Rhode Island & Connecticut. On an enlarged scale.

Surrounding the maps are the names of various railroads listed by states.

(630: 1858: St27rrr Book map)

584. 1862

Lloyd's new steel plate railroad map of the United States and Canadas from the latest surveys. Showing every railroad & station finished to September 1862. Engraved to accompany Lloyd's R. R. Weekly Guide. New York: J. T. Lloyd: 1862.

Scale: 69.1 Statute or American mi. = 1 degree.

Size: 94.5 x 125.3 cm.

Colored.

(630: 1862: L775rrs Large)

585. 1872

Weather map. U.S. War Department. Signal Service, U.S. Army. Division of Telegrams and Reports for the Benefit of Commerce. Washington: J. L. Kervand, lith.: 1872.

Lithographer: J. L. Kervand.

Size: 36.5 x 56.8 cm.

Colored.

Three copies for Friday, 31 May 1872: 7:35 A.M.; 4:35 P.M.; and 11:35 P.M.

There are also yellow flimsies filled-in in manuscript.

(630: 1872: Un42wem Small)

586. 1873-1876

Weather map. U.S. War Department. Signal Service, U.S. Army, Division of Telegrams and Reports for the Benefit of Commerce and Agriculture. All observations made at the same moment of actual (not local) time. 1873-1876. Washington: N. Peters, photo-lithographer: 1873-1876.

Scale: 4 cm. = 250 mi.

Cartographer: Prof. Guyot.

Size: 36.1 x 56.9 cm.

Colored.

Distributed by the Philadelphia Board of Marine Underwriters.

(630: 1873-1876: G999unm Large)

587. 1879

Map of the United States and territories, showing the extent of public surveys, Indian and military reservations, land grant R.R., rail roads, canals, and other details . . . compiled from the official surveys of the U.S. Department of the Interior General Land Office . . . by C. Roeser. New York: J. Bien: 1879.

Scale: 1 in. = 30 mi.

Cartographer: C. Roeser.

Lithographer: J. Bien.

Size: 122.3 x 197 cm.

Colored.

Inset: Territory of Alaska.

(630: 1879: Un38ust Large)

588. 1883-1901

New shippers railway map of the United States . . . showing all railroads, each in a separate color, and all railroad stations in large plain type. Chicago: Rand, McNally & Co.: 1883-1901. 8 sections.

Scale: 1 in. = 8 mi.

Size of each: 300 x 159 cm.

(630: 1883-1901: R157rrs Roller)

589. 1883

Outline map [of the United States] with dials showing standard railway time. Copyrighted 1883 by W. F. Allen. New York: 1883.

Cartographer: W. F. Allen.

Engraver: American Bank Note Company.

Size: 15.6 x 22 cm.

Colored.

(630: 1883: AL51rrt Book map)

590. 1884

Colton's railroad & commercial map of the United States & Canada. East of the 100th meridian of longitude. Compiled, engraved and published by G. W. & C. B. Colton & Co. New York: Colton: 1884.

Scale: 1 in. = 20 statute mi.

Cartographers: G. W. and C. B. Colton.

Engravers: G. W. and C. B. Colton.

Size: 174.9 x 146.8 cm.

Colored.

Insets: Map of the railroads of New England; vicinity of New York; and general map of the United States, showing the railroad routes across the continent.

Incomplete: wanting western half.

(635: 1884: C672rcm Large)

591. [ca. 1885]

Map showing the rail road and canal lines used in the determination of elevations by James T. Gardner, geographer. [Ca. 1885].

Cartographer: James T. Gardner.

Size: 38 x 56.1 cm.

(630: [ca. 1885]: G173rrc Small)

592. [1888]

Statistical map of the United States of America. Prepared by James S. Cowden and James D. Holman . . . under the direction of the Hon. John C. Black, U.S. Pension Bureau, Commissioner of Pensions. July 1888. [Washington: Government Printing Office: 1888].

Cartographers: James S. Cowden, James D. Holman, and John C. Black.

Size: 48.6 x 77.2 cm.

(630: [1888]: Un38pbs Large)

593. [1889]

Annual change of the magnetic declination for the epoch January, 1890. U.S. Coast and Geodetic Survey. T. C. Mendenhall, superintendent. N.p.: [1889].

Scale: 1:10,000,000.

Cartographers: T. C. Mendenhall and C. A. Schott.

Size: 36.6 x 51.6 cm.

At head of map: "Coast & Geodetic Survey Report for fiscal year 1887-'88."

(630: 1889: M521amd Small)

594. 1889

General railway map of the United States, the Dominion of Canada and Mexico. Every subscriber to the official railway guide for 1889, at $5.00 per annum, will be furnished with a copy of this map on application to the National Railway Publication Co., 41 Bond Street, New York. New York: American Bank Note Company: 1889.

Size: 90.8 x 125.2 cm.

Colored.

(630: 1889: N196grm Large)

595. [1890]

Isogonic chart for the epoch 1890. U.S. Coast and Geodetic Survey. T. C. Mendenhall, superintendent. N.p.: [1890].

Scale: 1:13,700,000.

Cartographers: T. C. Mendenhall and C. A. Schott.

Size: 24.5 x 52.6 cm.

Colored.

At head of map: "Coast & Geodetic Survey Report for fiscal year 1888-'89."

(630: [1890]: M521isc Small)

596. [1890]

Isogonic chart of the United States for the epoch 1890. U.S. Coast and Geodetic Survey. T. C. Mendenhall, superintendent. N.p.: [ca. 1890].

Scale: 1:7,000,000.

Cartographers: T. C. Mendenhall and C. A. Schott.

Size: 55.5 x 71.6 cm.

Colored.

At head of map: "Coast & Geodetic Survey Report for fiscal year 1888-'89."

(630: [1890]: M521ich Large)

597. [1890]

Magnetic meridians of the United States for January 1890. U.S. Coast and Geodetic Survey. T. C. Mendenhall, superintendent. N.p.: [1890].

Scale: 1:10,000,000.

Cartographers: T. C. Mendenhall and C. A. Schott.

Size: 35.6 x 57.6 cm.

Colored.

At head of map: "Coast & Geodetic Survey Report for fiscal year 1888-'89."

(630: [1890]: M521mme Small)

598. 1890

Map of the United States and territories, with adjacent parts of Canada and Mexico also part of the West India Islands. Showing the extent of public surveys, Indian and military reservations, rail roads, canals, and other details. Compiled from the official surveys for the U. S. Department of the Interior General Land Office and other authentic sources . . . by A. F. Dinsmore. 1890. Baltimore: Isaac Friedenwald: 1890.

Scale: 1 in. = 40 statute mi.

Cartographer: A. F. Dinsmore.

Size: 158.4 x 204.2 cm.

Colored.

Inset: Territory of Alaska.

(630: 1890: Un38cmw Large)

599. 1898

United States. Compiled under the direction of Henry Gannett . . . by Henry King. U.S. Department of the Interior. Geological Survey. [Washington]: 1898. 3 sheets.

Scale: 1:2,500,000.

Cartographers: Henry Gannett and Henry King.

Size: 122.6 x 192.2 cm.

Colored.

Inset: Alaska.

(630: 1898: Un38geo Large)

600. 1899

United States base map. Compiled by Henry Gannett. U.S. Geological Survey. J. W. Powell, director. [Washington]: 1899. Edition of Dec. 1896, reprinted Sept. 1899.

Scale: 9.1 cm. = 300 statute mi.

Cartographers: Henry Gannett and J. W. Powell.

Engraver: U.S. Geological Survey.

Size: 44.6 x 71.9 cm.

Colored.

(630: 1899: Un38bas Large)

601. 1899

United States contour map. Compiled by Henry Gannett. U.S. Geological Survey. J. W. Powell, director. [Washington]: 1899. Edition of Dec. 1896, reprinted Sept. 1899.

Scale: 9.1 cm. = 300 statute mi.

Cartographers: Henry Gannett and J. W. Powell.

Engraver: U.S. Geological Survey.

Size: 44.5 x 71.9 cm.

Colored.

(630: 1899: Un38con Large)

602. [1899]

United States relief map. Compiled by Henry Gannett. U.S. Geological Survey. J. W. Powell, director. [Washington]: 1899. Edition of Dec. 1896, reprinted September 1899.

Scale: 9.1 cm. = 300 statute mi.

Cartographers: Henry Gannett and J. W. Powell.

Engraver: U.S. Geological Survey.

Size: 44.4 x 71.9 cm.

Colored.

(630: [1899]: Un38rm Large)

603. [ca. 1901]

Distribution of the magnetic declination in the United States in 1900. Compiled by Henry Gannett. U.S. Geological Survey. [Washington]: U.S. Geological Survey: [ca. 1901].

Scale: 2.4 cm. = 100 mi.

Cartographer: Henry Gannett.

Lithographer: U.S. Geological Survey.

Size: 44.5 x 71.8 cm.

Colored.

See: U.S. Geological Survey, *17th annual report,* part I, pl. II.

(630: [ca. 1901]: G153mde Large)

604. [ca. 1902]

Normal annual precipitation in the United States, 1870-1901. N.p.: [ca. 1902].

Scale: 1 in. = 200 mi.

Cartographer: Alfred J. Henry.

Engraver: U.S. Geological Survey.

Size: 26 x 40.1 cm.

Colored.

(630: [ca. 1902]: H388nap Small)

605. 1904

Climatic charts of the United States. U.S. Department of Agriculture. Weather Bureau. Washington: Weather Bureau: 1904.

Size of paper: 48.3 x 61.1 cm.

Colored.

Contains title page and 26 charts.

(630: 1904: C611use Large)

606. 1906

United States. U.S. Geological Survey. Charles D. Walcott, director. [Washington]: 1906. Edition of May 1906.

Scale: 9.5 cm. = 600 mi.

Cartographer: Charles D. Walcott.

Engraver: U.S. Geological Survey.

Size: 26.5 x 40 cm.

Colored.

(630: 1906: Un38usa Small)

607. 1907

United States including territories & insular possessions; showing the extent of public surveys, Indian, military & forest reservations, railroads, canals, national parks & other details. Compiled under the direction of I. P. Berthrong. U.S. General Land Office, Department of the Interior. Corrected to 30 June 1907. New York: Sackett & Wilhelms Co.: 1907.

Scale: 1 in. = 37 mi.

Cartographers: M. Hendges and I. P. Berthrong.

Engraver: R. F. Bartle Co.

Size: 151 x 210 cm.

(630: 1907: Un38tsr Roller)

608. 1908

National forests with related projects and data. Base map compiled by the drafting division of the U.S. Department of Agriculture, General Land Office. Gifford Pinchot, Forester. Reissued with the national forests and related projects and data by the Office of Geography in the Forest Service. Information furnished by the U.S. Reclamation Service, U.S.

Weather Bureau, and Inland Waterways Commission. [Washington]: 1908. Edition of 1 July 1908. Map of 4 sheets.

Scale: 1 in. = 37 mi.

Size: 150 x 208.2 cm.

Colored.

Insets: Index map of most of the northern hemisphere; Panama Canal; Puerto Rico; Hawaiian Islands; Tutuila group of the Samoan Island; Guam or Cuajan Island; Philippine Islands; and Alaska with the Aleutian Islands.

(630: 1908: Un38nfa Large)

609. 1909

Lines of the Bell Telephone Companies, United States and Canada, 1 July 1909. American Telephone and Telegraph Company. Boston: Heliotype Co.: 1909.

Scale: 1 in. = 40 mi.

Size: 136 x 198 cm.

(630: 1909: AmllteL Roller)

610. 1910

United States. Original compiled under the direction of Henry Gannett. U.S. Geological Survey, George Otis Smith, director. 1910. [Washington]: 1910. First published in 1890. 3 parts.

Scale: 1:2,500,000.

Cartographers: Henry Gannett and George Otis Smith.

Size: 132.4 x 194.7 cm.

Colored.

Inset: Alaska.

(630: 1910: Un38int Large)

611. 1910

United States including territories and insular possessions; showing the extent of public surveys, Indian, military & forest reservations, railroads, canals, national parks. . . . U. S. General Land Office, Department of the Interior. Compiled under the direction of I. P. Berthrong. Corrected to 30 June 1910. New York: Sackett & Wilhelms Co.: 1910.

Scale: 1 in. = 37 mi.

Cartographers: M. Hendges and I. P. Berthrong.

Engraver: R. F. Bartle Co.

Size: 148 x 207 cm.

(630: 1910: Un38cnp Roller)

612. 1911

United States. U.S. Geological Survey. George Otis Smith, director. [Washington]: U.S. Geological Survey: 1911. Edition of March 1911.

Scale: 3 cm. = 200 mi.

Cartographer: George Otis Smith.

Size: 19.1 x 29.5 cm.

Colored.

(630: 1911: Un38usg Small)

613. 1911

United States base map. Compiled by Henry Gannett. U.S. Geological Survey. George Otis Smith, director. [Washington]: 1911. Edition of November 1911.

Scale: 1:7,000,000.

Cartographers: Henry Gannett and George Otis Smith.

Engraver: U.S. Geological Survey.

Size: 44.6 x 72 cm.

Colored.

(630: 1911: Un38bas Large)

614. 1911

United States contour map. Compiled by Henry Gannett. U.S. Geological Survey. George Otis Smith, director. [Washington]: 1911. Edition of Nov. 1911.

Scale: 1:7,000,000.

Cartographers: Henry Gannett and George Otis Smith.

Engraver: U.S. Geological Survey.

Size: 44.6 x 72.2 cm.

Colored.

(630: 1911: Un38con Large)

615. 1913

United States including territories and insular possessions; showing the extent of public surveys, Indian, military & forest reservations, railroads, canals, national parks. . . . U. S. General Land Office, Department of the Interior. Compiled under the direction of I. P. Berthrong. New York: Sackett & Wilhelms: 1913.

Scale: 1 in. = 37 mi.

Cartographers: I. P. Berthrong and M. Hendges.

Engraver: R. F. Bartle Co.

Size: 150 x 208 cm.

(630: 1913: Un38imf Roller)

616. 1914

United States. Original compilation under the direction of Henry Gannett. U.S. Geological Survey. George Otis Smith, director. [Washington]: U.S. Geological Survey: 1914. First published in 1890. 2 sheets.

Scale: 1:2,500,000.

Cartographers: Henry Gannett and George Otis Smith.

Engraver: U.S. Geological Survey.

Size: 122.3 x 194.4 cm.

Colored.

Inset: Alaska.

Also map of type-of-farming areas in the United States, 1930.

(630: 1914: Un38mus Large)

617. [ca. 1916]

Map of the United States showing locations of National Guard mobilization training camps; National Army cantonment camps; Reserve Officers training camps and aviation sites. Issued by the Rock Island Lines. Chicago: Rand McNally & Co.: [ca. 1916].

Size: 49.1 x 90.7 cm.

Colored.

(630: [ca. 1916]: R157ng Small)

618. 1916

United States. Department of the Interior. Geological Survey. United States. [Washington]: U.S. Geological Survey: 1916. Edition of 1911, reprinted 1916.

Scale: 1 in. = 250 mi.

Engraver: U.S. Geological Survey.

Size: 19.1 x 29.5 cm.

Colored.

(630: 1916: Un38rmr Small)

619. 1916

United States contour map. U.S. Department of the Interior. U.S. Geological Survey. George Otis Smith, director. [Washington]: 1916. Edition of 1916.

Scale: 1:7,000,000.

Cartographer: George Otis Smith.

Size: 44.7 x 72.2 cm.

Colored.

(630: 1916: Un38cm Small)

620. [1930]

Type-of-farming areas in the United States, 1930. U.S. Department of Commerce. Bureau of the Census. Prepared in cooperation with the Bureau of Agricultural Economics, Department of Agriculture. [Washington: 1930].

Size of paper: 76.2 x 104.3 cm.

Colored.

Contains agricultural notes on three sides.

(630: [1930]: Un38tfa Small)

621. [1939]

Indian tribes, reservations and settlements in the United States. U.S. Department of the Interior. 1939. Compiled and drawn by Sam Attahvich, Comanche. [Washington: 1939].

Cartographer: Sam Attahvich.

Size: 36.5 x 58.5 cm.

Inset: Alaska.

Provenance: Presented by Alban W. Hoopes, July 1939.

(630: [1939]: Un38str Small)

622. 1939

National forests, state forests, national parks, national monuments and Indian reservations. 1939. U.S. Department of Agriculture Forest Service. [Washington]: U.S. Geological Survey: 1939.

Scale: 1:7,000,000.

Engraver: U.S. Geological Survey.

Size: 44.5 x 71.8 cm.

Colored.

Insets: Puerto Rico, Alaska.

On right side of map are lists of forest supervisors, headquarters, experimental stations, etc.

(630: 1939: Un38nfs Large)

623. 1952

Pleistocene eolian deposits of the United States, Alaska and parts of Canada, compiled by the National Research Council Committee for the Study of Eolian Deposits. Geological Society of America. [Washington]: 1952.

Scale: 1:2,500,000.

Size: 121.5 x 99 cm.

Photograph.

(630: 1952: G293pLe Roller)

EASTERN UNITED STATES Nos. 624-625

624. 1858

Charts of the lights on the Atlantic & Gulf coasts of the United States. Published by the order and under the direction of the U.S. Light House Board. [New York]: J. Bien: 1858. Series 1 and 2 (12 sheets).

Scale: 1 in. = 10 mi.

Lithographer: J. Bien.

Size of paper: 117.7 x 67.1 cm.

Colored.

(635.2: 1858: Un48Lhb Large)

625. 1959

Glacial map of the United States east of the Rocky Mountains. Published by the Geological Society of America. 1959. First ed. 2 sheets.

Scale: 1:1,750,000.

Cartographers: Richard F. Flint, Roger B. Colton, Richard P. Goldthwaite, and H. B. Willman.

Lithographer: Williams & Heintz Litho. Corp.
Size: 87.5 x 102.7 cm.

(630: 1959: F643gLr Large)

NORTHEASTERN UNITED STATES AND CANADA Nos. 626-633

626. [1775]

A map of the present seat of war on the borders of Canada. [1775].
Scale: 2.5 cm. = 10 English mi.
Engraver: Aitken.
Size: 15.5 x 38.2 cm.
In: *Pennsylvania Magazine,* vol. 1 (1775), opposite p. 463.
Wheat: 89

(050: P383)

627. [1776]

A general map of the northern British colonies in America. Which comprehends the province of Quebec, the government of Newfoundland, Nova-Scotia, New-England and New-York. . . . Regulated by the astronomic and trigonometric observations of Major Holland, and corrected from Governor Pownall's late map 1776. [Title at top of sheet]: The seat of war, in the northern colonies, containing the province of Quebec, Newfoundland, Nova-Scotia, New-England, New-York, New-Jersey, Pennsylvania, &ca. London: Robert Sayer & John Bennett: 1776 [14 August 1776].
Scale: 2½ in. = 150 British statute mi.
Cartographers: S. J. Holland and T. Pownall.
Size: 48.1 x 66.3 cm.
Colored.
Originally in *American military atlas;* removed from binding 5 November 1962.

(635: [1776]: P876nbc Large)

628. [1789]

A map of the northern and middle states; comprehending the western territory and the British dominions in North America. Compiled from the best authorities. [1789].
Scale: [2.5 cm. = ca. 1° 40′ latitude].
Cartographer: Amos Doolittle.
Engraver: Amos Doolittle.
Size: 31.8 x 40.1 cm.
Provenance: Presented by Jacob Snider, Jr., from the library of John Vaughan, 18 March 1842.
In: Jedidiah Morse, *The American geography,* 1st ed., opposite p. 33.
Wheat: 149

(917.3: M83)

629. [1797]

Map of the northern part of the United States of America. [1797]. Fragment.

Scale: [2.5 cm. = ca. 100 mi.].

Cartographer: Abraham Bradley, Jr.

Engraver: B. Callender.

Size: 18.3 x 19 cm.

In: Jedidiah Morse, *The American gazetteer,* opposite "New England."

Wheat: 158

(917.3: M83am)

630. [1817]

Northern Provinces of the United States. Drawn and engraved for Thomson's New General Atlas. [Edinburgh: 1817].

Scale: 2¼ in. = 80 mi.

Size: 49.5 x 59.8 cm.

Colored.

Inset: Vignette of "The Great Falls of Niagara" at top of map.

Provenance: Presented by Fred B. Rogers, 7 October 1970.

(632.5: [1817]: Un38np Large)

631. [1840/41]

Profile with the spirit level, of the due north line from the monument at the source of the River St. Croix to the St. John, surveyed in 1840 & 1841 under the direction of Major J. D. Graham . . . one of the commissioners for surveying and exploring the N. E. boundary of the U.S. U.S. Topographical Engineers. Washington: Hass lith.: [1841].

Scale: 1 in. = 3 mi. *Horizontal scale of feet:* 1:200,000.

Cartographers: J. D. Graham and Wilson McC. Fairfax.

Lithographer: P. Haas.

Size: 32.2 x 76.4 cm.

Provenance: Presented by J. D. Graham, 19 July 1844. Major Graham wrote 28 March 1844 that he was presenting

> a profile or vertical Section with the Spirit level, of the country traversed by the due north line from the monument at the source of the river St. Croix, to the river St. John, derived from surveys executed under my direction in the years 1840 & 1841.
>
> In presenting this document, I beg to call the attention of the Society to the strong contrast, which will appear on a comparison, between the aspect of the country traversed by this due north line as there developed, and that exhibited by the profile of Colonel Bouchette, the British surveyor under the 5th article of the Treaty of Ghent, derived from his surveys of that line and his barometric measurements upon it in the years 1817 & '18.
>
> The direction of the meridian line whose profile or vertical section is now offered to the notice of the Society, was obtained from numerous astronomical observations, fully verified, with a transit instrument having a telescope of 43 inches focal length. It will be perceived on reference thereto that the meridian, thus traced, does not meet with any highland or elevation whatever in passing Mars Hill, but on the contrary that it passes the base of that eminence considerably to the eastward, and at a *depression* nowhere less that 12 feet below the level of the base of the monument which was erected to mark the source of the River St. Croix. . . . It may be well to remark that the base of this monument is surrounded (unless at seasons of extreme drought) by the water constituting the extreme source of that river [Archives. J. D. Graham to APS, 28 March 1844. Minutes. *Proc.,* vol. 4, p. 53. Donation Book].

Librarian George Ord was notified 22 June 1844 that W. R. Palmer had received these maps for the Society [Archives. W. R. Palmer to G. Ord, 22 June 1844].

See: U.S. Congress, 27th, 3rd session. House of Representatives.

(641: [1840/41]: Un 38scj Large)

632. 1843

Map of the boundary lines between the United States and the adjacent British provinces, from the mouth of the River St. Croix to the intersection of the parallel of 45 degrees of north latitude with the River St. Lawrence near St. Regis showing the lines as respectively claimed by the United States and Great Britain under the treaty of 1783, as awarded by the king of the Netherlands, and as settled in 1842 by the Treaty of Washington. Compiled by T. J. Lee and W. M. C. Fairfax under the direction of J. D. Graham, Corps of Topographical Engineers. March 1843. U.S. Army. Corps of Topographical Engineers. [Washington]: 1843. Map G, no. 3.

Scale: 1 in. = 16 mi.

Cartographers: T. J. Lee, W. McC. Fairfax, and J. D. Graham.

Engraver: W. J. Stone.

Size: 58.3 x 70.7 cm.

Colored.

Inset: Rouse's Point and its vicinity on Lake Champlain.

Provenance: Presented by J. D. Graham, 22 June 1844.

(632.5: 1843: Un38bLb Large)

633. 1979

Bouguer gravity map of northeastern United States and southeastern Canada. Onshore and offshore. Compiled by C. T. Hildreth during 1977 and 1978. Regional map Number 1 (western sheet). New England seismotectonic study. The University of the State of New York, the State Museum and Education Department. Coordinator, Patrick J. Barosh. [Albany, N.Y.]: 1979. 2 sheets.

Scale: 1:1,000,000.

Cartographers: C. T. Hildreth and Patrick J. Barosh.

Size: 137.5 x 101.7 cm.

Colored.

Inset: Alternative map for the Boston area.

There is also an eastern sheet. *Size:* 137.5 x 100.5 cm. *Insets:* Index map to sources of information; and Alternative map for the area east of Magdalen Islands.

(632.5: 1979: H548bgr Large)

NEW ENGLAND Nos. 634-636

634. 1776

Charts of the coast and harbours of New England . . . from the surveys taken by Samuel Holland. [London]: 1776. 7 pieces.

Scale: Various.

Cartographers: Samuel Holland and J. F. W. Des Barres.

Sizes: Various.

See: Samuel Holland. *Atlas* . . . London: 1776.

(640.1: 1776: D452cne Large)

635. [1794]

[Southern shore of New England from New York City to Barnstable, with map of Long Island. This is one section of a large topographical map intended to cover the entire United States and possibly more.] [1794. 1961 reprint].

Engraver: Eliza Colles.

Scale: 2.5 cm. = 10 mi.

Size: 14.2 x 9.8 cm.

In: Christopher Colles, *A survey of the roads,* p. 80.

Wheat: 155

(526.8: C68s)

636. 1916

The New England commercial and route survey, showing good auto roads, good roads, steep decline, electric railways, state capitals, state boundaries, court house, etc. Boston: The Bullard Co.: 1916.

Scale: 1 in. = 9 mi.

Size of paper: 110.8 x 116.2 cm.

Insets: Boston and vicinity; and Bar Harbour and vicinity.

Provenance: Presented by Mrs. A. I. Hallowell, November 1983.

On same sheet as: A map of Cape Cod and its approaches . . .1916.

(640.1: 1916: N423crs Large)

MAINE Nos. 637-650

637. [1795]

Maine. [1795].

Scale: 2.5 cm. = 44 mi.

Size: 18.7 x 15.5 cm.

Provenance: Presented by the Estate of Richard Gimbel, 1974.

In: Joseph Scott, *The United States gazetteer,* opposite "Maine."

Wheat: 173

(Paine 68: Sco3u)

638. 1795

A map of the district of Maine drawn from the latest surveys and other best authorities. 1795.

Scale: 2.5 cm. = 18 English statute mi.
Cartographer: Osgood Carleton.
Engraver: Doolittle.
Size: 50 x 40.7 cm.
Inset: A map of those parts of the country most famous for being harassed by the Indians.
Provenance: Presented by Jeremy Belknap, 15 October 1784.
In: James Sullivan, *The history of the district of Maine,* frontispiece.
Wheat: 171

(974.1: Su5)

639. [1796]

Province of Maine. [1796].
Scale: 2.5 cm. = 50 mi.
Engraver: Amos Doolittle.
Size: 19 x 14.5 cm.
Provenance: Presented by Suzanne Wister Eastwick and Joseph L. Eastwick, July 1975.
In: Mathew Carey, *Carey's American pocket atlas,* opposite p. 22.
Wheat: 174

(917.3: C25c)

640. 1799

The province of Maine from the best authorities. 1799.
Scale: 2.5 cm. = 35 mi.
Size: 26.8 x 18.5 cm.
In: John Payne, *A new and complete system of universal geography,* vol. 4, opposite p. 253.
Wheat: 178

(910: P29)

641. 1815

Map of the district of Maine from the latest and best authorities. By Moses Greenleaf Esqr. 1815. Boston: Cummings & Hilliard: 1815 [30 October 1815].
Scale: 1:506,880.
Cartographer: Moses Greenleaf.
Engraver: W. B. Annin.
Size: 103.3 x 66.8 cm.
Colored.
Provenance: Presented by Robert M. Gardner, 19 June 1818.
See: Moses Greenleaf, *Statistical view of the district of Maine.* Boston: Cummings: 1816.

(641: 1815: G843mdm Large)

642. 1851

Richmond's island harbor, [Maine] (Harbor of refuge No. [blank]) from a trigonometrical survey under the direction of A. D. Bache. U.S. Coast and Geodetic Survey. [Washington]: 1851.

Scale: 1:20,000.

Cartographers: A. D. Bache, J. Lambert, and John R. Barker.

Engravers: S. Siebert, J. V. N. Throop, and W. Smith.

Size: 35.6 x 43.7 cm.

Provenance: Presented by the Treasury Department through A. D. Bache, 21 March 1851.

(641: 1851: Un38rih Small)

643. 1854

York River Harbor, Maine. U.S. Coast Survey. A. D. Bache, Supdt. [Washington]: 1854.

Scale: 1:20,000.

Cartographers: F. Fairfax, A. D. Bache, T. D. Cram, and A. W. Longfellow.

Engravers: A. Maedel, J. Young, and C. A. Knight.

Size: 31.7 x 40.1 cm.

Provenance: Presented by A. D. Bache, 6 March 1857.

(641: 1854: Un38yrh Small)

644. 1876

Eastport harbor, Maine . . . U.S. Coast Survey. This chart is based upon a survey made under the direction of A. D. Bache. [Washington]: 1876.

Scale: 1:40,000.

Cartographers: A. D. Bache, C. O. Boutelle, W. H. Dennis, S. A. Gilbert and W. H. Richardson.

Size: 65.9 x 45.7 cm.

Provenance: Presented by the U.S. Coast and Geodetic Survey, 14 December 1885.

Spencer C. McCorkle wrote 14 December that many U.S. Coast and Geodetic Survey charts were seldom issued since "bouys &c. on the bars & in the rivers & harbors, are constantly changing (at least some of them) & were corrected every year or so" [Archives. U.S. Coast and Geodetic Survey to H. Phillips, 14 December 1885]. However, on 21 December he sent 29 charts. These were "all that can be spared at present." As soon as a new edition of the charts for the Delaware River in front of Philadelphia was issued, the Society would receive a copy. He enclosed a listing of the charts he sent: 302A Eastport Harbor Me. 103 Mt. Desert Id. 104 Penobscot Bay. 325 Portland harbor. 337 Boston harbor. 353 Narragansett Bay. 359a Harbor of New London & Thames R. 369 New York bay and harbor. 121 Sandy Hook to Barnegat Lt Hs. 122 Barnegat Lt Hs to Absecon Lt. 377 Cape May city & vicinity topl map. 379 Delaware breakwater. 126b Delaware River Cherry Id Flats to Bridenberg. 126c Delaware River Bombay Hook Lt to Cherry Id. 131 Entrance to Chesapeake Bay. 384 Patapsca River & Baltimore harbor. 391a Washington & Georgetown harbor. 404b Norfolk harbor. 431 Charleston harbor. 156 Savannah to Sapelo Id. 455a Jacksonville to Hibernia Fla. 469 Key West harbor. 177 Tampa Bay. 188 Mobile Bay. 510 No. 8 New Orleans to Sonias' Plantation. 30 Galveston bay Texas. 621 San Francisco entrance Cal. 617 Monterey harbor. 652 Pugets Sound Washington Terry [Archives. U.S. Coast . . . Survey List; 1885].

(641: 1876: Un38ehm Large)

645. [1885]

Coast chart no. 3. Mount Desert Island, Frenchman's and Blue Hill bays and approaches. Maine. U.S. Coast and Geodetic Survey. [Washington: 1885].

Scale: 1:80,000.

Cartographers: A. D. Bache, A. Lindenkohl, C. Junken, and H. Lindenkohl.

Engravers: J. Enthoffer, E. A. Maedel, J. G. Thompson. F. Courtenay, and H. M. Knight.

Size: 96.1 x 74 cm.

Provenance: Presented by U.S. Coast and Geodetic Survey through Spencer C. McCorkle, 14 December 1885.

(641: [1885]: Un38fbh Large)

646. 1885

Coast chart no. 4. Penobscot bay, Maine. U.S. Coast and Geodetic Survey. [Washington]: 1885.

Scale: 1:80,000.

Cartographer: A. D. Bache.

Size: 96 x 74.5 cm.

Provenance: Presented by U.S. Coast and Geodetic Survey through Spencer C. McCorkle, 14 December 1885.

(641: 1885: Un38pbm Large)

647. [1885]

Portland harbor, Maine. From a trigonometrical survey under the direction of A. D. Bache. 1871. U.S. Coast Survey. [Washington: 1885].

Scale: 1:20,000.

Cartographers: A. D. Bache, A. W. Longfellow, C. O. Boutelle, P. Witzel, and A. Balbach.

Engravers: G. B. Metzroth and A. Petersen.

Size: 72.6 x 64.5 cm.

Provenance: Presented by U.S. Coast and Geodetic Survey through Spencer C. McCorkle, 14 December 1885.

(641: [1885]: Un38phm Large)

648. 1898

Maine. [From: *Atlas of the world,* p. 9]. Rand, McNally & Co.: 1898.

Scale: 1 in. = 26 mi.

Size: 32 x 23.5 cm.

Colored.

Provenance: Presented by Mrs. A. I. Hallowell, November 1983.

On same sheet is: New Hampshire, *Atlas of the world.* 1898.

(641: 1898: M288nap Small)

649. 1901

Map of Mount Desert Island, Maine. Compiled by Waldron Bates, Edward L. Rand and Herbert Jacques, 1896, and revised 1901. Boston: Geo. H. Walker: 1901.

Scale: 1:40,000.

Cartographers: W. Bates, E. L. Rand, H. Jacques, and William H. Munroe.

Size: 76.5 x 61.5 cm.

Colored.

Nomenclature revised and corrected—adapted from the map dated June 1893, compiled for the *Flora of Mt. Desert Island.* For text see: Edward L. Rand and John Howard Redfield, *Flora of Mt. Desert Island, Maine.* Cambridge: Wilson: 1894.

(641: 1901: B311mdi Large)

650. 1913

Railroad map of Maine. Prepared under the direction of, and presented by Elmer P. Spofford, Frank Keizer, and John A. James. New York: George F. Cram: 1913 [30 June 1913].

Scale: 1 in. = 8 mi.

Cartographers: E. P. Spofford, Frank Keizer, and John A. James.

Engraver: George F. Cram.

Size: 103 x 79 cm.

Colored.

Inset: Vicinity of Portland.

(641: 1913: M285rrc Book map)

NEW HAMPSHIRE Nos. 651-660

651. [1768]

An accurate map of New Hampshire in New England, from a late survey. [London: R. Sayer and T. Jeffery: 1768].

Size: 32.5 x 27.7 cm.

Original in T. Jefferys, *General topography of North America and the West Indies.* London: R. Sayer and T. Jefferys: 1768.

(642: [1768]: Sa97nh Small)

652. 1784

A topographical map of the state of New Hampshire, surveyed under the direction of Samuel Holland Esqr Surveyor General of lands for the northern district of North America. London: William Faden: 1784 [1 March 1784].

Scale: 1¾ in. = 12 mi.

Cartographer: Samuel Holland.

Size: 53.1 x 35.7 cm.

Facsimile.

Wheat: 180

(642: 1784: H723tnh Small)

653. 1791

A new map of New Hampshire. 1791.
Scale: 2.5 cm. = 12 mi.
Cartographer: Jeremy Belknap.
Engraver: S. Hill.
Size: 38.3 x 25.7 cm.
Provenance: Presented by Jeremy Belknap, 1784-1792.
See: Jeremy Belknap, *History of New Hampshire,* vol. 2, frontispiece.
Wheat: 183

(641: 1791: B411nh Small)

654. [1795]

Newhampshire. [1795].
Scale: 2.5 cm. = 24 mi.
Size: 18.9 x 15.5 cm.
Provenance: Presented by the Estate of Richard Gimbel, 1974.
In: Joseph Scott, *The United States gazetteer,* opposite "New-Hampshire."
Wheat: 186

(Paine 68: Sco3u)

655. [1796]

The state of New Hampshire. [1796].
Scale: 2.5 cm. = 26 American mi.
Cartographer: Samuel Lewis.
Engraver: Seymour.
Size: 19 x 14.6 cm.
Provenance: Presented by Suzanne Wister Eastwick and Joseph L. Eastwick, July 1975.
In: Mathew Carey, *Carey's American pocket atlas,* opposite p. 18.
Wheat: 187

(917.3: C25c)

656. 1799

The state of New Hampshire compiled chiefly from actual surveys. 1799.
Scale: 2.5 cm. = 17 mi.
Cartographer: A. Anderson.
Engraver: C. Tiebout.
Size: 29.5 x 18.5 cm.
In: John Payne, *A new and complete system of universal geography,* vol. 4, opposite p. 229.
Wheat: 190

(910: P29)

657. 1816

Map of New Hampshire by Philip Carrigain. 1816. [Concord: 1816].

Cartographer: Philip Carrigain.

Size: 70.4 x 41.7 cm.

Insets: View of the White Mountains from Shelburne; the cap of the White Mountains.

(642: 1816: C232mnh Large)

658. 1850

Map of New Hampshire & Vermont: published by Thomas, Cowperthwait & Co. [Philadelphia]: Thomas, Cowperthwait & Co.: 1850.

Scale: 1 in. = 15 mi.

Size: 38 x 30.8 cm.

Colored.

(642: 1850: C832nhv Small)

659. 1878

Atlas accompanying the report on the geology of New Hampshire. C. H. Hitchcock, state geologist. Geological Survey of New Hampshire. 1878. New York: J. Bien: 1878. 15 sheets.

Scale: 1 in. = 2½ mi.

Cartographer: C. H. Hitchcock.

Engraver: Julius Bien.

Size: 54 x 65 cm.

Colored.

Provenance: Presented by the state of New Hampshire, 4 October 1878.

See: New Hampshire, Geological Survey, *Report* . . .

(642: 1878: N466geo Extra-oversize)

660. 1898

New Hampshire. [From: *Atlas of the world,* p. 10]. Rand, McNally & Co.: 1898.

Scale: 1 in. = 15.5 mi.

Size: 32 x 23.5 cm.

Colored.

Provenance: Presented by Mrs. A. I. Hallowell, November 1983.

On same sheet as: Maine, *Atlas of the world.* 1898.

(642: 1898: N428nap Small)

VERMONT Nos. 661-668

661. 1793

A map of the state of Vermont. 1793.

Scale: 2.5 cm. = 12½ mi.

Cartographer: J. Whitelaw.

Engraver: Callendar.

Size: 35.7 x 25.8 cm.

In: Samuel Williams, *The natural and civil history of Vermont,* frontispiece.
Wheat: 194

(974.3: W67)

662. [1795]

State of Vermont. [1795].
Scale: 2.5 cm. = 2.2 mi.
Size: 19.1 x 15.9 cm.
Provenance: Presented by the Estate of Richard Gimbel, 1974.
In: Joseph Scott, *The United States gazetteer,* opposite "Vermont."
Wheat: 196

(Paine 68: Sco3u)

663. 1796

A correct map of the state of Vermont from actual survey; exhibiting the country and town lines, rivers, lakes, ponds, mountains, meetinghouses, mills, public roads &c. By James Whitelaw, 1796. New Haven: A. Doolittle: 1796.
Scale: 1:253,440.
Cartographer: James Whitelaw.
Engraver: Amos Doolittle.
Size: 114.5 x 76.8 cm.
Colored.
Wheat: 200

(643: 1796: W589msv Large)

664. [1796]

Vermont, from actual survey. [1796].
Scale: 2.5 cm. = 24 American mi.
Engraver: Seymour.
Size: 19.3 x 14.6 cm.
Provenance: Presented by Suzanne Wister Eastwick and Joseph L. Eastwick, July 1975.
In: Mathew Carey, *Carey's American pocket atlas,* opposite p. 12.
Wheat: 197

(917.3: C25c)

665. [1799]

Vermont from the latest authorities. [1799].
Scale: 2.5 cm. = 18 mi.
Cartographer: Alexander Anderson.
Engraver: Cornelius Tiebout.
Size: 23 x 18.7 cm.
In: John Payne, *A new and complete system of universal geography,* vol. 4, opposite p. 229.
Wheat: 201

(910: P29)

666. [1830]

Geographical, statistical, and historical map of Vermont. N.p.: [1830].

Scale: 1 in. = 15 mi.

Cartographer: F. Lucas, Jr.

Engraver: Young and Delleker.

Size: 31 x 24 cm. *Size of paper:* 44.5 x 54.9 cm.

Colored.

On three sides are data concerning Vermont. Franklin is a place name.

(643: [1830]: L965ver Small)

667. 1898

Vermont. [From: *Atlas of the world,* p. 11]. Rand, McNally & Co.: 1898.

Scale: 1 in. = 14 mi.

Size: 31 x 24 cm.

Colored.

Provenance: Presented by Mrs. A. I. Hallowell, November 1983.

On same sheet is: Massachusetts, *Atlas of the world,* 1898.

(642: 1898: V598mas Small)

668. 1910-1911.

State of Vermont. U.S. Geological Survey. George Otis Smith, director. Compiled in 1910-1911. [Washington]: 1910-1911.

Scale: 1:500,000.

Cartographers: R. B. Marshall, A. F. Hassan, and George Otis Smith.

Size: 53.5 x 32.2 cm.

International map of the world. U.S. section.

(643: 1910-1911: Un38int Small)

MASSACHUSETTS Nos. 669-696

669. [1775]

A new and accurate map of the present seat of war in North America, from a late survey. [London: 1775].

Scale: 69½ English statute mi. = 1 degree.

Cartographer: Thomas Kitchin (?).

Size: 28.8 x 37.8 cm.

Colored.

"For the *London Magazine.*" London: 1775.

(640.1: [1775]: K644wna Small)

670. [1782]

Map of the island of Nantucket. [London: Davies: 1782].

Cartographer: James Tupper.
Size: 19.5 x 24.7 cm.
Original in M. G. J. de Crèvecoeur, *Letters from an American farmer.* London: Davies: 1782.
(644: [1782]: T838nm Small)

671. [1793]
A map of the county of Worcester. [1793].
Scale: 2.5 cm. = 4 mi.
Cartographers: Charles Baker and John Peirce.
Engraver: Seymour.
Size: 36 x 31.3 cm.
Provenance: Presented by John Vaughan, 20 December 1839.
In: Peter Whitney, *The history of the county of Worcester,* frontispiece.
Wheat: 211
(974.4: W61)

672. [1795]
Massachusetts. [1795].
Scale: 2.5 cm. = 30 mi.
Size: 15.3 x 18.3 cm.
Provenance: Presented by the Estate of Richard Gimbel, 1974.
In: Joseph Scott, *The United States gazetteer,* opposite "Massachusetts."
Wheat: 216
(Paine 68: Sco3u)

673. [1796]
Massachusetts. [1796].
Scale: 2.5 cm. = 27 mi.
Engraver: W. Barker.
Size: 14.9 x 19.2 cm.
Provenance: Presented by Suzanne Wister Eastwick and Joseph L. Eastwick, July 1975.
In: Mathew Carey, *Carey's American pocket atlas,* opposite p. 26.
Wheat: 217
(917.3: C25c)

674. 1799
The state of Massachusetts from the best authorities. 1799.
Scale: 2.5 cm. = 20 American mi.
Cartographer: A. Anderson.
Engraver: C. Tiebout.
Size: 19 x 23.5 cm.
In: John Payne, *A new and complete system of universal geography,* vol. 4, opposite p. 235.

Wheat: 222

(910:P29)

675. 1806

Chart of the harbors of Salem, Marblehead, Beverly and Manchester. From a survey taken in the years 1804, 5 & 6 by Nath[anie]l Bowditch . . . assisted by Geo. Burchmore & Wm. Ropes 3rd. N.p.: 1806 [27 June 1806].

Scale: 2 in. = 1¼ statute mi.

Cartographers: Nathaniel Bowditch, George Burchmore, and William Ropes III.

Engravers: Hooker & Fairman.

Size: 54.1 x 65.2 cm.

Provenance: Presented and endorsed by Nathaniel Bowditch, 20 October 1809.

(644: 1806: B672smb Large)

676. 1813

Map of the American coast, from Lynhaven Bay to Narragansett Bay, by John Melish. Philadelphia: Melish: 1813.

Scale: 1 in. = 20 mi.

Cartographer: John Melish.

Engraver: H. S. Tanner.

Size: 40.3 x 46.3 cm.

Colored.

Originally in: John Melish, *Military and topographical atlas.* Philadelphia: Palmer: 1813.

(635: 1813: M485ac Small)

677. 1825

Survey across the Isthmus of Cape Cod state of Massachusetts and town of Sandwich of a proposed canal between Buzzard's and Barnstable bays. 1825. Surveyed by the U.S. Army Corps of Topographical Engineers under the direction of Major Per[r]ault. [Washington]: 1825. 3 sheets.

Scale: 1 in. = 880 yd.

Cartographers: Major Perrault, Frederick Searle, W. B. Thompson, and John Farley.

Provenance: Presented by General Daniel Parker, 1 January 1833.

No. 1. Survey across isthmus. 26.4 x 47.4 cm. No. 2. Profile of proposed route. 27.5 x 46.7 cm. No. 3. Survey of a valley & ponds. 20.8 x 35.8 cm.

(644: 1825: Un31bb Small)

678. 1828

Plan of a survey for the proposed Boston and Providence rail-way. By James Hayward. Jan. 1828. [Boston]: 1828.

Scale: 1 in. = 1 mi.

Cartographer: James Hayward.

Engravers: Annin and Smith.

Size: 18.8 x 108 cm.

Map made to accompany report of the commissioners.

(640.1: 1828: H333bpr Small)

679. 1836

A map of the extremity of Cape Cod including the townships of Provincetown & Truro: with a chart of their sea coast and of Cape Cod harbor . . . Executed under the direction of J. D. Graham . . . during . . . years 1833, '34, & '35. U.S. Army. Corps of Topographical Engineers. Reduced from original . . . by Wash: Hood: 1836. [Washington]: 1836.

Scale: 1:10,560.

Cartographers: J. E. Johnston, A. A. Humphreys, J. N. Macomb, W. R. Palmer, and J. D. Graham.

Engraver: W. J. Stone.

Size: 148 x 178.3 cm.

Provenance: Presented by James Duncan Graham, 19 July 1844; one copy was presented by Graham, 20 December 1839.

See: U.S. Congress. 25th Congress. 2nd Session. *Report*. . . .

(644: 1836: Un38ccp Large)

680. 1836

A map of the town of Taunton, with a view of the public buildings. Published by Saml. O. Dunbar. Taunton: S. O. Dunbar: 1836.

Scale: 1 in. = 170 rods.

Cartographer: Samuel O. Dunbar.

Lithographer: Pendleton.

Size: 77.8 x 56.7 cm.

Inset: Map of the village of Taunton.

Provenance: Presented by the Reverend E. Neville, 5 April 1839.

There are engravings of 13 public buildings on the map.

(644.981: 1836: D912tpb Large)

681. 1838

Map of the Island of Nantucket, including Tuckernuck. Surveyed by William Mitchell. 1838. Boston: E. W. Bouvé: 1838.

Scale: 1:47,520.

Cartographer: William Mitchell and E. W. Bouvé.

Lithographer: E. W. Bouvé.

Size: 60.2 x 85.7 cm.

Provenance: Presented by Nathan Dunn, 20 September 1839.

(644: 1838: M695ntm Large)

682. 1841

Chart of Cape Cod harbor and the adjacent coast of Provincetown and Truro. Reduced from the original of Major James D. Graham . . . U.S. Army. Corps of Topographical

Engineers. Published under the patronage of the Boston Marine Insurance Company by I. W. P. Lewis. [Washington]: 1841.

Scale: 1:21,120.

Cartographers: James D. Graham and I. W. P. Lewis.

Engraver: W. J. Stone.

Size: 71.2 x 85.1 cm.

Provenance: Presented and endorsed by J. D. Graham, 1 April 1842.

(644: 1841: Un38cch Large)

683. [ca. 1842]

Cape Cod harbour and adjacent coast, reduced from the survey of Major J. D. Graham, U.S. Topographical Engineers. U.S. Army. 1833, 34 & 35. [Washington: ca. 1842].

Scale: 1 in. = 1 mi.

Cartographer: J. D. Graham.

Engravers: Hooker and Harrison.

Size: 22.3 x 30.9 cm.

Provenance: Presented by J. D. Graham, 19 July 1844.

(644: [ca. 1842]: Un31cch Small)

684. 1846

Harbor of New Bedford founded upon a trigonometrical survey under the direction of A. D. Bache. U.S. Coast and Geodetic Survey. [Washington]: 1846.

Scale: 1:40,000.

Cartographers: A. D. Bache, C. M. Eakin, H. L. Whiting, and G. S. Blake.

Engravers: Sherman & Smith.

Size: 44.7 x 35.8 cm.

Provenance: Presented by the U. S. Treasury Department through A. D. Bache, 21 August 1846.

(644: 1846: Un38hnb Small)

685. 1847

The harbor of Holmes' Hole; [and] the harbor of Tarpaulin Cove founded upon a trigonometrical survey under the direction of A. D. Bache. U.S. Coast and Geodetic Survey. [Washington]: 1847. 2 maps on one sheet.

Scale: 1:20,000.

Cartographers: A. D. Bache, C. M. Eakin, W. M. Boyce, G. S. Blake, and M. I. McClery.

Engravers: Sherman and Smith.

Size: 43.9 x 35.9 cm.

Both maps marked "Harbor of Refuge No. [blank]."

(644: 1847: Un33hhtc Small)

686. 1848

Edgartown harbor founded upon a trigonometrical survey under the direction of A. D. Bache. U.S. Coast and Geodetic Survey. [Washington]: 1848.

Scale: 1:20,000.

Cartographers: A. D. Bache, C. M. Eakin, H. L. Whiting, C. H. Davis, M. I. McClery, and John Robertson.

Size: 43.7 x 35.9 cm.

Provenance: Presented by J. R. Ingersoll through A. D. Bache, 21 July 1848.

(644: 1848: Un33eh Small)

687. 1848

Nantucket harbor from a trigonometrical survey under the direction of A. D. Bache. U.S. Coast and Geodetic Survey. [Washington]: 1848.

Scale: 1:20,000.

Cartographers: A. D. Bache, C. M. Eakin, H. L. Whiting, C. H. Davis, J. M. Wampler, and J. Robertson.

Engraver: F. Dankworth, O. A. Lawson, J. H. Young, J. Knight, and W. Smith.

Size: 35.8 x 44 cm.

Provenance: Presented by the U.S. Treasury Department, 5 April 1850.

(644: 1848: Un33nh Small)

688. 1850

The harbor of Hyannis from a trigonometrical survey under the direction of A. D. Bache. U.S. Coast and Geodetic Survey. [Washington]: 1850.

Scale: 1:30,000.

Cartographers: A. D. Bache, C. M. Eakin, W. M. Boyce, J. N. Maffitt, and Wm. Luce.

Engravers: Sherman and Smith.

Size: 46.9 x 35.8 cm.

Provenance: Presented by A. D. Bache, 27 December 1850.

(644: 1850: Un33hh Small)

689. 1851

Plano del puerto de Holmes' Hole levantado por el teniente de la Armada de los Estados-Unidos G. S. Blake en 1847. Madrid: 1851.

Scale: 3¼ in. = 1 sea mi.

Cartographers: G. S. Blake and F. Bregante.

Size: 20.9 x 32.6 cm.

Provenance: Presented by the Madrid Dirección de Hidrografia through Don Severiano Moreledo, 7 January 1853.

(644: 1851: B58lhh Small)

690. 1853

Wellfleet harbor Massachusetts from a trigonometrical survey under the direction of A. D. Bache. U.S. Coast and Geodetic Survey. [Washington]: 1853.

Scale: 1:50,000.

Cartographers: A. D. Bache, T. J. Cram, H. L. Whiting, J. B. Glück, C. H. McBlair, W. M. C. Fairfax, J. J. Ricketts, and M. C. Gritzner.

Engravers: S. Siebert, E. F. Woodward, and S. E. Stull.

Size: 35.6 x 44.3 cm.

Provenance: Presented "pursuant of an Act of Congress, and by the direction of the Treasury Department" through A. D. Bache, 10 October 1854.

(644: 1853: Un33wh Small)

691. 1855

Gloucester harbor Massachusetts from a trigonometrical survey under the direction of A. D. Bache. U.S. Coast and Geodetic Survey. [Washington]: 1855.

Scale: 1:20,000.

Cartographers: A. D. Bache, C. O. Boutelle, H. L. Whiting, H. S. Stellwagen, L. Daser, and W. T. Martin.

Engravers: G. B. Metzeroth and J. L. Hazzard.

Size: 44.7 x 35.1 cm.

Provenance: Presented by A. D. Bache, 6 March 1854.

(644: 1855: Un33gh Small)

692. 1855

Newburyport harbor, Massachusetts. From a trigonometrical survey under the direction of A. D. Bache. U.S. Coast and Geodetic Survey. [Washington]: 1855.

Scale: 1:20,000.

Cartographers: C. O. Boutelle, A. D. Bache, A. W. Longfellow, M. Woodhull, and C. Mahon.

Engravers: S. Siebert, A. Rollé, W. Smith, E. Yeager, J. Young, and H. C. Evens.

Size: 42.5 x 62.4 cm.

Provenance: Presented by the U.S. Treasury Department through A. D. Bache, 6 March 1857.

(644: 1855: Un38nph Large)

693. 1855

Salem harbor, Massachusetts. From a trigonometrical survey under the direction of A. D. Bache. U.S. Coast and Geodetic Survey. [Washington]: 1855.

Scale: 1:25,000.

Cartographers: A. D. Bache, C. O. Boutelle, H. L. Whiting, C. H. McBlair, C. Mahon, and C. Farquahar.

Engravers: F. Danksworth, H. M. Knight, J. Knight, S. E. Stull, and J. Young.

Size: 52.7 x 70.3 cm.

Provenance: Presented by the U.S. Treasury Department through A. D. Bache, 6 March 1857.

(644: 1855: Un38shm Large)

694. 1898

Massachusetts. [From: *Atlas of the world,* p. 12]. Rand, McNally & Co.: 1898.

Scale: 1 in. = 15.5 mi.

Size: 23 x 31.6 cm.

Colored.

Provenance: Presented by Mrs. A. I. Hallowell, November 1983.

On same sheet as: Vermont, *Atlas of the world.* 1898.

(644: 1898: M385ver Small)

695. 1916

A map of Cape Cod and its approaches, showing through roads, good roads, local roads, private roads, churches, schools, cemeteries, town halls, light houses, post office, altitudes (173), gradual decline, steep decline . . . Boston: The Bullard Co.: 1916.

Size of paper: 110.8 x 116.2 cm.

Colored.

Insets: Maps of Boston and vicinity; Nantucket; Martha's Vineyard.

Provenance: Presented by Mrs. A. I. Hallowell, November 1983.

On reverse is: The New England commercial and route survey. 1916.

(644: 1916: M386ccb Large)

696. 1935

A prospect of Harvard University and Radcliffe College, Cambridge, Massachusetts. Boston: Merrymount Press: 1935.

Scale: 1 in. = 200 ft.

Cartographer: Edwin J. Schruers.

Size: 82.2 xd 61.4 cm.

Colored.

(644: 1935: H266sch Extra-Oversize)

BOSTON Nos. 697-710

697. [1775]

A new and correct plan of the town of Boston, and provincial camp. [1775].

Scale: [2.5 cm. = 1/4 mi.].

Engraver: Aitken.

Size: 26 x 18.9 cm.

Inset: [Map of the country around Boston].

In: *Pennsylvania Magazine,* vol. 1 (1775), opposite p. 291.

Wheat: 238

(050: P383)

698. [1775]

A new plan of Boston harbour from an actual survey. [1775].

Scale: [2.5 cm. = 1 3/4 mi.].

Engraver: Clownes [Caleb Lownes].

Size: 25.5 x 18.9 cm.

In: *Pennsylvania Magazine,* vol. 1 (1775), opposite p. 241.

Wheat: 239

(050: P383)

699. 1777

Plan of Boston in New England with its environs, including Milton, Dorchester, Roxbury, Brookline, Cambridge, Medford, Charlestown, parts of Malden and Chelsea. With the military works constructed in those places in the years 1775 and 1776. [Signed] : Henry Pelham. London: 1777 [2 June 1777].

Scale: 1:50,000.

Cartographer: Henry Pelham.

Engraver: Francis Jukes.

Size: 99.3 x 69.3 cm.

Provenance: Presented by Edward Hudson, 5 January 1827.

(644.917: 1777: P366bne Large)

700. [1789]

A map of the seat of the late war at Boston in the state of Massachusetts. [1789].

Engraver: [James Trenchard?].

Size: 17.2 x 25.5 cm.

In: *Columbia Magazine,* vol. 3, opposite p. [385].

Wheat: 208

(050: C72)

701. [ca. 1795]

[A plan of the town and chart of the harbor of Boston.] N.p.: [ca. 1795].

Size: 26.1 x 34 cm.

(644.917: [ca. 1795]: B651pth Large)

702. 1818

Sketch of the action on the heights of Charles Town 17 June 1775, between his majesty's troops under the command of M. Genl. Howe and a large body of American rebels. Philadelphia: Harrison Hall: 1818.

Cartographers: Henry de Bernière and H. Dearborn.

Size: 33.3 x 50.2 cm.

On map: "N.B. the parts in red are corrections of the original by Maj. Gen. Dearborn." However, on this printed map there are no red lines.

(644: 1775: D325chm.a Small)

703. 1819

Map of Boston and its vicinity. From actual survey by John G. Hales. Boston: Hales. Philadelphia: Melish: 1819.

Scale: 1:63,360.

Cartographer: John G. Hales.

Engraver: Edwin Gillingham.

Size: 65.2 x 80.7 cm.

Colored.

Provenance: Presented by John Vaughan, 15 October 1819.

(644.917: 1819: H133bav Large)

704. 1831

Plan of Mount Auburn, by Alex[ande]r L. Wadsworth. November 1831. Boston: Pendleton: 1831.

Scale: 1 in. = 7½ rods.

Cartographer: A. L. Wadsworth.

Lithographer: Pendleton.

Size: 55.1 x 44.6 cm.

Colored.

Autograph note at bottom: "It is supposed that 3 to 4000 lots may be laid out without removing the trees & shrubbery so as to mar the beauty of the spot."

See: Joseph Story, *Address delivered on the dedication of the cemetery.* Boston: Buckingham: 1831; and James Mease, *Gingerbread age.* New York: Rinehart: 1957.

(644.917: 1831: W193pma Large)

705. 1878

Boston harbor, Massachusetts. From a trigonometrical survey under the direction of A. D. Bache. U.S. Coast and Geodetic Survey. [Washington: 1878.] Edition of 1878.

Scale: 1:40,000.

Cartographers: A. D. Bache, J. B. Glück, H. L. Whiting, J. S. Williams, S. A. Gilbert, C. H. Davis, and A. Boschke.

Size: 71.8 x 91.1 cm.

Provenance: Presented by the U.S. Coast and Geodetic Survey through Spencer C. McCorkle, 14 December 1885.

(644.917: 1878: Un38bhm Large)

706. 1898

Boston, [Massachusetts]. [From: *Atlas of the world*, p. 13]. Rand, McNally & Co.: 1898.

Scale: 3 in. = 1 mile.

Size: 31.8 x 23.5 cm.

Colored.

Inset: Outline map of Boston.

Provenance: Presented by Mrs. A. I. Hallowell, November 1983.

On same sheet as: Rhode Island, *Atlas of the world.* 1898.

(644: 1898: M385bos Small)

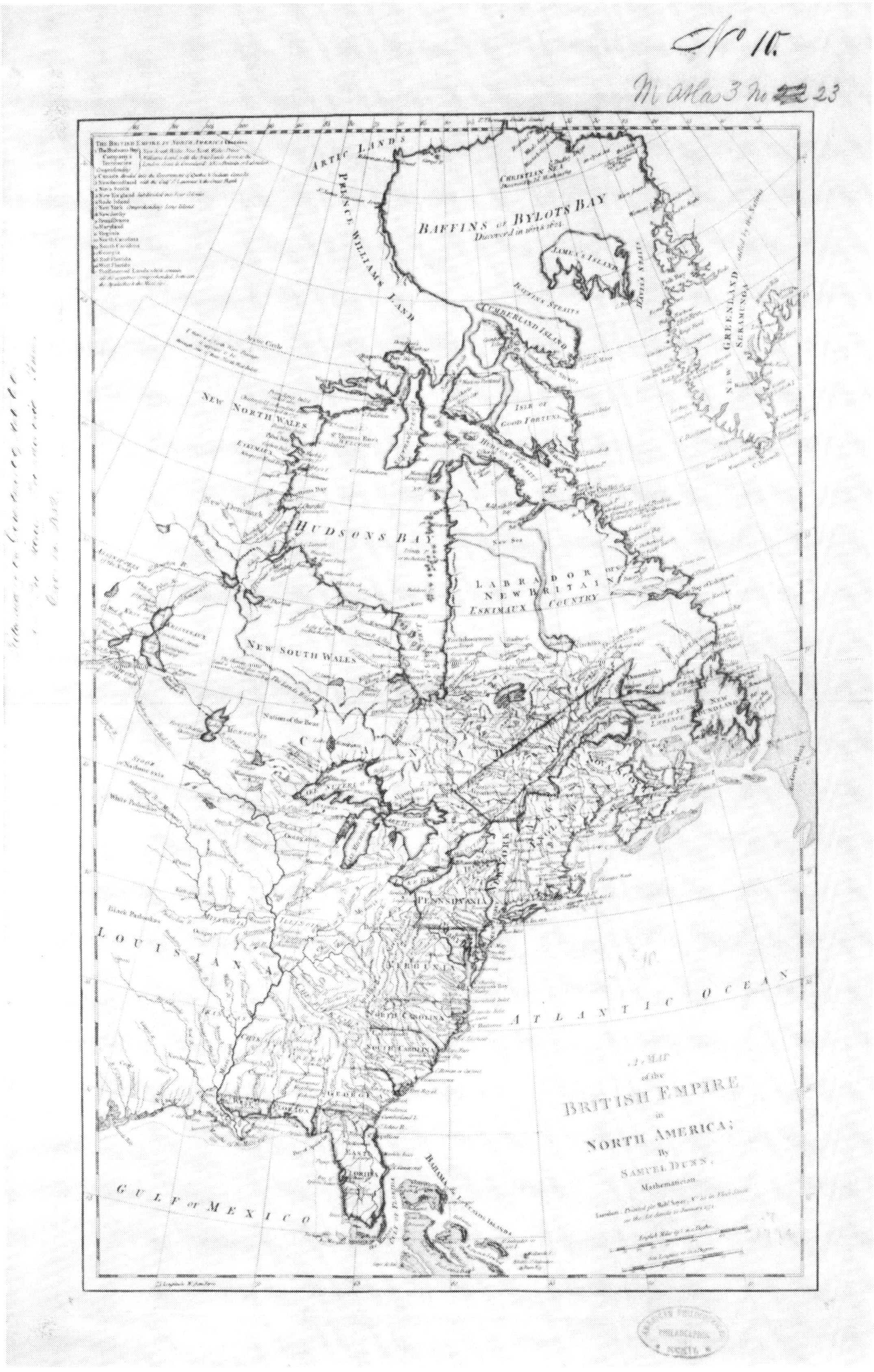

Samuel Dunn. A map of the British Empire in North America. 1774. [No. 453]

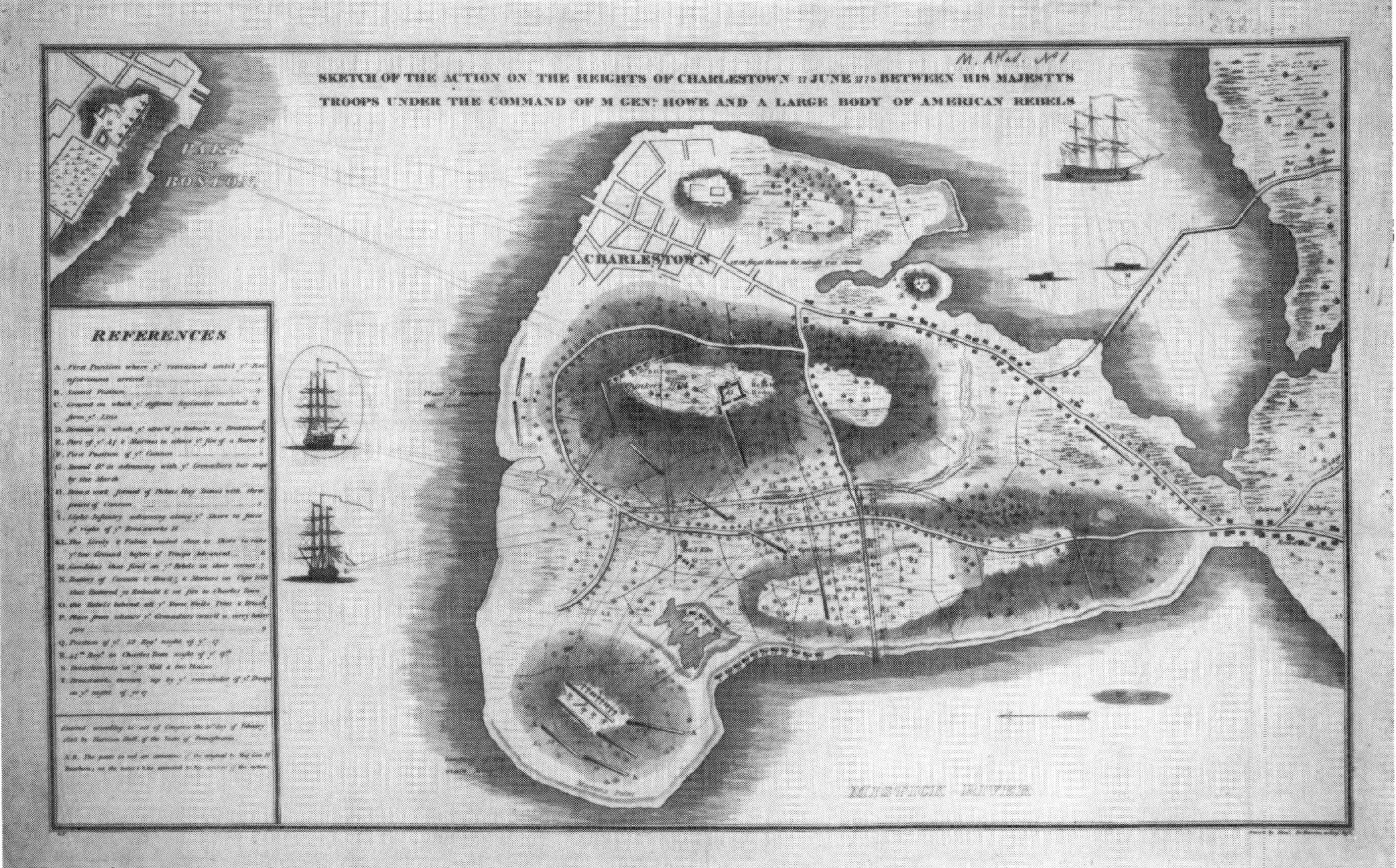

Henry de Bernière and H. Dearborn. Sketch of the action on the heights of Charles Town 17 June 1775, between his majesty's troops under the command of M. Genl. Howe and a large body of American rebels. 1818. [No. 702]

707. 1903

Boston, Massachusetts, and vicinity. U.S. Geological Survey, Charles D. Walcott, director. H. M. Wilson, geographer in charge. Surveyed in 1898-1900. [Washington]: 1903. Edition of October 1903.

Scale: 1:62,500.

Cartographers: H. M. Wilson, Frank Sutton, Charles D. Walcott, et al.

Size: 44.5 x 66.2 cm.

Colored.

(644: 1903: Un38top Large)

708. 1912

Boston. Compiled, engraved and published by the U.S. Geological Survey. George Otis Smith, director. North K 19. Washington: 1912.

Scale: 1:1,000,000.

Cartographer: George Otis Smith.

Size: 46.4 x 52.7 cm.

Colored.

(644.917: 1912: Un38usg Large)

709. [n.d.]

Old Boston compiled from the Book of Possessions, by George Lamb. Published by the Trustees of the Boston Public Library. [Boston]: Heliotype Printing Co.: n.d. 10 sheets of maps and title page.

Cartographer: George Lamb.

Size of sheets: 89 x 66 cm.

(644.917: [n.d.]: B615bbp Large)

710. [n.d.]

Plan of the town of Boston.

Size of paper: 21.9 x 17.2 cm.

(644.917: [n.d.]: B651pb Small)

RHODE ISLAND Nos. 711-718

711. [1794]

[Western Narragansett shore. 1794].

Scale: 2.5 cm. = 2½ mi.

Engraver: Doolittle.

Size: 15 x 9 cm.

Provenance: Presented by B. McMahon, 19 January 1816.

In: Ezra Stiles, *A history of three of the judges,* opposite p. 345.

Wheat: 249

(920.042: St55h)

712. 1795

A map of the state of Rhode Island; taken mostly from surveys by Caleb Harris. Drawn by Harding Harris. Engraved for Carter & Wilkinson, Providence, 1795. Providence: Mowbray Co.: 1969.

Scale: 69½ mi. = 1 degree.

Cartographers: Harding Harris and Caleb Harris.

Engraver: Samuel Hill.

Size: 54.7 x 41.3 cm.

Provenance: Presented by Rhode Island Historical Society, 28 October 1969.

Copyright Rhode Island Historical Society, 1969. Reproduction.

Wheat: 251

(645: 1795: H234rich Large)

713. [1795]

Rhode Island. [1795].

Scale: 2.5 cm. = 7½ mi.

Size: 15.6 x 18.4 cm.

Provenance: Presented by the Estate of Richard Gimbel, 1974.

In: Joseph Scott, *The United States gazetteer,* opposite "Rhode Island."

Wheat: 252

(Paine 68: Sco3u)

714. [1796]

Rhode Island. [1796].

Scale: 2.5 cm. = 8½ mi.

Engraver: W. Barker.

Size: 19 x 14.6 cm.

Provenance: Presented by Suzanne Wister Eastwick and Joseph L. Eastwick, July 1975.

In: Mathew Carey, *Carey's American pocket atlas,* opposite p. 34.

Wheat: 253

(917.3: C25c)

715. [1799]

Rhode Island. [1799].

Engraver: W. Barker.

Size: 24 x 18.7 cm.

In: John Payne, *A new and complete system of universal geography,* vol. 4, opposite p. 258.

Wheat: 255

(910: P29)

716. 1832

Chart of Narragansett Bay. Surveyed in 1832 by Capt. Alex. S. Wadsworth, Lieuts. Thos. R. Gedney, Charles Wilkes, Jr. and George S. Blake. U.S. Navy. [Washington: 1832].

Scale: 1 in. = 2000 ft.

Cartographers: Alexander S. Wadsworth, Thomas R. Gedney, Charles Wilkes, Jr., and George S. Blake.

Engraver: W. J. Stone.

Size: 138.6 x 110.3 cm.

Provenance: Presented by Colonel Watmough, prior to 1838 [*Transactions*, n.s., vol. 5].

(645: 1832: Un36cnb Large)

717. 1885

Narragansett Bay. From a trigonometrical survey under the direction of Benjamin Peirce. U.S. Coast and Geodetic Survey. [Washington]: 1885.

Scale: 1:40,000.

Cartographers: Benjamin Peirce, E. Blunt, S. C. McCorkle, A. M. Harrison, C. Hosmer, H. G. Ogden, G. S. Blake, J. R. Goldsborough, W. P. Trowbridge, H. Mitchell, and F. P. Webber.

Size: 115.3 x 77.1 cm.

Provenance: Presented by the U.S. Coast and Geodetic Survey through Spencer C. McCorkle, 14 December 1885.

(645: 1885: Un36nbr Large)

718. 1898

Rhode Island. [From: *Atlas of the world*, p. 14]. Rand McNally & Co.: 1898.

Scale: 1 in. = 5 mi.

Size: 31.7 x 23 cm.

Colored.

Inset: City of Providence.

Provenance: Presented by Mrs. A. I. Hallowell, November 1983.

On same sheet as: Boston [Massachusetts], *Atlas of the world.* 1898.

(645: 1898: R344prv Small)

CONNECTICUT Nos. 719-749

719. [1774]

[Map showing continuation west of the latitude of the Connecticut coast line. 1774].

Scale: [2.5 cm. = ca. 45 mi.].

Size of plate: 15 x 25.5 cm.

Provenance: Presented by William Barton.

In: William Smith, *An examination of the Connecticut claim to lands in Pennsylvania*, opposite p. 31.

Wheat: 260

(974.8: Sm6)

720. 1780

Connecticut and parts adjacent. [By Bernard Romans, 1777]. Amsterdam: Cóvens and Mortier and Cóvens, junior: 1780. Reprinted by U.S. Geological Survey, 1937.

Scale: 1 in. = 6 mi.

Cartographer: Bernard Romans.

Engraver: K. Kleckhoff, 1780.

Size: 46.1 x 51.7 cm.

Colored.

This is one of a series of maps, depicting the 13 original states at the time of the ratification of the Constitution. Issued by the U.S. Constitutional Sesquicentennial Commission, 1937. It bears the title: Connecticut at the time of the ratification of the Constitution.

(646: 1780: R667cpa Large)

721. [1789]

From New York (4) to Stratford. [1789, reprinted 1961].

Size: 17.3 x 11.6 cm.

In: Christopher Colles, *A survey of the roads.*

Wheat: 265

(526.8: C68s)

722. [1789]

From New York (5) to Stratford. [1789, reprinted 1961].

Size: 17.3 x 11.5 cm.

In: Christopher Colles, *A survey of the roads.*

Wheat: 266

(526.8: C68s)

723. [1789]

From New York (6) to Stratford. [1789, reprinted 1961].

Size: 17.1 x 11.5 cm.

In: Christopher Colles, *A survey of the roads.*

Wheat: 267

(526.8: C68s)

724. [1789]

From New York (7) to Stratford. [1789, reprinted 1961].

Size: 17.1 x 11.5 cm.

In: Christopher Colles, *A survey of the roads.*

Wheat: 268

(526.8: C68s)

725. [1789]

From Stratford (15) to Poughkeepsie. [1789, reprinted 1961].

Size: 16.9 x 11.4 cm.

In: Christopher Colles, *A survey of the roads.*

Wheat: 269

(526.8: C68s)

726. [1789]

From Stratford (16) to Poughkeepsie. [1789, reprinted 1961].

Size: 17.1 x 11.4 cm.

In: Christopher Colles, *A survey of the roads.*

Wheat: 270

(526.8: C68s)

727. [1789]

From Stratford (17) to Poughkeepsie. [1789, reprinted 1961].

Size: 17.2 x 11.5 cm.

In: Christopher Colles, *A survey of the roads.*

Wheat: 271

(526.8: C68s)

728. [1791]

Bellows or Great Falls on the Connecticut River. [1791].

Size: 6.6 x 5.3 cm.

In: "A description of the surprising cataract, in the great River Connecticut," *American Museum,* vol. 9 (1791), pp. 253-255. Map is on p. 254. Woodcut.

Wheat: 182

(050: Am33)

729. [1794]

City of New Haven. Plate V. [1794].

Scale: 2.5 cm. = 55 rods.

Engraver: Doolittle.

Size: 15.1 x 9.1 cm.

Provenance: Presented by B. McMahon, 19 January 1816.

In: Ezra Stiles, *A history of three of the judges,* opposite p. 126.

Wheat: 275

(920.042: St55h [copy 2])

730. [1794]

Graves of the judges in New-Haven. Plate IV, no. 3 [1794].

Scale: 2.5 cm. = 8 ft.

Engraver: Doolittle (?).
Size: 7.4 x 14.1 cm.
Provenance: Presented by B. McMahon, 19 January 1816.
In: Ezra Stiles, *A history of three of the judges*, opposite p. 114.
Wheat: 276

(920.042: St55h [copy 2])

731. [1794]
Guilford. Plate III, no. 2 [1794. Contains also a long map of the Connecticut River at New Haven, Conn.].
Scale: 2.5 cm. = ca. 23 rods.
Engraver: Doolittle (?).
Size: 8.9 x 6 cm. and 3 x 15 cm.
Provenance: Presented by B. McMahon, 19 January 1816.
In: Ezra Stiles, *A history of three of the judges*, opposite p. 80.
Wheat: 277

(920.042: St55h [copy 2])

732. [1794]
Hadley. Plate IV, no. 2 [1794].
Scale: 2.5 cm. = 78 rods.
Engraver: Doolittle (?).
Size: 16.9 x 7.4 cm.
Inset: Hadley.
Provenance: Presented by B. McMahon, 19 January 1816.
In: Ezra Stiles, *A history of three of the judges*, opposite p. 114.
Wheat: 278

(920.042: St55h [copy 2])

733. [1794]
Hadley. Plate VIII. [1794].
Scale: 2.5 cm. = ca. 140 rods.
Engraver: Doolittle.
Size: 15 x 9 cm.
Inset: Mr. Russels house. The judges chamber 1665.
Provenance: Presented by B. McMahon, 19 January 1816.
In: Ezra Stiles, *A history of three of the judges*, opposite p. 202.
Wheat: 279

(920.042: St55h [copy 2])

734. [1794]
Judges cave. Plate II. [The second harbor of the judges, 1794].
Scale: 2.5 cm. = 8 rods.

Engraver: Doolittle.
Size of page: 16.5 x 9.8 cm.
Provenance: Presented by B. McMahon, 19 January 1816.
In: Ezra Stiles, *A history of three of the judges*, opposite p. 77.
Wheat: 283

(920.042: St55h [copy 2])

735. [1794]

Lodgments of the judges in & about New Haven from 1661 to 1664. Plate IV, no. 1 [1794].
Scale: 2.5 cm. = 5½ mi.
Engraver: Doolittle (?).
Size: 10.1 x 14.3 cm.
Provenance: Presented by B. McMahon, 19 January 1816.
In: Ezra Stiles, *A history of three of the judges*, opposite p. 114.
Wheat: 280

(920.042: St55h [copy 2])

736. [1794]

Map of New-Haven and its environs. Plate I. [1794].
Scale: 2.5 cm. = ⅘ mi.
Engraver: Doolittle (?).
Size: 17.5 x 23.2 cm.
Provenance: Presented by B. McMahon, 19 January 1816.
In: Ezra Stiles, *A history of three of the judges*, opposite p. 29.
Wheat: 281

(920.042: St55h [copy 2])

737. [1794]

Milford. Plate III, no. 2 [1794].
Scale: 2.5 cm. = ca. 23 rods.
Engraver: Doolittle (?).
Size: 6.6 x 6 cm.
Provenance: Presented by B. McMahon, 19 January 1816.
In: Ezra Stiles, *A history of three of the judges*, opposite p. 80.
Wheat: 282

(920.042: St55h [copy 2])

738. [1796]

Connecticut. [1796].
Scale: 2.5 cm. = 15 mi.
Engraver: W. Barker.
Size: 18.8 x 14.2 cm.

Provenance: Presented by Suzanne Wister Eastwick and Joseph L. Eastwick, July 1975.

In: Mathew Carey, *Carey's American pocket atlas,* opposite p. 40.

Wheat: 286

(917.3: C25c)

739. 1799

A new map of Connecticut from the best authorities. 1799.

Scale: 2.5 cm. = 12 mi.

Cartographer: A. Anderson.

Size: 18.8 x 23.6 cm.

In: John Payne, *A new and complete system of universal geography,* vol. 4, opposite p. 270.

Wheat: 292

(910: P29)

740. 1838

Map of the mouth of the Connecticut River and Saybrook harbor. Reduced from the original survey of J. W. Adams. U.S. Bureau of Topographical Engineers. By Wash. Hood. [Washington]: 1838.

Scale: 1 in. = 700 ft.

Cartographers: J. W. Adams, W. H. Swift, Wash. Hood, and M. H. Stansbury.

Engraver: W. J. Stone.

Size: 88.8 x 73.7 cm.

Provenance: Presented and endorsed by Hartman Bache, 26 June 1838; delivered 17 August 1838.

(646: 1838: Un38mcr Large)

741. 1846

New Haven harbor. Founded on a trigonometrical survey under the direction of F. R. Hassler. U.S. Coast and Geodetic Survey. [Washington]: 1846.

Scale: 1:30,000.

Cartographers: F. R. Hassler, James Ferguson, Edmund Blunt, C. M. Eakin, W. M. Boyce, J. Farley, G. S. Blake, and John B. Gluck.

Engravers: Sherman and Smith.

Size: 43.5 x 35.5 cm.

Provenance: Presented through A. D. Bache, 26 January 1847.

(646: 1846: Un38nhh Small)

742. 1848

The harbor of New London. Founded upon a trigonometrical survey under the direction of F. R. Hassler. U.S. Coast and Geodetic Survey. [Washington]: 1848.

Scale: 1:20,000.

Cartographers: F. R. Hassler, E. Blunt, F. H. Gerdes, J.B. Glück, and G. S. Blake.

Engravers: A. Rollé and J. Knight.

Size: 43.6 x 35.6 cm.

Provenance: Presented by the U.S. Coast and Geodetic Survey through Spencer C. McCorkle, 14 December 1885.

(646: 1848: Un38hnL Small)

743. 1848

Harbors of Black Rock and Bridgeport. Founded upon a trigonometrical survey under the direction of F. R. Hassler. U.S. Coast and Geodetic Survey. [Washington]: 1848.

Scale: 1:20,000.

Cartographers: F. R. Hassler, J. Ferguson, C. M. Eakin, G. S. Blake, M. I. McClery, and J. M. Wampler.

Engravers: S. T. Pettit, F. Danksworth, O. A. Lawson, A. Rollé, and J. Knight.

Size: 43.8 x 35.8 cm.

Provenance: Presented by Joseph R. Ingersoll through A. D. Bache, 21 July 1848.

(646: 1848: Un38brb Small)

744. 1848

Harbors of Sheffield Island and Cawkin's Island. From a trigonometrical survey under the direction of F. R. Hassler. U.S. Coast and Geodetic Survey. [Washington]: H. Benner: 1848.

Scale: 1:20,000.

Cartographers: F. R. Hassler, E. Blunt, C. M. Eakin, G. S. Blake, and M. I. McClery.

Engravers: S. T. Pettit and H. Benner.

Size: 35.8 x 44 cm.

Provenance: Presented by J. R. Ingersoll through A. D. Bache, 16 March 1849.

(646: 1848: Un38sch Small)

745. 1849

Harbors of Captain's Island east and Captain's Island west. (Harbors of refuge no. [blank]). From a trigonometrical survey under the direction of F. R. Hassler. U.S. Coast and Geodetic Survey. [Washington]: 1849.

Scale: 1:20,000.

Cartographers: F. R. Hassler, J. Ferguson, C. M. Eakin, G. S. Blake, W. Luce, and W. C. Barney.

Engravers: A. Rollé, W. Smith, and S. T. Pettit.

Size: 35.7 x 43.6 cm.

Provenance: Presented by A. D. Bache "by direction of the Treasury Department," 5 April 1850.

(646: 1849: Un38cih Small)

746. 1853

Mouth of Connecticut River. From a trigonometrical survey under the direction of F. R. Hassler and A. D. Bache. U.S. Coast and Geodetic Survey. [Washington]: 1853.

Scale: 1:20,000.

Cartographers: F. R. Hassler, A. D. Bache, W. H. Swift, E. Blunt, H. L. Whiting, J. R. Goldsborough, M. Woodhull, and A. Boschke.

Engravers: S. Siebert, O. A. Lawson, and W. Smith.

Size: 44.2 x 35.5 cm.

Provenance: Presented "pursuant of an Act of Congress, and by the direction of the Treasury Department" through A. D. Bache, 10 October 1854.

(646: 1853: Un38cr Small)

747. 1884

Thames River from New London to Norwich, Connecticut. U.S. Coast Survey. [Washington]: 1884.

Scale: 1:20,000.

Cartographers: J. E. Hilgard, C. O. Boutelle, C. Hosmer, H. G. Ogden, S. C. McCorkle, W. H. Dennis, R. Clover, and C. Junken.

Size: 105.5 x 47.8 cm.

(646: 1884: Un38trn Large)

748. 1898

Connecticut. [From: *Atlas of the world.*] Rand, McNally & Co.: 1898.

Scale: 1 in. = 9 mi.

Size: 23.5 x 31.8 cm.

Colored.

Provenance: Presented by Mrs. A. I. Hallowell, November 1983.

On same sheet as: New York, *Atlas of the world.* 1898.

(646: 1898: C762nap Small)

749. 1906

Preliminary geological map of Connecticut by H. E. Gregory and H. H. Robinson. Connecticut Geological and Natural History Survey. New York: Julius Bien: 1906.

Scale: 1:250,000.

Cartographers: H. E. Gregory and H. H. Robinson.

Size: 60.6 x 69.1 cm.

Colored.

See: Connecticut. Geological and natural history survey, *Bulletin,* no. 7.

(646: 1906: C742gnh Large)

MIDDLE ATLANTIC STATES Nos. 750-782

750. [1747]

[Coast of North America from Cape Hatteras to Boston harbor. 1747].

Cartographer: Lewis Evans (?).

Engraver: James Turner.

Size: 39.6 x 30.4 cm.

In: Board of general proprietors of the eastern division of New Jersey. *A bill in the chancery of New-Jersey,* following p. 124.

Wheat: 294

(347.8: N46b)

751. 1749

A map of Pensilvania, New-Jersey, New-York, and the three Delaware counties: by Lewis Evans. MDCCXLIX. Published by Lewis Evans March 25, 1749 according to act of Parliament. [Philadelphia]: 1749.

Scale: 69 English mi. = 1 degree.

Cartographer: Lewis Evans.

Engraver: L. Hebert.

Size: 64.4 x 49.1 cm.

Colored.

Provenance: Joseph Parker Norris gave a copy of this map to the American Philosophical Society on 16 June 1815.

Norris's copy has now disappeared. It was being used as late as 5 October 1832, for James Mease commented that there was a marginal note by Evans that "all great storms begin at the leeward." Somehow, this note strengthened Mease's theory that the present earth "was made at the creation from the ruins of another" (Minutes).

Wheat: 672

Contains comments about the origins of the northeast storms (suggested by B. Franklin?), lightning and electricity, navigation of the Delaware River, and other bays, etc.

Endorsement on back, in Benjamin Franklin's hand: "To Dr. John Mitchell from Mr. B. Franklin." The map was at one time in the collection of the Marquess of Bute, a friend of Mitchell's.

(640.2: 1749: Ev12tdc Large)

752. 1755

A general map of the middle British colonies, in America, viz Virginia, Màriland, Dèlaware, Pensilvania, New-Jersey, New-York, Connecticut, and Rhode Island: of Aquanishuonîgy, the country of the Confederate Indians; comprehending Aquanishuonîgy proper, their place of residence, Ohio and Tiiuxsoxrúntie their deer-hunting countries, Couxsaxráge and Skaniadarade, their beaver-hunting countries; of the lakes Erie, Ontario and Champlain, and part of New France; wherein is also shewn the antient and present seats of the Indian nations. By Lewis Evans. 1755. Published according to act of Parliament, by Lewis Evans, June 23, 1755, and sold by R. Dodsley, in Pall-Mall, London, & by the author in Philadelphia. [Philadelphia: Franklin]: 1755.

Scale: 5 cm. = 60 mi.

Cartographer: Lewis Evans.

Engraver: James Turner.

Size: 49.5 x 66.5 cm.

Colored.

Inset: A sketch of the remaining part of Ohio R. &c.

Wheat: 298

(635: 1755: Ev12bca Large)

753. 1766

A general map of the middle British colonies in America (viz) Virginia, Delaware, Pensilvania, New Jersey, New York, Connecticut & Rhode Island. Of Aquanishuonigy, the country of the Confederate Indians, comprehending Aquanishuonigy proper, their place of residence, Ohio & Tiiusoxruntie, their deer hunting countries, Couxsaxrage & Skaniadarade, their beaver hunting countries. Of the Lakes Erie, Ontario and Champlain and part of New France. 1766.

Size: 50 x 67.5 cm.

Colored.

Inset: A sketch of the remaining part of the Ohio River, etc.

Provenance: Presented by C. William Miller, May 1967.

This map is a variant of the Lewis Evans map of 1755. It is a proof by Meriden Gravure company, and was never published.

(635: 1766: Evvar Extra-oversize)

754. [1769?]

A map of that part of America where a degree of latitude was measured for the Royal Society by Charles Mason and Jere[my] Dixon. [London: 1769?].

Size: 16.6 x 9.6 cm.

Variation of map reproduced in Royal Society of London, *Philosophical Transactions,* 1769, vol. 58, p. 325.

For letter related to above map see: "Claims of Maryland, Pennsilvania . . . "

(635: [1769?]: M385aL Small)

755. [1771]

[A map of part of Pennsylvania, Delaware & Maryland intended to shew, at one view, the several places proposed for opening a communication between the waters of the Delaware & Chesopeak Bays. 1771].

Scale: 2.5 cm. = 7¾ mi.

Engraver: J. Smither.

Size of plate: 31.2 x 44.6 cm.

In: American Philosophical Society. *Transactions,* vol. 1, following p. 292.

Wheat: 300

(506.73: Am4t, vol. 1)

756. [1776]

A general map of the Middle British colonies, in America. Containing Virginia, Maryland, the Delaware Counties, Pennsylvania, and New Jersey. With the addition of New York, and of the greatest part of New England, as also of the bordering parts of the Province of Quebec. Improved from several surveys made after the late war, and corrected from Governor Pownall's late map 1776. [Title on top of sheet]: The seat of war in the middle British colonies, containing Virginia, Maryland, the Delaware Counties, &c. London: R. Sayer and J. Bennett: 1776 [15 October 1776].

Scale: 69½ British statute mi. = 1 degree.

Cartographer: T. Pownall.

Size: 48.8 x 66.5 cm.

Colored.

Inset: A sketch of the upper parts of Canada.

Originally in *American military atlas;* removed from binding 5 November 1962.

(635: [1776]: P876mbc Large)

757. 1776

A map of the middle British colonies in North America. First published by Mr. Lewis Evans, of Philadelphia, in 1755; and since corrected and improved, as also extended, with the addition of New England and bordering parts of Canada; from actual surveys now lying at the Board of Trade. By T. Pownall. London: J. Almon: 1776 [25 March 1776].

Scale: 7 cm. = 90 mi.

Cartographers: Lewis Evans and T. Pownall.

Engraver: James Turner.

Size: 49.3 x 82.4 cm.

Colored.

Inset: A sketch of the remaining part of Ohio R. &c.

(635: 1776: P876bc Large)

758. 1778

A new map of the western parts of Virginia, Pennsylvania, Maryland and North Carolina; comprehending the River Ohio, and all of the rivers, which fall into it; part of the River Mississippi, the whole of the Illinois River, Lake Erie; part of the Lakes Huron, Michigan &c. And all the country bordering on these lakes and rivers. By Thos. Hutchins. London: Hutchins: 1778 [1 November 1778].

Scale: 1 in. = 20 mi.

Cartographer: Thomas Hutchins.

Engraver: T. Cheevers.

Size: 90 x 110.3 cm.

Colored.

(635: 1778: H973vpm Large)

759. [1786]

A map of the country between Albermarle Sound and Lake Erie, comprehending the whole of Virginia, Maryland, Delaware and Pennsylvania, with parts of several other of the United States of America. Engraved for the Notes on Virginia. The country on the eastern side of the Allegany Mountains, is taken from Fry and Jefferson's map of Virginia, and Scull's map of Pennsylvania, which were constructed chiefly on actual survey; that on the western side of the Alleganey, is taken from Hutchins, who went over the principal water courses, with a compass and log-line, correcting his work by observations of latitude; additions have been made, where they could be made on sure ground. [Paris: 1786].

Cartographers: Fry, Jefferson, Scull, and Hutchins.

Engraver: S. J. Neele.

Size: 59.9 x 59.6 cm.

Colored.

The Society asked Thomas Jefferson in 1805 for a copy of his *Notes on Virginia.* Jefferson wrote 2 May 1805 and presented "a single copy remaining of the original edition printed at

Paris, the only one almost perfectly correct, & was never sold, a few copies only having been printed and given to my friends" (Archives).

In: Thomas Jefferson, *Notes on Virginia.*

(917.55: J35)

760. [1771]

[A map of part of Pennsylvania, Delaware and Maryland intended to shew, at one view, the several places proposed for opening a communication between the waters of the Delaware and Chesopeak Bays]. [1771].

Scale: 2.5 cm. = 7¾ mi.

Cartographer: W. T. [Thomas] Fisher.

Engraver: J. Smither.

Size of plate: 31.2 x 44.6 cm.

The idea of a canal linking the Delaware River and Chesapeake Bay was of long standing when the Society formed a committee to arrange for a survey of the general area. Merchants of Philadelphia financed the early surveys and on 16 June 1769, Thomas Gilpin, one of the "Merchants' appointees," exhibited his plan of a canal with the elevation between Chester River and Duck Creek, and he included an estimate "for flat bottomed boats, carrying 1,000 bushels, £8050; for shallops &c., £28,298." The Society also received an elevation of the land between the Bohemia and Apoquinimy Rivers with the locks necessary to carry vessels from the tidewater of the Delaware to that of the Chesapeake. "The Distance from Tide to Tide is 5 Miles & 107 perches" (APS Minutes).

W. T. Fisher (Thomas Fisher) delivered a map of the proposed canal 2 March 1770 which was

> intended to shew, at one View, the several places proposed for opening a communication between the waters of the Delaware and Chesapeake Bays . . . to be kept among the Society's papers for future use. In this Map is also delivered the different Roads proposed to be opened for Land Carriage

from the Susquehanna River. On 2 November 1770, the "Abstract of Canal papers read & ordered to be printed. Fisher's map ordered to be engraved" (APS Minutes).

The desire for this canal continued and it is interesting to note that as late as 16 December 1803 two of the maps belonging to the Society, the surveys by Anderson and Gilpin, were acknowledged as being in the possession of Benjamin Henry Latrobe. Latrobe had surveyed a route in 1799 and was named Engineer of the Chesapeake and Delaware Canal Company in 1804 when the first canal was dug.

In: American Philosophical Society. *Transactions,* vol. 1.

Wheat: 300

(506.73: Am4t, vol. 1)

760a. [1789]

[A map of part of Pennsylvania, Delaware and Maryland intended to shew, at one view, the several places proposed for opening a communication between the waters of the Delaware & Chesopeak Bays. 1789].

Scale: 2.5 cm. = 7¾ mi.

Engraver: J. Smither.

Size of plate: 31.2 x 44.6 cm.

In: American Philosophical Society. *Transactions,* vol. 1, 2nd ed. corrected (1789), opp. p. 367.

Wheat: 305

(506.73: Am4t)

761. 1789

A new map of the states of Pennsylvania, New Jersey, New York, Connecticut, Rhode Island, Massachusetts and New Hampshire, including Nova Scotia and Canada, from the latest authorities. New York: 1789.

Engraver: C. Tiebout.

Size: 27 x 40.3 cm.

Provenance: Presented by John Vaughan, May 1818; by the Estate of Richard Gimbel, 1974.

In: William Gordon, *History of the rise of the United States.* New York: Hodge, Allen and Campbell: 1789. Frontispiece, vol. 2.

Wheat: 150

(Paine 68: G65h)

762. 1794

A chart of Delaware Bay and River, containing a full and exact description of the shores, creeks, harbours, soundings, shoals, sands and bearings of the most considerable land marks, &c. &c. Faithfully copied from that published at Philadelphia. By Joshua Fisher, together with the tide table from the capes to Philadelphia . . . London: Laurie and Whittle: 1794 [12 May 1794].

Scale: 69½ English mi. = 1 degree.

Cartographer: Joshua Fisher.

Size: 46.8 x 68.6 cm.

Original in *North American pilot* . . . London: Laurie and Whittle: 1800. No. 10.

(640.2: 1794: F533db Large)

763. [1798]

A map of Pennsylvania, Delaware, New Jersey & Maryland, with the parts adjacent [1798].

Scale: 2.5 cm. = 28 mi.

Engraver: Thackara.

Size: 31 x 38.5 cm.

Provenance: Presented by Robert Proud, 18 June 1802.

In: Robert Proud, *The history of Pennsylvania,* vol. 2, frontispiece.

Wheat: 312

(974.8: P94)

764. 1798

A map of the middle states, shewing the situation of the Genesee lands & their connection with the Atlantic coast. 1798.

Scale: [2.5 cm. = ca. 35 mi].

Size: 26 x 24.7 cm.

In: Charles Williamson, *Description of the Genessee country,* following p. 37.

Wheat: 313

(974.7: W67)

765. 1799

A map of the middle states of North America, shewing the position of the Geneseo country comprehending the counties of Ontario & Steuben as laid off in townships of six miles square each. 1799.

Scale: 2.5 cm. = ca. 33 mi.

Engraver: Maverick.

Size: 37.3 x 39.8 cm.

In: Charles Williamson, *Description of the settlement of the Genesee country,* frontispiece.

Wheat: 316

(974.7: W67d)

766. 1800

A map exhibiting the different stage routs, between the cities of New-York, Baltimore, and parts adjacent. To which is added, as an historical companion, the operations of the British army, from their landing at Elk River in 1777, to their embarkation at Nevisink in 1778. By John Hills. Philadelphia: E. Savage: 1800.

Scale: 2.5 cm. = 3 mi.

Cartographer: John Hills.

Size: 28 x 152.1 cm.

Colored.

(635: 1800: H553nyb Large)

767. [ca. 1800]

[Map of Pennsylvania, Delaware and Maryland; showing proposed new roads and canals]. Plate VII, fig. 1. [Philadelphia: ca. 1800].

Scale: 1 cm. = 3 mi.

Size of paper: 32.7 x 41 cm.

Engraver: J. Smither.

See: American Philosophical Society. *Transactions,* n.s., vol. 1, pl. 7: "An abstract of sundry papers and proposals for improving the inland navigation of Pennsylvania and Maryland, by opening a communication between the tide-waters of Delaware or Susquehanna, or Chesopeake Bay . . . " See Maps 760 and 760a, above.

(635: [ca. 1800]: Sm67pdm Small)

768. [ca. 1815]

A new map of the seat of war. N.p.: [ca. 1815].

Scale: 1 in. = 42 mi.

Cartographer: S. Lewis.

Engraver: J. Bower.

Size: 34.2 x 46 cm.

Inset: Niagara River.

(640.2: [ca. 1815]: L585msw Small)

769. 1826

Map of reconnaissance between Baltimore and Philadelphia exhibiting the several routes of the mail-road contemplated by the resolution of Congress approved on the 4th of May 1826. U.S. War Department. [Washington: 1826].

Scale: 1 in. = 3 mi.

Cartographers: S. Bernard and William Tell Poussin.

Size: 22.3 x 79.6 cm.

(650: 1826: Un38rbp Large)

770. 1826

Map of the country between Washington & Pittsburgh referring to the contemplated Chesapeake & Ohio canal and its general route and profile. October 1826. Reduced from the general map, annexed to the report upon the contemplated canal, & drawn by Lieut. Farley . . . U.S. War Department. Georgetown, D.C.: 1826.

Scale: 1 in. = 8 mi.

Cartographer: John Farley.

Engravers: William Harrison and D. R. Harrison.

Size: 45.4 x 64.5 cm.

Colored.

Provenance: Presented through Daniel Parker, 4 January 1833.

(635: 1826: Un38wpo Large)

771. [ca. 1835]

A map showing the effect of a connection between the Delaware Division and the Delaware and Raritan canal. Philadelphia: Watson's Lith.: [1835].

Lithographer: Watson.

Size of paper: 49.2 x 65.8 cm.

(640.2: [ca. 1835]: W339drc Small)

772. [ca. 1848]

Geological map of the middle and western states. By James Hall. New York: Endicott Lithographer: [ca. 1848].

Scale: 1 in. = 30 mi.

Cartographer: James Hall.

Lithographer: Endicott.

Size: 58.4 x 82.4 cm.

Provenance: Presented and autographed by Charles B. Trego.

(635: [ca. 1848]: H143mws Large)

773. 1852

General chart of the coast from Gay Head to Cape Henlopen. From a trigonometrical survey under the direction of F. R. Hassler and A. D. Bache. U.S. Coast and Geodetic Survey. [Washington]: 1852.

Scale: 1:400,000.

Cartographers: F. R. Hassler and A. D. Bache.

Size: 78.3 x 95.3 cm.

Provenance: Presented "pursuant of an Act of Congress, and by the direction of the Treasury Department" through A. D. Bache, 10 October 1854.

Contains six views and five sections of the slope of the ocean bottom.

(635: 1852: Un38ghh Large)

774. 1867

Military maps illustrating the operations of the armies of the Potomac & James May 4th, 1864 to April 9th, 1865 including battlefields of the Wilderness, Spottsylvania, Northanna, Totopotomoy, Cold Harbor, the siege of Petersburg and Richmond. Battlefields of Five Forks, Jetersville & Sailor's Creek, Highbridge, Farmville & Appomatox Court-House. U.S. War Department. New York: J. Bien: 1867. 15 maps and title page.

Scales: 3 in. = 1 mi.; 1½ in. = 1 mi.

Lithographer: J. Bien.

Size of paper: 61.1 x 45.4 cm.

Colored.

Provenance: Presented by the U.S. War Department, 4 November 1870.

List of maps: 1. North Anna. 2. Cold Harbor. 3. Chancellorsville. 4. Harper's Ferry. 5. Fredericksburg. 6. High Bridge & Farmville. 7. Bermuda Hundred. 8. Jetersville & Sailors Creek. 9. Petersburg & Five Forks. 10. Totopotomoy. 11. Spottsylvania Court House. 12. The Wilderness. 13. Appomattox Court House. 14. Richmond. 15. Antietam.

(635: 1867: Un38apj Large)

775. 1869

Map of the region between Gettysburg, Pa., and Appomattox Court House, Va., exhibiting the connection between the campaign and battle-field maps . . . under the direction of Brig. & Bvt. Maj. Gen. A. A. Humphreys . . . by N. Michler. U.S. War Department. [Washington]: 1869.

Scale: 1:300,000.

Cartographers: N. Michler, John E. Weyss, C. Thompson, J. de la Camp, and A. A. Humphreys.

Size: 111.3 x 69.3 cm.

Colored.

(635: 1869: Un38cbf Large)

776. [ca. 1880]

Map of the Buffalo, New York and Philadelphia R. R. and its principal connections. New York: [ca. 1880].

Scale: 1 in. = 20 mi.

Cartographer: F. S. Cook.

Size: 53.9 x 84.5 cm.

Colored.

(640.2: [ca. 1880]: C772byp Large)

777. 1880

Post route map of the states of Pennsylvania, New Jersey, Delaware and Maryland and of the District of Columbia with adjacent parts of New York, Ohio, Virginia and West Virginia. Designed and constructed under the orders of the U.S. Post Office Department Postmaster General . . . by W. L. Nicholson. Washington: 1880. 1st ed. 1869.

Scale: 1 in. = 6 mi.

Cartographers: C. H. Poole and W. L. Nicholson.

Engraver: D. McClellan.

Size: 153 x 153 cm.

Colored.

(635: 1880: Un38pod Large)

778. 1881

Map of the Buffalo, Pittsburgh & Western R. R. & connections. N.p.: 1881.

Size: 57.1 x 60.4 cm.

Colored.

(640.2: 1881: B861wrr Large)

779. 1882

Map of the railroads of Pennsylvania and New Jersey and parts of adjoining states. 1882. Prepared from official data by J. A. Anderson. Philadelphia: Smith and Stroup: 1882.

Scale: 1 in. = 8 mi.

Cartographer: J. A. Anderson.

Size: 98.5 x 121.5 cm.

Colored.

Inset: Map showing the eastern shore of Maryland & Virginia.

(640.2: 1882: An21pnj Large)

780. 1884

Post route map of the state of New York, and parts of Vermont, Massachusetts, Connecticut, New Jersey and Pennsylvania. Also the adjacent portions of the Dominion of Canada, showing post offices, with the intermediate distances between them and mail routes in operation on 1st June 1884. [Washington]: 1884. 1st ed. issued 1868.

Scale: 1:380,160.

Size: 140.2 x 156.7 cm.

Colored.

Insets: Postal service of Long Island, with the principal mail connections of the city of New York; map of the Lake St. Frances to Montreal area, Canada.

(647: 1884: Un38prm Large)

781. 1892

New York-New Jersey. New York sheet. U.S. Geological Survey. J. W. Powell, director. Surveyed in 1888-1889. [Washington]: 1888. Edition of 1892.

Scale: 1:62,500.

Cartographers: Henry Gannett and J. W. Powell.

Size: 91.8 x 71.1 cm.

Colored.

(640.2: 1892: Un38nyj Large)

782. 1953

Bridgeport Quadrangle. New Jersey-Pennsylvania. Marcus Hook quadrangle. Pennsylvania-Delaware-New Jersey. U.S. Department of the Interior. Geological Survey. [Washington]: 1953. 2 sheets. Edition of 1953.

Scale: 1:24,000.

Size: 57.7 x 44.6 cm.

Colored.

Provenance: Presented by George Washington Corner.

(640.2: 1953: Un38pdn Large)

NEW YORK Nos. 783-843

783. [1724]

A map of the countrey of the Five Nations belonging to the province of New York and of the lakes near which the nations of the Far Indians live with part of Canada taken from the map of the Louisiane done by Mr. Delisle in 1718. [1724].

Scale: 2.5 cm. = 80 mi.

Cartographer: de l'Isle.

Size: 21.8 x 35.2 cm.

In: Cadwallader Colden, *Papers relating to an act of New York for encouragement of the Indian trade,* frontispiece.

Wheat: 317

(970.1: C67p)

784. 1762

A survey of Lake Champlain, including Lake George, Crown Point and St. John. Surveyed by order of . . . Jeffrey Amherst. By William Brassier. 1762. London: Robt. Sayer & Jn. Bennett: 1776 [5 August 1776].

Scale: 2 cm. = 5 mi.

Cartographers: William Brassier and Capt. Jackson.

Size: 66.2 x 47.9 cm.

Colored.

Inset: A particular plan of Lake George. Surveyed in 1756 by Capt. Jackson.

This was removed from: *The American military pocket atlas.* R. Sayer and J. Bennett: [1776].

(647: 1762: Bcja Large)

785. [1775]

A map of the present seat of war on the borders of Canada. [Philadelphia: Aitken: 1775].

Scale: 1 in. = 10 mi.
Size: 38.9 x 15.4 cm.
At top of map is: "for the Penna Magazine p. 463" [see: *Pennsylvania Magazine.* 1775, vol. 1].
Wheat: 89

(647: [1775]: Ai1lwbc Small)

786. 1778

The southern part of the province of New York: with part of the adjoining colonies. By Tho[ma]s Kitchin, sen[io]r hydrographer to His Majesty. [London]: R. Baldwin: 1778.
Cartographer: T. Kitchin.
Size: 24.7 x 18.4 cm.
At top of map is: "For the *London Magazine* 1778" [vol. 47, March 1778, p. 99].

(647: 1778: K644pny Small)

787. [1789]

From Albany (26) to Newborough. [1789, 1961 reprint].
Size: 17 x 11.5 cm.
In: Christopher Colles, *A survey of the roads.*
Wheat: 349

(526.8: C68s)

788. [1789]

From Albany (27) to Newborough. [1789, 1961 reprint].
Size: 17 x 11.3 cm.
In: Christopher Colles, *A survey of the roads.*
Wheat: 350

(526.8: C68s)

789. [1789]

From Albany (28) to Newborough. [1789, 1961 reprint].
Size: 17 x 11.4 cm.
In: Christopher Colles, *A survey of the roads.*
Wheat: 351

(526.8: C68s)

790. [1789]

From Albany (29) to Newborough. [1789, 1961 reprint].
Size: 17 x 11.4 cm.
In: Christopher Colles, *A survey of the roads.*
Wheat: 352

(526.8: C68s)

791. [1789]
From Albany (30) to Newborough. [1789, 1961 reprint].
Size: 17 x 11.5 cm.
In: Christopher Colles, *A survey of the roads.*
Wheat: 353

(526.8: C68s)

792. [1789]
From Albany (31) to Newborough. [1789, 1961 reprint].
Size: 17 x 11.4 cm.
In: Christopher Colles, *A survey of the roads.*
Wheat: 354

(526.8: C68s)

793. [1789]
From Albany (32) to Newborough. [1789, 1961 reprint].
Size: 17 x 11.8 cm.
In: Christopher Colles, *A survey of the roads.*
Wheat: 355.

(526.8:C68s)

794. [1789]
From Albany (33) to Newborough. [1789, 1961 reprint].
Size: 17 x 11.5 cm.
In: Christopher Colles, *A survey of the roads.*
Wheat: 356

(526.8: C68s)

795. [1789]
From New York (1) to Stratford. [1789, 1961 reprint].
Size: 17.3 x 11.4 cm.
In: Christopher Colles, *A survey of the roads.*
Wheat: 331

(526.8: C68s)

796. [1789]
From New York (2) to Slratford [sic]. [1789, 1961 reprint].
Size: 17.2 x 11.4 cm.
In: Christopher Colles, *A survey of the roads.*
Wheat: 332

(526.8: C68s)

797. [1789]

From New York (3) to Stratford. [1789, 1961 reprint].
Size: 17.2 x 11.4 cm.
In: Christopher Colles, *A survey of the roads.*
Wheat: 333

(526.8: C68s)

798. [1789]

From New York (8) to Poughkeepsie. [1789, 1961 reprint].
Size: 17.2 x 11.5 cm.
In: Christopher Colles, *A survey of the roads.*
Wheat: 334

(526.8: C68s)

799. [1789]

From New York (9) to Poughkeepsie. [1789, 1961 reprint].
Size: 17.2 x 11.5 cm.
In: Christopher Colles, *A survey of the roads.*
Wheat: 335

(526.8: C68s)

800. [1789]

From New York (10) to Poughkeepsie. [1789, 1961 reprint].
Size: 17.2 x 11.4 cm.
In: Christopher Colles, *A survey of the roads.*
Wheat: 336

(526.8: C68s)

801. [1789]

From New York (11) to Poughkee[p]sie. [1789, 1961 reprint].
Size: 17.2 x 11.5 cm.
In: Christopher Colles, *A survey of the roads.*
Wheat: 337

(526.8: C68s)

802. [1789]

From New York (12) to Poughkeepsie. [1789, 1961 reprint].
Size: 16.7 x 11.3 cm.
In: Christopher Colles, *A survey of the roads.*
Wheat: 338

(526.8: C68s)

803. [1789]
From New York (13) to Poughkeepsie. [1789, 1961 reprint].
Size: 17.2 x 11.4 cm.
In: Christopher Colles, *A survey of the roads.*
Wheat: 339
(526.8: C68s)

804. [1789]
From Poughkeepsie (14) to Albany. [1789, 1961 reprint].
Size: 17.2 x 11.5 cm.
In: Christopher Colles, *A survey of the roads.*
Wheat: 340
(526.8: C68s)

805. [1789]
From Poughkeepsie (21) to Albany. [1789, 1961 reprint].
Size: 17 x 11.3 cm.
In: Christopher Colles, *A survey of the roads.*
Wheat: 344
(526.8: C68s)

806. [1789]
From Poughkeepsie (22) to Albany. [1789, 1961 reprint].
Size: 17.2 x 11.5 cm.
In: Christopher Colles, *A survey of the roads.*
Wheat: 345
(526.8: C68s)

807. [1789]
From Poughkeepsie (23) to Albany. [1789, 1961 reprint].
Size: 17.2 x 11.4 cm.
In: Christopher Colles, *A survey of the roads.*
Wheat: 346
(526.8: C68s)

808. [1789]
From Poughkeepsie (24) to Albany. [1789, 1961 reprint].
Size: 17.3 x 11.7 cm.
In: Christopher Colles, *A survey of the roads.*
Wheat: 347
(526.8: C68s)

809. [1789]
From Poughkeepsie (25) to Albany. [1789, 1961 reprint].
Size: 17.1 x 11.4 cm.
In: Christopher Colles, *A survey of the roads.*
Wheat: 348
(526.8: C68s)

810. [1789]
From Stratford (18) to Poughkeepsie. [1789, 1961 reprint].
Size: 17.1 x 11.3 cm.
In: Christopher Colles, *A survey of the roads.*
Wheat: 341
(526.8: C68s)

811. [1789]
From Stratford (19) to Poughkeepsie. [1789, 1961 reprint].
Size: 17.1 x 11.5 cm.
In: Christopher Colles, *A survey of the roads.*
Wheat: 342
(526.8: C68s)

812. [1789]
From Stratford (20) to Poughkeepsie. [1789, 1961 reprint].
Size: 17.1 x 11.5 cm.
In: Christopher Colles, *A survey of the roads.*
Wheat: 343
(526.8: C68s)

813. [1795]
New York. Forts Erie, Schlosser, Niagara, Oswego and Schuyler are located. [1795].
Scale: 2.5 cm. = 50 mi.
Size: 15.8 x 9.3 cm.
Provenance: Presented by the Estate of Richard Gimbel, 1974.
In: Joseph Scott, *The United States gazetteer,* opposite "New-York." Incomplete.
Wheat: 366
(Paine 68: Sco3u)

814. [1796]
New York. [1796].
Scale: 2.5 cm. = 60 mi.
Engraver: W. Barker.
Size: 14.6 x 19.3 cm.

Provenance: Presented by Suzanne Wister Eastwick and Joseph L. Eastwick, July 1975.
In: Mathew Carey, *Carey's American pocket atlas,* following p. 46.
Wheat: 367

(917.3: C25c)

815. [1798]

Map of Ontario and Steuben counties [1798].
Scale: 2.5 cm. = ca. 7 mi.
Cartographer: Heslop.
Engraver: Fairman.
Size: 38 x 33.8 cm.
In: Charles Williamson, *Description of the settlement of the Genesee country,* following p. 37.
Wheat: 376

(974.7: W67d)

816. [1800]

Map of Morris's purchase or West Geneseo in the state of New York. Exhibiting part of the Lakes Erie and Ontario, the straights of Niagara, Chautauque Lake and all the principal waters, the boundary lines of the several tracts of lands purchased by the Holland Land Company, William and John Willink and others. Boundary lines of townships: boundary lines of New York and Indian Reservations: laid down by actual survey; also a sketch of part of Upper Canada. By Joseph & B. Ellicott. 1800. N.p.: 1800.
Cartographers: Joseph & B. Ellicott.
Size: 51.6 x 66.6 cm.; and 52.2 x 66.8 cm.
See: Ellis Paxson Oberholtzer, *Robert Morris, patriot and financier.* New York: Macmillan: 1903.

(647.934: 1800: EL52mpg Large)

817. 1800

The state of New-York from the best authorities. 1800.
Scale: 2.5 cm. = ca. 38 American mi.
Engraver: A. Anderson.
Size: 18.7 x 21.7 cm.
In: John Payne, *A new and complete system of universal geography,* vol. 4, opposite p. 296.

(910: P29)

818. 1802

A map of the state of New York. By Simeon De Witt. Albany: 1802 [16 October 1802].
Scale: 69 mi. = 1 degree.
Cartographer: Simeon De Witt.
Engraver: G. Fairman.
Size: 167.5 x 133.7 cm.
Inset: Continuation of the state westward from the preemption line.
Provenance: Presented by Simeon De Witt, 5 January 1827.

(647: 1802: D512snn Large)

819. 1804

Map of Morris's purchase or West Geneseo in the state of New York. Exhibiting part of the Lakes Erie and Ontario, the straights of Niagara, Chautauque Lake and all the principal waters, and boundary lines of the several tracts of lands purchased by the Holland Land Company, William and John Willink and others. Boundary lines of townships: boundary lines of New York and Indian Reservations: laid down by actual survey; also a sketch of part of Upper Canada. By Joseph & B. Ellicott. 1804. N.p.: 1804.

Scale: 1¼ in. = 6 mi.

Cartographers: Joseph and B. Ellicott.

Size: 52.2 x 67 cm.

See: Ellis Paxson Oberholtzer, *Robert Morris, patriot and financier.* New York: Macmillan: 1903.

(647.934: 1804: EL52mpg Large)

820. 1804

A map of the state of New York. By Simeon De Witt. Contracted from his large map of the state 1804. New York: 1804 [19 March 1804].

Scale: 69 mi. = 1 degree.

Cartographer: Simeon De Witt.

Size: 55.4 x 69.3 cm.

Colored.

Mutilated: lower left corner is missing.

(647: 1804: D512sny Large)

821. 1804

A map shewing the relative situation of a tract of land belonging to I. B. Church Esqr. New York: 1804.

Scale: 1¾ in. = 40 mi.

Cartographer: Jos. Fra[?].

Engraver: Peter Maverick.

Size: 38.9 x 48.6 cm.

(647: 1804: M455nys Small)

822. 1808

A map of the state of New York. Compiled from the latest authorities; including the turnpike roads now granted as also, the principal common roads connected therewith. By William McCalpin. Oxford: 1808.

Scale: 69 mi. = 1 degree.

Cartographer: William McCalpin.

Size: 41.6 x 51.2 cm.

Provenance: Presented by Robert Patterson.

(647: 1808: M125mny Small)

823. 1812

Map of the northern part of the state of New York. Compiled from actual survey by Amos Lay. New York: 1812 [16 July 1812].

Scale: 1:443,520.

Cartographer: Amos Lay.

Size: 76.2 x 126.2 cm.

Colored.

Provenance: Presented by Amos Lay, 20 May 1814.

(647: 1812: L455npn Large)

824. 1814

Plan of the siege of Plattsburg, and capture of the British fleet on Lake Champlain. The 11th Septr. 1814. [Philadelphia: ca. 1815].

Engraver: B. Tanner.

Size: 25 x 19.9 cm.

Provenance: Presented by Henry Schenck Tanner, 20 February 1818.

This is to accompany Tanner's engraving of MacDonough's victory on Lake Champlain, copied from the painting by Hugh Reinagle.

(647: 1814: T168psp Small)

825. 1816

Map of the north west part of the state of New York. Compiled from original surveys, by P. Tardieu. Paris: 1816.

Scale: 1 in. = 4½ mi.

Cartographer: P. Tardieu.

Engravers: P. A. F. Tardieu and Richomme.

Size: 59.6 x 72.9 cm.

Colored.

(647: 1816: T178sny Large)

826. 1820

Map of the Hudson between Sandy Hook and Sandy Hill with the post road between New York and Albany. New York: A. T. Goodrich and Co.: 1820.

Cartographer: Bridges.

Engraver: Rollinson.

Size: 252.6 x 22.8 cm.

(647: 1820: G633hr Small)

827. [ca. 1825]

Map and profile of the Erie Canal; commenced 1817, finished 1825. N.p.: [ca. 1825].

Scale: 1 in. = 6 mi.

Size: 22.7 x 116.2 cm.

Provenance: Presented by De Witt Clinton, 14 April 1826.

(647: [ca. 1825]: Er42ec Small)

828. [1835]

Survey for a ship canal around the Falls of Niagara. Made and drawn under the direction of Capt. W. G. Williams . . . by Lts. T. F. Drayton and J. G. Reed. U.S. Topographical Engineers. [Washington: 1835].

Vertical scale: 50 ft. = 1 mi.

Cartographers: W. G. Williams, T. F. Drayton, and J. G. Reed.

Size: 104.6 x 74.2 cm.

Insets: Engraving of Niagara Falls; Niagara River; U.S. north of the Ohio River and Upper Canada; Mouth of Four Mile creek with soundings adjacent.

Provenance: Presented through J. R. Ingersoll, 15 July 1836.

(647: [1835]: Un38cfn Large)

829. [1835]

Survey for a ship canal to connect Lakes Erie, & Ontario. Lockport route. Profiles levelled and surveyed under the direction of Capt. W. G. Williams, U.S. Topographical Engineers, assisted by Lieuts. Drayton and Reed. [Washington: 1835].

Cartographers: W. G. Williams, T. F. Drayton, and J. G. Reed.

Size of paper: 60.6 x 137.2 cm.

See also map: 647: [1835]: Un38cfn Large.

(647: [1835]: Un38cfn.1 Large)

830. [1835]

Survey for a ship canal to connect the Lakes Erie & Ontario. Lockport route. Surveyed under the direction of Capt. W. G. Williams by Lieuts. T. F. Drayton and J. G. Reed. U.S. Topographical Engineers. Map no. 3. [Washington: 1835].

Scale: 4 in. = 5280 ft.

Cartographers: W. G. Williams, T. F. Drayton, and J. G. Reed.

Size: 128.1 x 69.7 cm.

Inset: Tonnawanta Creek from the Niagara River.

Provenance: Presented by W. G. Williams through J. R. Ingersoll, 15 July 1836.

(647: [1835]: Un38eop Large)

831. [1836]

Survey for a ship canal around the Falls of Niagara. Made and drawn under the direction of Capt. W. G. Williams, U.S.T. Engineers by Lts. T. F. Drayton and J. G. Reed. U.S. Army Topographical Engineers. N.p.: [1836].

Scale: 12 in. = 1 mi. (horizontal).

Cartographers: W. G. Williams, T. F. Drayton, and J. G. Reed.

Size: 104.6 x 74.2 cm.

Insets: Niagara River segments; Mouth of Four Mile Creek; Great Lakes and St. Lawrence River to Quebec.

Provenance: Presented by J. R. Ingersoll, 15 July 1836.

With this are: 1) Profiles of the survey around Niagara Falls. Nos. 1, 2, 3. *Sizes:* various. 2) Plan of an artificial harbor & project for descent of ridge. *Insets:* Termination of line; and Section of lock. *Size:* 73 x 117.4 cm.

(647: [1836]: W678scn Large)

832. 1847

Oyster or Syosset Bay. Harbor of refuge no. [blank]. Founded upon a trigonometrical survey under the direction of F. R. Hassler. U.S. Coast and Geodetic Survey. [Washington]: 1847.

Scale: 1:30,000.

Cartographers: F. R. Hassler, E. Blunt, A. D. MacKay, F. H. Gerdes, G. S. Blake, and John B. Glück.

Engravers: Sherman and Smith.

Size: 43.7 x 35.6 cm.

Provenance: Presented by U.S. Treasury Department through A. D. Bache.

(647: 1847: Un31oyb Small)

833. 1849

Huntington Bay. Founded upon a trigonometrical survey under the direction of F. R. Hassler. U.S. Coast and Geodetic Survey. [Washington]: 1849.

Scale: 1:30,000.

Cartographers: F. R. Hassler, E. Blunt, F. H. Gerdes, G. S. Blake, and J. B. Glück.

Engraver: George G. Smith.

Size: 35.8 x 43.8 cm.

Provenance: Presented by the U.S. Treasury Department through A. D. Bache, 5 April 1850.

(647: 1849: Un33hb Small)

834. 1856

Chart of Buffalo harbor and head of Niagara River, with the outlet of Lake Erie. U.S. Bureau of Topographical Engineers. From the surveys of Capt. W. G. Williams. [Washington]: 1856.

Scale: 1:30,000.

Cartographers: W. G. Williams and I. C. Woodruff.

Engraver: W. H. Dougal.

Size: 54.5 x 50.1 cm.

Provenance: Presented by G. G. Meade, 19 July 1861.

(647.919: 1856: Un38cbh Large)

835. 1898

New York. [From: *Atlas of the world,* p. 16]. Rand, McNally & Co.: 1898.

Scale: 1 in. = 28 mi.

Size: 23.5 x 31.7 cm.

Colored.

Inset: Long Island and lower part of New York state.

Provenance: Presented by Mrs. A. I. Hallowell, November 1983.

On same sheet is: Connecticut, *Atlas of the world.* 1898.

(647: 1898: N429Lis Small)

836. 1901

Geologic map of New York, exhibiting the structure of the state so far as known. By Frederick J. H. Merrill. University of the State of New York. New York State Museum. New York: James B. Lyon, state printer, and Julius Bien & Co.: 1901. 12 sheets and wrapper.

Scale: 1 in. = 5 mi.

Cartographer: F. J. H. Merrill.

Size: ca. 52.5 x 42.5 cm.

Colored.

(647: 1901: N486geo Large)

837. 1901

Niagara River and vicinity. U.S. Geological Survey. Charles D. Walcott, director. [Washington]: 1901.

Scale: 1:62,500.

Cartographers: Henry Gannett, H. M. Wilson, Charles D. Walcott, Frank Sutton, J. H. Wheat, and W. W. Gilbert.

Size: 89.4 x 34 cm.

Colored.

(647: 1901: Un38nrv Large)

838. 1902

Albany and vicinity, New York. U.S. Geological Survey. Charles D. Walcott, director. Surveyed in 1891-1892 in cooperation with the State of New York. [Washington]: 1902. Edition of March 1902.

Scale: 1:62,500.

Cartographers: C. D. Walcott, Henry Gannett, H. M. Wilson, Frank Sutton, and J. H. Jennings.

Size: 91.7 x 68.6 cm.

Colored.

(647.912: 1902: Un38aav Large)

839: 1962

Correlation of the Cambrian rocks in New York state by Donald W. Fisher, state paleontologist. Geologic survey of the State Museum and Science Service, State Education Department of the University of the State of New York. Geological survey by John G. Broughton, state geologist. Albany: 1962. Map and chart series, no. 2, 1962.

Cartographers: Donald W. Fisher and John G. Broughton.

Size of paper: 89 x 114.4 cm.

(647: 1962: N486cam Small)

840. 1962

Geological map of New York; 1961. University of the State of New York. State Education Department. State Museum and Science Service. Geological Survey map and chart series nos. 5 and 15. 16 pieces. Albany: 1962.

Sizes: various.

Colored.

Provenance: Presented by the New York State Museum, February 1963.

(647: 1961: N486umg Extra-oversize)

841. 1971

Simple Bouguer gravity anomaly map of western New York. [Covers area of Niagara and Finger Lakes sheets of the 1961 and 1970 editions of the Geological Map of New York.] By Frank A. Revetta and William D. Diment. New York State Museum and Science Service. Geological Survey maps and charts series no. 17. The University of the State of New York. The State Education Department. Compiled 1971.

Scale: 1:250,000.

Cartographers: Frank A. Revetta and William D. Diment.

Size: 87.7 x 150.5 cm.

(647: 1971: R328bga Large)

842. 1974-1977

Preliminary brittle structure map of New York. Yngvar W. Isachsen and William G. McKendree. New York State Museum. Map and chart series nos. 31 a-g. Geological Survey of New York. The University of the State of New York. The State Education Department. [Albany, N.Y.]: 1974-1977.

Cartographers: Yngvar W. Isachsen and William G. McKendree.

Colored.

31a. Adirondack sheet. *Scale:* 1:250,000. *Size:* 72.1 x 110 cm.

31b. Hudson-Mohawk sheet. *Scale:* 1:250,000. *Size:* 72.5 x 112.5 cm.

31c. Lower Hudson. *Scale:* 1:250,000. *Size:* 73.5 x 32.2 cm.

31d. Niagara-Finger Lakes sheet. *Scale:* 1:250,000. *Size:* 76 x 114.5 cm.

31e. Preliminary brittle structure map of New York. *Scale:* 1:250,000. *Size:* 104.8 x 140.4 cm.

31f. Generalized map of recorded joint systems in New York. *Scale:* 1:1,000,000. *Size:* 51.9 x 67.8 cm.

31g. Sources of information for preliminary brittle structures map of New York and generalized map of recorded joint system in New York. *Scale:* 1:250,000. *Size:* 67.8 x 51.8 cm.

(647: 1974-1977: Is11bsm Large)

843. 1977

Quaternary geology of New York, Niagara sheet. By Ernest H. Muller. New York State Museum and Science Service. Geological Survey map and chart series number 28. [Albany, N.Y.]: 1977.

Scale: 1:250,000.

Cartographer: Ernest H. Muller.

Size: 74.9 x 77 cm.

Colored.

Inset: Moraines of western New York.

(647: 1977: M918qni Large)

NEW YORK CITY Nos. 844-857

844. 1810

Map of the ground occupied as a botanic garden, the property of Dr. Hosack, situate on the west side of the Middle Road, New York. Surveyed January 1810 by Wm. Bridges.

Cartographer: William Bridges.

Size: 9.7 x 32.5 cm.

Provenance: Presented by Mrs. Caroline Robbins, 16 November 1965.

Original in Columbia University; this photograph from published copy on file in the office of the Secretary of State of New York, Charters . . . , vol. 39, p. 82.

(B: H78.r)

845. [ca. 1830]

Plan of the city of New York. Drawn and engrav'd for D. Longworth, map and print seller. [New York]: D. Longworth: [ca. 1830].

Scale: 1 in. = 1000 ft.

Cartographer: J.A.

Engraver: P. R. Maverick.

Size: 37.9 x 55 cm.

Provenance: Presented by Daniel Parker, 4 January 1833.

(647.957: [ca. 1830]: M455pny Small)

846. 1834

City of New York. New York: J. H. Colton & Co.: 1834.

Engraver: S. Stiles & Co.

Size: 54.2 x 44 cm.

Colored.

Provenance: Presented by Suzanne Wister Eastwick and Joseph L. Eastwick, 1977.

(647.957: 1834: B94c Book map)

847. 1844

Map of New-York bay and harbor and the environs. Founded upon a trigonometrical survey under the direction of F. R. Hassler. U.S. Coast Survey. [Washington]: 1844.

Scale: 1:31,680.

Cartographers: F. R. Hassler, James Ferguson, Edmund Blunt, Thomas R. Gedney, C. Renard, and T. A. Jenkins.

Size: 167.1 x 160.7 cm.

Provenance: Presented by Joseph R. Ingersoll through A. D. Bache, 1 November 1844.

(647: 1844: Un38nyb Large)

848. 1845

Map of New-York bay and harbor and the environs. Founded upon a trigonometrical survey under the direction of F. R. Hassler. U.S. Coast Survey. [Washington]: 1845.

Scale: 1:84,480.

Cartographers: F. R. Hassler, James Ferguson, Edmund Blunt, C. Renard, T. A. Jenkins, and B. F. Sands.

Engravers: S. Siebert, A. Rollé, O. A. Lawson, F. Dankworth, and J. Knight.

Size: 62.7 x 89.6 cm.

Provenance: Presented by J. R. Ingersoll through A. D. Bache, 28 June, 8 December 1845.

(647: 1845: Un38nyh Large)

849. 1851

Hart & City Island. [And] Sachem's Head Harbor. From a trigonometrical survey under the direction of A. D. Bache. U.S. Coast and Geodetic Survey. [Washington]: 1851. 2 maps on 1 sheet.

Scale: 1:20,000 and 1:10,000.

Cartographers: A. D. Bache, E. Blunt, T. W. Werner, H. L. Whiting, G. S. Blake, J. R. Goldsborough, M. McClery, J. Lambert, and J. Robertson.

Engravers: R. T. Knight and S. T. Pettit.

Size: 35.7 x 43.7 cm.

Provenance: Presented by U.S. Treasury Department through A. D. Bache, 21 March 1851.

Both maps marked "Harbor of refuge no. [blank]."

(647: 1851: Un33hcsh Small)

850. 1851

Hell Gate and its approaches. From a trigonometrical survey under the direction of F. R. Hassler and A. D. Bache. U.S. Coast Survey. [Washington]: 1851.

Scale: 1:5,000.

Cartographers: F. R. Hassler, A. D. Bache, Edmund Blunt, H. L. Whiting, D. D. Porter, M. Woodhull, C. Mahon, and J. R. P. Mechlin.

Engravers: A. Rollé, F. Dankworth, G. McCoy, and J. Knight.

Size: 62.4 x 89.3 cm.

Provenance: Presented by U.S. Treasury Department through A. D. Bache, 21 March 1851.

(647: 1851: Un38hga Large)

851. 1851

Western part of the southern coast of Long Island. From a trigonometrical survey under the direction of F. R. Hassler and A. D. Bache. U.S. Coast Survey. [Washington]: 1851.

Scale: 1:80,000.

Cartographers: F. R. Hassler, A. D. Bache, Edmund Blunt, C. Renard, H. L. Dickens, T. R. Gedney, R. Bache, A. S. Baldwin, M. Woodhull, F. H. Gerdes, and J. J. Ricketts.

Engravers: F. Dankworth, O. A. Lawson, and J. Knight.

Size: 62.5 x 89.3 cm.

Provenance: Presented by the U.S. Treasury Department through A. D. Bache.

(647: 1851: Un38sLi Large)

852. 1853

Map of New York bay and harbor. John Hill in 1778, J. F. W. Des Barres in 1779, Capt. Leconte in 1819. U.S. Coast Survey. Philadelphia: Wagner & McGuigan, lith.: 1853.

Cartographers: John Hill, J. F. W. Des Barres, and Capt. Leconte.

Lithographers: Wagner & McGuigan.

Size: 52.7 x 42.6 cm.

(647: 1853: Un38Lec Small)

853. 1857

Preliminary chart of New York bay and harbor. From a trigonometrical survey under the direction of A. D. Bache. U.S. Coast Survey. New York: Hatch & Co.: 1857.

Scale: 1:80,000.

Cartographers: A. D. Bache, J. Ferguson, E. Blunt, S. A. Gilbert, A. S. Wadsworth, A. M. Harrison, H. Adams, H. L. Whiting, R. Wainwright, T. A. Craven, W. G. Temple, and A. Balbach.

Size: 77 x 67.5 cm.

Provenance: Presented by William Ricketts Palmer, April 1859.

(647: 1857: Un38pcy Large)

854. 1874

Bay and harbor of New York. U.S. Coast and Geodetic Survey. [Washington]: 1874.

Scale: 1:40,000.

Cartographers: J. Ferguson, E. Blunt, H. L. Whiting, S. A. Gilbert, A. M. Harrison, F. H. Gerdes, A. Boschke, F. W. Dorr, C. Rockwell, J. Mechan, R. Wainwright, T. A. Craven, H. Mitchell, and F. F. Nes.

Engravers: E. Molkow, W. A. Thompson, R. F. Bartle, H. C. Evans, H. M. Knight, E. A. Maedel, H. Lindenkohl, and A. Lindenkohl.

Size: 121.9 x 99.7 cm.

Provenance: Presented by the U.S. Coast and Geodetic Survey through Spencer C. McCorkle, 14 December 1885.

(647: 1874: Un38bhy Large)

855. [ca. 1885]

Central Park, New York City. Official map. By the courtesy of the Department of Parks of the City of New York. N.p.: [ca. 1885].

Size of paper: 42.2 x 187.2 cm.

(647.957: [ca. 1885]: N426cen Small)

856. 1901

New York City and vicinity. U.S. Geological Survey. Charles D. Walcott, director. [Washington]: 1901. Edition of January 1901.

Scale: 1:62,500.

Cartographers: H. M. Wilson and C. D. Walcott.

Size: 114.3 x 83.4 cm.

Colored.

(647.957: 1901: Un38nyc Large)

857. [ca. 1905]

Map of New York City. Directory Office. New York: [ca. 1905]. 2 pieces.

Scale: 1 in. = ¾ mi.

Engravers: Fisk and Rusell.

Size: 19.8 x 50.5 cm.

(647.957: [ca. 1905]: F523nyc Small)

NEW JERSEY Nos. 858-894

858. 1677

A mapp of New Jersey in America, by John Seller and William Fisher. London: Seller and Fisher: 1677.

Scale: 1:2 English leagues.

Cartographers: Seller and Fisher.

Size: 43.5 x 92.3 cm.

Colored.

Inset: New York [City].

Reproduced in collotype by the Meriden Gravure Company, 1958.

(648: 1677: Se57rnj Large)

859. [1747]

[Map of New Jersey north of the Raritan River to the North Mountain. 1747].

Scale: 2.5 cm. = 5 mi.

Cartographer: Lewis Evans (?).

Engraver: James Turner.

Size of plate: 28.3 x 61.5 cm.

In: Board of General Proprietors of the Eastern Division of New Jersey. *A bill in the chancery of New Jersey,* following p. 124.

Wheat: 397

(347.8: N46b)

860. [1747]

[Map of the northern parts of New Jersey and the adjacent parts of New York. 1747].

Scale: 2.5 cm. = 150 chains.

Cartographer: Lewis Evans.

Engraver: James Turner.

Size of plate: 42.5 x 34.5 cm.

In: Board of general proprietors of the eastern division of New Jersey. *A bill in the chancery of New Jersey*, following p. 124.

Wheat: 398

(347.8: EN46b)

861. 1778

The province of New Jersey, divided into east and west, commonly called the Jerseys. Second edition with considerable improvements. This map has been drawn from the surveys made in 1769 by order of the commissioners appointed to settle the partition line between the provinces of New York & New Jersey by Bernard Ratzer . . . and from another large survey of the northern parts . . . by Gerard Ban[c]ker . . . In this second edition great use has been made of several military surveys generously communicated by the officers of the British troops and of the regiments of Hesse and Anspach. [London]: W. Faden: 1778 [1 December 1778].

Scale: 1 degree = ca. 69½ British mi.

Cartographers: Bernard Ratzer, Gerard Bancker, and W. Faden.

Engraver: W. Faden.

Size: 78 x 56.6 cm.

Colored.

Facsimile made from a fine hand-colored copy of the 1778 map in the Ely Collection of the New Jersey Historical Society . . . printed in six colors on hand-fashioned paper by the Historic Publication Company of Montclair, New Jersey, from plates made for this edition in Milan, Italy.

(648: 1778: F123jer Large)

862. [1780]

New and accurate map of New Jersey, from the best authorities. [London: J. Hinton: 1780].

Size: 31.5 x 26.6 cm.

From *Universal Magazine*. London: John Hinton: 1780. June 1780, vol. 66, p. 281.

(648: [1780]: N426nj Small)

863. [1789]

From New-York (40) to Elizabeth Town. [1789, 1961 reprint].

Size: 17 x 11.5 cm.

In: Christopher Colles, *A survey of the roads;* also Christopher Colles, *The geographical ledger and systematized atlas.*

Wheat: 401

(526.8: C68s, and Pam., vol. 349, no. 4)

864. [1789]

From New-York (41) to Brunswick. [1789, 1961 reprint].

Size: 17.1 x 11.6 cm.

In: Christopher Colles, *A survey of the roads;* also Christopher Colles, *The geographical ledger and systematized atlas.*

Wheat: 402

(526.8: C68s, and Pam., vol. 349, no. 4)

865. [1789]

From New-York (42) to Brunswick. [1789, 1961 reprint].

Size: 17 x 11.3 cm.

In: Christopher Colles, *A survey of the roads;* also Christopher Colles, *The geographical ledger and systematized atlas.*

Wheat: 403

(526.8: C68s, and Pam., vol. 349, no. 4)

866. [1789]

From New-York (43) to Kingston. [1789, 1961 reprint].

Size: 17.1 x 11.6 cm.

In: Christopher Colles, *A survey of the roads;* also Christopher Colles, *The geographical ledger and systematized atlas.*

Wheat: 404

(526.8: C68s, and Pam., vol. 349, no. 4)

867. [1789]

From New-York (44) to Trenton. [1789, 1961 reprint].

Size: 17.3 x 11.3 cm.

In: Christopher Colles, *A survey of the roads;* also Christopher Colles, *The geographical ledger and systematized atlas.*

Wheat: 405

(526.8: C68s, and Pam., vol. 349, no. 4)

868. [1789]

From New-York (45) to Bristol. [1789, 1961 reprint].

Size: 17.2 x 11.6 cm.

In: Christopher Colles, *A survey of the roads;* also Christopher Colles, *The geographical ledger and systematized atlas.*

Wheat: 406

(526.8: C68s, and Pam., vol. 349, no. 4)

869. [1789]

From New-York (45*) to Cranberry. [1789, 1961 reprint].

Size: 17.2 x 11.6 cm.

In: Christopher Colles, *A survey of the roads.*

Wheat: 407

(526.8: C68s)

870. [1789]
From New-York (46*) to Allen town. [1789, 1961 reprint].
Size: 17.2 x 11.5 cm.
In: Christopher Colles, *A survey of the roads.*
Wheat: 408
(526.8: C68s)

871. [1789]
From New-York (47*) to the Blackhorse. [1789, 1961 reprint].
Size: 17.1 x 11.5 cm.
In: Christopher Colles, *A survey of the roads.*
Wheat: 409
(526.8: C68s)

872. [1789]
From New-York (48) to Mount Holly. [1789, 1961 reprint].
Size: 17.3 x 11.6 cm.
In: Christopher Colles, *A survey of the roads.*
Wheat: 410
(526.8: C68s)

873. [1789]
From New-York (49) to Philadelphia. [1789, 1961 reprint].
Size: 17.2 x 11.6 cm.
In: Christopher Colles, *A survey of the roads.*
Wheat: 411
(526.8: C68s)

874. [1789]
From New-York (50) to Philadelphia. [1789, 1961 reprint].
Size: 17.2 x 11.5 cm.
In: Christopher Colles, *A survey of the roads.*
Wheat: 412
(526.8: C68s)

875. [1795]
New Jersey. [1795].
Scale: 2.5 cm. = 16½ mi.
Size: 18.4 x 15.3 cm.
Provenance: Presented by the Estate of Richard Gimbel, 1974.
In: Joseph Scott, *The United States gazetteer,* opposite "Jersey."
Wheat: 415
(Paine 86: Sco3u)

876. [1796]

New Jersey. [1796].

Scale: 2.5 cm. = 25 American mi.

Engraver: A. Doolittle.

Size: 19.1 x 14.4 cm.

Provenance: Presented by Suzanne Wister Eastwick and Joseph L. Eastwick, July 1975.

In: Mathew Carey, *Carey's American pocket atlas,* following p. 52.

Wheat: 416

(917.3: C25c)

877. [1799]

State of New Jersey. [1799].

Engraver: W. Barker.

Size: 27.3 x 18.7 cm.

In: John Payne, *A new and complete system of universal geography,* vol. 4, opposite p. 320.

Wheat: 420

(910: P29)

878. [ca. 1820]

The Morris canal route. N.p.: [ca. 1820].

Scale: 2 cm. = 3 mi.

Engraver: J. A. Adams.

Size: 18.5 x 42.4 cm.

(648: [ca. 1820]: Ad11mcr Small)

879. 1828

A map of the state of New Jersey with part of adjoining states. Compiled under the patronage of the legislature of said state by Thomas Gordon. 1828. Trenton: Gordon: 1828 [20 March 1828].

Scale: 1:190,080.

Cartographer: Thomas Gordon.

Engraver: H. S. Tanner.

Size: 141 x 83 cm.

Colored.

Provenance: Presented by H. S. Tanner, 21 November 1828.

(648: 1828: G653snt Large)

880. 1836

Cape May roads, including Crow Shoal, Del[aware] Bay (September, 1836) and shewing the plan of an artificial harbor proposed for that place. U.S. Topographical Engineers. [Washington]: 1838.

Scale: 6 in. = 1 mi.

Cartographer: Hartman Bache.

Engraver: William J. Stone.

Size: 65 x 113.4 cm.

Provenance: Presented and endorsed by Hartman Bache, 26 June 1838.

See: U.S. 25th Congress. 2nd Session. Document no. 155.

(648: 1836: Un38cmr Large)

881. 1846

Little Egg Harbor. Founded upon a trigonometrical survey under the direction of F. R. Hassler. U.S. Coast and Geodetic Survey. [Washington]: 1846.

Scale: 1:30,000.

Cartographers: F. R. Hassler, C. Renard, G. M. Bache, B. F. Sands, and M. I. McClery.

Engravers: O. A. Lawson and J. Knight.

Size: 44 x 35.8 cm.

Provenance: Presented through A. D. Bache, 26 January 1847.

(648: 1846: Un33Leh Small)

882. 1879

Coast chart no. 22. From Barnegat Inlet to Absecon Inlet, New Jersey. U.S. Coast and Geodetic Survey. [Washington]: 1879.

Scale: 1:80,000.

Cartographers: C. Renard, J. Farley, F. W. Perkins, Charles M. Bache, H. M. De Wees, T. R. Gedney, T. S. Phelps, J. F. Moser, W. I. Vinal, C. Fendall, W. W. Harding, T. J. Lee, C. A. Schott, A. Lindenkohl, W. Lindenkohl, and C. Junken.

Engravers: H. C. Evans, E. A. Maedel, and F. Courtenay.

Size: 96.8 x 77.4 cm.

Inset: Diagram of charts of the New Jersey shoreline.

Provenance: Presented by the U.S. Coast and Geodetic Survey through Spencer C. McCorkle, 14 December 1885.

(648: 1879: Un38bai Large)

883. 1880

Cape May city and vicinity, New Jersey. U.S. Coast and Geodetic Survey. Surveyed by C. M. Bache . . . during the summer of 1879. [Washington]: 1880.

Scale: 1:10,000.

Cartographer: C. M. Bache.

Size: 59.5 x 97.2 cm.

Provenance: Presented by the U.S. Coast and Geodetic Survey through Spencer C. McCorkle, 14 December 1885.

(648: 1880: Un38cmc Large)

884. 1882

A topographical map of a part of northern New Jersey. Geological Survey of New Jersey. From surveys and levelings made and local surveys corrected by Geo. W. Howell and C. C. Vermeule, upon a projection made by the U.S. Coast and Geodetic Survey. [New York]: J. Bien: 1882.

Scale: 1:63,360.

Cartographers: George W. Howell and C. C. Vermeule.

Lithographer: Julius Bien.

Size: 89.6 x 90.1 cm.

Colored.

Letter from H. B. Kümmel identifies the map as a preliminary edition which is not in the New Jersey collection; 28 February 1903.

(648: 1882: N466nnj Large)

885. 1885

Coast chart no. 21. From Sandy Hook to Barnegat Inlet, New Jersey. U.S. Coast and Geodetic Survey. Aids to navigation corrected to 1885. [Washington]: 1885.

Scale: 1:80,000.

Cartographers: E. Blunt, C. Renard, J. Ferguson, J. Farley, W. S. Edwards, A. T. Mosman, F. W. Perkins, C. M. Bache, A. M. Harrison, Thomas R. Gedney, Geo. M. Bache, J. F. Moser, C. Fendall, W. I. Vinal, J. Renwick, C. A. Schott, T. C. Hilgard, H. Lindenkohl, P. Erichsen, and A. Lindenkohl.

Engravers: H. C. Evans, W. A. Thompson, J. G. Thompson, E. A. Maedel, and F. Courtenay.

Size: 97.9 x 76.1 cm.

Inset: Diagram of charts of the New Jersey shoreline.

Provenance: Presented by the U.S. Coast and Geodetic Survey through Spencer C. McCorkle, 14 December 1885.

(648: 1885: Un38shb Large)

886. 1887-1889

Atlas of New Jersey. Geological Survey of New Jersey. George H. Cook, state geologist; John C. Smock, assistant geologist; and, C. Clarkson Vermeule, topographer. New York: J. Bien, lith.: 1887-1889.

Map is in numbers: 1-10, 11-20, 21-28, 30-33, 35-38. Nos. 21-38 revised in 1903-1907.

Scale: 1 in. = 1 mi.

Cartographers: George H. Cook, John C. Smock, and C. Clarkson Vermeule.

Size: ca. 88 x 62 cm.

Colored.

1. Kittatinny valley & mountain. 2. Southwestern highlands. 3. Central highlands. 4. Northeastern highlands. 5. Vicinity of Flemington. 6. The valley of the Passaic. 7. Counties of Bergen, Hudson & Essex. 8. Vicinity of Trenton. 9. Monmouth shore. 10. Vicinity of Salem. 11. Vicinity of Camden. 12. Vicinity of Mount Holly. 13. Vicinity of Barnegat Bay. 14. Vicinity of Bridgeton. 15. Southern interior. 16. Egg Harbor and vicinity. 17. Peninsula of Cape May. 18. State map of New Jersey, geographic. 19. Relief map of New Jersey. 20. Geological map of New Jersey. 21-38. Atlas sheets 21-28, 31-33, 35-38 with original wrapper.

(648: 1887-1889: N466geo Large)

887. 1887

Map of Newbold, New Jersey. Laid out by Fowler and Lummis for Newbold Improvement Company. Philadelphia: F. Bourquin: 1887.

Scale: 1 in. = 200 ft.

Cartographers: Fowler and Lummis.

Lithographer: F. Bourquin.

Size: 64.7 x 57.8 cm.

Colored.

(648.957: 1887: N426mnb Small)

888. 1893

New Jersey. Bordentown sheet. U.S. Geological Survey. J. W. Powell, director. Surveyed in 1885. Edition of September 1893. [Washington]: 1893.

Scale: 1:62,500.

Cartographers: J. W. Powell, Henry Gannett, George H. Cook, and C. C. Vermeule.

Size: 44.5 x 34.2 cm.

Colored.

(648: 1893: Un33bs Small)

889. 1896

Relief map of New Jersey, 1896. Geological Survey of New Jersey. John C. Smock, state geologist. Based upon the topographical survey. N.p.: 1896. Name sheet in three pieces.

Scale: 1 in. = 4 mi.

Cartographers: John C. Smock and C. C. Vermeule.

Size: 110 x 61.8 cm.

Colored.

(648: 1896: N466ref Large)

890. 1899-1902

[Geologic and topographic maps of New Jersey]. Geological survey of New Jersey. John C. Smock, state geologist. C. C. Vermeule, topographer. Surveyed in 1884-1887. [Trenton?]: 1899-1902. Edition of 1899-1902.

Scale: 6.5 cm. = ½ mi.

Cartographers: John C. Smock and C. C. Vermeule.

Colored.

1. Taunton. 54 x 77.6 cm. 2. Amboy. 54 x 76.7 cm. 3. Navesink. 54 x 76.8 cm. 4. Long Branch. 54 x 76.9 cm. 5. Trenton. 54 x 77 cm. 6. Atlantic City. 54 x 78 cm. 7. New York Bay. 54 x 76.5 cm. 8. Shark River. 54 x 76.9 cm. 9. Jersey City. 54 x 76.5 cm. 10. Newark. 54.2 x 76.2 cm. 11. Morristown. 54 x 76.2 cm. 12. Hackensack. 54.2 x 76.6 cm. 13. Paterson. 54 x 76.2 cm. 14. Elizabeth. 54 x 76.4 cm. 15. Plainfield. 54 x 76.4 cm. 16. Camden. 54 x 77.3 cm. 17. Mount Holly. 54 x 77.3 cm. 18. Woodbury. 54 x 77.3 cm.

(648: 1899-1902: N466top Large)

891. 1900

New Jersey showing forest area and its relation to the principal watersheds. Geological Survey of New Jersey. John C. Smock, state geologist. C. C. Vermeule, topographer. [Trenton]: 1900. 7 sheets.

Scales: 1 in. = 5 mi., and 1 mi.

Cartographers: John C. Smock and C. C. Vermeule.

Sizes: Various.

Colored.

See: New Jersey. Geological Survey. Forests of New Jersey, *Annual report.* 1900.

(648: 1900: N466for Large)

892. [ca. 1905]

[Geological Survey of New Jersey]. Geological Survey of New Jersey. Henry B. Kümmel, state geologist. C. C. Vermeule, topographer. Surveyed in 1881-1883-1886-1887. Resurvey in 1904. N.p.: [ca. 1905]. 6 sheets. Edition of 1905.

Scale: 5¼ in. = 2 mi.

Cartographers: Henry B. Kümmel and C. C. Vermeule.

1. Dover-Stanhope. 54 x 76.2 cm. 2. Boonton. 54 x 76.2 cm. 5. Chester. 54 x 76.3 cm. 9. Pluckemin. 54 x 76.6 cm. 13. Somerville. 54 x 76.6 cm. 14. New Brunswick. 54 x 76.6 cm.

(648: [ca. 1905]: N466gsj Large)

893. [1955]

New Jersey. Sheet 32. Department of Conservation and Economic Development. Topographic Service. [1955].

Scale: 1 in. = 1 mi.

Cartographers: C. C. Vermeule and R. G. Blanchard.

Size: 83.6 x 60.6 cm.

Colored.

Provenance: Presented by Mrs. A. I. Hallowell, November 1983.

(648: [1955]: V591ccd Small)

894. [n.d.]

Plan of Barnegat City, situated at Barnegat Inlet, Long Beach, N[ew] J[ersey]. [Camden, N.J.]: W. T. Bailey & Bro.: n.d.

Size: 49.3 x 83.7 cm.

Inset: Picture of Barnegat lighthouse, bay, and inlet.

(648: [n.d.]: P376pbc Large)

PENNSYLVANIA Nos. 895-986

895. [1654/55]

Nova Suecia; eller the Swenska Revier [now Delaware River] in India Occidentalis. Drawn by Peter Lindeström, Royal Swedish Engineer, 1654 & 1655.

Scale: Milm. Germm. 15 in uno Gradu.

Cartographer: Peter Lindeström.

Size: 14.2 x 60.4 cm.

Reproduction of manuscript in Royal Swedish Archives. Size of original: 14.4 x 70 cm.

(635: [1654/55]: L645nso Large)

896. [1654/55]

Nova Suecia hodie dicta Pennsylvania Anno 1654 och 1655. Ar denna Nova Sueciae carta med dess riviers och lande situation, ock Beskaffenhet oftagen och tilt carts fôrd af P. Lindeström. [Philadelphia: 1834].

Cartographer: P. Lindeström.

Engraver: Fortif.

Size: 26.4 x 12.8 cm.

Original map in Campanius Holm, *Kort beskrifning om provincien Nya Swerige uti America. . . .* Stockholm: Wankijfs: 1702.

This is a duplicate of the map in Historical Society of Pennsylvania, *Memoirs,* vol. 3, p. 46, 1834.

(649: [1654/55]: L645ns Small)

897. [1687]

Map of the improved part of the province of Pennsilvania containing the three countyes of Chester, Philadelphia & Bucks as far as yet surveyed & laid out; ye divisions or distinctions made by ye different coullers, respects the settlements by way of townships. By Thomas Holme. London: Green & Thornton: [1687].

Scale: 4 English mi. or 32 furlongs.

Cartographer: Thomas Holme.

Engraver: F. Lamb.

Size: 82 x 140.5 cm.

Inset: City of Philadelphia.

Provenance: Presented by John Thornton and Robert Greene, mapsellers, London, 21 May 1824.

(649: [1687]: H733pen Framed. In map and print room)

898. [1702?]

Novae Sueciae tabula. ex Nic. Visscheri, del. [Stockholm?: 1702?].

Cartographer: Nikolaes Jamsz Visscher II.

Size: 13.1 x 8.2 cm. *Size of paper:* 21.7 x 28.7 cm.

See Historical Society of Pennsylvania, *Memoirs,* vol. 3, p. 78.

Original in: Thomas Campanius Holm, *Kort beskrifning om provincien Nya Swerige.* Stockholm: S. Wankijfs: 1702. P. 26.

Contains also a drawing of Trinity Fort.

(640.2: [1702?]: V828ns Small)

899. 1747

Plot of adjusted survey [of] Moravian road: Authorized: Court of Quarter-Sessions, New Town, June 11, 1747.

Cartographers: Thomas Craig, George Gray, and Robert Greeg.

Size: 26 x 34.4 cm.

Provenance: Presented by John Robert Connelly, December 1964.

Photostat.

Report of survey, 9 March 1748: "From the kings road near *Bethlehem* to the *Mahoning Creek* passing near the *Healing Waters* beyond the Blue Mountain." See: Northampton County Historical and Genealogical Society. *Publications,* vol. 2. "The Penn patents in the Forks of the Delaware."

See: "Independent conformation of the magnetic declination 1737-1748" by John Robert Connelly [Palmerton, Pa.: 1963].

(649: 1747: P376mor Small)

900. 1759

To the Honorable Thomas Penn and Richard Penn . . . map of the improved part of the province of Pennsylvania. By Nicholas Scull. Philadelphia: John Davis: 1759 [1 January 1759].

Scale: 69½ British mi. = 1 degree.

Cartographer: Nicholas Scull.

Engraver: J. Turner.

Size: 76.5 x 152.6 cm.

Colored.

Provenance: Presented by John Vaughan, 17 January 1806.

Wheat: 422

(649: 1759: Scu47tr Large)

901. [1765]

Plan of the battle near Bushy-Run gained by His Majesty's troops, commanded by Colonel Henry Bouquet over the Delaware, Shawanese, Mingoes, Wyandots, Mohikons, Miamies & Ottawas, on the 5th and 6th: of August, 1763. From an actual survey. [1765].

Scale: 2.5 cm. = 26 perches.

Cartographer: Thomas Hutchins.

Size: 28.9 x 33.2 cm.

In: William Smith, *An historical account of the expedition under the command of Henry Bouquet,* opposite p. 44.

Wheat: 424

(973.2: Sm6)

902. [ca. 1768]

Plan of the boundary lines between the provinces of Maryland and the three lower counties of Delaware, with part of the parallel of latitude which is the boundary between the provinces of Maryland and Pennsylvania. [The line was run by Mason and Dixon. London: ca. 1768]. 3 maps.

Scale: 1 in. = 4 mi.

Cartographers: Charles Mason and Jeremiah Dixon.

Sizes: 63.7 x 190.9 cm.; 50 x 64.7 cm.; and 37.5 x 66 cm.

Provenance: Presented and autographed by James Mease, 19 May 1826, another copy presented by Chew Family Trust, 15 November 1982.

Mease presented several "rare and interesting maps, of early date, of the division lines between Pennsylvania, Delaware and Maryland, accompanied by some explanatory notes." The gifts were listed as:

> 1. A plan of the Boundary line between the Province of Maryland, and the three lower Counties on Delaware, with facts (?) of the paralel [sic] of Latitude, which is the boundary between the provinces of Maryland and Pennsa.
>
> 2. A plan of the famous "West Line," or paralel [sic] of latitude, which is the boundary between the Provinces of Maryland and Pennsa. which was run by Mason and Dixon in 1762 & 8.
>
> 3. A map of that part of America, where a degree of latitude was measured for the Royal Society by Mason and Dixon in 1768 [Archives. J. Mease to APS, 19 May 1826; Minutes; Donation Book; *Transactions,* n.s., vol. 3].

See: Thomas D. Cope, "Degrees along the west line . . ." APS *Proceedings,* 1949.

Wheat: 497

(649: [ca. 1768]: Sm57madL Large)

903. 1770

To the Honorable Thomas Penn and Richard Penn . . . and to John Penn . . . map of the province of Pennsylvania. By William Scull. Philadelphia: James Nevil: 1770 [4 April 1770].

Scale: 1 in. = 10 mi.

Cartographer: William Scull.

Engraver: Henry Dawkins.

Size: 55.8 x 81.4 cm.

Colored.

William Scull, of West Jersey, sent 13 August 1768 to the American Society, through John Morgan, his proposals for publishing a "Map of the province of Pennsylvania from a more exact Survey and on a larger Scale than any yet extant." He asked

> That if agreeable a Committee be appointed by the Society to examine into the Accuracy of the performance, and Report accordingly.
>
> That if found by the Society worthy of Notice, they would grant him the Liberty of dedicating the Same to the Society in conjunction with the philosophical Society.
>
> That an Advertisement be printed in which he desires may be inserted the Minute of this Society as a Recommendation of his Undertaking to the public.
>
> That as it will be attended with considerable Expence, the Society to give him all the Assistance they are able in the Sale of his Map [APS Minutes].

Several members of the Society are thanked by Scull for their help.

Wheat: 425

(649: 1770: Scu47pn Large)

904. [1777]

A map of Pennsylvania exhibiting not only the improved parts of that province, but also its extensive frontiers: laid down from actual surveys and chiefly from the late map of W. Scull published in 1770. La Pensilvanie en trois feuilles. Traduite des meilleures cartes anglaises. Paris: Le Rouge: [1777].

Scale: 69½ English mi. = 1 degree.

Cartographer: W. Scull.

Size: 66.4 x 131.3 cm.

Colored.

See: Georges Louis Le Rouge, *Atlas Amériquain septentrional.* Paris: Le Rouge: 1778-[ca. 1792]. No. 14.

(649: [1777]: P376par Large)

905. [1788]

To the patrons of the Colombian Magazine this map of Pennsylvania is dedicated by their obliged and obedt. servants. The Proprietors. [1788].

Scale: 2.5 cm. = 15½ mi.

Size: Top right corner fragment of map is all that remains.

In: *Columbian Magazine,* vol. 2, opposite p. 3.

Fragment.

Wheat: 426

(050: C72)

906. [1789]

From New-York (46) to Frankford. [1789, 1961 reprint].

Size: 17.1 x 11.5 cm.

In: Christopher Colles, *A survey of the roads;* also Christopher Colles, *The geographical ledger and systematized atlas.*

Wheat: 427

(526.8: C68s, and Pam., vol. 349, no. 4)

907. [1789]

From New-York (47) to Philadelphia. [1789, 1961 reprint].

Size: 17.1 x 11.6 cm.

In: Christopher Colles, *A survey of the roads;* also Christopher Colles, *The geographical ledger and systematized atlas.*

Wheat: 428

(526.8: C68s, and Pam., vol. 349, no. 4)

908. [1789]

From Philadelphia (51) to Annapolis, Maryland. [1789, 1961 reprint].

Size: 17.1 x 11.5 cm.

In: Christopher Colles, *A survey of the roads.*

Wheat: 429

(526.8: C68s)

909. [1789]

From Philadelphia (52) to Annapolis, Maryl[an]d. [1789, 1961 reprint].

Size: 17.1 x 11.5 cm.

In: Christopher Colles, *A survey of the roads.*
Wheat: 430

(526.8: C68s)

910. [1789]

From Phila[delphia] (53) to Annapolis, Maryl[an]d. [1789, 1961 reprint].
Size: 17.6 x 11.4 cm.
In: Christopher Colles, *A survey of the roads.*
Wheat: 480

(526.8: C68s)

911. [1789]

From Philadelphia (54) to Annapolis, Maryland. [1789, 1961 reprint].
Size: 17 x 11.5 cm.
In: Christopher Colles, *A survey of the roads.*
Wheat: 481

(526.8: C68s)

912. [1791]

District of Pennsylvania . . . A map of Pennsylvania, & the parts connected therewith, relating to the roads and inland navigation, especially as proposed to be improved by the late proceedings of the Assembly. (Copied from his larger map.) To the Legislature and the Governor of Pennsylvania this map is respectfully inscribed by Reading Howell. [Philadelphia: 1791. 13 June 1791].
Scale: 1 in. = 12 mi.
Cartographer: Reading Howell.
Engraver: J. Trenchard.
Size: 46.3 x 66.7 cm.

(649: [1791]: H833cfm Large)

913. [ca. 1791]

A map exhibiting a general view of the roads and inland navigation of Pennsylvania, and part of the adjacent states. By John Adlum and John Wallis. N.p.: [ca. 1791].
Scale: 1 degree = 69 mi.
Cartographers: John Adlum and John Wallis.
Size: 84.1 x 93.5 cm.
Colored.
Inset: Perspective view of part of a canal with locks.
Mutilated.
Wheat: 432

(649: [ca. 1791]: Ad1lgrn Large)

914. 1792

A map of the state of Pennsylvania, by Reading Howell. 1792. To Thomas Mifflin governor, the Senate, and House of Representatives of the Commonwealth of Pennsylvania, this map is respectfully inscribed by the author. [Philadelphia]: 1792 [1 August 1792].

Scale. 1 degree – 69 mi.

Cartographer: Reading Howell.

Size: 94.5 x 160.3 cm.

Provenance: Presented by Reading Howell, February 1793.

Trimmed.

Wheat: 433

(649: 1792: H833cpa Large)

915. [1795]

Pennsylvania. [1795].

Scale: 2.5 cm. = 45 mi.

Size: 15.5 x 18.8 cm.

Provenance: Presented by the Estate of Richard Gimbel, 1974.

In: Joseph Scott, *The United States gazetteer,* opposite "Pennsylvania."

Wheat: 444

(Paine 68: Sco3u)

916. [1796]

Pennsylvania. [1796].

Scale: 2.5 cm. = 50 mi.

Engraver: W. Barker.

Size: 14.7 x 19.3 cm.

Provenance: Presented by Suzanne Wister Eastwick and Joseph L. Eastwick, July 1975.

In: Mathew Carey, *Carey's American pocket atlas,* opposite p. 58.

Wheat: 445

(917.3: C25c)

917. [1797]

The state of Pennsylvania. From the latest surveys. New York: J. Reid: [1797].

Scale: 1 degree = 69½ American mi.

Engraver: D. Martin.

Size: 32.2 x 45.1 cm.

See: John Reid, publisher, *The American atlas.* New York: J. Reid: 1796, no. 12.

Wheat: 451

(649: [1797]: M365paa Small)

918. 1800

The state of Pennsylvania from the latest surveys. 1800.

Scale: 2.5 cm. = 20 American mi., 69½ = 1 degree.

Size: 18.3 x 26.6 cm.

In: John Payne, *A new and complete system of universal geography,* vol. 4, opposite p. 329.
(910: P29)

919. [1803]

A correct plan of the town of Kittanning, in Armstrong county, Pennsylvania; laid out in the year 1803. Printed by James Alexander: [1803].

Size of paper: 32.4 x 19.5 cm.

(649.946: [1803]: K654acp Small)

920. [ca. 1810]

A draft of the city of Germany. Founded on the 25th day of September A.D. 1810. By Samuel F. Conover. Situated in the county of Somerset, state of Pennsylvania. [Philadelphia?: ca. 1810].

Scale: 1 in. = 200 ft.

Cartographer: Reading Howell.

Engraver: Henry S. Tanner.

Size: 58.6 x 69.3 cm.

Colored.

Provenance: Presented by Dr. Conover, 3 May 1811.

(649.776: [ca. 1810]: C743gsc Large)

921. [ca. 1811]

A map of the state of Pennsylvania by Reading Howell. 1811. Philadelphia: Kimber & Conrad & Johnson & Warner: [ca. 1811].

Scale: 1 in. = 10 mi.

Cartographer: Reading Howell.

Engraver: F. Vallance.

Size: 54.5 x 85.9 cm.

Inset: Schuylkill permanent bridge.

(649: [ca. 1811]: H833pkv Large)

922. [ca. 1811]

A map of the state of Pennsylvania, by Reading Howell. Revised 1811. Philadelphia: Kimber & Conrad and Johnson & Warner: [1811].

Scale: 1 degree = 69 mi.

Cartographer: Reading Howell.

Size of paper: 96.4 x 163.5 cm.

Colored.

(649: [1811]: H833pre Large)

923. 1814

Plan of the town of York Haven. N.p.: 1814.

Engravers: Harrison & Frederick.

Size: 60.1 x 73.8 cm.

Top half of this map is missing.

(649.997: 1814: H243tyh Large)

924. [ca. 1815]

Specimen of the county maps to be constructed by virtue of an act of legislature directing the formation of a map of Pennsylvania. N.p.: [ca. 1815].

Cartographer: John Melish.

Size: 2 cm. = 3 mi.

Colored.

In: John Melish, "Geographical intelligence . . . " [ca. 1815].

(Pam., vol. 45, no. 8.)

925. [ca. 1816]

Plan of Pittsburgh and adjacent country. Surveyed by W[illia]m Darby. Philadelphia: Darby: [ca. 1816].

Scale: 3 1/4 cm. = 2 furlongs.

Cartographer: William Darby.

Engraver: H. S. Tanner.

Size: 50.9 x 43.3 cm.

Colored.

Provenance: Presented, with manuscript notes, by George Merrick, 4 January 1830; Reverend Robert Patterson presented an 1816 plan of Pittsburgh by William Darby, 16 May 1817.

(649.963: [ca. 1816]: D242pac Small)

926. 1818

A map shewing the situation of Susquehanna county, Pennsylvania. Silver Lake, Pennsylvania. N.p.: Robert H. Rose: 1818.

Cartographer: Robert H. Rose.

Engraver: J. Bower.

Size of plate: 21.3 x 18.5 cm. *Size of paper:* 40 x 25.4 cm.

Map is part of a broadside advertising a tract of land in Susquehanna County, Pennsylvania.

(649.778: 1818: R727scp Small)

927. 1821

Map of York & Adams counties, [Pennsylvania]. Published by D. Small and W. Wagner. 1821. N.p.: D. Small & W. Wagner: 1821 [23 June 1821].

Scale: 1 in. = 3 mi.

Engraver: W. Wagner.

Size: 39.5 x 59 cm.

Colored.

(649.797: 1821: S27ya Large)

928. 1822

[Geological map of the north central and south central sections of Pennsylvania]. [Philadelphia?:] John Melish: 1822 [7 June 1822]. 2 book maps.

Size of paper: 66 x 67 cm.; and 67 x 67.1 cm.

Colored.

Provenance: Presented by the heirs of Martin H. Boyé, ca. 1880.

Incomplete. Mutilated.

Contains statistical table, showing prominent features of each county.

(649: 1822: M485geo Book maps [2])

929. 1824

Map of Lancaster County, Pennsylvania. By Joshua Scott. Lancaster: 1824 [1 January 1824].

Scale: 2½ cm. = 1 mi.

Cartographer: Joshua Scott.

Size: 91.4 x 118.3 cm.

Colored.

(649.748: 1824: Sco37Lcp Large)

930. [1826]

Map of Pennsylvania, constructed from the county surveys authorized by the state; and other original documents, by John Melish. Corrected and improved to 1826. [Philadelphia]: J. Melish: [1826].

Scale: 1 in. = 6 mi.

Cartographer: John Melish.

Engraver: B. Tanner.

Size: 129 x 186.6 cm.

Colored.

Provenance: Presented by Henry S. Tanner, 15 February 1828.

Tanner noted that the map was corrected to 1825. He presented the map "as a mark of my high respect for an Institution, equally distinguished for its liberality and extensive usefulness."

Contains statistical tables.

(649: [1826]: M485csp Large)

931. [ca. 1830]

Plan of the town of Port Clinton, at the forks of the Schuylkill. Philadelphia: C. G. Childs: [ca. 1830].

Scale: 1 in. = 200 ft.

Lithographer: C. G. Childs.

Size: 64.8 x 46.5 cm.

(649.964: [ca. 1830]: C432cfs Large)

932. [1831]

Plan and profile of the Danville and Pottsville Railroad. 1831. Philadelphia: Kennedy & Lucas, lith.: [1831].

Scale: 1 in. = 1¼ mi.

Lithographers: Kennedy & Lucas.

Size: 46.8 x 64.5 cm.

See: *Report of the superintendent of the Danville and Pottsville Railroad.* Pam. vol. 715, no. 5.

(649: [1831]: D192pdr Large)

933. 1832

Map of Fayette County, Pa. From actual surveys by F. Lewis. N.p.: 1832.

Scale: 1 in. = 2 mi.

Cartographer: F. Lewis.

Engraver: S. M. Lewis.

Size: 40.5 x 49 cm.

Colored.

(649.731: 1832: L585fcp Small)

934. [1832]

Map of Pennsylvania, constructed from the county surveys authorized by the state; and other original documents, by John Melish. Corrected and improved to 1832. Philadelphia: B. Tanner: [1832].

Scale: 1 in. = 6 mi.

Cartographers: John Melish, James Finlayson, John G. Melish, and Edward Paguenaud.

Engraver: B. Tanner. Seal of the state, etc., drawn by F. Kearny.

Size: 130.5 x 188.8 cm.

Colored.

(649: [1832]: M485cst Large)

935. 1837

A map of the canals & rail roads of Pennsylvania and New Jersey and the adjoining states. By H. S. Tanner. 1837. Philadelphia: H. S. Tanner: 1837.

Cartographer: H. S. Tanner.

Engraver: H. S. Tanner.

Size: 52.6 x 70.6 cm.

Colored.

Removed from Pam. vol. 715, no. 12.

(649: 1837: T158rcc Large)

936. 1839

Map of Pennsylvania. Philadelphia: Spencer & Co.: 1839.

Cartographers: Ely & Hammond.

Engravers: Ely & Hammond.

Size: 55 x 81.7 cm.

Colored.

Insets: Philadelphia; Scale of mountains; Profile of the Pennsylvania Canal; Harrisburg; Lehigh and Schuylkill coal regions; Pittsburgh; William Penn's treaty with the Indians.

(649: 1839: Sp37pyv Extra-Oversize)

937. [ca. 1840]

Plan of the Philipsburg estate in Centre, Clearfield & Cambria counties, Pennsylvania. Philadelphia: Watson's lith.: [ca. 1840].

Scale: ¾ in. = 1 mi.

Lithographer: Watson.

Size: 55.8 x 80.7 cm.

Map shows coal fields.

(649.763: [ca. 1840]: P536ccc Large)

938. 1844

Industrial and political map of Pennsylvania, exhibiting the population, the principal resources of each county, and the valuation of property: the several congressional districts, and the vote for presidential electors in 1840. Compiled from the returns of the sixth census and from other official sources. [Philadelphia]: 1844.

Size of paper: 41.2 x 65 cm.

Colored: "The map is so colored as to represent the several congressional districts."

Provenance: Presented by C. Gibbons.

(649: 1844: P376cen Book map)

939. [ca. 1850]

Conestoga navigation. E. Colman, esq., proprietor. [Philadelphia]: P. S. Duval: [ca. 1850].

Scale: 1¾ in. = ¾ mi.

Cartographer: J. Scott, engineer.

Engraver and lithographer: J. Probst and P. S. Duval.

Size: 21.5 x 74.6 cm.

Colored.

(649: [ca. 1850]: Sco37cn Small)

940. [ca. 1850]

New Philadelphia [Schuylkill County, Pennsylvania]. N.p.: [ca. 1850].

Lithographer: Cephus G. Childs.

Size: 36.8 x 41.6 cm.

(649.957: [ca. 1850]: C432np Small)

941. [1851]

Map of Pennsylvania railroad with its connections. Showing the different routes, projected or constructed between the seaboard & the western states. Philadelphia: Friend & Aub, Lith.: [1851].

Lithographers: Friend & Aub.

Size: 59.5 x 142.3 cm.

Contains profile of Pennsylvania Rail Road.

See: Pennsylvania Railroad, *Fourth annual report.* 1851. Pam. vol. 730, no. 15.

(635: [1851]: M322prr Large)

942. [1852]

Map of the projected railway from Harrisburg to Pittsburg[h]. With proposed extensions to Cleveland, Cincinnati and St. Lewis, in connexion with the public works of Pennsylvania, Ohio, Indiana and Illinois. Philadelphia: J. T. Bowen, lithographer: [1852].

Cartographers: J. A. Sheaff and E. F. Gay.

Lithographer: J. T. Bowen.

Size: 40.8 x 103.8 cm.

Colored.

(635: [1852]: R137hpc Large)

943. [ca. 1853]

Map of the Lackawanna & Pittsburg[h] R. R. and connections. N.p.: [ca. 1853].

Size: 48.3 x 75.2 cm.

Colored.

(635: [ca. 1853]: L115Lpr Large)

944. 1855

Pennsylvania. New York: J. H. Colton & Co.: 1855.

Size of paper: 35 x 42.7 cm.

Colored.

Provenance: Presented by Suzanne Wister Eastwick and Joseph L. Eastwick, July 1975.

(649: 1855: C72p Book map)

945. 1856

Map of the canals and rail roads for transporting anthracite coal from the several coal fields to the city of New York. Drawn under the direction of J. Dutton Steele by W. Lorenz. Baltimore: Hanckel & Son: 1856.

Scale: 1 in. = 6 mi.

Cartographers: J. Dutton Steele and W. Lorenz.

Lithographers: Hanckel & Son.

Size: 83.7 x 106.5 cm.

Colored.

(649: 1856: L885can Book map)

946. 1857

A geological and topographical map of Pennsylvania and New Jersey, compiled from the official reports of the geological surveys and from various other sources public and private. Constructed and engraved by W. Williams. Philadelphia: Charles De Silver: 1857.

Scale: 1 in. = 8 mi.

Cartographer: W. Williams.

Engraver: W. Williams

Size: 95.7 x 134 cm.

Colored.

Inset: Colored diagram showing geological folds within the earth.

(649: 1857: P386pnj Large)

947. [ca. 1860]

Franklin township, Orange; [including] Fell [in Luzerne county]. N.p.: [ca. 1860].

Scale: 1½ in. = 1 mi.

Size: 33 x 27 cm.

Colored.

Locates individual properties.

(649.733: [ca. 1860]: P376twp Small)

948. [ca. 1860]

Young's additions to Pottsville and Port Carbon, with part of their adjoining coal lands. The whole for sale in tracts or lots to suit purchasers. N.p.: [ca. 1860].

Size: 73.4 x 53.2 cm.

Colored.

(649.773: [ca. 1860]: Y79ppca Large)

949. 1862

A new county map of Pennsylvania, and adjoining states, showing the route of the railroads, etc. Published by R. L. Barnes. Philadelphia: R. L. Barnes: 1862.

Scale: 1 in. = 10 mi.

Size: 65.7 x 95.2 cm.

Colored.

Contains the autograph of S. Bradford.

(649: 1862: B261roa Book map)

950. [ca. 1865]

Battle field of Gettysburg. U.S. Corps of Engineers. Survey ordered by Brevet Major General A. A. Humphreys . . . and conducted under Brevet Major General G. K. Warren. [Washington: ca. 1865].

Scale: 1:12,000.

Cartographers: G. K. Warren and A. A. Humphreys.

Size: 73.4 x 69.9 cm.

Map reduced from one on a scale of 1 in. = 200 ft., deposited in the archives of the Chief of Engineers.

(649.934: [ca. 1865]: Un38gfb Large)

951. 1865

Outline map of Venango Co., Pa., from actual surveys. New York: F. W. Beer & Co.: 1865.

Scale: 1 in. = 480 rods.

Lithographer: Ferd. Mayer & Co.

Size: 54.1 x 77.8 cm.

Colored.

(649: 1865: Ou95vcp Large)

952. 1871

Map of the bituminous coal region of north western Pennsylvania and the proposed railroad connections with Buffalo & Western New York. Prepared by N. F. Jones and published by Thos. King, Smethport, Pa., 1867. Revised edition 1871. [Smethport, Pa.]: Thos. King: 1871.

Scale: 1 in. = 8 mi.

Cartographer: N. F. Jones.

Lithographer: T. Sinclair.

Size: 68.1 x 48.8 cm.

Colored.

(649: 1871: Si67coL Large)

953. 1871

Map of the first and second anthracite coal fields of Pennsylvania. By Strauch & Cochran. [New York: American Photo-Lith. Co.]: 1871.

Scale: 1 cm. = 1 mi.

Cartographers: Strauch & Cochran.

Size: 32.5 x 65.5 cm.

Colored.

Contains a list of collieries in above coal fields.

(649: 1871: St87ant Large)

954. 1872

[Geological map of a part of Bedford County, Pennsylvania]. N.p.: 1872.

Scale: 1 in. = 1 mi.

Cartographer: J. P. Lesley.

Size: 29.5 x 20.2 cm.

Colored.

(649.716: 1872: L565bcp Small)

955. 1875-1885

Topographical map (in 10 foot contour lines) of the south mountains in Adams, Franklin, Cumberland & York counties. Second Geological Survey of Pennsylvania. Surveyed and drawn by Ambrose E. Lehman, from 1875 to 1885. Under the superintendence until 1877 of Dr. Persifor Frazer. [Harrisburg, Pa.]: Lane S. Hart: [ca. 1885]. 3 sheets.

Scale: 1 in. = 1600 ft.

Cartographers: Persifor Frazer, Ambrose E. Lehman, George M. Lehman, and James E. Maull.

Lithographer: Julius Bien & Co.

Size: Each sheet, ca. 69 x 57 cm.

(649: 1875-1885: P386afy Large)

956. 1876

Maps of the battle field of Gettysburg July 1, 2, 3, 1863 . . . Compiled and added for the government by John D. Bachelder, 1st-3rd. day's battle. U.S. War Department. Boston: J. B. Bachelder: 1876. New York: Endicott & Co.: 1876. 3 maps: one for each day of the battle.

Scale: 1:12,000.

Cartographer: John D. Bachelder.

Engravers: Julius Bien and Louis E. Neuman.

Sizes: 73.7 x 70.2 cm.; 73.9 x 70 cm.; and 73.8 x 69.9 cm.

Colored.

Provenance: Presented by the U.S. Government, 15 February 1878.

(649.934: 1876: Un38bfg Large)

957. 1882

Delaware River from Cherry Island flats to Bridesburg. U.S. Coast and Geodetic Survey. J. E. Hilgard, superintendent. [Washington]: 1882.

Scale: 1:40,000.

Cartographers: S. C. McCorkle, R. M. Bache, C. T. Iardella, H. L. Marindin, H. B. Mansfield, and J. E. Hilgard.

Size: 68.7 x 106.4 cm.

Provenance: Presented by U.S. Coast and Geodetic Survey through Spencer C. McCorkle, 14 December 1885.

Note: Preliminary edition by photo-lithography, to be superceded eventually by an engraved edition.

(649: 1882: Un38cfb Large)

958. 1883-1884

Reference survey maps [of the]anthracite district: Mahonoy/Shamokin basin in Schuylkill county; Mahonoy-Shamokin in Schuylkill, Northumberland and Columbia counties: Wyoming basin, Luzerne County. Second Geological Survey of Pennsylvania. J. P. Lesley, state geologist. Charles A. Ashburner, geologist in charge. New York: J. Bien, lith.: 1883-1884. 4 maps.

Scale: 1 in. = 400 ft.

Cartographers: J. P. Lesley, Charles A. Ashburner, Frank A. Hill, H. E. Parrish, O. B. Harden, and T. J. Williams.

Lithographer: Julius Bien & Co.

Sizes: ca. 61 x 73.5 cm.

See: Pennsylvania. Second Geological Survey, *Report* . . . 1883-1884.

(649: 1883-1884: P37ref Large)

959. 1884

Geological map of Berks county, compiled from the surveys of F. Prime, E. V. D'Invilliers, R. H. Sanders . . . J. P. Lesley, state geologist. Geological Survey of Pennsylvania. [Harrisburg, Pa.]: Lane S. Hart, state printer: 1884.

Scale: 1 in. = 2 mi.

Cartographers: F. Prime, E. V. D'Invilliers, R. H. Sanders, and J. P. Lesley.

Size: 51.2 x 65.6 cm.

Colored.

See: Pennsylvania. Second Geological Survey, *Report of progress;* D'Invilliers . . . , *Geology of the south mountain belt of Berks County* .

(649.716: 1884: P386gmb Large)

960. 1885

A map of Huntingdon County. Compiled by E. B. Harden . . . and geologically colored by J. P. Lesley, from a manuscript field map of the Juniata Valley made by A. A. Henderson, in 1840 . . . C. A. Ashburner and C. E. Billin, in 1875 . . . J. H. Dewes in Tuscarora valley in 1875 . . . C. C. Billin, in 1877 . . . R. H. Sanders, in 1876 . . . by E. B. Harden, 1881 . . . I. C. White in 1883. Second Geological Survey of Pennsylvania. N.p.: 1885.

Scale: 1 in. = 2 mi.

Cartographers: J. P. Lesley, A. A. Henderson, C. A. Ashburner, C. E. Billin, J. H. Dewes, R. H. Sanders, E. B. Harden, and I. C. White.

Size: 62.7 x 49.3 cm.

Colored.

See: Pennsylvania. Second Geological Survey, *Report of progress.* T3, 1885.

(649.739: 1885: P386hcp Large)

961. 1889

A geological and topographical map of the New Boston and Morea coal lands, in Schuylkill County, Pennsylvania. By Benj. Smith Lyman . . . Aided by Amos P. Brown and J. S. Elverson. [Philadelphia]: 1889.

Scale: 1 in. = 400 ft.; or 1:4,800.

Cartographers: B. S. Lyman, Amos P. Brown, and J. S. Elverson.

Size: 66.2 x 114 cm.

Provenance: Presented by Benjamin Smith Lyman, 15 February 1889.

(649.773: 1889: L995sch Large)

962. 1893

Map of the railroad lines in the Pennsylvania system. 1892. [Philadelphia]: 1893.

Scale: 1:37 mi.

Size: 25.9 x 74 cm.

Colored.

Insets: Maps of the New York, Philadelphia, and Norfolk Railroad lines; Grand Rapids and Indiana Railroad.

(635: 1893: P368pr Small)

963. 1894

Maps of the railroad lines in the Pennsylvania system. 1893. [Philadelphia]: 1894.

Scale: 1:37 mi.

Size: 25.9 x 78.4 cm.

Colored.

Insets: Maps of the New York, Philadelphia, and Norfolk Railroad lines; Grand Rapids and Indiana Railroad.

(635: 1894: P368pr Small)

964. [1904]

Pennsylvania railroad to the World's Fair, St. Louis, 1904. N.p.: [1904].

Size: 16.2 x 39.2 cm.

Colored.

(635: [1904]: P386sL Small)

965. 1909

The Rand-McNally new commercial and statistical map of Pennsylvania. Complete index of all post offices and railroads, with latest official census. Chicago: Rand, McNally & Co.: 1909.

Scale: 1 in. = 8.5 statute mi.

Size: 64.9 x 95.6 cm. *Size of paper:* 141.6 x 114 cm.

Colored.

Insets: Eleven maps of Pennsylvania showing: the yield of wheat, manufactured products, lumber and timber products, hay and forage, tobacco, rye per square mile; the earliest exploration and settlements in Pennsylvania; the coal fields of Pennsylvania, and the production of coal by counties; Philadelphia and vicinity; relief map of Pennsylvania; Pittsburgh and vicinity.

On reverse: Rand-McNally new official railroad map of the United States and Dominion of Canada.

Scale: 1 in. = 83 statute mi.

Size: 71.7 x 98.4 cm. *Size of paper:* 141.6 x 114 cm.

Colored.

Insets: Eleven insets of the U.S. showing yield of wheat, corn, oats, hogs, sheep, cattle, horses and mules per square mile; coal fields; annual rainfall; the world; the U.S. with Puerto Rico and the Philippine Islands.

(649: 1909: R157rmp Large)

966. [ca. 1910]

Geology of the Broad Top coal field, Bedford, Fulton and Huntingdon counties, by James H. Gardner. Topographical and Geological Survey of Pennsylvania. Plate IV. N.p.: [ca. 1910].

Scale: 1:24,000.

Cartographer: James H. Gardner.

Size: 105.6 x 101.1 cm.

Colored.

Inset: Coal fields of Pennsylvania.

Provenance: Presented by Richard R. Hill.

(649: [ca. 1910]: P386cfb Large)

967. 1910

Railroad map of Pennsylvania. Published by the Department of Internal Affairs of Pennsylvania. Drawn and compiled by J. Sutton Wall. N.p.: 1910.

Scale: 1 in. = 6 mi.

Cartographer: J. Sutton Wall.

Size: 86.8 x 141.4 cm.

Colored.

Provenance: Presented by Francis S. McIlhenny.

(649: 1910: P376rrm Large)

968. [1912]

Map of New Sweden; 1638-1655. Philadelphia: S. J. Clarke Pub. Co.: [1912].

Cartographer: Gregory S. Keen.

Size: 16.3 x 9.6 cm.

In: Ellis Paxson Oberholtzer, *Philadelphia, a history of the city and its people.* Philadelphia: S. J. Clarke Pub. Co.: [1912]. 4 vols. See vol. 1, facing p. 12.

(974.811:Ob2p Reading Room)

969. 1915

Map of the location [of] Union and Confederate corps and division hospitals on Gettysburg battlefield, July 1863. U.S. Gettysburg National Park Commission. N.p.: 1915.

Scale: 1 in. = 2,000 ft.

Cartographers: John P. Nicholson, E. B. Cope, and E. M. Hewitt.

Size: 46.2 x 47.7 cm.

Blueprint.

(649.934: 1915: Un33gp Small)

970. 1916

Map of the battlefield of Gettysburg from original surveys by the engineers of the commission. By the authority of the Hon. Newton D. Baker, Secretary of War, July 1st. 1916, under the direction of the Gettysburg National Park Commission. U.S. Army. Corps of Engineers. Philadelphia: John T. Palmer Company: 1916.

Scale: 6 in. = 4200 ft.

Cartographer: S. A. Hammond.

Size: 117 x 84.6 cm.

Colored.

(649.934: 1916: Un38ace Large)

971. 1916-1919

Perspective view of Gettysburg National Military Park. U.S. Gettysburg National Park Commission. [Washington: 1916-1919].

Cartographers: John P. Nicholson and E. B. Cope.

Size: 43 x 53.5 cm.

Colored.

(649.934: 1916-1919: Un38gpn Small)

972. [ca. 1922]

Pennsylvania showing counties in different colors - townships - cities - boroughs - villages - postoffices - steam and electric railways, with stations and distances between stations and other features - complete index to all places on map showing location and population according to latest official census. Published by National Map Co. Indianapolis, Ind.: National Map Co.: [ca. 1922].

Scale: 1 in. = 5¼ mi.

Size: 91.5 x 149 cm.

Colored.

Inset: Congressional and state senatorial districts.

(649: [ca. 1922]: P384cou Large)

973. 1924-1926

Composite map showing the Mason-Dixon Line, compiled by Thomas D. Cope from U.S. Geological Survey maps: Pennsylvania-Delaware-Coatesville quadrangle; Pennsylvania-Delaware-West Chester quadrangle; Pennsylvania-Delaware-New Jersey quadrangle; Pennsylvania-New Jersey-Philadelphia quadrangle; Pennsylvania-McCalls Ferry quadrangle; Pennsylvania-Quarryville quadrangle. [Washington]: 1924-1926. 8 pieces.

Size: 215.3 x 44.3 cm.

Colored.

Provenance: Presented by Thomas D. Cope.

See: Thomas D. Cope, Collection of reprints on Mason and Dixon (974.5:C79).

Shows lines run by [William?] Talbot in 1683, Benjamin Chambers in 1688, and John Taylor in 1732-33.

Contains also: Blueprint of the states of Pennsylvania, New Jersey, Delaware, Maryland and West Virginia (22.5 x 29.5 cm.). Title: Mason and Dixon's Line established by agreement made in 1732. Located & milestones put in from 1763 to 1767 . . .

Also: Photostat of original drawing of map showing the area between the Delaware and the Susquehanna Rivers. 44 x 26 cm.

(649: 1924-1926: M385cj Small)

974. [1937]

Map of Penn's manors superimposed on a modern map by Fred G. Gorman, January 9, 1937. Highway map of southeastern Pennsylvania. Compiled from records of Pennsylvania Department of Highways, the Montgomery County Commissioners, the U.S. Geological Survey. [Philadelphia: 1937].

Scale: 1 in. = 2 mi.

Cartographer: Fred J. Gorman.

Size: 40.5 x 49.6 cm.

Provenance: Presented by Arthur Bloch, 3 December 1952.

(649: [1937]: G653pmp Small)

975. [1943]

A map of some of the south and east bounds of Pennsylvania in America being partially inhabited [by William Penn?]. London: John Thornton and John Seller: [1681]. Meriden Gravure Co.: 1943.

Scale: 1 in. = 6 mi.

Cartographer: William Penn.

Size: 41.7 x 51.1 cm.

Colored.

Facsimile reproduced in collotype by the Meriden Gravure Company from the original in the John Carter Brown Library: 1943.

With "a description at the end of it; and some proposals." Accompanied by "A note on the William Penn map of Pennsylvania" [London: 1681].

(649: [1943]: P376sep Large)

976. [1949?]

Plans of Benjamin Chambers' and John Taylor's boundary line of southern Pennsylvania. 1688.

Scale: 1 in. = 200 ft.

Cartographers: Benjamin Chambers, John Taylor, and Fred J. Gorman.

Size of paper: ca. 39 x 448.3 cm.

Colored.

Provenance: Presented by Fred J. Gorman, 30 June 1949.

Photograph, ca. 1949.

(649: [1949?]: C344spa Large)

977. 1950

Birdsboro quadrangle. Pennsylvania. Berks County. U.S. Geological Survey. [Washington]: 1950.

Scale: 1:24,000.

Size: 57.9 x 44.4 cm.

Colored.

Provenance: Presented by George Washington Corner.

(649.716: 1950: Un38bibo Large)

978. 1953

Oil and gas fields of Pennsylvania. Geological Survey of Pennsylvania. Fourth series. Compiled by Thomas H. Jones and Lillian Heeren. N.p.: 1953.

Scale: 1 in. = 6 mi.

Cartographers: Thomas H. Jones and Lillian Heeren.

Size: 69 x 118.2 cm.

Colored.

Pipelines shown on reverse side of this map.

(649: 1953: P376gas Large)

979. 1955

Media quadrangle. Pennsylvania. U.S. Geological Survey. [Washington]: 1955.

Scale: 1:24,000.

Size: 57.7 x 44.5 cm.

Colored.

Provenance: Presented by George Washington Corner.

(649.753: 1955: Un38med Large)

980. 1956

Lansdowne quadrangle. Pennsylvania. U.S. Geological Survey. [Washington]: 1956.

Scale: 1:24,000.

Size: 57.7 x 44.5 cm.

Colored.

Provenance: Presented by George Washington Corner.

(649.749: 1956: Un38Lan Large)

981. 1956

A map for the Welcome Society of Pennsylvania, shewing the most travelled roads from sundry places to W[illia]m Penn's manor of Pennsbury. Ao. Di. 1935. [Philadelphia]: 1956.

Cartographer: Wm. M. Campbell.

Size: 16.5 x 24.8 cm.

Inset: From Holme's map begun in 1681 shewing some of the proprietary's manors, including Pennsbury.

Provenance: Presented by William Jacob Robbins, November 1956.

(649: 1956: C142wsp Small)

982. 1957

Boyertown quadrangle. Pennsylvania. U.S. Geological Survey. [Washington]: 1957.

Scale: 1:24,000.

Size: 57.8 x 44.4 cm.

Colored.

Provenance: Presented by George Washington Corner.

(649.716: 1957: Un38boy Large)

983. 1959

Combination atlas map of Montgomery County, Pennsylvania. Compiled, drawn & published from personal examinations and surveys: J. D. Scott. Philadelphia: Thos. Hunter: 1877. Reprinted in 1959. 41 pages of maps, cover, title page, introduction, and index.

Cartographer: J. D. Scott.

Size of paper: 28 x 21.5 cm.

Colored.

Provenance: Presented by the Montgomery County Planning Commission, October 1959.

Re-issued in honor of the 175th anniversary of Montgomery County, September 1959. Contains list of townships and boroughs.

(649.755: 1959: Sco37mc Small)

984. 1960

Geologic map of Pennsylvania. Topographic and Geologic Survey of Pennsylvania. Fourth series. Carlisle Gray, state geologist. Washington: Williams & Heintz Map Corporation: 1960.

Scale: 1:250,000.

Cartographers: Carlyle Gray, R. R. Conlin, D. M. Hoskins, M. N. Shaffner, A. A. Sokolow, D. B. McLaughlin, A. R. Geyer, A. S. Cate, V. C. Shepps, W. S. Lytle, J. M. Bergsten, J. T. Miller, G. H. Wood, Jr., H. H. Arndt, T. M, Kehn, and A. E. van Olden.

Size: 144.7 x 203.4 cm.

Colored.

(649: 1960: P37togr Large)

985. 1961

Subsurface structure of plateau region of north-central and western Pennsylvania on top of Oriskany formation. Geological Survey of Pennsylvania. Fourth series. N.p.: 1961.

Scale: 14 cm. = 25 mi.

Cartographers: Addison S. Cate and C. R. Fettke.

Size: 103.8 x 129.8 cm.

See: Geological Survey of Pennsylvania, *Bulletin.* 1953, pl. 3.

(649: 1961: P37ori Large)

986. [n.d.]

Plan of Keim's addition to Port Clinton. Philadelphia: Childs: [n.d.]

Scale: 1 in. = 200 ft.

Lithographer: Childs.

Size: 33.3 x 49.6 cm.

Endorsement: "Ask W. J. Duane for a map drawn by Christian Brobst." See map 649: [ca. 1835]: B81msc.

On reverse: manuscript map in pencil of central Pennsylvania showing Danville & Catawissa, etc.

(649.964: [n.d.]: C432kpc Small)

PHILADELPHIA Nos. 987-1040

987. [1683]

Portraiture of the city of Philadelphia in the province of Pennsylvania in America. By Thomas Holme, surveyor. London: Andrew Sowle: [1683].

Scale: ca. ¾ in. = 528 ft.

Cartographer: Thomas Holme.

Size: 30 x 45 cm.

Inset: Peninsula between Delaware and Schuylkill Rivers.

Provenance: Presented by General Parker from the library of Zaccheus Collins, 1831.

(649.962: [1683]: H733sph Small)

988. [1756]

A plan of the city of Philadelphia; the Battery; & the State House. Also, a small copy of "An East Prospect of the city . . ." Engraved and published by T. Jefferys. London: [1756].

Cartographer: Nicholas Scull.

Engraver: Thomas Jefferys.

Size: 50 x 93.5 cm.

(649.962: [1756]: Scu47ep Framed. Map and Print room.)

989. [1774]

[Map of Philadelphia. 1774].

Size: 49.5 x 49.8 cm.

Provenance: Presented by John Reed, March 1816.

In: John Reed, *An explanation of the map of the city and liberties of Philadelphia.* Philadelphia: 1774.

(912.748: R25)

990. [1774]

. . .Map of the city and liberties of Philadelphia, with the catalogue of purchasers, is humbly dedicated by their most obedient humble servant, John Reed. [Philadelphia]: T. Man: [1774].

Scale: 1 in. = 100 perches.

Cartographer: John Reed.

Engraver: James Smither.

Size: 75.4 x 150.5 cm.

Inset: Philadelphia; vignettes of the State House, Pennsylvania Hospital, and the house of employment and the alms house.

See: John Reed, *An explanation of the city and liberties of Philadelphia.* Philadelphia: 1774.

Wheat: 457

(649.962: 1774: R257cLp Large)

991. [1774]

[Map of the western part of Philadelphia. 1774].

Size: 33.5 x 16 cm.

In: John Reed, *An explanation of the city and liberties of Philadelphia.* Philadelphia: 1774.

(912.748: R25)

992. [1774]

To the Honourable House of Representatives of the freemen of Pennsylvania this map of the city and liberties of Philadelphia with the catalogue of purchasers is humbly dedicated by their most obedient humble servant Iohn Reed. [1774].

Scale: 2.5 cm. – 100 perches.

Cartographer: John Reed?

Engraver: James Smither.

Size: 75.5 x 49.7 cm.

Provenance: Presented by Joseph Reed, 15 March 1816.

In: John Reed, *An explanation of the map of the city.* [1774].

Wheat: 457

(912.748: R25 and 975.7: R14h)

993. [1777]

Map of Philadelphia and parts adjacent. By N. Scull and G. Heap. N.p.: [1777].

Scale: 1 in. = 8 furlongs.

Cartographers: N. Scull and G. Heap.

Size: 33.9 x 29.5 cm.

See: *Gentleman's Magazine,* 1777, vol. 47, opp. p. 573, for American edition of map.

Wheat: 455

(649.962: [1777]: Scu47pha Small)

994. [ca. 1777]

Philadelphie, par Eas[t]burn. Paris: Le Rouge: [ca. 1777].

Scale: 1 in. = 900 English ft.

Cartographer: B. Eastburn.

Size: 21.2 x 35.1 cm.

Colored.

(649.962: [ca. 1777]: Ea82ppp Small)

995. [1777]

Seat of war in the environs of Philadelphia. By Thomas Kitchin. [London]: R. Baldwin: [1777].

Cartographer: Thomas Kitchin.

Size: 19.2 x 25 cm.

At top of map is: "For the *London Magazine,* 1777."

(649.962: [1777]: K644swp Small)

996. 1779

A plan of the city and environs of Philadelphia, with the works and encampments of His Majesty's forces. Under the command of Lieutenant General Sir William Howe, K.B. London: Wm. Faden: 1779 [1 January 1779].

Scale: 3 in. = 1 mi.

Engraver: William Faden.

Size: 51.7 x 47.3 cm.

Provenance: Presented by Dr. John W. Reps.

Reproduction by Historic Urban Plans: Ithaca, New York: 1965.

(649.962: 1779: F123how Large)

997. [1792]

[Plan of thirty-six lots between Locust and Spruce Streets on Twelfth, in Philadelphia. 1792].

Scale: 2.5 cm. = 85 ft.

Size: 11.8 x 6 cm.

In: *Dunlap's American Daily Advertiser.* 10 March 1792.

Wheat: 460

(071: D92)

998. 1796

Map of Philadelphia. From "Plan of the City of Philadelphia and environs." [Philadelphia]: 1796.

Cartographer: John Hills.

Size: 16 x 21.4 cm.

Photograph from original in the Library of Congress.

See: Philip Lee Phillips, *A descriptive list of maps and views of Philadelphia.* [Philadelphia: 1926].

Wheat: 468

(649.962: 1796: H553phi Small)

999. [1796]

Plan of the city of Philadelphia. [1796].

Scale: 2.5 cm. = 90 perches.

Size: 24.3 x 36 cm.

In: Thomas Stephens, *Stephens' Philadelphia directory, for 1796,* frontispiece.

Wheat: 467

(917.4811: P53d)

1000. [1797]

Plan of the city of Philadelphia. [1797].

Scale: 2.5 cm. = 90 perches.

Engraver: J. Bowes.

Size: 24.3 x 36 cm.

In: Cornelius W. Stafford, *The Philadelphia Directory for 1797,* frontispiece.

Wheat: 470

(917.4811: P53d)

1001. [ca. 1802]

To the citizens of Philadelphia, this plan of the city and its environs is respectfully dedicated by the editor. [Philadelphia: 1802].

Scale: 1 in. = 75 perches.

Cartographer: P. C. Varle.

Engraver: Scott.

Size of paper: 45.8 x 61 cm.

Colored.

Insets: City Hall, State House, and Congress Hall; Library Company of Philadelphia library; First United States Bank.

Trimmed.

(649.962: [1802]: V428ppv Large)

1002. 1807

Plan of the city of Philadelphia. London: James Cundee: 1807 [7 March 1870].

Scale: 1 in. = 100 perches.

Size: 15.9 x 24.8 cm.

See: Charles William Janson, *The stranger in America.* London: J. Cundee: 1807, p. 175.

(649.962: 1807: C934ph Small)

1003. 1808

Plan of the city of Philadelphia and environs. Surveyed by John Hills in the summers of 1801, 2, 3, 4, 5, 6, & 7. Philadelphia: John Hills: 1808.

Scale: Distance between the circles is one mile.

Cartographer: John Hills.

Engraver: William Kneass.

Size: 100.6 x 100.6 cm. In 10 circles, outermost circle 99 cm. diameter.

Contains engravings of the Philadelphia Water Works and Cape Henlopen lighthouse.

(649.962: 1808: H553cmp Large)

1004. 1811

To the citizens of Philadelphia, this new plan of the city and its environs. Taken from actual survey . . .[and] drawn under the direction of J. A. Paxton by William Strickland. [Philadelphia]: C. P. Harrison: 1811 [1 January 1811].

Scale: 1 in. = 1,000 ft.

Cartographer: William Strickland.

Engraver: W. Harrison.

Size: 45.8 x 49.7 cm.

Colored.

To accompany: *Paxton's stranger's guide.*

(649.962: 1811: P286npp Large)

1005. [ca. 1816]

This new map of the city of Philadelphia for the use of fire men and others is most respectfully dedicated to the citizens and members of the engine & hose companies, by . . . John A. Paxton. [Philadelphia: ca. 1816].

Cartographer: W. Strickland.

Size: 35.4 x 61.2 cm. *Size of paper:* 47.5 x 61.2 cm.

Contains: a listing of engine & hose houses; a vignette of the Fair Mount Water Works; an extension of the northwestern end of the map of the city to include Fair Mount; and cut away diagrams of two fire plugs.

(649.962: [ca. 1816]: P286feh Large)

1006. 1819

Map of Philadelphia County, constructed by virtue of an act of the legislature of Pennsylvania. By John Melish. 1816. Philadelphia: Melish: 1819 [14 April 1819].

Scale: 1 in. = 1¼ mi.

Cartographer: J. Melish.

Engravers: Tanner, Vallance, Kearny, & Company.

Size: 43.4 x 49 cm.

Colored.

(649.762: 1819: M485pc Small)

1007. [ca. 1820]

To the hose companies of Philadelphia this plan of the city is inscribed. The first hose company instituted in 1803. Philadelphia: W. Kneass: [ca. 1820].

Cartographer: William Strickland.

Engraver: W. Kneass.

Size: 37.9 x 38.7 cm.

Colored.

Inset: Engraving of Center Square water works.

(649.962: [ca. 1820]: K73hcp Small)

1008. 1824

Plan of the city of Philadelphia, compiled from actual surveys by F. Drayton, 1824. Philadelphia: 1824.

Scale: 1 in. = 2,000 ft.

Cartographer: F. Drayton.

Engraver: R. M. Filbert.

Size: 28.7 x 23.9 cm.

Colored.

(649.962: 1824: D792pcp Small)

1009. [1825]

Philadelphia & environs. Philadelphia: H. S. Tanner: [1825].

Scale: 1 in. = 2 mi.

Size: 17.2 x 10.5 cm.

Provenance: Presented by Henry S. Tanner, July 1825.

Autograph inscription: George Merrick from H. S. Tanner, esqr., Phila.

Published in *A general outline of the United States.* Philadelphia: H.S. Tanner: 1825, p. 122.

(649.962: [1825]: T158pae Small)

1010. [ca. 1826]

A map of Philadelphia and adjacent country. Philadelphia: H. S. Tanner: [ca. 1826].

Scale: 1 in. = 2 mi.

Size: 28 x 37.2 cm.

Colored.

Provenance: Presented by H. S. Tanner, 5 March 1830.

(649.962: [ca. 1826]: T158pcc Small)

1011. 1840

Philadelphia. [London]: Society for the Diffusion of Useful Knowledge: 1840.

Scale: 2 in. = 3,000 ft.

Size: 28.8 x 37.5 cm.

Colored.

Contains vignettes of the Second U.S. Bank and the Merchants' Exchange.

(649.962: 1840: So17ukp Small)

1012. [1843]

Map of Washington Square, Walnut Street, Philadelphia. [Philadelphia]: M. Schmitz & P. Kereven: 1843.

Scale: 1 in. = 30 ft.

Cartographers: John B. Colahan and M. Schmitz.

Lithographer: T. Sinclair.

Size: 47 x 47 cm. *Size of paper:* 63.4 x 85.6 cm.

Provenance: Presented by Thomas Sinclair, 26 May 1843.

The botanical and English names of the trees on this map were collected with great care, and their position given by Patrick Kereven, chief gardener of Wash[i]ng[ton] & Indepen[den]ce Squares.

(649.962: [1843]: Sch17ws Large)

1013. 1847

A plan of the city and environs of Philadelphia. Surveyed by N. Scull and G. Heap. London: W. Faden: 1777. Philadelphia: Thomas Fisher: 1847.

Scale: 1:47,520.

Cartographers: N. Scull and G. Heap.

Engraver: W. Faden.

Size: 62.7 x 45.4 cm.

Colored.

Provenance: Presented by Thomas Fisher, 19 February 1847.

Copy of Faden's map originally in W. Faden, *The North American Atlas.* Republished by Thomas Fisher, 1847.

Contains vignette of the State House.

(649.962: 1847: F123cvp Large)

1014. 1849

Map of the city of Philadelphia, together with all the surrounding districts, including Camden, N. J., from official records, plans of the district surveyor & original surveys by J. C. Sydney. 1849. Philadelphia: Smith & Wistar: 1849.

Scale: 1 in. = 450 ft.

Cartographer: J. C. Sydney.

Lithographer: P. S. Duval.

Size: 171 x 173.5 cm.

Colored.

Provenance: Presented 12 July 1849.

(649.962: 1849: Si17phL Large)

1015. 1849

Philadelphia. Published by A. McElroy, Philadelphia 1849. Philadelphia: A. McElroy and P. S. Duval: 1849.

Scale: 1 in. = 1,000 ft.

Engraver: N. Friend.

Size: 43 x 35.3 cm.

(649.962: 1849: F913phi Large)

1016. 1854

An east prospect of the city of Philadelphia, taken by George Heap from the Jersey shore, under the direction of Nicholas Scull. 1754. Philadelphia: E. H. Coggins: 1854.

Scale: 4 furloughs or ½ a mile.

Cartographers: George Heap and Nicholas Scull.

Engraver: Gerard Vandergucht.

Lithographer: L. N. Rosenthal.

Size: 20 x 45.6 cm. *Size of paper:* 60.5 x 96.1 cm.

Insets: The battery, and the State House.

Provenance: John Vaughan presented "An East Prospect . . ." 1752, 2 November 1810.

(649.962: 1854: Scu47ep Large)
[Lithograph copy]

1017. 1854

An outline of the newly consolidated city [of Philadelphia] showing the boundaries of the wards, according to the act, passed by the Legislature, Jany. 31st, 1854. Philadelphia: R. L. Barnes: 1854.

Lithographers: Friend & Aub.

Size: 64.5 x 50.4 cm.

Colored.

(649.962: 1854: Ou85ph Small)

1018. 1855

Philadelphia. No. 21. New York: J. H. Colton & Co.: 1855.

Scale: 1 in. = 650 yd.

Cartographer: George W. Colton.

Size: 40.5 x 31.6 cm.

Colored.

See: George W. Colton, *Colton's atlas of the world, illustrating physical and political geography.* New York: J. H. Colton & Co.: 1856.

(649.962: 1855: C672mca Small)

1019. 1855

Scott's map of the consolidated city of Philadelphia, from the latest records and actual surveys. Philadelphia: Scott & Moore: 1855.

Engraver: E. Herrlein.

Size: 164.5 x 189.5 cm.

Inset: Map of Philadelphia County. Also contains business directory of the subscribers to the map, engravings of the Independence Hall complex, Merchants' Exchange, Laurel Hill cemetery, and Girard College.

Colored.

Provenance: Purchased "for the use of the Society" in 1855.

(649.962: 1855: Sco37cm Large)

1020. 1891

Philadelphia water front. Delaware River, Pennsylvania. U.S. Coast and Geodetic Survey. [Washington]: 1891.

Scale: 1:9600.

Cartographers: F. R. Hassler, C. M. Eakin, S. C. McCorkle, F. W. Perkins, W. M. Boyce, R. M. Bache, J. Hergesheimer, H. L. Marindin, E. O. Kendall, J. E. Hilgard, J. Lock, C. A. Schott, E. Goodfellow, J. B. Baylor, E. Smith, and C. M. Thomas.

Size: 129.4 x 66 cm.

(649.962: 1891: Un38pdr Large)

1021. 1891

Philadelphia water front. U.S. Coast and Geodetic Survey. Schuylkill River, Pennsylvania. [Washington]: 1891.

Scale: 1:9,600.

Cartographers: F. R. Hassler, C. M. Eakin, R. M. Bache, S. C. McCorkle, W. M. Boyce, P. J. Hergesheimer, H. L. Marindin, E. O. Kendall, John Locke, C. A. Schott, E. Goodfellow, J. B. Baylor, J. E. Hilgard, E. Smith, and C. M. Thomas.

Size: 129.2 x 65.9 cm.

(649.962: 1891: Un38pwf Large)

1022. [1893]

Plan showing improvements in the 21st, 22nd, and 34th wards, and Fairmount Park, north of Girard Avenue, Philadelphia, 1893. Philadelphia: Century Lith. Co.: 1893.

Cartographer: Chester E. Albright.

Lithographer: Century Lithograph Company.

Size: 63.4 x 113.5 cm.

Insets: View of proposed Falls bridge over Schuylkill river; Memorial Hall; Country Club House, Gentlemen's driving park; Wissahickon Inn; Park Boulevard from City Hall; Manheim; Midvale Avenue, under P.& R.R.R.

Provenance: Presented by Margah Toogood Flood, September 1958.

(649.962: [1893]: AL11wfp Extra-Oversize)

1023. 1896

New map of the city of Philadelphia from the latest city surveys. Prepared for Gopsill's directories. Philadelphia: J. L. Smith: 1896.

Scale: 3 in. = 1 mi.

Size: ca. 101 x 72 cm.

Colored.

(649.962: 1896: P536cdp Large)

1024. 1896

Philadelphia and vicinity. Pennsylvania and New Jersey. U.S. Geological Survey. Charles D. Walcott, director. [Washington]: 1894. Edition of June 1896.

Scale: 1:62,500.

Cartographers: Henry Gannett, C. D. Walcott, H. M. Wilson, Frank Sutton, J. H. Jennings, R. D. Cummin, W. H. Lovell, E. B. Clark, and Robert Muldrow.

Size: 91.5 x 71.3 cm.

Colored.

(649.962: 1896: Un38pav Large)

1025. 1901

New map of the city of Philadelphia from the latest city surveys. Prepared for Gopsill's directories. Philadelphia: J. L. Smith: 1901.

Scale: 3 in. = 1 mi.

Size: ca. 101 x 72 cm.

Colored.

(649.962: 1901: P536cdp Large)

1026. 1902

Philadelphia and vicinity. Pennsylvania and New Jersey. U.S. Geological Survey. Charles D. Walcott, director. [Washington]: 1902. Edition of February 1902.

Scale: 1:62,500.

Cartographers: Henry Gannett, C. D. Walcott, H. M. Wilson, Frank Sutton, J. H. Jennings, R. D. Cummin, W. H. Lovell, E. B. Clark, and Robert Muldrow.

Size: 91.5 x 71.3 cm.

Colored.

(649.962: 1902: Un38pac Large)

1027. 1906

New map of the city of Philadelphia from the latest city surveys. Prepared for Gopsill's directories. Philadelphia: J. L. Smith: 1906.

Scale: 3 in. = 1 mi.

Size: ca. 101 x 72 cm.

Colored.

(649.962: 1906: P536cdp Large)

1028. [1908]

The Philadelphia Parkway as planned for the Fairmount Park Association. Horace Trumbauer, C. C. Zantzinger, and Paul P. Cret. [Philadelphia: 1908].

Cartographers: Horace Trumbauer, C. C. Zantzinger, and Paul P. Cret.

Size: 66.7 x 66.7 cm.

See: Philadelphia Fairmount Park Art Association, *Report of the Commission. . . .* Philadelphia: 1908.

(649.962: [1908]: T778ppf Large)

1029. 1910

Map of Philadelphia, Camden and vicinity. Compiled from city plans and personal surveys. Philadelphia: C. E. Howe Company: 1910.

Scale: 1 in. = 2,000 ft.

Cartographer: Elvino V. Smith.

Size: 133.2 x 104.2 cm.

Colored.

(649.962: 1910: P536pcv Large)

1030. 1913-1914

New map of Philadelphia and vicinity. Philadelphia: J. L. Smith: 1913-1914.

Scale: 1 in. = 1 mi.

Size: 167.2 x 110.8 cm.

Contains autograph of Francis J. Packard.

(649.962: 1913-1914: Sm57ph Book map)

1031. 1928

Philadelphia, by the City Planning Division, Bureau of Engineer & Surveys. No. 9. 1 January 1928.

Size: 41 x 124 cm.

Blueprint.

(649.962: 1928: P536cpd Large)

1032. 1932

General plan of Girard College, Philadelphia, Pa., 1932.

Cartographers: Day & Zimmermann.

Size: 29.7 x 106.5 cm.

Provenance: Presented by Girard College, May 1967.

Photograph in two pieces of architects' plan, 24 April 1930, and revised 20 January 1932.

(649.962: 1932: D332gir Large)

1033. 1934

A map of old Philadelphia on which are indicated buildings of architectural interest that were built during the 18th. and early 19th. centuries and are still standing Ao. Di. 1932, two hundred and fifty years after the founding of the city by William Penn. Philadelphia: J. L. Smith Co.: 1934.

Cartographer: William M. Campbell.

Size: 52.2 x 77.1 cm.

Colored.

Inset: Philadelphia from the map made by John Reed in 1774; also contains pictures of Independence Hall, William Penn, B. Franklin, Stephen Girard, and William Penn and the Indians.

Made by a committee of the Philadelphia chapter of the American Institute of Architecture.

(649.962: 1934: Am31aia Large)

1034. 1940

Roads and bridle paths, Wissahickon Valley, Fairmount Park, Philadelphia, Pennsylvania. Friends of the Wissahickon, Inc. Philadelphia: 1940.

Scale: 1 in. = 1,000 ft.

Cartographer: Ralph Weston Jones.

Size: 83.5 x 47.7 cm.

Colored.

Provenance: Presented by the Friends of the Wissahickon, Inc., February 1940.

(649.962: 1940: W759fwp Large)

1035. 1941

Map of the Schuylkill River from Fairmount to the Falls, showing the roads and residences which existed in the adjacent territory before 1850. Prepared and rendered for the Modern Club, 1941. By Henry B. McIntire. 1941. N.p.: 1941.

Cartographer: Henry B. McIntire.

Size: 24.8 x 47.2 cm. *Size of photograph:* 27 x 73.7 cm.

Two panels in memoriam: Warwick James Price.

(649.962: 1941: M185srf Large)

1036. 1947

Preliminary plans for Old City Area, by the Philadelphia City Planning Commission. [Philadelphia]: 1947.

Size: 40.6 x 51.2 cm.

Includes the present Independence National Historic Park.

(649.962: 1947: In24cpc Small)

1037. 1957

A map of Philadelphia; published November 1, 1762 by Matthew Clarkson and Mary Biddle. Republished November, 1858, by Joseph H. Bonsall and Samuel L. Smedley from whose lithograph, in the possession of the Library Company of Philadelphia, this facsimile is reproduced in March, 1957, by the Society of Colonial Wars in the Commonwealth of Pennsylvania. N.p.: 1957.

Scale: 1 in. = 400 ft.

Cartographers: Matthew Clarkson and Mary Biddle.

Size: 49 x 65.7 cm.

Provenance: Presented by Thomas D. Cope, 23 May 1957.

Facsimile.

(649.962: 1957: C562cbp Large)

1038. [n.d.]

Map of Philadelphia and parts adjacent with a perspective view of the state-house. N.p.: [1750].

Scale: 1 in. = 8 furlongs.

Cartographers: N. Scull and George Heap.

Size: 50 x 29.8 cm.

Mutilated. Facsimile?

(649.962: n.d.: Scu47ph Small)

1039. [n.d.]

Plan of Glenwood Cemetery. Ridge Road and Islington Lane. Philadelphia: T. Sinclair: [n.d.].

Scale: 1 in. = 40 ft.

Lithographer: T. Sinclair.

Size: 61 x 81.8 cm.

(649.962: [n.d.]: G473pgc Large)

1040. [n.d.]

Washington Square east urban renewal area - illustrative site plan. Prepared by Wright, Andrade, and Amenta. . . . N.p.: n.d.

Cartographers: Wright, Andrade, and Amenta.

Size of paper: 35.8 x 40.8 cm.

Provenance: Presented by Richard Harrison Shryock, February 1962.

(649.962: n.d.: P536rea Small)

DELAWARE Nos. 1041-1052

1041. [ca. 1786]

[Map of Chesapeake Bay, Delaware Bay and Potomac River]. N.p.: [ca. 1786].

Size of paper: 51.9 x 35.5 cm.

See: John Churchman below, No. 1042.

(651: [ca. 1786]: C412bay Small)

1042. 1786

To the American Philosophical Society this map of the peninsula between Delaware & Chesopeak bays with the said bays and shores adjacent drawn from the most accurate surveys. By John Churchman. N.p.: 1786.

Scale: 1 in. = 10 mi.

Cartographer: John Churchman.

Size: 57.2 x 43 cm.

Colored.

Date set by Col. Lawrence Martin of the U.S. Library of Congress. Received from the Library of Congress in March 1938, as a duplicate exchange.

Wheat: 477

(651: 1786: C472dcb Large)

1043. [1795]

Delaware. [1795].

Scale: 2.5 cm. = 16 mi.

Size: 18.5 x 15.5 cm.

Provenance: Presented by the Estate of Richard Gimbel, 1974.

In: Joseph Scott, *The United States gazetteer,* opposite "Delaware."

Wheat: 484

(Paine 68: Sco3u)

1044. [1796]

Delaware. [1796].

Scale: 2.5 cm. = 15 mi.

Engraver: A. Doolittle.

Size: 19.3 x 14.5 cm.

Provenance: Presented by Suzanne Wister Eastwick and Joseph L. Eastwick, July 1975.

In: Mathew Carey, *Carey's American pocket atlas,* following p. 82.

Wheat: 485

(917.3: C25c)

1045. [ca. 1830]

Map of the country contiguous to the Chesapeake & Delaware Canal. [Philadelphia: H. S. Tanner: ca. 1830].

Scale: 1 in. = 8 mi.

Cartographer: H. S. Tanner.

Engraver: H. S. Tanner.

Size: 21 x 26.2 cm.

(651: [ca. 1830]: T158cdc Small)

1046. 1833

Delaware River, from Bombay Hook light to Cherry Island flats. U.S. Coast and Geodetic Survey. J. E. Hilgard, superintendent. [Washington]: 1883.

Scale: 1:40,000.

Cartographers: J. E. Hilgard, R. M. Bache, J. A. Sullivan, S. C. McCorkle, A. T. Mosman, C. T. Iardella, H. B. Mansfield, and H. L. Marindin.

Size: 68.7 x 106.2 cm.

Provenance: Presented by the U.S. Coast and Geodetic Survey through Spencer C. McCorkle, 14 December 1885.

Preliminary edition by photo-lithography, to be superseded eventually by an engraved edition.

(651: 1833: Un38dbc Large)

1047. [1833]

Harbour of the Delaware breakwater from the Atlantic Ocean. Diagram copied from the original survey. Philadelphia: Childs & Inman, lith.: [1833].

Scale: 1 in. = 1,000 ft.

Cartographers: E. Morris and W. Strickland.

Lithographers: Childs & Inman.

Size: 54.3 x 40.7 cm.

Provenance: Presented and endorsed by W. Strickland, 8 June 1833.

Strickland wrote that this was a lithographic copy of his plan of the Delaware breakwater which he was "now executing" [Donation Book; APS, *Transactions*, n.s., vol. 4].

At top of the map is a picture of the breakwater area.

(651: [1833]: St87deb Large)

1048. [1834]

Plan of the town & fort of Christina, besieged by the Dutch, 1655. [Philadelphia: 1834].

Size: 12.6 x 12.1 cm.

Original in Thomas Campanius Holm, *Kort beskrifning om provincien Nya Swerige.* Stockholm: S. Wankijfs 1702, p. 26.

Duplicate of map in Historical Society of Pennsylvania, *Memoirs*, vol. 3, p. 84. 1834.

(651: [1834]: C462tc Small)

1049. 1848

Map of Delaware bay and river. Founded upon a trigonometrical survey under the direction of F. R. Hassler and A. D. Bache. U.S. Coast Survey. [Washington]: 1848.

Scale: 1:80,000.

Cartographers: F. R. Hassler, A. D. Bache, J. Ferguson, E. Blunt, C. M. Eakin, F. H. Gerdes, J. J. S. Hassler, J. E. Johnstone, W. M. Boyce, H. L. Whiting, G. D. Wise, B. F. Sands, G. S. Blake, T. R. Gedney, G. M. Bache, W. M. C. Fairfax, and J. B. Dale.

Engravers: F. Dankworth, O. A. Lawson, and J. Knight.

Size: 192.7 x 89.9 cm.

Inset: Part of the Delaware River continued from Frankford to Trenton.

Provenance: Presented by A. D. Bache, 5 January 1849.

(651: 1848: Un38dbr Large)

1050. 1883

Delaware breakwater. U.S. Coast and Geodetic Survey. [Washington]: 1883.

Scale: 1:20,000.

Cartographers: J. E. Hilgard, A. T. Mosman, W. I. Vinal, E. Hergesheimer, and G. C. Hanus.

Size: 52.5 x 68.3 cm.

Provenance: Presented by U.S. Coast and Geodetic Survey through Spencer C. McCorkle, 14 December 1885.

(651: 1883: Un38dbs Large)

1051. [n.d.]

[Delaware Bay and River to Philadelphia.] N.p.: n.d.

Scale: 1 in. = 3¼ mi.

Size: 38.7 x 82.2 cm.

Wheat: 478(?)

(651: n.d.: D372dba Large)

1052. [n.d.]

Map of the state of Delaware. N.p.: Rand McNally & Co.: n.d.

Scale: 6.4 cm. = 4 mi.

Size: ca. 151 x 73.5 cm.

Colored.

Inset: Georgetown, Delaware.

(651: n.d.: M323deL Large)

MARYLAND Nos. 1053-1077

1053. 1731

Map showing boundary lines between Maryland and Pennsylvania. N.p.: 1731.

Engraver: John Senex.

Size: 22.7 x 15.2 cm.

Photograph from the Historical Society of Pennsylvania.

See: APS, *Proceedings,* vol. 98, part 6, fig. 1. T. D. Cope, "Maryland-Pennsylvania boundary survey."

Wheat: 474

(652: 1731: Se57mpb Small)

1054. [1789]

From Annapolis (64) to Alexandria. [1789, 1961 reprint].

Size: 17.2 x 11.5 cm.

In: Christopher Colles, *A survey of the roads.*

Wheat: 508

(526.8: C68s)

1055. [1789]

From Annapolis (65) to Alexandria. [1789, 1961 reprint].

Size: 17.3 x 11.5 cm.

In: Christopher Colles, *A survey of the roads.*

Wheat: 509

(526.8: C68s)

1056. [1789]

From Annapolis (62) to Bladensburg. [1789, 1961 reprint].

Size: 17.1 x 11.5 cm.

In: Christopher Colles, *A survey of the roads.*

Wheat: 506

(526.8: C68s)

1057. [1789]

From Annapolis (63) to Bladensburg. [1789, 1961 reprint].

Size: 17.2 x 11.6 cm.

In: Christopher Colles, *A survey of the roads.*

Wheat: 507

(526.8: C68s)

1058. [1789]

From Philadelphia (56) to Annapolis, M[arylan]d. [1789, 1961 reprint].

Size: 17 x 11.4 cm.

In: Christopher Colles, *A survey of the roads.*

Wheat: 500

(526.8: C68s)

1059. [1789]
From Philadelphia (57) to Annapolis, M[arylan]d. [1789, 1961 reprint].
Size: 17.1 x 11.5 cm.
In: Christopher Colles, *A survey of the roads.*
Wheat: 501
(526.8: C68s)

1060. [1789]
From Philadelphia (58) to Annapolis, M[arylan]d. [1789, 1961 reprint].
Size: 17.2 x 11.3 cm.
In: Christopher Colles, *A survey of the roads.*
Wheat: 502
(526.8: C68s)

1061. [1789]
From Philadelphia (59) to Annapolis, M[arylan]d. [1789, 1961 reprint].
Size: 17.1 x 11.5 cm.
In: Christopher Colles, *A survey of the roads.*
Wheat: 503
(526.8: C68s)

1062. [1789]
From Philadelphia (60) to Annapolis, M[arylan]d. [1789, 1961 reprint].
Size: 17 x 11.4 cm.
In: Christopher Colles, *A survey of the roads.*
Wheat: 504
(526.8: C68s)

1063. [1789]
From Philadelphia (61) to Annapolis, M[arylan]d. [1789, 1961 reprint].
Size: 17 x 11.3 cm.
In: Christopher Colles, *A survey of the roads.*
Wheat: 505
(526.8: C68s)

1064. [1789]
From Philadelphia (55) to Maryl[an]d. [1789, 1961 reprint].
Size: 17.2 x 11.5 cm.
In: Christopher Colles, *A survey of the roads.*
Wheat: 499
(526.8: C68s)

1065. 1795

Map of the state of Maryland, laid down from an actual survey of all the principal waters, public roads, and divisions of the counties therein; describing the situation of the cities, towns, villages, houses of worship and other public buildings, furnaces, forges, mills, and other remarkable places; and of the federal territory; as also a sketch of the state of Delaware; shewing the probable connexion of the Chesapeake and Delaware bays. By Dennis Griffith, June 20, 1794. Philadelphia: J. Vallance: 1795 [6 June 1795].

Scale: 1 degree = 68⁹⁄₁₀ mi.

Cartographer: Dennis Griffith.

Engravers: Thackara & Vallance.

Size: 75.3 x 132.6 cm.

Inset: Plan of the city of Washington and territory of Columbia.

Wheat: 511

(652: 1795: G873msm Large)

1066. [1795]

Maryland. [1795].

Scale: 2.5 cm. = 28 mi.

Size: 15.5 x 18.6 cm.

Provenance: Presented by the Estate of Richard Gimbel, 1974.

In: Joseph Scott, *The United States gazetteer,* opposite "Maryland."

Wheat: 512

(Paine 68: Sco3u)

1067. [1796]

Maryland. [1796].

Scale: 2.5 cm. = 26 mi.

Engraver: W. Barker.

Size: 14.8 x 19.4 cm.

Inset: Maryland west of Port Cumberland.

Provenance: Presented by Suzanne Wister Eastwick and Joseph L. Eastwick, July 1975.

In: Mathew Carey, *Carey's American pocket atlas,* following p. 90.

Wheat: 513

(917.3: C25c)

1068. 1799

A map of the head of Chesapeake Bay and Susquehanna River. Shewing the navigation of the same with a topographical description of the surrounding country from an actual survey. By C. P. Hauducoeur. N.p.: 1799.

Scale: 1:250 poles.

Cartographer: C. P. Hauducoeur.

Engraver: Allardice.

Size: 55.9 x 75 cm.

Inset: Plan of the town of Havre de Grace.

Wheat: 520

(652: 1799: H293cbs Large)

1069. 1799

The states of Maryland and Delaware from the latest surveys. 1799.

Scale: 2.5 cm. = 221½ mi.

Cartographer: Anderson.

Engraver: Scoles.

Size: 18.6 x 23.5 cm.

Inset: Continuation of the Potomac River from Fort Cumberland.

In: John Payne, *A new and complete system of universal geography,* vol. 4, opposite p. 377.

Wheat: 518

(910: P29)

1070. 1801

Warner & Hanna's plan of the city and environs of Baltimore . . .; 1801. N.p.: 1801.

Scale: 1 in. = 40 perches.

Engraver: Francis Shallus.

Size: 48 x 72.6 cm.

Colored.

Insets: View of the market space canal; New assembly room.

Provenance: Priestman's library was put up for sale 17 June 1831, and the Society purchased $70.00 of books which were not received until 5 May 1837. This map was probably included in the purchase.

Autographed by Charles Priestman, 1 March [18]27, Philadelphia.

(652.915: 1801: Sh17ceb Large)

1071. [1813]

Map of the seat of war in North America. [Philadelphia: Palmer: 1813].

Scale: 2 cm. = 50 mi.

Cartographer: J. Melish.

Engraver: H. S. Tanner.

Size: 37 x 56.2 cm.

Colored.

Original in: John Melish, *Military and topographical atlas of the United States.* Philadelphia: Palmer: 1813.

One copy contains handwritten names of points in New York State and the St. Lawrence Valley.

(632.5: [1813]: M485na Small)

1072. [1831]

Plan and section of an intended railway from the Columbia rail-road in the Great Chester valley to the Susquehanna River near the town of Port Deposit in the state of Maryland. Laid down from surveys made under the direction of J. Edgar Thomson. [Philadelphia: 1831].

Scale: 1 in. = 1 mi.

Cartographer: J. E. Thomson.

Lithographers: Childs and Inman.

Size: 31.2 x 74.4 cm.

Colored.

Inset: General map of the country between Baltimore and Philadelphia.

(652: [1831]: T368csp Large)

1073. 1846

The harbor of Annapolis. Founded upon a trigonometrical survey under the direction of A. D. Bache. U.S. Coast and Geodetic Survey. [Washington]: 1846.

Scale: 1 in. = 1 mi.

Cartographers: A. D. Bache, James Ferguson, Ferd. H. Gerdes, George M. Bache, and R. D. Cutts.

Engravers: J. H. Young and O. Lawson.

Size: 35.8 x 43.9 cm.

Provenance: Presented by the U.S. Coast and Geodetic Survey through A. D. Bache, 21 August 1846.

(652: 1846: Un33ha Small)

1074. 1849

Mouth of the Chester River. (Harbor of refuge, no. [blank].) From a trigonometrical survey under the direction of A. D. Bache. U.S. Coast and Geodetic Survey. [Washington]: H. Benner: 1849.

Scale: 1:40,000.

Cartographers: A. D. Bache, J. Ferguson, J. E. Johnson, H. L. Whiting, J. C. Neilson, W. P. McArthur, Wm. Luce, R. D. Cutts, and C. Mahon.

Engravers: F. Dankworth, O. A. Lawson, and S. T. Pettit.

Size: 43.7 x 35.8 cm.

Provenance: Presented by A. D. Bache "by direction of the Treasury Department," 5 April 1850.

(652: 1849: Un32cr Small)

1075. 1885

Patapsco River and Baltimore harbor. From a trigonometrical survey under the direction of F. R. Hassler and A. D. Bache. U.S. Coast and Geodetic Survey. [Washington]: 1885.

Scale: 1:60,000.

Cartographers: F. R. Hassler, A. D. Bache, J. Ferguson, J. B. Glück, H. L. Whiting, R. D. Cutts, G. M. Bache, E. A. Webber, and J. W. Donn.

Size: 44.4 x 68.6 cm.

Colored.

Provenance: Presented by U.S. Coast and Geodetic Survey through Spencer C. McCorkle, 14 December 1885.

(652: 1885: Un38prb Large)

1076. 1902

Map of Garrett County showing geological formations and agricultural soils. Maryland Geological Survey, U.S. Geological Survey and U.S. Bureau of Soils. Baltimore: Hoen, lith.: 1902.

Scale: 1:62,500.

Cartographers: Wm. Bullock Clark, Charles D. Walcott, Milton Whitney, G. C. Martin, and Clarence W. Dorsey.

Lithographer: Hoen.

Size: 103.7 x 86 cm.

Colored.

Inset: Vertical section of the area.

On back of map: Manuscript map of the world, possibly demonstrating the prevailing ocean currents or winds. *Size of paper:* 90.5 x 108.2 cm.

(652: 1902: M365geo Large)

1077. 1937

To the American Philosophical Society this map of the peninsula between Delaware & Chesopeak bays with the said bays and shores adjacent drawn from the most accurate surveys. By John Churchman. N.p.: 1786. U.S Geological Survey: 1937.

Scale: 1 in. = 10 mi.

Cartographer: John Churchman.

Size: 57 x 43.4 cm.

Colored.

This is one of the series of maps depicting the 13 original states at the time of the ratification of the Constitution. Issued by the U.S. Constitution Sesquicentennial Commission.

Title: Delaware at the time of the ratification of the Constitution, from a 1787 original in the Library of Congress.

Copy 1: Contains autograph inscription of Col. Lawrence Martin: "To the American Philosophical Society with Lawrence Martin's compliments. 1937 facsimile of map dedicated to the Society in 1786. This is from the second, or 1787 edition. It is only a proof. Fine copy will be presented as soon as off the press."

Copy 2: Reprinted by U.S. Geological Survey from a 2nd edition of the original map. 1937.

Wheat: 479

(651: 1937: C472cbs Large)

DISTRICT OF COLUMBIA Nos. 1078-1080

1078. 1792

Plan of the city of Washington in the territory of Columbia, ceded by the states of Virginia and Maryland to the United States of America, and by them established as the seat of their government, after the year MDCCC. Philadelphia: 1792.

Scale: 1 in. = 100 poles.

Cartographer: James Thackara.

Engravers: Thackara and Vallance.

Size of paper: ca. 51.5 x 71.5 cm.

Provenance: Presented by George Washington to the Earl of Buchan, 22 April 1793; subsequently by Buchan to the American Philosophical Society, 17 July 1793.

Contains autograph of Washington.

Wheat: 531

(653: 1792: T328pcw Large)

1079. 1857

Map of Washington City, District of Columbia, seat of the Federal Government. Respectfully dedicated to the Senate and the House of Representatives of the United States of North America. By A. Boschke. [Washington]: 1857.

Scale: 1 in. = 500 ft.

Cartographer: A. Boschke.

Lithographer: J. Bien.

Size: 148 x 143 cm.

Colored.

Border contains vignettes of places of interest in the city.

(653: 1857: B651wdc Book map)

1080. 1882

Washington and Georgetown harbors, District of Columbia. U.S. Coast and Geodetic Survey. [Washington]: 1882.

Scale: 4 in. = 1 mi.

Cartographers: S. T. Abert and A. and H. Lindenkohl.

Size: 46.3 x 72.3 cm.

Provenance: Presented by U.S. Coast and Geodetic Survey through Spencer C. McCorkle, 14 December 1885.

(653: 1882: Un38wgh Large)

SOUTHERN STATES Nos. 1081-1093

1081. 1776

A general map of the southern British colonies, in America. Comprehending North and South Carolina, Georgia, East and West Florida, with the neighboring Indian countries. From the modern surveys of Engineer de Brahm, Capt. Collet, Mouzon & others; from the large hydrographical survey of the coasts of East and West Florida. By B. Romans. 1776. [Title at top of sheet]: The seat of war, in the southern British colonies, comprehending North and South Carolina, Georgia, East and West Florida, &ca. London: R. Sayer and J. Bennett: 1776 [15 October 1776].

Scale: 1 degree = 69½ British statute mi.

Cartographers: B. Romans, De Brahm, Capt. Collet, and Mouzon.

Size: 50 x 64.5 cm.

Colored.

Insets: Plan of Charleston; Plan of St. Augustine.

Originally in *American military pocket atlas;* removed from binding 5 November 1962

(636.5: 1776: R667bcm Large)

1082. [1776]

A new map of North & South Carolina, & Georgia. [1776].

Scale: 2.5 cm. = 110 mi.

Engraver: Aitken.

Size: 15.7 x 21.8 cm.

In: *Pennsylvania Magazine,* vol. 2 (1776), opposite p. 268.

Wheat: 489

(050: P383)

1083. 1788

A map of the states of Virginia, North Carolina, South Carolina and Georgia comprehending the Spanish provinces of East and West Florida exhibiting the boundaries as fixed by the late treaty of peace between the United States and the Spanish dominions compiled from late surveys and observations. 1788.

Scale: 2.5 cm. = 95 mi.

Cartographer: Joseph Purcell.

Engraver: Amos Doolittle.

Size: 31 x 36.4 cm.

In: Jedidiah Morse, *American geography,* opposite p. 1.

Wheat: 491

(917.3: M83)

1084. 1789

New map of the states of Georgia, South and North Carolina, Virginia and Maryland, including the Spanish provinces of West and East Florida, from the latest surveys. New York: 1789.

Engraver: C. Tiebout.

Size: 30.3 x 37 cm.

Provenance: Presented by Estate of Richard Gimbel, 1974.

In: William Gordon, *History of the rise of the United States.* New York: Hodge, Allen and Campbell: 1789, vol. 1, front.

Wheat: 492

(Paine 68: G65h, and 973.3: G65)

1085. [1797]

Map of the southern parts of the United States of America. [1797].

Scale: 2.5 cm. = 68 mi.

Cartographer: Abraham Bradley, Jr.

Engraver: S. Hill.
Size: 19.6 x 39.1 cm.
In: Jedidiah Morse, *The American gazetteer,* opposite "Southern States."
Wheat: 496
(917.3: M83am)

1086. 1813
A map of the southern section of the United States including the Floridas & Bahama Islands shewing the seat of war, in that department. Philadelphia: Melish: 1813.
Scale: 1 in. = 60 mi.
Cartographer: John Melish.
Engraver: H. S. Tanner.
Size: 40.7 x 52.8 cm.
Colored.
Originally in: John Melish, *Military and topographical atlas.* Philadelphia: Palmer: 1813, no. 5.
(636.5: 1813: M485fb Small)

1087. 1813
A map of the southern section of the United States including the Floridas & Bahama Islands, shewing the seat of war, in that department. Philadelphia: John Melish: 1813 [March 1813].
Cartographer: John Melish.
Engraver: H. S. Tanner.
Size: 40.5 x 52.3 cm.
Colored.
(650: 1813: M485fbi Small)

1088. [1817]
Southern provinces of the United States. No. 57. [London: 1817].
Scale: 1 in. = 47 British statute mi.
Engraver: Hewitt.
Size: 49.6 x 59.2 cm.
Colored.
Inset: Characteristic scenery of the Hudson River.
Drawn and engraved for Thompson's *New general atlas,* 1817.
(636.5: [1817]: H493sps Large)

1089. [1822]
The seat of war of the revolution in the southern states; shewing the principal movements of the hostile armies. Philadelphia: [1822].
Scale: 1 in. = 30 mi.
Cartographer: H. S. Tanner.

Engraver: H. S. Tanner.

Size: 45 x 38.4 cm.

Colored.

See: William Johnson, *Sketches of the life and correspondence of Major General Greene.* Charleston: Miller: 1822.

(650: [1822]: T158swr Small)

1090. [1826]

Map of reconnaissance exhibiting the country between Washington and New Orleans with the routes examined in reference to a contemplated national road between these two cities. U.S. War Department. N.p.: [1826].

Scale: 1 in. = 10 mi.

Size: 50.6 x 69.3 cm.

Colored.

Provenance: Presented by General Parker, 4 January 1833.

To accompany Document No. 156. 19th Congress. 1st Session. House of Representatives. *Road from Washington to New Orleans.* Washington: 1826. Compiled from Tanner's atlas.

(636.5: [1826]: Un38wno Large)

1091. 1865

Military map showing the marches of the United States forces under command of Maj. Gen. W. T. Sherman U.S.A. during the years 1863, 1864, 1865 . . . Drawn by Capt. William Kosak . . . and John B. Muller. U.S. Army. Corps of Topographical Engineers. St. Louis, Mo.: 1865.

Scale: 1 in. = 25 mi.

Cartographers: William Kosak and John B. Muller.

Engravers: H. C. Evans and F. Courtenay.

Size: 54.5 x 94 cm.

Colored.

(636.5: 1865: Un38mfs Large)

1092. 1900

West Indian hurricane of September 1-10, 1900. U.S. Department of Agriculture. Weather Bureau. Storm bulletin no. 1, 1900. Willis L. Moore, chief. Washington: 1900. 6 maps on one sheet.

Cartographer: Willis L. Moore.

Size of sheet: 61 x 48.5 cm.

Colored.

Descriptive matter at bottom of sheet.

(630: 1900: Un38hur Small)

1093. 1937

A general map of the southern British colonies, in America. Comprehending North and South Carolina, Georgia, East and West Florida, with the neighbouring Indian countries. From the modern surveys of . . . de Brahm, Capt. Collet, Mouzon & others; and from the large hydrographical survey of the coasts of East and West Florida. By B. Romans. 1776. London: R. Sayer and J. Bennett: 1776. Reprinted by U.S. Geological Survey, 1937, under the title: Georgia at the time of the ratification of the Constitution.

Scale: 1 degree = 69½ British mi.

Cartographer: B. Romans.

Size: 43.6 x 55.5 cm.

Colored.

Insets: The settled portion of Georgia (from Abraham Bradley's 1796 map of the United States); the correct southern boundary of Georgia in 1787.

See: U.S. Geological Survey, *Bulletin* 817. Washington: 1930.

(658: 1937: R667cgf Large)

VIRGINIA Nos. 1094-1138

1094. 1775

A map of the most inhabited part of Virginia, containing the whole province of Maryland, with part of Pensilvania, New Jersey and North Carolina. Drawn by Joshua Fry and Peter Jefferson in 1775. . . . London: R. Sayer and Thomas Jefferys: 1775.

Scale: 1 in. = 10 mi.

Cartographers: Joshua Fry, Peter Jefferson Will[ia]m Byrd, William Dandridge, Richard Fitzwilliams, Alexander Irving, Will[ia]m Mayo, William Churton, and Daniel Weldon.

Size: 78.3 x 123.3 cm.

Colored.

(654: 1755: F943vmc Large)

1095. [1776]

Map of the maritime parts of Virginia exhibiting the seat of war, and of Ld. Dunmore's depredations. In that colony. [1776].

Scale: 2.5 cm. = 10 mi.

Cartographer: P. S. Du Simitière.

Engraver: Robert Aitken.

Size: 24.9 x 27.6 cm.

See: *Pennsylvania Magazine,* vol. 2 (1776), opposite p. 184.

Wheat: 540

(654: [1776]: D852mpv Small)

1096. [1789]

From Annapolis (70) to Bowling-green Ord[inar]y. [1789, 1961 reprint].

Size: 17.1 x 11.6 cm.

In: Christopher Colles, *A survey of the roads.*

Wheat: 550

(526.8: C68s)

1097. [1789]

From Annapolis (66) to Dumfries. [1789, 1961 reprint].
Size: 17.3 x 11.5 cm.
In: Christopher Colles, *A survey of the roads.*
Wheat: 546

(526.8: C68s)

1098. [1789]

From Annapolis (67) to Fredericksburg, Va. [1789, 1961 reprint].
Size: 17.2 x 11.5 cm.
In: Christopher Colles, *A survey of the roads.*
Wheat: 547

(526.8: C68s)

1099. [1789]

From Annapolis (68) to Fredericksburg. [1789, 1961 reprint].
Size: 17.1 x 11.4 cm.
In: Christopher Colles, *A survey of the roads.*
Wheat: 548

(526.8: C68s)

1100. [1789]

From Annapolis (73) to Hanover & Newcastle. [1789, 1961 reprint].
Size: 17.2 x 11.5 cm.
In: Christopher Colles, *A survey of the roads.*
Wheat: 553

(526.8: C68s)

1101. [1789]

From Annapolis (72) to Hanover Court-House. [1789, 1961 reprint].
Size: 17.2 x 11.4 cm.
In: Christopher Colles, *A survey of the roads.*
Wheat: 552

(526.8: C68s)

1102. [1789]

From Annapolis (71) to Head Lynchs Ord[inar]y. [1789, 1961 reprint].
Size: 17.1 x 11.6 cm.
In: Christopher Colles, *A survey of the roads.*
Wheat: 551

(526.8: C68s)

1103. [1789]
From Annapolis (74) to New Kent Court-house. [1789, 1961 reprint].
Size: 17.2 x 11.5 cm.
In: Christopher Colles, *A survey of the roads.*
Wheat: 554
(526.8: C68s)

1104. [1789]
From Annapolis (75) to New Kent, Court-house. [1789, 1961 reprint].
Size: 17.2 x 11.6 cm.
In: Christopher Colles, *A survey of the roads.*
Wheat: 555
(526.8: C68s)

1105. [1789]
From Annapolis (69) to Todd's Ordinary. [1789, 1961 reprint].
Size: 17.1 x 11.5 cm.
In: Christopher Colles, *A survey of the roads.*
Wheat: 549
(526.8: C68s)

1106. [1789]
From Annapolis (76) to Williamsburgh. [1789, 1961 reprint].
Size: 17.1 x 11.5 cm.
In: Christopher Colles, *A survey of the roads.*
Wheat: 556
(526.8: C68s)

1107. [1789]
From Annapolis (77) to Williamsburgh. [1789, 1961 reprint].
Size: 17.2 x 11.6 cm.
In: Christopher Colles, *A survey of the roads.*
Wheat: 557
(526.8: C68s)

1108. [1789]
From Annapolis (78) to York. [1789, 1961 reprint].
Size: 17.2 x 11.4 cm.
In: Christopher Colles, *A survey of the roads.*
Wheat: 558
(526.8: C68s)

1109. [1789]
From Annapolis (79) to York. [1789, 1961 reprint].
Size: 17.2 x 11.5 cm.
In: Christopher Colles, *A survey of the roads.*
Wheat: 559

(526.8: C68s)

1110. [1789]
From Williamsburgh (80) to Ayletts Warehouse. [1789, 1961 reprint].
Size: 17.1 x 11.7 cm.
In: Christopher Colles, *A survey of the roads.*
Wheat: 560

(526.8: C68s)

1111. [1789]
From Williamsburg (81) to Aylett's Ware-houses. [1789, 1961 reprint].
Size: 17.1 x 11.6 cm.
In: Christopher Colles, *A survey of the roads.*
Wheat: 561

(526.8: C68s)

1112. [1789]
From Williamsburg (85) to Hooe's ferry. [1789, 1961 reprint].
Size: 17.2 x 12.1 cm.
In: Christopher Colles, *A survey of the roads.*
Wheat: 565

(526.8: C68s)

1113. [1789]
From Williamsburg (86) to Hooe's ferry. [1789, 1961 reprint].
Size: 17.7 x 11.6 cm.
In: Christopher Colles, *A survey of the roads.*
Wheat: 566

(526.8: C68s)

1114. [1789]
From Williamsburg (83) to Port-Royal. [1789, 1961 reprint].
Size: 17 x 11.6 cm.
In: Christopher Colles, *A survey of the roads.*
Wheat: 563

(526.8: C68s)

1115. [1789]

From Williamsburg (84) to Port-royal. [1789, 1961 reprint].
Size: 17.2 x 11.5 cm.
In: Christopher Colles, *A survey of the roads.*
Wheat: 564

(526.8: C68s)

1116. [1789]

From Williamsburg (82) to Sneed's Ordinary. [1789, 1961 reprint].
Size: 17 x 11.7 cm.
In: Christopher Colles, *A survey of the roads.*
Wheat: 562

(526.8: C68s)

1117. [1795]

Virginia. [1795].
Scale: 2.5 cm. = 68 mi.
Size: 15.3 x 18.5 cm.
Provenance: Presented by the Estate of Richard Gimbel, 1974.
In: Joseph Scott, *The United States gazetteer,* opposite "Virginia."
Wheat: 569

(Paine 68: Sco3u)

1118. [1796]

Virginia. [1796].
Scale: 2.5 cm. = 65 mi.
Engraver: W. Barker.
Size: 14.7 x 19.1 cm.
Provenance: Presented by Suzanne Wister Eastwick and Joseph L. Eastwick, July 1975.
In: Mathew Carey, *Carey's American pocket atlas,* opposite p. 94.
Wheat: 570

(917.3: C25c)

1119. 1799

The state of Virginia from the best authorities. 1799.
Scale: 2.5 cm. = 48 American mi.
Engraver: A. Anderson.
Size: 19 x 25.2 cm.
In: John Payne, *A new and complete system of universal geography,* vol. 4, opposite p. 385.
Wheat: 574

(910: P29)

1120. [1809]

Frederick, Berkeley & Jefferson counties in the state of Virginia. Executed A.D. 1809 by Chas. Varlé. [Richmond: 1809].

Scale: 1 in. = 2¼ mi.

Cartographer: Charles Varlé.

Engraver: Benjamin Jones.

Size: 84.9 x 61.8 cm.

Colored.

Inset: Plan of Winchester.

(654: [1809]: V428fbj Large)

1121. [ca. 1824]

University of Virginia ground plan. N.p.: [ca. 1824].

Scale: 1 in. = 50 ft.

Cartographer: Peter Maverick.

Size: 43.3 x 48.9 cm.

(654: [ca. 1824]: M446unv Small)

1122. [1826]

Map of the state of Virginia. Constructed in conformity to law, from the late surveys, authorized by the legislature, and other original and authentic documents, by Herman Boyé. [Philadelphia: 1826].

Scale: 1 degree = 60 geographical mi. = 69.08 statute mi.

Cartographer: Herman Boyé.

Engravers: H. S. Tanner and B. Tanner.

Size: 155.5 x 233.3 cm.

Colored.

Insets: View of Richmond from the west; University of Virginia.

Provenance: Presented by the State of Virginia, 17 August 1827.

On the map: "Mason & Dixon's line . . . " and "the author has examined a variety of rare and original documents, in the library of the Am: Phil: Society, from which he has collected and incorporated with the map, such facts as cannot fail to enhance its value."

(654: [1826]: B691svr Large)

1123. 1861

Map of the battle fields of Manassas and the surrounding region showing the various actions of the 21st July, 1861, between the armies of the Confederate States and the United States, surveyed and drawn by W. G. Atkinson. U.S. Army Corps of Topographical Engineers. N.p.: 1861.

Cartographer: W. G. Atkinson.

Size: 34.5 x 34 cm.

Provenance: Presented and inscribed by Capt. Abbot, 6 February 1863.

(654: 1861: Un38mbm Small)

1124. 1861

Part of the map of the military department of S.E. Virginia and Fort Monroe compiled in the Bureau of Topographical Engineers of the War Department. August, 1861. [Washington: 1862].

Scale: 1:200,000.

Cartographers: Herman Boyé, L. V. Buckholz, and H. L. Abbot.

Engraver: J. Schedler.

Size: 67.2 x 103.8 cm.

(654: 1861: Un38mon Large)

1125. 1861

Part of the map of the military department of S.E. Virginia and Fort Monroe, showing the approaches to Richmond and Petersburg, compiled in the Bureau of Topographical Engineers of the War Department, 1861, with additions and corrections from the map of the siege of Yorktown and the campaign maps of the army of the Potomac, compiled by Capt. H. L. Abbot. 1862. [Washington: 1862].

Scale: 1:80,000.

Cartographer: H. L. Abbot.

Engraver: J. Schedler.

Size: 73.4 x 123.9 cm.

(654: 1861: Un38vir Large)

1126. [1862]

Campaign maps, Army of the Potomac. Map no. 1. Yorktown to Williamsburg, prepared by command of Maj. Gen. George B. McClellan . . . A. A. Humphreys Brig. Gen. and Chief of Topographical Engineers. Compiled by Capt. H. L. Abbot. [Washington: 1862]. Map in 4 pieces.

Scale: 1 in. = 1 mi.

Cartographers: H. L. Abbot and A. A. Humphreys.

Size of paper: 91.2 x 60.1 cm.

Provenance: Presented and autographed by H. L. Abbot.

Photograph of a manuscript map.

(654: [1862]: Un38wil Large)

1127. [1862]

Campaign maps. Army of the Potomac. Map no. 2. Williamsburg to White House. Prepared by command of Maj. Gen. George B. McClellan . . . U.S. War Department. Compilation, under the direction of Brig. Gen. A. A. Humphreys, by Capt. H. L. Abbot. [Washington: 1862].

Scale: 1 in. = 1 mi.

Cartographers: A. A. Humphreys, H. L. Abbot, W. R. Palmer, N. Bowen, F. W. Dorr, J. W. Donn, F. A. Churchill, et. al.

Engraver: W. H. Dougal.

Size: 67.2 x 59.8 cm.

Provenance: Photographed and inscribed copy presented by H. L. Abbot.

(654: [1862]: Un38wht Large)

1128. [1862]

Campaign map, Army of the Potomac. Map no. 3. White House to Harrisons Landing. Prepared by command of Major General George B. McClellan, commanding Army of the Potomac. Compiled under the direction of Brig. Gen A. A. Humphreys, by Capt. H. L. Abbot. [Washington: 1862].

Scale: 1 in. = 1 mi.

Cartographers: A. A. Humphreys and H. L. Abbot.

Engraver: W. H. Dougal.

Size: 86.8 x 71.1 cm.

Provenance: Presented by H. L. Abbot.

(654: [1862]: Un38har Large)

1129. 1862

Central Virginia, compiled in the Bureau of Topographical Engineers of the War Department for military purposes. July, 1862. U.S. War Department. Washington: J. F. Gedney: 1862.

Scale: 1:350,000.

Size: 65.7 x 78.8 cm.

(654: 1862: Un38cen Large)

1130. [1862]

Map of eastern Virginia, compiled from the best authorities, and printed at the Coast Survey Office, A. D. Bache, Supdt. 1862. U.S. Coast and Geodetic Survey. Compiled by W. L. Nicholson. [Washington: 1862].

Scale: 1 in. = 15 mi.

Cartographers: A. D. Bache and W. L. Nicholson.

Lithographer: Charles G. Krebs.

Size: 55.8 x 47.2 cm.

Colored.

Provenance: Presented by A. D. Bache, 3 October 1862.

(654: [1862]: Un38evb Large)

1131. 1862

Official plan of the siege of Yorktown, Va. Conducted by the Army of the Potomac under command of Maj. Gen. George B. McClellan, April 5th to May 3rd 1862. Prepared under the direction of Brig. Gen J. G. Barnard by Lieut. Henry L. Abbot. U.S. Topographical Engineers. [Washington]: 1862.

Scale: 2½ in. = 800 yd.

Cartographers: Henry L. Abbot, J. G. Barnard, C. B. Comstock, N. J. Hall, O. G. Wagner, N. Bowen, F. W. Dorr, and J. W. Donn.

Size: 43 x 73 cm.

Provenance: Presented and endorsed by H. L. Abbot.

(654: 1862: Un38pyu Small)

1132. 1863

Portions of the military departments of Virginia, Washington middle, & the Susquehanna. U.S. War Department. Engineering Department. [Washington]: 1863.

Scale: 1:200,000.

Cartographer: Dennis Callahan.

Size: 68.4 x 103.5 cm.

Colored.

(654: 1863: Un38vws Large)

1133. 1863

Upper Potomac from McCoy's Ferry to Conrad's Ferry and adjacent portions of Maryland and Virginia, compiled from county maps and maps prepared by Col. J. N. McComb, with additions and corrections by Lt. Col. D. H. Strother. U.S. War Department. 1863. [Washington]: 1863.

Scale: 1:126,720.

Cartographers: J. N. McComb and D. H. Strother.

Lithographer: J. F. Gedney.

Size: 78.6 x 89.6 cm.

(654: 1863: Un38mcc Large)

1134. [1864]

The approaches from Washington, to Richmond. From surveys supplied by officers of the Army. Home Insurance Company of New York. [New York: 1864].

Scale: 1 in. = 7 mi.

Cartographer: J. H. Higginson.

Size: 48.5 x 63.5 cm.

Colored.

Inset: Strategic points of East Tennessee, northern Alabama and Georgia, from the U.S. Official Military Map, 1864.

(654: [1864]: H753wr Small)

1135. 1864

Portions of Virginia and North Carolina, embracing Richmond & Lynchburg, Va., and Goldsboro & Salisbury, N.C., compiled in the Engineering Bureau, War Department, for military purposes. New York: J. Schedler: 1864.

Scale: 1:350,000.

Size: 82.6 x 102.3 cm.

(654: 1864: Un38vnc Large)

1136. 1881

Map of the vicinity of Yorktown, Virginia. N.p.: 1881.

Size: 44.8 x 65.5 cm. *Size of paper:* 73 x 86.5 cm.

Colored.

Inset: Yorktown.

On three sides of the map are likenesses of famous Revolutionary War persons and pictures of famous buildings.

(654: 1881: M322ytv Small)

1137. [1882]

Norfolk harbor. Elizabeth River, Virginia. Topography by C. M. Bache. U.S. Coast and Geodetic Survey. 1882. [Washington]: 1883.

Scale: 1:10,000.

Cartographers: J. W. Donn, B. A. Colonna, C. M. Bache, Eugene Ellicott, E. B. Thomas, and C. M. Chester.

Size: 70.8 x 85.5 cm.

Provenance: Presented by the U.S. Coast and Geodetic Survey through Spencer C. McCorkle, 14 December 1885.

(654: [1882]: Un38nhv Large)

1138. [1885]

Coast Chart no. 31. Chesapeake Bay. Sheet no. 1. York River, Hampton Roads, Chesapeake entrance. From a trigonometrical survey under the direction of A. D. Bache. U.S. Coast and Geodetic Survey. [Washington: 1885].

Scale: 1:80,000.

Cartographers: A. D. Bache, F. Blunt, J. Farley, J. J. S. Hassler, J. Seib, J. W. Donn, J. Mechan, J. J. Almy, W. W. Harding, W. M. C. Fairfax, A. Strausz, G. D. Wise, and L. D. Williams.

Engravers: J. Knight, A. Sengteller, H. S. Barnard, and J. C. Kondrup.

Size: 63 x 96.2 cm.

Provenance: Presented by U.S. Coast and Geodetic Survey through Spencer C. McCorkle, 14 December 1885.

(654: [1885]: Un38cyh Large)

WEST VIRGINIA Nos. 1139-1144

1139. 1864

Map of the oil district of West Virginia. Compiled & drawn by Peter F. Stout. Philadelphia: R. L. Barnes: 1864.

Scale: 1½ in. = 4 mi.

Cartographer: P. F. Stout.

Size: 35.8 x 44.6 cm.

Colored.

(655: 1864: St77wvo Small)

1140. [ca. 1900]

Map of Coal River, in Kanawha & Boone counties, West Virginia. Geological Survey of West Virginia. Cincinnati, O.: Ehrgott & Krebbs: [ca. 1900].

Size: 46 x 61 cm.

Inset: Map of West Virginia and surrounding states.

(655: [ca. 1900]: W529crb Small)

1141. [1917]

Map of West Virginia showing coal, oil, gas, iron ore and limestone areas. Geological Survey of West Virginia. From the topographical sheets of the U.S. Geological Survey. Baltimore: Hoen & Co.: [1917].

Scale: 1 in. = 8 mi.

Cartographers: I. C. White, Ray V. Hennen, D. B. Reger, and R. C. Tucker.

Lithographer: A. Hoen & Co.

Size: 82.8 x 111.4 cm.

Colored.

Contains listing by county of coal mines.

(655: [1917]: W529Lim Large)

1142. [1917]

Map of West Virginia showing railroads. Geological Survey of West Virginia. Base from government and other surveys by Ray V. Hennen and R. C. Tucker. Baltimore: Hoen & Co.: [1917].

Scale: 1 in. = 8 mi.

Cartographers: Hennen and Tucker.

Lithographer: A. Hoen & Co.

Size: 83 x 94.8 cm.

Colored.

(655: [1917]: W529rwv Large)

1143. 1921

Map of West Virginia showing coal, oil, gas, iron ore and limestone areas. Geological Survey of West Virginia. Baltimore: Hoen, lith.: 1921.

Scale: 1 in. = 8 mi.

Cartographers: I. C. White, Ray V. Hennen, D. B. Reger, and R. C. Tucker.

Lithographer: Hoen.

Size: 83 x 94.3 cm.

Colored.

Accompanied by: List of coal mines in West Virginia; 1 July 1921.

(655: 1921: W529geo Small)

1144. 1937

Relief map of West Virginia, prepared by the Geological Survey of West Virginia. Paul H. Price, state geologist. Data from U.S. Geological Survey topographical quadrangles surveyed in cooperation with the West Virginia Geological Survey. 1937. Baltimore: A. Hoen & Co.: 1937.

Scale: 1 in. = 8 mi.

Cartographer: Paul H. Price.

Lithographer: A. Hoen & Co.

Size: 78.4 x 88.5 cm.
Colored.

(655: 1937: W529rmw Extra-oversize)

NORTH CAROLINA Nos. 1145-1156

1145. [1771]

Plan of the entrance into Cape Fear harbour, North Carolina. [London: 1771].
Cartographer: J. S. Speer.
Size: 25.2 x 36.6 cm.
See: Joseph Smith Speer, *West-India pilot.* London: Hooper: 1771.

(656: [1771]: C222cfh Small)

1146. [1795]

North-Carolina. [1795].
Scale: 2.5 cm. = 62 mi.
Size: 15.5 x 18.8 cm.
Provenance: Presented by the Estate of Richard Gimbel, 1974.
In: Joseph Scott, *The United State gazetteer,* opposite "North-Carolina."
Wheat: 583

(Paine 68: Sco3u)

1147. [1796]

North Carolina. [1796].
Scale: 2.5 cm. = 50 mi.
Engraver: W. Barker.
Size: 14.8 x 19.4 cm.
Provenance: Presented by Suzanne Wister Eastwick and Joseph L. Eastwick, July 1975.
In: Mathew Carey, *Carey's American pocket atlas,* following p. 102.
Wheat: 584

(917.3: C25c)

1148. 1798

To navigators this chart. being an actual survey of the sea coast and inland navigation from Cape Henry to Cape Roman is most respectfully inscribed by Price and Strother. New-Bern, North Carolina: W. Johnston: 1798.
Scale: 1 degree = 20 leagues.
Cartographers: Price and Strother.
Engraver: W. Johnston.
Size: 36.8 x 98.3 cm.
Wheat: 590

(656: 1798: P936hr Large)

1149. 1800

North Carolina from the best authorities. 1800.

Scale: 2.5 cm. = ca. 24 American mi.

Engraver: A. Anderson.

Size: 19 x 38.5 cm.

In: John Payne, *A new and complete system of universal geography,* vol. 4, opposite p. 414.

(910: P29)

1150. 1808

To David Stone and Peter Brown, esqrs. this first actual survey of the state of North Carolina, taken by the subscribers is respectfully dedicated by . . . Jona. Price, John Strother. 1808. Philadelphia: C. P. Harrison: 1808.

Scale: 1 degree = 69½ mi.

Cartographers: John Strother and Jonathan Price.

Engraver: W. Harrison.

Size of paper: ca. 70.5 x 150.5 cm.

Colored.

Provenance: Received 19 February 1808.

Trimmed.

(656: 1808: P936snc Large)

1151. 1837

[Core Sound, North Carolina]. Surveyed under the direction of J. Kearney. U.S. Army. Topographical Engineers. 1837. [Washington]: 1837. Nos. 1-4: 1. Harbor Island bar. 62.3 x 32.5 cm.; 2. Drum shoal. 25.3 x 32.6 cm.; 3. Piney Point shoal. 52.4 x 32.6 cm.; 4. Bells Point shoal. 52.4 x 32.6 cm.

Scale: 1:15,840.

Cartographers: J. Kearney, T. J. Lee, L. Sitgreaves, and A. M. Mitchell.

Engraver: W. J. Stone.

Size: 54.5 x 33.7 cm.

Provenance: Presented by J. D. Graham, 18 June 1841.

In: U.S. Congress. 25th Congress. House of Representatives. 2nd Session. Document 445.

(656: 1837: Un31acs Small)

1152. 1837

Core sound, North Carolina, surveyed, under the direction of Lieut. Col. J. Kearney, by T. J. Lee, L. Sitgreaves, A. M. Mitchell and Wash: Hood. U.S. Bureau of Topographical Engineers. Washington: 1837.

Scale: 16,000 ft.

Cartographers: J. Kearney, T. J. Lee, L. Sitgreaves, A. M. Mitchell, and Wash. Hood.

Engraver: W. J. Stone.

Size: 42.7 x 88.4 cm.

Provenance: Presented by J. D. Graham, 18 June 1841.

In: U.S. Congress. 25th Congress. House of Representatives. 2nd Session. Document 445.

(656: 1837: Un38csn Large)

1153. 1850

Pasquotank River. From a trigonometrical survey under the direction of A. D. Bache. U.S. Coast and Geodetic Survey. [Washington]: 1850.

Scale: 1:60,000.

Cartographers: A. D. Bache, W. M. Boyce, J. C. Neilson, W. P. McArthur, and H. Adams.

Engravers: W. Smith and S. T. Pettit.

Size: 35.7 x 43.7 cm.

Provenance: Presented by A. D. Bache, 27 December 1850.

(656: 1850: Un37pr Small)

1154. 1911

Map of North Carolina. Prepared by North Carolina Geological and Economic Survey from base map compiled by the U.S. Geological Survey. Washington: Norris Peters: 1911.

Scale: 1 in. = 11 mi.

Cartographer: Joseph Hyde Pratt.

Size: 50.3 x 126.4 cm.

Colored.

(656: 1911: Un38ges Large)

1155. 1916

State of North Carolina. In cooperation with the state of North Carolina, Joseph Hyde Pratt, State Geologist. U.S. Geological Survey. George Otis Smith, Director . . . compiled in 1909 and 1910. [Washington]: 1916.

Scale: 1:500,000.

Cartographers: Joseph Hyde Pratt, George Otis Smith, R. B. Marshall, and A. F. Hassan.

Engraver: U.S. Geological Survey.

Size: 63.9 x 166.2 cm.

(656: 1916: Un38ncs Large)

1156. 1926

Proposed Great Smoky Mountains National Park, North Carolina-Tennessee. U.S. Department of the Interior. Geological survey. [Washington]: U.S. Geological Survey: 1926.

Scale: 1:125,000.

Engraver: U.S. Geological Survey.

Size: 61.6 x 75.1 cm.

Colored.

(635: 1926: Un35gsm Small)

SOUTH CAROLINA Nos. 1157-1169

1157. [1795]

South Carolina. [1795].
Scale: 2.5 cm. = 40 mi.
Size: 15.5 x 18.2 cm.
Provenance: Presented by the Estate of Richard Gimbel, 1974.
In: Joseph Scott, *The United States gazetteer,* opposite "South-Carolina."
Wheat: 602

(Paine 68: Sco3u)

1158. [1796]

South Carolina. [1796].
Scale: 2.5 cm. = 45 American mi.
Engraver: Doolittle.
Size: 14.6 x 19.1 cm.
Provenance: Presented by Suzanne Wister Eastwick and Joseph L. Eastwick, July 1975.
In: Mathew Carey, *Carey's American pocket atlas,* following p. 110.
Wheat: 603

(917.3: C25c)

1159. 1796

The state of South Carolina: from the best authorities. 1796. New York: J. Reid: 1796.
Scale: 1 degree = 69½ American mi.
Engraver: Tanner.
Size: 38.1 x 43 cm.
On reverse: Georgia, from the latest authorities. *Size:* 22.3 x 38.7 cm.
In: John Reid, *The American atlas.* New York: J. Reid: 1796, no. 17.
Wheat: 605

(657: 1796: T158ssc Small)

1160. 1799

The state of South Carolina from the best authorities. 1799.
Scale: 2.5 cm. = 35 mi.
Engraver: Scoles.
Size: 18.7 x 21.5 cm.
In: John Payne, *A new and complete system of universal geography,* vol. 4, opposite p. 429.
Wheat: 608

(910: P29)

1161. [1802]

[Map of South Carolina]. [Charleston: 1802].

Scale: 1 degree = 69½ mi.

Cartographer: T. Coram.

Engraver: J. Akin.

Size: 41.8 x 47.7 cm.

Colored.

Original in John Drayton, *View of South Carolina, as respects her natural and civil concerns.* Charleston: Young: 1802.

(657: [1802]: D792sc Small)

1162. [ca. 1802]

Sketch of the Santee canal. [Charleston: ca. 1802].

Scale: 2¼ in. = 5 mi.

Engraver: James Akin.

Size: 32.5 x 14.7 cm.

Duplicate of map in John Drayton, *View of South Carolina, as respects her natural and civil concerns.* Charleston: Young: 1802.

(657: [ca. 1802]: Ak51scc Small)

1163. [1809]

A plan of Charles Town, from a survey of Edw[ar]d Crisp in 1704. N.p.: [1809].

Scale: 1 in. = 40 perches or 660 ft.

Cartographer: Edward Crisp.

Engraver: James Akin.

Size: 23.1 x 29.1 cm.

In: David Ramsay, *History of South Carolina.* Charleston: Longworth: 1809, vol. 2.

(657.976: [1809]: C873csc Small)

1164. [1818]

[South Carolina]. N.p.: [1818].

Cartographer: John Wilson.

Size of paper: 38.8 x 49.9 cm.

Contains manuscript note: "Sketched by John Wilson to accompany his report, 1818."

(657: [1818]: W699sc Small)

1165. [1822]

A map of South Carolina, constructed and drawn from the district surveys, ordered by the legislature: By John Wilson. The astronomical observations by Professor Geo: Blackburn & I. M. Elford. Philadelphia: H. S. Tanner: [1822. 10 April 1822].

Scale: 1 in. = 6 mi.

Cartographer: John Wilson.

Engraver: H. S. Tanner.

Size: 112.2 x 149.8 cm.

Colored.

Provenance: Presented by H. S. Tanner, 15 August 1823; and another presented by John L. Wilson, governor of South Carolina, 15 August 1823.

Inset: Charleston harbour.

(657: [1822]: W699scm Large)

1166. [1825]

Charleston harbour and the adjacent coast and country, South Carolina. U.S. Bureau of Topographical Engineers. Surveyed at intervals in 1823, 1824 and 1825 by Hartman Bache, James D. Graham, C. M. Eakin and W. M. Boyce. [Washington: 1825].

Scale: 4 in. = 1 mi.

Cartographers: Hartman Bache, J. D. Graham, C. M. Eakin, W. M. Boyce, and S. Wragg.

Engraver: W. J. Stone.

Size: 158.8 x 135.4 cm.

Provenance: Presented by Hartman Bache, 3 April 1835; and presented by Colonel Long, 7 February 1862.

Hartman Bache told the Society that the survey for this map took place in 1823-1825. He described this survey and gave "evidence of its minuteness and accuracy." The members then asked Bache to write his comments for use at a future meeting [Minutes; Donation Book; *Transactions.*, n.s., vol. 5].

In: U.S. Congress. *A collection of maps, etc.* published by order of Congress. Washington: 1843, no. 102.

(657: [1825]: Un38chc Large)

1167. 1855

Preliminary chart of Charleston harbor and its approaches. U.S. Coast Survey. From a trigonometrical survey under the direction of A. D. Bache. Hydrography by the party under the command of J. N. Maffitt. 1855. [Washington]: 1855.

Scale: 1:30,000.

Cartographers: A. D. Bache, C. O. Boutelle, S. A. Gilbert, J. N. Maffitt, J. J. Ricketts, and A. Boschke.

Engravers: F. Dankworth, J. Knight, E. Yeager, W. Smith, and G. McCoy.

Size: 86.4 x 60.7 cm.

Provenance: Presented by A. D. Bache, 6 March 1857.

(657: 1855: Un38pcc Large)

1168. 1870

Charleston harbor and its approaches. U.S. Coast Survey. From a trigonometrical survey under the direction of A. D. Bache. Hydrography by the party under the direction of C. O. Boutelle. 1870. [Washington]: 1870.

Scale: 1:30,000.

Cartographers: A. D. Bache, C. O. Boutelle, S. A. Gilbert, J. Seib, and W. S. Edwards.

Size: 77.5 x 80.8 cm.

Provenance: Presented by the U.S. Coast and Geodetic Survey through Spencer C. McCorkle, 14 December 1885.

(657: 1870: Un38scc Large)

1169. 1911

Official map of South Carolina, issued by the State Department of Agriculture, Commerce and Industries. 1911. New York: E. C. Bridgman: 1911.

Scale: 2.5 cm. = 10 mi.

Size: 57 x 73.1 cm.

Colored.

Contains printed material extolling the value of South Carolina for living. Also picture of the capitol at Columbia.

(657: 1911: So87oms Large)

1169a. 1983

General highway map of Oconee County, South Carolina. Contains notes by Margaret Mills Seaborn and outline of William Bartram's path taken in 1776. N.p.: 1983.

Scale: 1 in. = 2 mi.

Cartographer: Margaret Mills Seaborn.

Size: 60.3 x 44.7 cm.

Colored.

Provenance: Presented by Mrs. M. M. Seaborn, 3 July 1986.

(657: 1983: So82oco Large)

GEORGIA Nos. 1170-1183

1170. [1771]

Plan du Port de Gouadaquini now called Jekil Sound in the province of Georgia in North America. N.p.: [1771].

Scale: 1 degree = 60 mi.

Size: 26.8 x 37.5 cm.

On map: Jekil Island called Pallavona in 1721.

See: Joseph Smith Speer, *West-India pilot.* London: Hooper: 1771.

(658: [1771]: J384pjs Small)

1171. 1779

Siège de Savanah, fait par les troupes du roi, aux ordres de Monsieur le conte d'Estaing, vice-amiral de France en Septembre et Octobre 1779.

Scale: 10 cm. = 1,200 toises.

Photograph.

Size: 32.1 x 46 cm.

Provenance: Presented by Morton Deutsch, April 1964.

(658.972: 1779: Un38sav Large)

1172. [1795]

Georgia. [1795].

Scale: 2.5 cm. = 115 mi.

Size: 15.3 x 19 cm.
Provenance: Presented by the Estate of Richard Gimbel, 1974.
In: Joseph Scott, *The United States gazetteer,* opposite "Georgia."
Wheat: 612

(Paine 68: Sco3u)

1173. [ca. 1795]
The state of Georgia. N.p.: [ca. 1795].
Scale: 1 in. = 20 mi.
Engraver: Gridley.
Size: 44.4 x 37.5 cm.
Colored.

(658: [ca. 1795]: G873gfc Small)

1174. [1796]
Georgia. [1796].
Scale: 2.5 cm. = 95 mi.
Engraver: W. Barker.
Size: 14.8 x 19 cm.
Provenance: Presented by Suzanne Wister Eastwick and Joseph L. Eastwick, July 1975.
In: Mathew Carey, *Carey's American pocket atlas,* following p. 114.
Wheat: 613

(917.3: C25c)

1175. 1796
Georgia, from the latest authorities. [New York: J. Reid: 1796].
Scale: 1 degree = 69½ American mi.
Engraver: Benjamin Tanner.
Size: 38.7 x 22.2 cm.
In: John Reid, *The American Atlas.* New York: J. Reid: 1796.
On reverse: The state of South Carolina: from the best authorities.
Wheat: 616

(657: 1796: T158ssc Small)

1176. 1799
Georgia from the latest authorities. 1799.
Scale: 2.5 cm. = ca. 50 mi.
Engraver: I. Scoles.
Size: 20.1 x 38.5 cm.
In: John Payne, *A new and complete system of universal geography,* vol. 4, opposite p. 438.
Wheat: 619

(910: P29)

1177. 1814

Map of the seat of war among the Creek Indians. From the original drawing in the War Department. Philadelphia: J. Melish: 1814 [8 November 1814].

Scale: 1 in. = 20 mi.

Cartographer: J. Melish.

Engraver: H. S. Tanner.

Size: 42.6 x 49.5 cm.

Colored.

This was originally the property of Benjamin Smith Barton.

(660: 1814: M485ci Small)

1178. 1861

Sketch of the Atlantic coast of the United States, from Savannah River to St. Mary's River, embracing the coast of the state of Georgia. U.S. Coast Survey. A. D. Bache, superintendent. [Washington]: 1861.

Scale: 1:200,000.

Cartographers: A. Lindenkohl, A. D. Bache, J. R. Butts, R. Mills, and Col. Campbell.

Inset: Atlantic coast of North Florida.

Size: 57.2 x 98.5 cm.

Trimmed.

(658: 1861: Un38acg Large)

1179. 1864

Northwestern Georgia (with portions of the adjoining states of Tennessee and Alabama) being part of the department of the Cumberland. U.S. War Department. Bureau of Topographical Engineers. January, 1863. New York: J. Bien: [1864].

Scale: 1:350,000.

Cartographers: Dennis Callahan and William E. Merrill.

Lithographer: J. Bien.

Size: 97.4 x 85.8 cm.

(658: 1864: Un38gta Large)

1180. 1875

Map illustrating the military operations in front of Atlanta, Georgia, from the passage of Peach Tree Creek, July 19th, 1864, to the commencement of the movement upon the enemy's lines of communication, south of Atlanta, August 26th, 1864. U.S. War Department. New York: Graphic Co.: 1875.

Scale: 2 in. = 1 mi.

Size: 73 x 78.3 cm.

Colored.

Contains diagrams of defensive works of each side.

(658: 1875: Un38apt Large)

1181. 1877

Maps illustrating the military operations of the Atlanta campaign . . . exhibiting the works of the U.S. and Confederate forces. 1864. New York: American Photo-Lith. Co.: [1877]. Maps in 5 pieces: No. 1. Tennessee River to Oostanaula River. 51.6 x 88.5 cm.; No. 2. Resaca to Ackworth. 70.7 x 63.7 cm.; No. 3. Kingston & Cassville to Dallas and Marietta. 57.2 x 79.9 cm.; No. 4. Pine, Lost and Kennesaw Mountains south to include Atlanta and environs. 62.6 x 66.6 cm.; No. 5. Chattahoochee River to Jonesboro and Lovejoy's Station. 64.3 x 67 cm.

Scale: 2¼ in. = 3 mi.

Colored.

(658: 1877: Un38mmg Large)

1182. 1885

Coast chart no. 56. From Savannah to Sapelo Island, Georgia. 1876. Surveys made between the years 1850 and '69 under the direction of A. D. Bache. U.S. Coast and Geodetic Survey. [Washington]: 1885.

Scale: 1:80,000.

Cartographers: A. and H. Lindenkohl, A. D. Bache, Benjamin Peirce, C. O. Boutelle, C. P. Bolles, E. O. C. Ord, A. W. Longfellow, A. W. Evans, F. P. Webber, H. L. Whiting, A. M. Harrison, W. H. Dennis, C. Hosmer, C. Fendall, C. Rockwell, H. L. Du Val, J. H. Moore, C. M. Fauntleroy, C. Junken, and J. E. Hilgard.

Engravers: E. Hergesheimer, A. Sengteller, E. A. Maedel, H. S. Barnard, and H. M. Knight.

Size of paper: 104.8 x 81.5 cm.

Provenance: Presented by U.S. Coast and Geodetic Survey through Spencer C. McCorkle, 14 December 1885.

Trimmed.

(658: 1885: Un38ssg Large)

1183. 1911-1912

State of Georgia. U.S. Geological Survey. George Otis Smith, director. Compiled in 1911-1912. [Washington]: U.S. Geological Survey: 1911-1912.

Scale: 1:500,000.

Cartographers: G. O. Smith, R. B. Marshall, and A. F. Hassan.

Engraver: U.S. Geological Survey.

Size: 104.5 x 86.4 cm.

Advance sheet. Subject to correction.

(658: 1911-1912: Un38int Large)

FLORIDA Nos. 1184-1199

1184. [1775]

Entrances of Tampa Bay. Page LXVIIII of the appendix. [1775].

Scale: 1 in. = 3⅓ geometric mi.

Cartographer: B. Romans.

Size: 9.6 x 15.3 cm.

Provenance: Presented by John Vaughan, 17 January 1806.

In: B. Romans, *A concise natural history of East and West Florida; containing an account of the natural produce of all the southern part of British America in the three kingdoms of nature, particularly the animal and vegetable.* New York: 1775.

Wheat: 623

(917.59: R66)

1185. [1775]

Pensacola bar. Page LXXXIV of the appendix. [1775].

Scale: 2.5 cm. = 2 geometric mi.

Cartographer: Bernard Romans (?).

Size: 14.3 x 9.6 cm.

Provenance: Presented by John Vaughan, 17 January 1806.

In: Bernard Romans, *A concise natural history of East and West Florida,* appendix.

Wheat: 624

(917.59: R66)

1186. [1791]

A map of the coast of East Florida from the River St. John southward near to Cape Canaveral. [1791].

Scale: [1½ in. = 10 mi.].

Cartographer: William Bartram.

Size: 25 x 14.7 cm.

Provenance: Presented by William Bartram, 20 July 1792.

In: William Bartram, *Travels through North & South Carolina, Georgia, East & West Florida, the Cherokee country, the extensive territories of the Muscogulges, or Creek Confederacy, and the country of the Chactaws . . .* Philadelphia: James & Johnson: 1791.

Wheat: 628

(917.3: B28)

1187. [1819]

New chart of the Gulf of Florida, Bahama Banks and part of the Gulf of Mexico, including St. Nicholas and Santaren Channels. From the most approved Spanish and other modern surveys. By Richard Patten. Published . . . 1817, with additions to 1819. New York: [1819].

Cartographer: Richard Patten.

Engravers: J. Ridley and W. Sim.

Size: 57.2 x 86.2 cm.

Provenance: Presented and autographed by C. F. Heazlitt, 22 January 1847.

(659: [1819]: P276fbb Large)

1188. [ca. 1820]

Plan of lands in East Florida purchased by Messrs. John Forbes & Company, from the Indians, supposed to contain 1,200,000 acres. [Savannah?: ca. 1820].

Scale: 1½ in. = 8 geographical mi.

Engraver: C. C. Wright.

Size: 32.7 x 27.6 cm.

(659: [ca. 1820]: F663Lef Small)

1189. 1820

Plan of the town of Colinton in East Florida. [New York?]: 1820.

Scale: 1 in. = 500 ft.

Cartographer: I. McKennon.

Engraver: [Charles? or William?] Rollinson.

Size: 34.1 x 34.5 cm.

See map No. 1188.

(659.923: 1820: R657tcf Small)

1190. 1821

Map of Florida. Philadelphia: Wm. Darby & B. Tanner; 1821 [21 March 1821].

Scale: 1 in. = 40 mi.

Size: 47 x 39.4 cm.

Insets: Espiritu Santo; Pensacola; Mobile Bays.

(659: 1821: D242flo Small)

1191. [1822]

Pensacola harbor and bar, Florida. Surveyed in 1822 by Maj. James Kearney, assisted by Lieuts. Thompson, Turnbull and Butler. U.S. Bureau of Topographical Engineers. [Washington: 1822].

Scale: 1:15,840.

Cartographers: James Kearney, Lts. Thompson, Turnbull, and Butler.

Engraver: W. J. Stone.

Size: 147.6 x 120.4 cm.

Provenance: Presented by Hartman Bache, 20 January 1837; delivered by J. D. Graham, 18 June 1841.

(659: [1822]: Un38phb Large)

1192. [1829]

Map of the territory of Florida. From its northern boundary to lat: 27°, 30′ N. connected with the delta of the Mississippi. Annexed to the report of the board of internal improvement . . . relating to the canal contemplated to connect the Atlantic with the Gulf of Mexico. U.S. Topographical Engineers. N.p.: [1829].

Scale: 1 in. = 11 mi.

Cartographer: W. H. Swift.

Engraver: W. J. Stone.

Size: 67.7 x 166.5 cm.

Insets: 15 surveys of sounds and bays; Gulf of Mexico with surrounding territory.

Provenance: Presented and autographed by William Tell Poussin, 11 June 1830.

Poussin sent Secretary John Vaughan a copy of the report "on the contemplated canal across the Peninsula of Florida" with a map. The map, he assured Vaughan, "was constructed with the greatest care and record [sic] all facts of actual survey" (Archives. Poussin to Vaughan; 11 June 1830. *Transactions,* n.s., vol. 4).

(659: [1829]: Un38fdm Large)

1193. [1839]

Map of the seat of war in Florida, compiled by orders of Bvt. Brigr. Genl. Z. Taylor, principally from the surveys and reconnaissances of the officers of the United States Army. By Capt. John Mackay and Lieut. J. E. Blake. U.S. Bureau of Topographical Engineers. [Washington: 1839].

Scale: 2 cm. = 9 mi.

Cartographers: John Mackay and J. E. Blake.

Engraver: W. J. Stone.

Size: 103.2 x 73.8 cm.

Colored.

(659: [1839]: Un38swf Large)

1194. 1841

The Bahama banks and gulf of Florida. By Edmund Blunt. New York: E. & G. W. Blunt: 1841.

Cartographer: Edmund Blunt.

Engraver: R. M. Gaw.

Size: 97.8 x 125.8 cm.

Inset: Chart of Key West.

Provenance: Presented and autographed by Edmund and G. W. Blunt, 16 October 1841.

The Blunts "spent a great deal of money in keeping up our charts to the latest surveys. . . . Our Chart, & our Coast pilot, contained a description of the shoal on which the Florida was short time since lost; & particular directions respecting the Coasts of New Jersey." The Blunts also wrote that their "new chart of the Bahama Bank" was the most dangerous section for American vessels and it differed from the British charts [Archives. E. Blunt to J. Vaughan, 16 October 1841].

(659: 1841: B561bbf Large)

1195. [1855]

Key West harbor and its approaches. U.S. Coast and Geodetic Survey. From a trigonometrical survey under the direction of A. D. Bache. Hydrography by the party under the command of Lieut. John Rodgers. 1855. [Washington: 1855].

Scale: 1:50,000.

Cartographers: A. D. Bache, J. E. Hilgard, I. H. Adams, E. K. Knorr, and John Rodgers.

Engravers: E. Yeager, E. F. Woodward, and H. M. Knight.

Size: 61.8 x 86.6 cm.

Inset: Sub-sketch of Key West harbor.

Provenance: Presented by A. D. Bache, 6 March 1857; by the U.S. Coast and Geodetic Survey through Spencer C. McCorkle, 14 December 1885.

(659: [1855]: Un38kwh Large)

1196. 1884

St. John's River, Florida, from Jacksonville to Hibernia. U.S. Coast and Geodetic Survey. 1884. [Washington]: 1884.

Scale: 1:40,000.

Cartographers: A. Lindenkohl, H. G. Ogden, J. B. Baylor, and C. M. Chester.

Engravers: R. F. Bartle, F. Courtenay, E. A. Maedel, and J. G. Thompson.

Size: 81.5 x 53.1 cm.

Provenance: Presented by the U.S. Coast and Geodetic Survey through Spencer C. McCorkle, 14 December 1885.

(659: 1884: Un38sjr Large)

1197. 1885

Coast chart no. 177. Tampa Bay, Florida. U.S. Coast and Geodetic Survey. [Washington]: 1885.

Scale: 1:80,000.

Cartographers: H. G. Ogden, W. I. Vinal, J. Hergesheimer, E. B. Thomas, H. B. Mansfield, J. M. Hawley, R. Platt, A. Braid, E. Smith, and C. M. Chester.

Size: 83.1 x 98.9 cm.

Provenance: Presented by U.S. Coast and Geodetic Survey through Spencer C. McCorkle, 14 December 1885.

(659: 1885: Un38tbf Large)

1198. [1911]

State of Florida. U.S. Department of the Interior. Compiled from the official records of the General Land Office and other sources, under the direction of I. P. Berthrong. 1911. Washington: Eckert Litho. Co.: [1911].

Scale: 1 in. = 12 mi.

Cartographer: I. P. Berthrong.

Lithographers: Eckert Lith. Co. and Wm. Bauman, Jr.

Size: 102.7 x 81 cm.

Colored.

Inset: Western panhandle of Florida.

(659: [1911]: Un38dif Large)

1199. 1913

A new sectional map of Florida. U.S. Department of Agriculture. Buffalo, N.Y.: Wm. P. Northrup: 1913.

Scale: 1 in. = 10 mi.

Size: 110.3 x 79.4 cm.

Colored.

Inset: Panhandle of western Florida.

(659: 1913: Un38fLo Large)

ALABAMA Nos. 1200-1208

1200. [1775]

Mobile bar. [1775].

Scale: 1 in. = ca. 2 mi.

Cartographer: B. Romans.

Size: 9 x 15 cm.

Provenance: Presented by John Vaughan, 17 January 1806.

At top of page is: "Page LXXXV of the Appendix."

In: Bernard Romans, *A concise natural history of East and West Florida; containing an account of the natural produce of all the southern part of British America in the three kingdoms of nature, particularly the animal and vegetable.* New York: 1775.

Wheat: 487

(917.59: R66)

1201. 1818

Map of Alabama constructed from the surveys in the General Land Office and other documents. By John Melish. Philadelphia: Melish: 1818 [29 October 1818].

Scale: 1 in. = 15 mi.

Cartographer: John Melish.

Size: 68.1 x 48.5 cm.

Colored.

(663: 1818: M485sma Large)

1202. [ca. 1830]

Map of four townships granted to the French emigrants in Alabama. N.p.: [ca. 1830].

Size: 19.2 x 24.9 cm.

Map shows Black Warrior or Tuskaloosa River.

(663: [ca. 1830]: AL11tfe Small)

1203. 1851

Entrance to Mobile Bay. U.S. Coast Survey. From a trigonometrical survey under the direction of A. D. Bache. [Washington]: 1851.

Scale: 1:40,000.

Cartographers: A. D. Bache, F. H. Gerdes, R. H. Fauntleroy, W. E. Greenwell, C. P. Patterson, and J. J. Ricketts.

Engravers: R. T. Knight, S. T. Pettit, and W. Smith.

Size: 70.5 x 48.1 cm.

Provenance: Presented by A. D. Bache, 6 April 1852; received 21 May 1852.

(663: 1851: Un38emb Large)

1204. 1856

Mobile Bay, Alabama. U.S. Coast Survey. From a trigonometrical survey under the direction of A. D. Bache. [Washington]: 1856.

Scale: 1:80,000.

Cartographers: A. D. Bache, C. M. Eakin, F. H. Gerdes, W. E. Greenwell, B. F. Sands, J. Alden, C. P. Patterson, C. Mahon, and J. J. Ricketts.

Engravers: S. Siebert, F. Dankworth, J. Knight, J. H. Goldthwaite, J. V. N. Throop, and H. M. Knight.

Size: 95.7 x 73 cm.

Provenance: Presented by A. D. Bache, 6 March 1857; by the U.S. Coast and Geodetic Survey through Spencer C. McCorkle, 14 December 1885.

(663: 1856: Un38mba Large)

1205. 1885

Mobile Bay, Alabama. From a trigonometrical survey under the direction of A. D. Bache, Superintendent of the Survey of the Coast of the United States. Triangulation by C. M. Eakin & F. H. Gerdes. Topography by W. E. Greenwell. Hydrography by the parties under the command of Lieuts. B. F. Sands, J. Alden & C. P. Patterson. Aids to navigation corrected to 1885. N.p.: 1885.

Scale: 1:80,000.

Cartographers: A. D. Bache, C. M. Eakin, F. H. Gerdes, W. E. Greenwell, B. F. Sands, J. Alden, and C. P. Patterson.

Engravers: S. Siebert, E. Dankworth, J. Knight, and J. H. Goldthwait.

Size: 96.5 x ca. 71 cm.

(663: 1885: B124mob Extra-oversize)

1206. [1890]

Maps of Muscle Shoals canal, Tennessee River, showing proposed intercoastal canal . . . U.S. War Department. Corps of Engineers. Made under the direction of Lt. Col. J. W. Barlow. Washington: Norris Peters Co.: [1890]. 4 maps.

Cartographers: Julius Shutting, John H. Morgan, and J. W. Barlow.

Sizes: 13.9 x 39; 22.7 x 35.4; 30.5 x 45.5; and 29.8 x 268.6 cm.

Colored.

(663: [1890]: Un38msc Large)

1207. [1895]

Map of the state of Alabama. U.S. Department of the Interior. Compiled from the official records of the General Land Office and other sources under the direction of Harry King. 1895. New York: Julius Bien: [1895].

Scale: 1 in. = 12 mi.

Cartographers: Harry King, Daniel O'Hare, and Robert H. Morton.

Lithographer: Julius Bien.

Size: 82.2 x 53.5 cm.

Colored.

(663: [1895]: Un38aLa Large)

1208. [1915]

State of Alabama. U.S. Department of the Interior. Compiled from official records of the General Land Office, U.S. Geological Survey and other sources. Under the direction of I. P. Berthrong. 1915. [New York: 1915].

Scale: 1 in. = 12 statute mi.

Cartographers: I. P. Berthrong, Daniel O'Hare, and George A. Daidy.

Size: 81.7 x 53.4 cm.

Colored.

Insets: Chattanooga and vicinity, showing the Chickamauga and Chattanooga National Park; Birmingham-Bessemer; Mobile; Shiloh and vicinity.

(636: [1915]: Un38aLa Large)

MISSISSIPPI Nos. 1209-1214

1209. 1850

Cat and Ship Island harbors. U.S. Coast Survey. From a trigonometrical survey under the direction of A. D. Bache. New York: Sherman & Smith: 1850.

Scale: 1:40,000.

Cartographers: A. D. Bache, F. H. Gerdes, J. E. Hilgard, W. E. Greenwell, C. P. Patterson, and Charles Mahon.

Engravers: Sherman & Smith and J. Knight.

Size: 42.6 x 85 cm.

Provenance: Presented by A. D. Bache, 27 December 1850.

(664: 1850: Un38csi Large)

1210. [1890]

State of Mississippi. U.S. Department of the Interior. Compiled from the official records of the General Land Office and other sources under supervision of A. F. Dinsmore. 1890. [Washington: 1890].

Scale: 1 in. = 12 mi.

Cartographers: A. F. Dinsmore and Robert H. Morton.

Size: 82.9 x 52.9 cm.

Colored.

(664: [1890]: Un38smd Large)

1211. 1910-1911

State of Mississippi. U.S. Geological Survey. George Otis Smith, director. [Washington]: 1910-1911.

Scale: 1:500,000.

Cartographers: G. O. Smith, R. B. Marshall, and A. F. Hassan.

Size: 111.8 x 67.6 cm.

(664: 1910-1911: Un38int Large)

1212. 1911

Mississippi. Rand, McNally & Co's new Business Atlas map of Mississippi. Chicago: Rand, McNally: 1911.

Scale: 1 in. = 13 statute mi.

Size: 67 x 47.8 cm.

Colored.

(664: 1911: M655bam Large)

1213. 1915

State of Mississippi. U.S. Department of the Interior. Compiled from official records of the General Land Office, U.S. Geological Survey, and other sources; under the direction of I. P. Berthrong. Washington: Eckert Lithographing Co.: 1915.

Scale: 1:760,320.

Cartographers: A. F. Dinsmore and I. P. Berthrong.

Lithographer: Eckert Lithographing Co.

Size: 83.9 x 52.5 cm.

Colored.

Inset: Jackson.

(664: 1915: Un38som Small)

1214. 1916

Mississippi. U.S. Department of the Interior. Geological Survey. George Otis Smith, director. [Washington]: 1916.

Scale: 1:1,000,000.

Cartographers: George Otis Smith, R. B. Marshall, and A. F. Hassan.

Size: 57.2 x 35 cm.

(664: 1916: Un38mgs Small)

SOUTH CENTRAL UNITED STATES No. 1215

1215. 1820

A map of Louisiana and Mexico. Carte de la Louisiane et du Mexique, dressée par P. Tardieu, fils ainé, gravée par P. A. F. Tardieu, père. Paris: P. A. F. Tardieu and C. Picquet: New York: Anthony Girard: 1820.

Cartographer: P. Tardieu.

Engravers: P. A. F. Tardieu and Richomme.

Size: 106.5 x 79.4 cm.

Colored.

(633.6: 1820: L935cLm Large)

LOUISIANA Nos. 1216-1224

1216. 1803

Plan of New Orleans & its environs. New Orleans: 1803.

Cartographer: J. L. Boqueta de Woiseri.

Size: 70.8 x 48.5 cm.

Colored.

Insets: Esplanade or Place of Arms; Louisiana Purchase; map of New Orleans & larger environs.

Only three copies of this map are known.

(666.957: 1803: B641noe Large)

1217. 1806

Carte générale du territoire d'Orléans comprenant aussi la Floride occidentale et une portion du territoire du Mississippi. Dressée d'après les observations les plus récentes par Bmi. Lafon. Nouvelle Orléans. 1806.

Scale: 1 degree = 25 common leagues.

Cartographer: B. Lafon.

Size of paper: 87.5 x 131 cm.

Provenance: Presented by Benjamin Lafon, 16 February 1816.

Trimmed.

(666: 1806: L135nof Large)

1218. 1816

A map of the state of Louisiana, with part of the Mississippi territory, from actual survey. By Wm. Darby. Philadelphia: John Melish: 1816 [8 April 1816].

Scale: 1:10 English mi.

Cartographer: William Darby.

Engraver: Samuel Harrison.

Size: 80.2 x 114 cm.

Colored.

Provenance: Presented by William Darby, 17 May 1816.

See: William Darby, *A geographical description of Louisiana.* Philadelphia: Melish: 1816.

(666: 1816: D242sLm Large)

1219. 1816

Plan of the city and environs of New Orleans. Taken from actual survey. By B. Lafon. N.p.: 1816.

Cartographer: B. Lafon.

Size: 50.4 x 54.2 cm.

Colored.

Trimmed.

(666.957: 1816: L135pno Large)

1220. 1834

Topographical map of New Orleans and its vicinity. Embracing a distance of twelve miles up, and eight and three quarter miles down the Mississippi River and part of Lake Pontchartrain representing all public improvements existing and projected, and important establishments, accompanied by a statistical table containing the most accurate illustrations prefaced by a splendid view of New Orleans, &. Compiled from actual surveys and the best authorities by Charles F. Zimpel. New Orleans: 1834 [4 September 1834].

Scale: 1 in. = 6 mi.

Cartographer: Charles F. Zimpel.

Size: 155.2 x 160.1 cm.

Insets: Plan of the banks of the Mississippi and adjacent country, extending from Caraby's district to A. Ducros' plantation being a continuation of the above map, on a scale of 2⅜ inches to the mile; Lake Pontchartrain; portrait of New Orleans; plan of New Orleans; Defeat of the British Army on the 8 of January 1815. Important buildings are depicted across the top of the map.

Provenance: Presented by Cash, Esq., 21 October 1836.

(666.957: 1834: Z79tmno Large)

1221. 1850

A new map of Louisiana, with its canals, roads & distances from place to place along the stage & steam boat routes. Philadelphia: Thomas, Cowperthwait & Co.: 1850.

Scale: 1 in. = 30 mi.

Size: 29.4 x 36.5 cm.

Colored.

Inset: New Orleans.

(666: 1850: C832Lcr Small)

1222. [1896]

Map of the state of Louisiana. U.S. Department of the Interior. Compiled from the official records of the General Land Office and other sources under the direction of Harry King. 1896. Baltimore: Friedenwald Co.: [1896].

Scale: 1 in. = 12 mi.

Cartographers: Harry King and Daniel O'Hare.

Size: 76.3 x 72.5 cm.

Colored.

(666: [1896]: Un38smL Large)

1223. [1915]

Southern Louisiana, showing waterways, etc. in the New Orleans, La., engineering district. U.S. War Department. Corps of Engineers. Compiled from U.S. Engineers, Coast and Geodetic, U.S. Geological, U.S. Land, Railroad surveys & other data. U.S. Survey Office, Detroit, Mich.: [1915].

Scale: 1 in. = 6 mi.

Cartographers: A. H. Guillot, R. P. Howell, Jr., Lansing H. Beach, and Edward H. Schulz.

Size: 77.5 x 138.4 cm.

Colored.

Insets: United States; the world.

(666: [1915]: Un38sLw Large)

1224. [1916]

State of Louisiana. U.S. Department of the Interior. Compiled from official records of the General Land Office and other sources under the direction of I. P. Berthrong. 1916. Washington: Eckert Lithog. Co.: [1916].

Scale: 1 in. = 12 mi.

Cartographers: I. P. Berthrong, A. F. Dinsmore, and Wm. Bauman, Jr.

Lithographer: Eckert Lithographing Co.

Size: 69.9 x 72.2 cm.

Colored.

Inset: City of New Orleans.

(666: [1916]: Un38Ldi Large)

TEXAS Nos. 1225-1235

1225. [1840]

Map of the river Sabine from its mouth on the Gulf of Mexico in the sea to Logan's Ferry in lat. 31° 58′ 24″ north. Shewing the boundary between the United States and the Republic of Texas between said points, as marked and laid down by survey in 1840, under the direction of the commissioners appointed for that purpose under the first article of the convention signed at Washington, April 25, 1838. Surveyed in 1840. Washington: [1840].

Scale: 1 in. = 1 mi.

Cartographers: J. D. Graham, Thomas J. Lee, George G. Meade, P. J. Pillans, D. C. Wilber, and A. B. Gray.

Lithographer: P. Haas.

Size: 74.1 x 426.3 cm.

Provenance: Presented by James D. Graham, 28 August 1842.

See note on map 1227 for Graham's comments on this map.

(668: [1840]: Un38sgm Large)

1226. [1841]

Map of the river Sabine from Logan's Ferry to 32nd. degree of north latitude. Shewing the boundary between the United States of America and the Republic of Texas between said points, as laid down by survey in 1841, under the direction of the joint commission appointed for that purpose under the 1st article of the convention signed at Washington, 25th. day of April 1838. [Washington: 1841].

Scale: 10½ in. = 2 mi.

Cartographer: A. B. Gray.

Size: 47.2 x 67.1 cm.

See: U.S. Congress, *A collection of maps, etc., published by order of Congress.* Washington: 1843, No. 173.

(668: [1841]: Un38sLf Large)

1227. [1841]

Part of the boundary between the United States and Texas, from Sabine River, northward, to the 36th. mile mound. Drawn from the notes of survey and copied for the use of the commissioners, under the convention of the 25th. April 1838. By J. Edmd. Blake. U.S. Army Corps of Topographical Engineers. [Washington: 1841]. 3 maps.

Scale: 2 in. = 1 mi.

Cartographers: J. Edmund Blake, T. R. Conway, A. B. Gray and James Kearney.

Size: ca. 69 x 53 cm.

Provenance: Presented by James D. Graham, 28 August 1842.

A^2 from Sabine River, northward to the 36th mile mound.

B^2 North of Sabine River from the 36th to the 72nd. mile mound.

C^2 North of Sabine River from 72nd. mile mound to Red River.

See: United States Congress, *A collection of maps, etc. pub. by order of Congress.* Washington: 1843, Nos. 174-176.

Graham wrote that he was forwarding some maps and charts

> illustrating the demarcation of that portion of the boundary between the United States and the Republic of Texas, included between the mouth of the river Sabine, and the point where the due north line from 32° of N. latitude on that river, intersects the Red River. Also a chart of the entrance of the river Sabine. These maps and charts are from actual surveys executed, as particularly stated on the respective sheets, in the years 1840, and 1841. There are in all, ten sheets. The river Sabine from its mouth, to Logan's Ferry, consists of five sheets which I have had mounted in one continuous map for more convenient reference. This part of the series is printed in colours so as to define the line of the boundary with great minuteness.
>
> The latitudes and longitudes of several points, and the dip and declination of the magnetic needle, as expressed on these five sheets, are from actual observations made by myself while attached to the Commission on the part of the United States for marking out the said boundary [Archives. J. D. Graham to A. D. Bache, 28 August 1842. Minutes. *Proceedings,* vol. 2, p. 214. Donation book].

See also: map No. 1225.

(668: [1841]: Un38pbt Large)

1228. [1841]

Sabine Pass and mouth of the river Sabine in the sea. Surveyed under the direction of Maj. J. D. Graham . . . by Lieut. T. J. Lee . . . and P. J. Pillans, Texan army, attached to the joint commission for marking the boundary between the United States and Texas. U.S. Topographical Engineers. 1840. [Washington: 1841].

Scale: 2 in. = 1 mi.

Cartographers: J. D. Graham, T. J. Lee, and P. J. Pillans.

Size: 65.7 x 46.4 cm.

See: U.S. Congress, *A collection of maps, etc., published by order of Congress.* Washington: 1843, No. 177.

(668: [1841]: Un38spm Large)

1229. 1844

Map of Texas and the country adjacent: compiled in the Bureau of the Corps of Topographical Engineers. U.S. War Department. From the best authorities. For the State Department, under the direction of Colonel J. J. Abert. By W. H. Emory. 1844. [Washington]: 1844.

Scale: 2.5 cm. = 100 mi.

Cartographers: J. J. Abert and W. H. Emory.

Size: 35.9 x 55.7 cm.

Provenance: Presented by Joseph R. Ingersoll, 20 June 1845.

(668: 1844: Un38tca Large)

1230. 1853

Galveston entrance, Texas. U.S. Coast Survey. From a trigonometrical survey under the direction of A. D. Bache. [Washington]: 1853.

Scale: 1:40,000.

Cartographers: A. D. Bache, R. H. Fauntleroy, J. S. Williams, J. M. Wampler, T. A. Craven, A. S. Baldwin, A. Fornaro, and A. Balbach.

Engravers: H. C. Evans and S. E. Stull.

Size: 34.5 x 43.5 cm.

Provenance: Presented by A. D. Bache.

(668: 1853: Un38etg Small)

1231. 1884

Preliminary chart of Galveston Bay, Texas. U.S. Coast Survey. A. D. Bache Supdt. [Washington]: 1884.

Scale: 1:200,000.

Cartographers: A. D. Bache, R. H. Fauntleroy, R. D. Cutts, J. M. Wampler, T. A. Craven, E. J. De Haven, E. M. Hughes, J. M. Mansfield, and C. W. Howell.

Size: 48.7 x 42.5 cm.

Provenance: Presented by the U.S. Coast and Geodetic Survey through Spencer C. McCorkle, 14 December 1885.

(668: 1884: Un38gbt Large)

1232. 1900

Map of Texas and parts of adjoining territories. U.S. Geological Survey. Compiled by and under the direction of Robert T. Hill. [Washington]: 1900.

Scale: 1 in. = 25 mi.

Cartographers: Robert T. Hill, Henry Selden, and Willard D. Johnson.

Size: 79.7 x 90.1 cm.

Colored.

(668: 1900: Un38tth Large)

1233. 1903

Railroad and county map of Texas. 1903. Railroad Commission of Texas. St. Louis, Mo.: Woodward & Tiernan: 1903.

Scale: 1 in. = 25 mi.

Size: 79.5 x 89.4 cm.

Colored.

(668: 1903: T318rcm Large)

1234. 1904

Geological map of a portion of west Texas, showing parts of Brewster, Presidio, Jeff. Davis, and El Paso counties, and south of the Southern Pacific R.R. By Benjamin F. Hill and J. A. Udden. University of Texas Mineral Survey. 1904. Buffalo, N.Y.: Matthews-Northrup Works: 1904.

Cartographers: C. P. Scrivener, Benjamin F. Hill, and J. A. Udden.

Size: 40.5 x 128.2 cm.

Colored.

(668: 1904: T318tws Large)

1235. 1904

[Topographical survey of the Terlingua quadrangle, Texas, in cooperation with the University of Texas Mineral Survey]. William B. Phillips, Director. U.S. Geological Survey. Charles D. Walcott, Director. [Washington]: 1904. 3 maps. Editions of 1902 and 1904.

Scale (2 maps): 1:125,000.; 1:50,000.

Cartographers: Charles D. Walcott, William B. Phillips, E. M. Douglas, Arthur Stiels, Fred McLaughlin, and J. E. Blackburn.

Engraver: U.S. Geological Survey.

Size: 44.5 x 39 cm.; 16.6 x 48.5 cm.

Colored.

(668: 1904: Un38tsm Small)

KENTUCKY Nos. 1236-1253

1236. 1784

This map of Kentucke, drawn from actual observations, is inscribed with the most profound respect, to the Honorable the Congress of the United States of America; and to his Excell[en]cy George Washington. By John Filson. Philadelphia: T. Rook: 1784. 3 facsimiles.

Scale: 1 in. = 10 mi.

Cartographer: John Filson.

Engraver: Henry D. Purcell.

Sizes: 49 x 44.7 cm.; 48 x 44.6 cm.; 44.7 x 50 cm.

One facsimile was by the Kentucky Geological Survey, ser. VI, 1929. Source: Original in the private library of W.R. Jillson . . . Pub. date 20 December, 1929.

Another facsimile is of the New-York Historical Society's copy of the map. Reprinted by the Kentucky Geological Survey, 15 January, 1931. Written on the scroll on facsimile is: "New York Historical Society from John Pintard. 1809."

One facsimile presented by R.T. Durrett.

There is also a reprint in Z.F. Smith's *History of Kentucky*. 1886.

See: John Filson, *The discovery, settlement and present state of Kentucke.*

Wheat: 632

(661: 1784: F483prf Small)

1237. [ca. 1785]

Carte de Kentucke, d'après les observations actuelles: dediée a l'honorable Congrès des Etats-Unis de l'Amérique et à son excellence George Washington, commandant en chef des ses armées par leur très humble serviteur, John Filson. [Paris: Buisson: 1785].

Scale: 3.5 cm. = 20 mi.

Cartographer: John Filson.

Engraver: André.

Size: 34.7 x 31.6 cm.

Removed from John Filson, *Histoire de Kentucke* . . . trad. par M. Parraud. Paris: Buisson: 1785.

(661: [ca. 1785]: F483prf Small)

1238. 1793

A map of the state of Kentucky, drawn from the best authorities. 1793.

Scale: 2.5 cm. = 42½ British statute mi.

Engraver: T. Condor.

Size: 15 x 19.1 cm.

In: Gilbert Imlay, *A topographical description,* following p. 400.

Wheat: 640

(917.3:Im5 [London edition])

1239. [1795]

Kentucky. [1795].

Scale: 2.5 cm. = ca. 57 mi.

Size: 15 x 18.5 cm.

Provenance: Presented by the Estate of Richard Gimbel, 1974.

In: Joseph Scott, *The United States gazetteer,* opposite "Kentucky."

Wheat: 643

(Paine 68: Sco3u)

1240. 1795

Map of the state of Kentucky; with adjoining territories. New York: Smith, Reid and Wayland: 1795.

Engraver: A. Anderson.

Size: 37.4 x 44.3 cm.

In: *The American atlas.* New York: J. Reid: 1796, no. 15.

Wheat: 646

(661: 1795: An21msk Large)

1241. [1796]

Kentuckey [sic; 1796].

Scale: 2.5 cm. = ca. 57 mi.

Engraver: W. Barker.

Size: 14.7 x 19 cm.

Provenance: Presented by Suzanne Wister Eastwick and Joseph L. Eastwick, July 1975.

In: Mathew Carey, *Carey's American pocket atlas,* following p. 98.

Wheat: 644

(917.3: C25c)

1242. [1796]

Plan of Franklinville, in Mason County, Kentucky. [1796].

Scale: 2.5 cm. = 3,200 ft.

Engraver: Tanner.

Size: 15.4 x 11.5 cm.

In: William Winterbotham, *An historical view of the United States,* vol. 3, opposite p. 145.

Wheat: 648

(917: W73)

1243. [1796]

Plan of Lystra, in Nelson-County, Kentucky. [1796].

Scale: 2.5 cm. = ca. 800 ft.

Engraver: Tanner.

Size: 16 x 11.5 cm.

In: William Winterbotham, *An historical view of the United States,* vol. 3, opposite p. 144.

Wheat: 649

(917: W73)

1244. 1800

The state of Kentucky and the adjoining territories from the best authorities. 1800.

Engraver: Scoles.

Size: 18.6 x 21.7 cm.

In: John Payne, *A new and complete system of universal geography,* vol. 4, opposite p. 404.

(910: P29)

1245. 1818

A map of the state of Kentucky, from actual survey. Also part of Indiana and Illinois, compiled principally from returns in the Surveyor-General's office. By Luke Munsell. Frankfort: Luke Munsell: 1818 [16 December 1818].

Scale: 1 degree = 69.5 mi.

Cartographer: Luke Munsell.

Engraver: H. Anderson.

Size: 98.7 x 228 cm.

Colored.

Provenance: Presented by Luke Munsell, 15 January 1819.

There are two engraved scenes from pictures painted by Thomas Sully.

(661: 1818: M925msk Large)

1246. 1850

A new map of Kentucky, with its roads & distances from place to place along the stage & steam boat routes. [Philadelphia]: Thomas Cowperthwait: 1850.

Scale: 1 in. = 30 mi.

Size: 29.6 x 35.5 cm.

Colored.

Insets: Lexington & vicinity; Falls of the Ohio; Maysville and vicinity.

(661: 1850: C832ken Small)

1247. [1859]

Map of eastern Kentucky, shewing the western outcrop of its coal field as determined by the surveys of 1858 and 1859, by Jos. Lesley, Jr., Philadelphia: J. Rehn: [1859].

Scale: 1 in. = 18 mi.

Cartographer: Joseph Lesley, D. Downes, and J. P. Lesley, Jr.

Lithographer: J. Rehn.

Size: 25 x 24.5 cm.

Provenance: Presented by George B. Wood, 6 January 1860.

(661: [1859]: L565ken Small)

1248. 1886

Agricultural map of the Jackson Purchase. Geological Survey of Kentucky. John R. Proctor, Director. [New York?]: Julius Bien & Co.: 1886.

Scale: 1:300,000.

Cartographers: John R. Proctor, R. H. Loughridge, and J. B. Hoening.

Lithographer: Julius Bien & Co.

Size: 33.6 x 51 cm.

Colored.

(661: 1886: K414ajp Small)

1249. [ca. 1891]

[County maps of Kentucky.] Geological Survey of Kentucky. John R. Proctor, J. B. Hoening, Charles J. Norwood, Directors. N.p.: [ca. 1891]. 12 maps: 1. Boyle & Mercer counties. 49.6 x 39.1 cm.; 2. Clinton county. 39 x 33.1 cm.; 3. Crittenden county. 78.8 x 81.5 cm.; 4. Green, Taylor & Adair counties. 54.7 x 49.1 cm.; 5. Henry, Shelby and Oldham counties. 55.3 x 57.4 cm.; 6. Lincoln county. 42.2 x 37.7 cm.; 7. Madison county. 42.3 x 38.6 cm.; 8. Meade and Breckenridge counties. 59.4 x 55.5 cm.; 9. Montgomery and Clark counties. 45.1 x 53.2 cm.; 10. Trigg and Christian counties. 47.7 x 61.1 cm. On same sheet are maps of Taylor county [30.4 x 39.4 cm.] and Green county [34.7 x 25.7 cm.]; 11. Warren county. 51.4 x 46.6 cm.; 12. Washington and Marion counties. 54.4 x 44.4 cm.

Cartographers: John R. Proctor, J. C. Fales, W. M. Linney, J. B. Hoening, R. H. Loughridge, H. R. Ayres, L. M. Sellier, A. M. Miller, J. W. Norwood, F. H. Bagby, F. Julius Fohs, S. C. Jones, J. F. McAdoo, and W. T. Knott.

Lithographer: Julius Bien.

Colored.

(661: [ca. 1891]: 1891: K414geo Large)

1250. 1891

Map of Meade County showing locations of gas wells. Geological Survey of Kentucky. 1891. [New York]: Julius Bien & Co.: 1891.

Scale: 1 in. = 2 mi.

Lithographer: J. Bien & Co.

Size: 34.8 x 40.2 cm.

Colored.

To accompany report by R. H. Loughridge.

(661: 1891: K414mgw Small)

1251. 1905

Map of the Big Sandy Valley, by J. B. Hoening. Geological Survey of Kentucky, Charles J. Norwood, Director. Compiled from surveys made by the Kentucky and U.S. Geological Surveys and from data from other authentic sources. N.p.: 1905.

Scale: 1:253,440.

Cartographers: J. B. Hoening and Charles J. Norwood.

Size: 52.5 x 36.5 cm.

(661: 1905: K414bsv Small)

1252. [1905]

Soil map of Webster county. Geological Survey of Kentucky. Charles J. Norwood, Director. Compiled and drawn by J. W. Norwood. Delineation of soils by S. C. Jones. Baltimore: [1905].

Scale: 1:126,720.

Cartographers: J. W. Norwood and S. C. Jones.

Lithographer: A. Hoen & Co.

Size: 27.6 x 41.2 cm.

Colored.

(661: [1905]: K414swc Small)

1253. 1916

Map of upper Cumberland River and adjacent territory. By L. M. Sellier. Geological Survey of Kentucky. Baltimore: A. Hoen & Co.: 1916.

Scale: 1:63,360.

Cartographer: L. M. Sellier.

Lithographer: A. Hoen & Co.

Size: 48.8 x 102 cm.

(661: 1916: K414ucr Small)

TENNESSEE Nos. 1253a-1262

1253a. [1795]

A map of the Tennassee Government formerly part of North Carolina, taken chiefly from surveys by Gen[era]l D. Smith & others. [Philadelphia: Carey: 1795].

Scale: 1 in. = 22 mi.

Cartographer: D. Smith.

Engraver: J. Scott.

Size: 24 x 51.5 cm.

Provenance: Presented by Mathew Carey, 18 October 1805.

The *Transactions* (vol. 6) state that Carey presented the "Materials from which Guthrie's Geography were compiled." This is one of those maps. See: William Guthrie, *The general atlas for Carey's edition of his geography improved.* Philadelphia: Carey: 1795 [1 May 1795].

This map contains numerous manuscript notations.

(662: [1795]: Sm5.2 Large)

1254. [1795]

S[outh] W[est] Territory. [1795].

Scale: 2.5 cm. = 66 mi.

Size: 15.5 x 9.5 cm.

Provenance: Presented by the Estate of Richard Gimbel, 1974.

Incomplete.

In: Joseph Scott, *The United states gazeteer,* opposite "Territory South of the Ohio."

Wheat: 651

(Paine 68: Sco3u)

1255. [1796]

Tennassee [sic]: lately the S[outh] W[ester]n Territory. [1796].

Scale: 2.5 cm. = ca. 70 mi.

Engraver: W. Barker.

Size: 14.8 x 19.3 cm.

Provenance: Presented by Suzanne Wister Eastwick and Joseph L. Eastwick, July 1975.

In: Mathew Carey, *Carey's American pocket atlas,* following p. 106.

Wheat: 652

(917.3: C25c)

1256. 1799

A map of the Tennassee [sic] government from the latest surveys. 1799.

Scale: 2.5 cm. = 30 mi.

Size: 17.8 x 39 cm.

In: John Payne, *A new and complete system of universal geography,* vol. 4, opposite p. 422.

Wheat: 657

(910: P29)

1257. [ca. 1830]

Tennessee. [Philadelphia: ca. 1830].

Scale: 1 in. = 40 mi.

Cartographer: Samuel Lewis.

Engraver: Henry S. Tanner.

Size: 21.7 x 27.6 cm.

Colored.

(662: [ca. 1830]: L585mst Small)

1258. 1850

A new map of Tennessee, with its roads and distances from place to place along the stage & steam boat routes. [Philadelphia]: Thomas Cowperthwait & Co.: 1850.

Scale: 1 in. = 30 mi.

Size: 28.9 x 39 cm.

Colored.

Insets: Environs of Nashville; Environs of Knoxville.

(662: 1850: C832ten Small)

1259. [1863].

Battlefield of Chattanooga, with the operations of the National Forces under the command of Maj. Gen. U. S. Grant, during the battles of November 23, 24, & 25, 1863. U.S. War Department. [Washington]: Lithographic Association of New York: [1863].

Scale: 1:47,520.

Cartographers: W. F. Smith, F. W. Dorr, J. W. Donn, Major Morhardt, Capts. Ligowsky, McDowell, Jenny, Preston, and Lts. Boeckh and Dahl.

Lithographer: Charles G. Krebs.

Size: 46.4 x 40.4 cm.

Colored.

Provenance: Presented by the U.S. Coast Survey, 20 January and 21 April 1865; presented by C. H. Hart, 15 March 1865.

Printed for the benefit of the Sanitary Commission by the Lithographic Association of New York.

(662: [1863]: Un38cwb Small)

1260. [1864]

Topographical map of the approaches and defenses of Knoxville, E. Tennessee, shewing the positions occupied by the U.S. & Confederate forces during the siege. U.S. War Department. Department of Engineers. Surveyed by direction of Capt. O. M. Poe during Dec., Jan. and Feb. 1863-64. New York: Graphic Co.: [1864].

Scale: 6⅜in. = 1 mi.

Cartographer: O. M. Poe.

Size: 66.3 x 75.5 cm.

Colored.

(662.946: [1864]: Un38kdt Large)

1261. 1874

Battlefield in front of Franklin, Tennessee, where the U.S. forces consisting of the 4th and 23rd Corps and the Cavalry Corps M.D.M., all under the command of Maj. Gen'l. J. M. Schofield, severely repulsed the Confederate Army, commanded by Lt. Gen'l. Hood, November 30th, 1864. U.S. War Department. Compiled under the direction of Col. W. E. Merrill. New York: American Photo-Lithograph Co.: 1874.

Scale: 6 in. = 1 mi.

Cartographers: W. E. Merrill, James R. Willett, and T. J. L. Remington.

Size: 69.8 x 44.9 cm.

(662: 1874: Un38bft Large)

1262. 1875

Map of the battlefield of Chattanooga. U.S. War Department. Department of Engineers. Prepared to accompany report of Maj. Gen. U. S. Grant. By direction of Brig. Gen. W. F. Smith. 1864. New York: Graphic Company: 1875.

Scale: 2.9 cm. = ½ mi.

Cartographers: W. F. Smith, F. W. Dorr, J. W. Donn, Maj. Morhardt, Capts. Ligowsky and McDowell, Lts. Boeckh and O. H. Dahl, Capts. W. L. B. Jenny and McElfatrick.

Size: 72.4 x 66.1 cm.

Colored.

(662: 1875: Un38bcg Large)

ARKANSAS Nos. 1263-1264

1263. [1914]

State of Arkansas. U.S. Department of the Interior. Compiled from the official records of the General Land Office and other sources under the direction of I. P. Berthrong. 1914. Washington: Eckert Lithographing Co.: [1914].

Scale: 1 in. = 12 mi.

Cartographer: I. P. Berthrong.

Lithographer: Eckert Lithographing Co.

Size: 58.7 x 71 cm.

Colored.

Insets: City of Little Rock; City of Hot Springs.

(665: [1914] Un38saw Large)

1264. 1916

Arkansas. U.S. Department of the Interior. Geological Survey. George Otis Smith, Director. Compiled in 1912-1913. [Washington]: 1916.

Scale: 1:1,000,000.

Cartographers: George Otis Smith, R. B. Marshall, and A. F. Hassan.

Size: 41.3 x 47 cm.

(665: 1916: Un38aws Small)

OKLAHOMA Nos. 1265-1268

1265. 1837

Plan of Cherokee lands, Ind[ian] Ter[ritory]: surveyed under instructions from Isaac McCoy. N.p.: 1837.

Scale: 1 in. = 20 mi.

Cartographers: J. C. McCoy and Isaac McCoy.

Size: 27.4 x 50.5 cm.

See: Isaac McCoy, *Remarks on the practicability of Indian reform* . . . U.S. 25th Congress, 2nd Session, Doc. no. 120. Second ed. New York: Gray and Bunce: 1829.

(633: 1837: M135pcL Small)

1266. 1902

Indian territory. U.S. Department of the Interior. Geological Survey. Surveyed under the direction of Charles H. Fitch, 1895-1899. Edition of July 1902. [Washington]: 1902.

Scale: 1:500,000.

Cartographer: Charles H. Fitch.

Size: 82.6 x 75.1 cm.

Colored.

(667: 1902: Un38ito Large)

1267. [1907]

State of Oklahoma. U.S. Department of the Interior. Compiled from the official records of the General Land Office and other sources under the direction of I. P. Berthrong. 1907. Washington: A. B. Graham Co.: 1907.

Scale: 1 in. = 12 mi.

Cartographers: I. P. Berthrong, Charles J. Helm, A. F. Dinsmore, and M. Hendges.

Size: 55 x 108.4 cm.

Colored.

Insets: McAlester; Lawton; Muskogee; Guthrie.

(667: [1907]: Un38soo Large)

1268. 1914

Railroad map of Oklahoma. Published by the state. Oklahoma City: Corporation Commission: 1914.

Size: 65.7 x 129.2 cm.

Colored.

(667: 1914: Okl4raa Book map)

OLD NORTHWEST Nos. 1269-1273

1269. 1788

A plan of the several villages in the Illinois country, with part of the River Mississippi &c. London: 1788.

Scale: 2.5 cm. = 10 mi.
Cartographer: Thomas Hutchins.
Engraver: [John Norman?].
Size: 18.3 x 12.7 cm.
Provenance: Presented by John Vaughan, 16 June 1815.
In: Thomas Hutchins, *A topographical description of Virginia,* opposite p. 41.
Wheat: 665

(917.3: H97T)

1270. [1793]

A map of the western part of the territories belonging to the United States of America. Drawn from the best authorities. [London: 1793].
Scale: 2.5 cm. = ca. 110 British statute mi.
Engraver: T. Condor.
Size: 34.9 x 33 cm.
In: Gilbert Imlay, *A topographical description,* frontispiece.

(917.3: Im5)

1271. [1795]

N[orth] W[est] Territory. [1795].
Scale: 2.5 cm. = 140 mi.
Size: 18.3 x 15.4 cm.
Provenance: Presented by the Estate of Richard Gimbel, 1974.
In: Joseph Scott, *The United States gazetteer,* opposite "Territory North-West of the Ohio."
Wheat: 674

(Paine 68: Sco3u)

1272. [1796]

N[orth] W[est] Territory. [1796].
Scale: 2.5 cm. = 130 mi.
Engraver: W. Barker.
Size: 19 x 14.8 cm.
Provenance: Presented by Suzanne Wister Eastwick and Joseph L. Eastwick, July 1975.
In: Mathew Carey, *Carey's American pocket atlas,* following p. 84.
Wheat: 675

(917.3: C25c)

1273. [1803]

Ohio River [by Andrew Ellicott]. Plates A and B. [Philadelphia: 1803].
Cartographer: Andrew Ellicott.
Engraver: A. Lawson.
Sizes of plates: 32.7 x 46.5; 31.2 x 46.4 cm.
See: Andrew Ellicott, *Journal* . . . Philadelphia: Dobson: 1803.

(635: [1803]: EL52ohr Small)

GREAT LAKES AREA Nos. 1274-1301

1274. [ca.1814]

A correct map of the seat of war. Published by John Conrad. Philadelphia: John Conrad; Baltimore: Fielding Lucas, Jr.: [ca. 1814].

Scale: 1 in. = 42 mi.

Cartographer: Samuel Lewis.

Engraver: S. Harrison.

Size: 40.2 x 58 cm.

Colored.

(635: [ca. 1814]: H243cmw Small)

1275. [ca. 1820]

Map exhibiting the country between Lakes Michigan and Erie, and the contested boundary lines. By David H. Burr. New York: P. A. Mesier: [ca. 1820].

Cartographer: David H. Burr.

Lithographer: P. A. Mesier.

Size: 44.6 x 162.1 cm.

Colored.

Remarks on map refer to map by Thomas Hutchins, 1778.

(670: [ca. 1820]: B941Lme Large)

1276. [1836]

Survey for a ship canal to connect the lakes Erie & Ontario. Lockport route. Surveyed under the direction of Capt. W. G. Williams, U.S. Top. Engrs. by Lieuts. T. F. Drayton & J. G. Reed. Map. no. 3. N.p.: [1836].

Scale: Horizontal scale — 4 in. = 5,280 ft.

Cartographers: W. G. Williams, T. F. Drayton, and J. G. Reed.

Size: 129.5 x 69.4 cm.

Inset: Creek flowing into Niagara River.

Provenance: Presented by J. R. Ingersoll, 15 July 1836.

With this is a profile of the Lockport route.

(670: [1836]: W678eon Large)

1277. [1847]

Map of that part of the mineral lands adjacent to Lake Superior, ceded to the United States by the treaty of 1842 with the Chippewas. Comprising that district lying between Chocolate River and Fond du Lac, under the superintendency of Gen. John Stockton . . . Projected and drawn under the direction of Lieut.-Col. George Talcott . . . by A. B. Gray, assisted by John Seib. Compiled from surveys and explorations made by direction of the U.S. War Department . . . ; from U.S. surveys made by Dr. Houghton and Wm. A. Burt Esq.; from Nicollet's map of the upper Mississippi River; and Bayfield's charts of Lake Superior. Washington: C. B. Graham Lith. Co.: [1847].

Scale: 8.5 cm. = 20 mi.

Cartographers: Nicollet, Bayfield, George Talcott, A. B. Gray, John Sieb, Dr. Houghton, and William A. Burt.

Lithographer: C. B. Graham Lith. Co.

Size: 90.1 x 115.5 cm.

Colored.

Insets: Lake Superior; Vignette of Camp Gray, October 1st, 1844, near Talcott Harbor.

See accompanying report of 1845.

(626.2: [1845]: Un38mLs Large)

1278. 1852

Kelley's and Bass Islands showing the harbors of refuge in their vicinity being an extract from the survey of the west end of Lake Erie. U.S. War Department. Survey of the north and north western lakes. Under the orders of Lieut.-Col. Kearney, 1849. Washington: 1852.

Scale: 1:50,000.

Cartographers: John Lambert, J. N. Macomb, Woodruff, Gunnison, Scammon, W. F. Smith, Burgess, Peter, Forster, and Kearney.

Engraver: W. H. Dougal.

Size: 50.8 x 62.9 cm.

Provenance: Presented by G. G. Meade, 19 July 1861.

(626.5: 1852: Un398kbi Large)

1279. [1852]

West end of Lake Erie and Detroit River from surveys made under the direction of . . . Lieut. J. N. Macomb, I. C. Woodruff, J. W. Gunnison, E. P. Scammon, R. W. Burgess, J. F. Peter, W. H. Warner, and J. H. Forster under Lt.-Col. James Kearney, 1849. U.S. War Department. Survey of the northern and north western lakes. [Washington: 1852].

Scale: 1:120,000.

Cartographers: John Lambert, J. N. Macomb, I. C. Woodruff, J. W. Gunnison, E. P. Scammon, R. W. Burgess, J. F. Peter, W. H. Warner, J. H. Forster, and James Kearney.

Engraver: W. Smith.

Size: 93.2 x 71.5 cm.

Provenance: Presented by Hartman Bache, 1 April 1853.

(626.5: [1852]: Un38edr Large)

1280. 1853

Lake Erie compiled from surveys made under the direction of the U.S. War Department. Survey of the northern and north western lakes. Bureau of Topographical Engineers. 1849. Reduced for engraving by John Lambert . . . under the direction of Capt. T. J. Lee. Washington: 1853.

Scale: 1:400,000.

Cartographers: John Lambert and T. J. Lee.

Size: 61.3 x 99.5 cm.

Provenance: Presented by G. G. Meade, 19 July 1861.

(626.5: 1853: Un38Les Large)

1281. [1854]

Head of Green Bay and entrance to Fox River, Wisconsin . . . surveyed in 1845 by Capt. W. G. Williams and Lieut. J. W. Gunnison with corrections to date by Lieut. W. F. Raynolds, under the orders of Capt. J. N. Macomb. U.S. War Department. Survey of the northern and north western lakes. 1853. [Washington]: 1854.

Scale: 1:30,000.

Cartographers: John Lambert, F. Herbst, W. G. Williams, J. W. Gunnison, W. F. Raynolds, and J. N. Macomb.

Engraver: W. H. Dougal.

Size: 49.5 x 67.5 cm.

Provenance: Presented by G. G. Meade, 19 July 1861.

(626.3: [1854]: Un38hgb Large)

1282. [1855]

Preliminary chart of Eagle Harbor Lake Superior . . . Surveyed and drawn under the direction of Capt. J. N. Macomb by Lieut. W. F. Raynolds assisted by J. U. Mueller, D. F. Henry and J. Wallace. U.S. War Department. Survey of the northern and north western lakes. 1855. [Washington: 1855].

Scale: 1:5,000.

Cartographers: O. M. Poe, H. Gillman, J. N. Macomb, W. F. Raynolds, J. U. Mueller, D. F. Henry, and J. Wallace.

Engraver: H. Dougal.

Size: 51.7 x 73.6 cm.

Inset: Eagle Harbor.

Provenance: Presented by G. G. Meade, 19 July 1861.

(626.2: [1855]: Un38ehs Large)

1283. [1856]

Straits of Mackinac with the approaches thereto from Lakes Huron and Michigan and the entrance by the detour passage to the St. Mary's River from trigonometrical surveys under . . . Lieut.-Col. James Kearney in 1849, and of Capt. J. N. Macomb in 1851, 1852, 1853 & 1854. U. S. War Department. Survey of the northern and north western lakes. [Washington: 1856].

Scale: 1:120,000.

Cartographers: A. Boschke, James Kearney, and J. N. Macomb.

Engraver: Selmar Siebert.

Size: 55 x 103 cm.

Provenance: Presented by G. G. Meade, 19 July 1861.

(626.3: [1856]: Un38smm Large)

1284. [1857]

Chart of St. Clair flats reduced from the original surveys of Capt. Geo[rge] G. Meade under the direction of Lieut.-Col. J. Kearney . . . . Showing also the improvement at the mouth of the South Pass now being effected by the United States. U.S. War Department. Bureau of Topographical Engineers. 1857. [Washington: 1857].

Scale: 1:32,000.

Cartographers: George G. Meade and J. Kearney.

Engraver: W. H. Dougal.

Size: 67.6 x 64.2 cm.

Provenance: Presented by G. G. Meade, 19 July 1861.

(626.4: [1857]: Un38scf Large)

1285. [1857]

North end of Lake Michigan including the Beaver Island group from trigonometrical surveys under . . . Captain J. N. Macomb in 1854 & 1855. U.S. War Department. Survey of the northern and north western lakes. [Washington: 1857].

Scale: 1:120,000.

Cartographers: J. U. Müller and J. N. Macomb.

Engraver: W. H. Dougal.

Size: 60 x 90.5 cm.

Provenance: Presented by G. G. Meade, 19 July 1861.

(626.3: [1857]: Un38mbg Large)

1286. [1857]

Preliminary chart of Tawas Harbor [Harbor of Refuge, Lake Huron]. U. S. War Department. Survey of the northern and north western lakes. Surveyed under the direction of Capt. J. N. Macomb, G. W. Lamson, H. Gillman and A. G. Lamson. 1856. [Washington: 1857].

Scale: 1:16,000.

Cartographers: J. N. Macomb, G. W. Lamson, H. Gillman, and A. G. Lamson.

Engraver: W. H. Dougal.

Size: 62.2 x 73 cm.

(626.4: [1857]: Un38thh Large)

1287. [1857]

Maumee Bay surveyed and drawn under the direction of Capt. Geo. G. Meade in 1857. U.S. War Department. Survey of the northern and north western lakes. [Washington: 1858].

Scale: 1:30,000.

Cartographers: G. G. Meade, J. H. Foster, W. H. Hearding, and P. C. Rabout.

Engraver: W. H. Dougal.

Size: 58.2 x 68.1 cm.

Provenance: Presented by G. G. Meade, 19 July 1861.

(626.5: [1858]: Un38mbe Large)

1288. [1858]

Preliminary chart of Agate Harbor Lake Superior surveyed and drawn under the direction of Capt. J. N. Macomb . . . by Lieut. W. F. Raynolds assisted by J. Mueller, D. F. Henry and J. Wallace in 1855. U.S. War Department. Survey of the lakes. Published under the supervision of Capt. G. G. Meade 1858. [Washington: 1858].

Scale: 1:10,000.

Cartographers: C. P. Rabout, J. N. Macomb, W. F. Raynolds, J. Mueller, D. F. Henry, J. Wallace, and G. G. Meade.

Engraver: W. H. Dougal.

Size: 37.5 x 49.9 cm.

Provenance: Presented by G. G. Meade, 19 July 1861.

(626.2: [1858]: Un38cah Small)

1289. [1858]

Preliminary chart of Eagle River Lake Superior surveyed and drawn under the direction of Capt. J. N. Macomb, by Lieut. W. F. Raynolds assisted by J. Mueller, D. F. Henry and J. Wallace in 1855. U.S. War Department. Survey of the lakes. Published under the supervision of Capt. G. G. Meade 1859. [Washington: 1859].

Scale: 1:10,000.

Cartographers: C. P. Rabaut, J. N. Macomb, W. F. Raynolds, J. Mueller, D. F. Henry, J. Wallace, and G. G. Meade.

Engraver: W. H. Dougal.

Size: 47.5 x 34.9 cm.

Provenance: Presented by G. G. Meade, 19 July 1861.

(626.2: [1859]: Un38eLs Small)

1290. [1859]

Preliminary chart of Ontanogon Harbor Lake Superior . . . U.S. War Department. Survey of the northern and north western lakes. Surveyed and drawn under the direction of Capt. J. N. Macomb by Lieut. W. F. Raynolds assisted by J. U. Mueller, D. F. Henry and J. Wallace. Reduced for publication, under the direction of Capt. G. G. Meade, 1859. [Washington: 1859].

Scale: 1:16,000.

Cartographers: Joshua Barney, J. N. Macomb, W. F. Raynolds, J.U. Mueller, D. F. Henry, J. Wallace, and G. G. Meade.

Engraver: W. H. Dougal.

Size: 36.7 x 53.1 cm.

Provenance: Presented by G. G. Meade, 19 July 1861.

(626.2: [1859]: Un38soh Small)

1291. [1860]

General chart of Lake Huron projected from trigonometrical surveys under the orders of Capt. G. G. Meade and from other reliable information, 1860. U.S. War Department. Surveys of the northern and north western lakes. [Washington: 1860].

Scale: 1:400,000.

Cartographers: G. G. Meade and J. U. Mueller.

Engraver: W. H. Dougal.

Size: 90.3 x 67.3 cm.

(626.4: [1860]: Un38gch Large)

1292. [1860]

Harbors of refuge Presqu'ile, False Presqu'ile, and Middle Island, Lake Huron. U.S. War Department. Survey of the northern and north western lakes. Projected from a trigonometrical survey executed under the orders of Capt. G. G. Meade, 1858. [Washington: 1860].

Scale: 1:40,000.

Cartographer: G. G. Meade.

Engraver: W. H. Dougal.

Size: 69.4 x 99.8 cm.

(626.4: [1860]: Un38ppm Large)

1293. 1860

Saginaw Bay and part of Lake Huron projected from a trigonometrical survey under the orders of Capt. J. N. Macomb in 1856, and of Capt. G. G. Meade in 1857 & 1858. U. S. War Department. Survey of the northern and north western lakes. [Washington: 1860].

Scale: 1:120,000.

Cartographers: J. U. Mueller, J. N. Macomb, and G. G. Meade.

Engraver: W. H. Dougal.

Size: 74 x 110 cm.

(626.4: 1860: Un38sbh Large)

1294. [1860]

Thunder Bay, Lake Huron projected from a trigonometrical survey executed under the orders of Capt. Geo. G. Meade in 1858. U. S. War Department. Survey of the northern and north western lakes. [Washington: 1860].

Scale: 1:40,000.

Cartographer: Joshua Barney and George G. Meade.

Engraver: W. H. Dougal.

Size: 74.1 x 88.4 cm.

(626.4: [1860]: Un38tbh Large)

1295. 1861

Chart of the lights on the lake coast of the United States. Published by the order and under the direction of the U.S. Light House Board. Prepared at the office of the survey of the northern and north western lakes under the supervision of Capt. Geo. G. Meade. N.p.: 1861. 5 pieces.

Scale: 1:600,000.

Cartographer: Geo. G. Meade.

Sizes: Various.

Colored.

(670: 1861: Un48Lhb Large)

1296. [1861]

South end of Lake Huron and head of the St. Clair River projected from a trigonometrical survey, executed under the orders of Capt. Geo. G. Meade in 1859. U.S. War Department. Survey of the northern and north western lakes. [Washington: 1861].

Scale: 1:120,000.

Cartographers: Joshua Barney and G. G. Meade.

Engraver: William H. Dougal.

Size: 69 x 68.2 cm.

Inset: Head of St. Clair River showing the anchorage ground.

(626.4: [1861]: Un38sLh Small)

1297. 1862

Preliminary chart of Grand Island and its approaches, Lake Superior, projected from a trigonometrical survey, executed under the orders of Capt. Geo[rge] G. Meade, 1859. U.S. War Department. Survey of the northern and north western lakes. [Washington: 1862].

Scale: 1:25,000.

Cartographers: George G. Meade and Joshua Barney.

Engraver: W. H. Dougal.

Size: 92.9 x 72.3 cm.

(626.2: 1859: Un38gis Large)

1298. [ca.1869]

Section of the lake, river and canal navigation for Lake Superior to the Gulf of St. Lawrence. [ca. 1869].

Scales: Horizontal, 1 in. = 60 mi. Vertical, 1 in. = 400 ft.

Cartographer: E. Blackwell.

Size: 23 x 85.4 cm.

Colored.

For text, see: APS, *Transactions.* Ser. 2, vol. 13, 1869.

With this is a water color painting of "Sketch from the Mountain of Montreal showing part of the plain extending south east & southwards to the Green Mountain Range in Eastern townships of Canada and Vermont & the Adirondack Mountain regions of New York. To illustrate Mr. Blackwell's paper on the basin of the St. Lawrence." *Size of paper:* 36.3 x 93.1 cm.

(670: [ca. 1869]: B561Lsg Large)

1299. 1916

Coast Chart No. 5. Lake Superior from Ontonagon to Oronto Bay and Outer Island prepared under the direction of Colonel Mason M. Patrick . . . 1915-1916 from U.S. Lake Surveys made between the years 1868 and 1905. U.S. War Department. Army. Survey of the northern and northwestern lakes. Issued August 9, 1916. [Washington]: 1916.

Scale: 1:120,000.

Cartographer: Mason M. Patrick.

Engraver: G. F. Penner.

Size: 71.1 x 80.9.

Colored.

Inset: Ontonagon harbor, Michigan.

(626.2: 1916: Un38soo Large)

1300. 1916

General chart of the northern and northwestern lakes including New York state canals and Lake Champlain prepared from the latest U.S. lake survey . . . and first issued under the direction of Col. G. J. Lydecker, and Major Charles Keller . . . in 1906-1907. U.S. War Department. Army Corps of Engineers. Survey of the northern and north western lakes. Issued June 26, 1916. Catalogue No. O. [Washington]: 1916. Fifth edition.

Scale: 1:1,200,000.

Cartographers: G. J. Lydecker and Charles Keller.

Size: 78.1 x 125.7 cm.

Colored.

(620: 1916: Un38Lny Large)

1301. 1916

Lake Ontario prepared from a trigonometrical survey in 1873-1875 and first issued under the direction of Major C. B. Comstock, U.S. War Department. Survey of the northern and north western lakes. Revised from results of surveys by the U.S. Lake Survey and the Canadian Government . . . Issued 7 October 1916. Catalogue No. 2. [Washington]: 1916. Sixth edition (in colors).

Scale: 1:400,000.

Cartographers: Edward Molitor and C. B. Comstock.

Size: 51.9 x 87.5 cm.

Colored.

(626.6: 1916: Un38wLo Large)

OHIO Nos. 1302-1317

1302. [1765]

A topographical plan of that part of the Indian-country through which the army under the command of Colonel Bouquet marched in the year 1764. [1765].

Scale: 2.5 cm. = 6 mi.

Cartographers: Thomas Hutchins, Guy Johnson, and Bernard Ratzer.

Engraver: Henry Dawkins (?).

Size: 36.2 x 49.3 cm.

Inset: A general map of the country on the Ohio and Muskingham shewing the situation of the Indian-towns with respect to the army under the command of Colonel Bouquet.

In: William Smith, *An historical account of the expedition,* frontispiece.

Wheat: 658

(973.2: Sm6)

1303. 1785

A map of the federal territory [Ohio] from the western boundary of Pennsylvania to the Scioto River; laid down from the latest informations and divided into townships and fractional parts of townships agreeably to the Ordinance of the Honle. Congress passed in May 1785. N.p.: 1785.

Scale: 2.5 cm. = 10 mi.

Size: 65.9 x 48.9 cm.

Colored.

To accompany Manasseh Cutler, *Explanation of the map of the federal lands confirmed by the treaties of 1784.* Salem: 1787.

Wheat: 662

(672: 1785: Un38fed Large)

1304. [1787]

A plan of the rapids in the River Ohio. [1787].

Scale: 2.5 cm. = 80 yd.

Cartographer: Thomas Hutchins.

Engraver: John Cheevers.

Size: 14.7 x 18.5 cm.

Provenance: Presented by John Vaughan, 16 June 1815.

In: Thomas Hutchins, *A topographical description of Virginia,* opposite p. 8.

Wheat: 664

(917.3:H97t [London edition])

1305. [1787]

Plan of the remains of some ancient works on the Muskingum. [1787].

Scale: 2.5 cm. = 10 chains.

Cartographer: Jona. Heart.

Size: 17.2 x 21.6 cm.

In: *Columbian Magazine,* vol. 1, following p. 424.

Wheat: 663

(050: C72)

1306. [1788]

A plan of Campus Martius, at the city of Marietta, territory of the United States, N.W. of the River Ohio. [1788].

Scale: 2.5 cm. = 40 ft.

Size of plate: 17.5 x 15.2 cm.

In: *Columbian Magazine,* vol. 2, opposite p. 646.

Wheat: 668

(050: C72)

1307. [1789]

Indian works, on Huron River or Bald Eagle Creek. [Ohio, 1789].

Size of page: 10.6 x 19.5 cm.

In: *Columbian Magazine,* vol. 3, opposite p. 543.

Wheat: 669

(050: C72)

1308. [1793]

A plan of the rapids of the Ohio. [1793].

Scale: 2.5 cm. = 800 yd.

Engraver: T. Condor.

Size: 15 x 18.4 cm.

In: Gilbert Imlay, *A topographical description,* opposite p. 51.

Wheat: 672

(917.3: Im5 [London edition])

1309. 1815

A map of the state of Ohio, from actual survey. By B. Hough & A. Bourne. Chillicothe: B. Hough & A. Bourne. Philadelphia: J. Melish: 1815 [1 May 1815].

Scale: 1 degree = 69½ mi.

Cartographers: B. Hough and A. Bourne.

Engraver: H. S. Tanner.

Size: 125.7 x 114.7 cm.

Colored.

Provenance: Presented by John Melish, 6 October 1815.

(672: 1815: H813mso Large)

1310. 1826

State of Ohio. Chart of the entrance of Sandusky-Bay. U.S. Topographical Engineers. Reduced from the original survey of Lieut. C[ampbell] Graham of 1826, under the direction of Col. Jno. J. Abert . . . by Wash: Hood. [Washington]: 1838.

Scale: 4 in. = 1 mi.

Cartographers: John J. Abert, Campbell Graham, and Wash. Hood.

Engraver: W. J. Stone.

Size: 54 x 41.7 cm.

Inset: Map of Sandusky Bay.

Provenance: Presented and endorsed by J. D. Graham, 18 June 1841; and Hartman Bache, 17 August 1838.

See: U.S. 25th Congress. 2nd Session. Document no. 399.

(672: 1826: Un38csb Small)

1311. 1833

Ohio. By David H. Burr. New York: J. H. Colton & Co.: 1833.

Scale: 1 in. = 15 mi.

Cartographer: David H. Burr.

Engraver: S. Stiles & Co.

Size: 47.5 x 57.3 cm.

Colored.

Inset: A plan of Cincinnati.

(672: 1833: B941moc Large)

1312. 1872

Climatological map of the state of Ohio. By Lorin Blodget. N.p.: 1872.

Cartographer: Loren Blodget.

Size: 37 x 27.7 cm.

Colored.

See: H. F. Walling and O. W. Gray, *New topographical atlas of the state of Ohio.* Cincinnati: Stedman, Brown & Lyon: 1872.

(672: 1872: B621cmo Small)

1313. [1878]

State of Ohio. U.S. Department of the Interior. Compiled from the official records of the General Land Office and other sources by C. Roeser. New York: Julius Bien: [1878].

Scale: 1 in. = 10 mi.

Cartographer: C. Roeser.

Lithographer: Julius Bien.

Size: 63.5 x 72.6 cm.

Colored.

(672: [1878]: Un38gLo Large)

1314. 1883

A geological and topographical sketch of the Hinckley coal tracts, in Brookfield Township, Trumbull County, Ohio. By Benj. Smith Lyman. New York: J. Bien: 1883.

Scale: 1:18,000.

Cartographer: Benjamin Smith Lyman.

Lithographer: Julius Bien.

Size: 44.8 x 33.2 cm.

(672: 1883: L995hct Small)

1315. 1900

Topographic sheet of Ohio-Kentucky. Cincinnati Quadrangle. U.S. Geological Survey. Charles D. Walcott, Director. [Washington]: 1900.

Scale: 1:62,500.

Cartographers: Jno. H. Renshawe, Geo. T. Hawkins, W. J. Peters, Chas. E. Cooke, and Charles D. Walcott.

Engraver: U.S. Geological Survey.

Size: 44.4 x 69.5 cm.

Colored.

(672: 1900: Un38cqo Large)

1316. 1910

State of Ohio. U.S. Department of the Interior. Compiled from the official records of the General Land Office and other sources under the direction of I. P. Berthrong. Washington: Eckert: 1910.

Scale: 1 in. = 12 mi.

Cartographers: I. P. Berthrong and A. F. Dinsmore.

Lithographer: Eckert Lithograph. Co.

Size: 54.4 x 60.5 cm.

Colored.

(672: 1910: Un38soo Large)

1317. 1914

Railroad map of Ohio. Published by the state. Columbus, Ohio: Columbus Lithog. Co.: 1914.

Scale: 1¼ in. = 10 mi.

Lithographer: Columbus Lithograph Company.

Size of paper: 82.3 x 72.3 cm.

Colored.

(672: 1914: Oh36rrc Book map)

INDIANA Nos. 1318-1320

1318. 1886

State of Indiana. U.S. Department of the Interior. Compiled from the official records of the General Land Office and other sources by Geo. U. Mayo. Washington: Graham: 1886.

Scale: 1 in. = 12 mi.

Cartographers: Geo. U. Mayo and M. Hendges.

Lithographer: A. B. Graham Co.

Size: 63.3 x 44.5 cm.

Colored.

(673: 1886: Un38soi Large)

1319. 1909-1910

State of Indiana. U.S. Geological Survey. George Otis Smith, Director. [Washington]: U.S. Geological Survey: 1909-1910.

Scale: 1:500,000.

Cartographers: George Otis Smith, R. B. Marshall, and A. F. Hassan.

Engraver: U. S. Geological Survey.

Size: 91.8 x 59.7 cm.

(673: 1909-1910: Un38int Large)

1320. 1916

State of Indiana. U.S. Department of the Interior. Compiled from official records of the General Land Office, U.S. Geological Survey and other sources under the direction of I. P. Berthrong. Washington: 1916.

Scale: 1:760,320.

Cartographers: I. P. Berthrong, A. F. Dinsmore, and Thomas O. Wansleben.

Lithographer: Eckert Lithographic Company.

Size: 62.9 x 43.9 cm.

Colored.

(673: 1916: Un38inb Small)

ILLINOIS Nos. 1321-1328

1321. [1778]

A plan of the several villages in the Illinois country, with part of the River Mississippi &c. By Thos. Hutchins. [London: J. Almon: 1778].

Scale: 1 in. = 10 mi.

Cartographer: Thomas Hutchins.

Size: 18.1 x 12.9 cm.

See: *A topographical description of Virginia, Pennsylvania, Maryland and North Carolina,* by Thos. Hutchins. London: J. Almon: 1778.

(674: [1778]: H973vmr Small)

1322. 1836

The tourist's pocket map of the state of Illinois, exhibiting its internal improvements; roads, distances, etc. By J. H. Young. Philadelphia: S. Augustus Mitchell: 1836.

Scale: 1 in. = 30 mi.

Cartographer: J. H. Young.

Engraver: E. F. Woodward.

Size: 39.2 x 32.5 cm.

Colored.

Inset: Map of the lead mine region east of the Mississippi River.

(674: 1836: Y79stLL Book map)

1323. [1838]

Diagram of the state of Illinois. No. 2. (A) [Washington: 1838].

Scale: 1 in. = 18 mi.

Engraver: W. J. Stone.

Size of paper: 66.1 x 39.6 cm.

Colored.

See: U.S. 25th Congress. 3rd Session. Senate document 17, no. 10.

(674: [1838]: Un38dsi Large)

1324. 1853

New sectional [pocket] map of the state of Illinois. Compiled from the United States surveys. Also exhibiting the internal improvements, distances between towns, villages & post offices, the outlines of prairies, woodlands, marshes, & the lands donated to the state by the Gen[era]l Gov[ernmen]t for the purpose of internal improvements. By J. M. Peck, John Messenger and A. J. Mathewson. New York: J. H. Colton & Co.: 1853.

Scale: 1 in. = 10 mi.

Cartographers: J. M. Peck, John Messenger, and A. J. Mathewson.

Engraver: S. Stiles & Co.

Size: 103.6 x 69.4 cm.

Colored.

Insets: Vicinity of Alton and St. Louis; vicinity of Galena, the lead region.

(674: 1853: C672sec Book map)

1325. [1854]

Chart of the harbor & river of Chicago, from surveys made in Aug[us]t 1854 under the superintendence of Brevet Lt. Col. Graham. Sheet G. No. 12. Chicago: H. Acheson: [1854].

Scale: 1 in. = 200 ft.

Cartographer: J. D. Graham.

Lithographer: H. Acheson.

Size: 55.7 x 72.1 cm.

Provenance: Presented by J. D. Graham, 10 March 1858.

Accompanies J.D. Graham's report (no. 32) to the Chief Topographical Engineer dated 19 August 1854. Published for the Board of Trade of Chicago.

Graham wrote that he was presenting several charts:

1. A chart, (marked G. No. 12) of Chicago harbor, shewing its condition and the facilities of entrance produced by dredging on the bar, under my direction in the summer of 1854. Before the dredging, the bar was very much in the same condition as it is shewn in the next succeeding map marked G No. 38 [map no. 1325].
2. A map marked G No. 38 of Chicago harbor and bar, shewing their condition on the 9th of April 1856. The re-formation at the bar, at the harbor entrance, between August 1854 and April 1856, is here clearly shewn. The positions of the block-house and other buildings of Old Fort Dearborn in 1856, now no longer existing, are shewn on this map. They have all yielded to the progress of improvement in the rapidly increasing City of Chicago[map no. 1326].
3. Map G. No. 43 Chicago harbor, shewing its condition and the remnant of the bar on the 15th of September 1856 after the dredging (done for the second time under my direction) between April 22d and September 15th 1856.
4. Map G No. 44. Chicago harbor & Bar, shewing the effect of the violent gales of wind in October 1856, in diminishing the depth of water on the bar. This map shews the actual condition of the harbor & bar on the 14th of November 1856, from careful survey ending on that day.

All of the above mentioned maps based on careful triangulations made under my direction, and the harbors mentioned are all on Lake Michigan.

I will add that these maps are all referred to in my reports on lake harbor improvements which I have sent to the Society's library . . .

P.S.: Maps 43 and 44 shew the position to be occupied by the new light house now being constructed for Chicago harbor. It is immediately to the north of the extremity of the north harbor-pier [Archives. J. D. Graham to C. B. Trego, 10 March 1858. Minutes. *Proceedings*, vol. 6, p. 309. Donation book for 16 April 1858].

(674: [1854]: Un38chr Large)

1326. [1856]

Chicago Harbor and Bar, Illinois. U.S. Topographical Engineers. From survey made between March 29th & April 9th, 1856. Under the direction of Brevet Lt. Col. J. D. Graham. Map G no. 38. Chicago: Ed. Mendel: [1856].

Scale: 1 in. = 200 ft.

Cartographer: J. D. Graham.

Lithographer: Ed. Mendel.

Size: 58.6 x 77 cm.

Provenance: Presented by J. D. Graham, 10 March 1858.

See: Map no. 1325 for Graham's comments on this map.

(674: [1856]: Un38chb Large)

1327. 1911

State of Illinois, U.S. Department of the Interior. Compiled from the official records of the General Land Office and other sources under the direction of I. P. Berthrong. Washington: Eckert: 1911.

Scale: 1 in. = 12 mi.

Cartographers: Charles J. Helm and I. P. Berthrong.

Lithographer: Eckert Litho. Co.

Size: 83.6 x 49 cm.

Colored.

(674: 1911: Un38gLo Large)

1328. 1915

State of Illinois. U.S. Department of the Interior. Geological Survey. George Otis Smith, Director. [Washington]: U.S. Geological Survey: 1915.

Scale: 1:1,000,000.

Cartographer: George Otis Smith.

Engraver: U.S. Geological Survey.

Size: 63 x 36.6 cm.

(674: 1915: Un38ssi Small)

MISSISSIPPI RIVER Nos. 1329-1343

1329. [1803]

[Section of the Mississippi River. By Andrew Ellicott]. No. 2. [Philadelphia: 1803].

Cartographer: Andrew Ellicott.

Engraver: Alexander Lawson.

Size: 21.1 x 42.3 cm.

Original in: Andrew Ellicott, *Journal* . . . Philadelphia: Dobson: 1803.

(630: [1803]: EL52mr Small)

1330. [1807]

Map of the Mississippi River from its source to the mouth of the Missouri: laid down from the notes of Lieut. Z. M. Pike, by Anthony Nav. Reduced, and corrected by the astronomical observations of Mr. Thompson at its source: and of Capt. M. Lewis, where it receives the waters of the Missouri. [1807].

Scale: 1 in. = 25 mi.

Cartographers: Anthony Nav, Z. M. Pike, Mr. Thompson, and M. Lewis.

Engraver: Francis Shallus.

Size: 23.5 x 77.5 cm.

Provenance: Presented by Thomas Jefferson, 1807.

See: Zebulon M. Pike's *Account of a voyage up the Mississippi River.* Washington: 1807, and *Account of expeditions to the source of the Mississippi.* Philadelphia: 1810.

(633: [1807]: K584mrm Large)

1331. [1837]

Map of the Des Moines Rapids of the Mississippi River. No. 1. U.S. Corps of Engineers. Drawn by Lt. M. C. Meigs and Henry Kayser. Surveyed by Lt. R. E. Lee, M. C. Meigs, Henry Kayser and J. S. Morehead. September 1837. [Washington: 1837].

Scale: 5 in. = 1 mi.

Cartographers: R. E. Lee, J. S. Morehead, M. C. Meigs, and Henry Kayser.

Engraver: W. J. Stone.

Size of paper: 59.4 x 141.2 cm.

In: *A collection of maps, etc., published by order of Congress.* Washington: 1843. Senate. Documents, no. 139. 25th Congress. 2nd Session.

(678: [1837]: Un38dmr Large)

1332. [1837]

Map of the Rock Island Rapids of the Mississippi River. U.S. Corps of Engineers. No. 2. Surveyed by Lt. R. E. Lee, etc. in Sept[embe]r and Oct[obe]r 1837. [Washington: 1837].

Scale: 1 in. = 16 chains.

Cartographers: R. E. Lee, M. C. Meigs, J. S. Morehead, and Henry Kayser.

Engraver: W. J. Stone.

Size of paper: 53.3 x 170.6 cm.

Provenance: Presented by J. T. Abert.

In: U.S. 25th Congress. 2nd Session. Senate. Documents, no. 139.

(674: [1837]: Un38rir Large)

1333. 1878

Mississippi River, Louisiana. From New Orleans to Soniat Plantation, including Carrollton, Jefferson and Kennerville. U.S. Coast Survey Sheet no. 8. [Washington]: 1878.

Scale: 1:20,000.

Cartographers: C. P. Patterson, C. H. Boyd, and A. McCrackin.

Size: 58 x 83.7 cm.

Inset: Diagram of sheets.

Provenance: Presented by U.S. Coast and Geodetic Survey through Spencer C. McCorkle, 14 December 1885.

(666: 1878: Un38mrs Large)

1334. [1893]

Atlas illustrating report of March 20, 1909, of Board on examination and survey of Mississippi River created by Act of Congress, approved March 2, 1907. [1893]. 52 maps.

Size: 63.7 x 101.5 cm.

Colored.

Accompanies U.S. 61st Congress. 1st Session. House of Representatives. Documents, no. 50.

(630: [1893]: P18lmsra Large)

1335. 1893-1907

Survey of the Mississippi River. Published by the Mississippi River Commission. 1893-1907. Nos. 101-144.

Scale: 1 in. = 1 mi.

Cartographer: C. W. Clark.

Sizes: 31.5 x 57 and 32 x 57.5 cm.

Traces the course of the river from Wisconsin to its junction with the Ohio River.

(630: 1893-1907: P181msr Large)

1336. [1893]

Survey of the Mississippi River [New Orleans sections] made under the direction of the Mississippi River Commission. Chart no. 76. Projected from a trigonometrical survey made by the U.S. Coast Survey in 1874. N.p.: [1893]. 4 sheets.

Scale: 1:10,000.

Cartographers: Carl F. Palfrey, J. G. Warren, Geo. A. Zinn, C. H. Boyd, James A. Paige, E. J. Thomas, A. T. Morrow, W. S. Williams, W. H. Roper, W. G. Comber, G. H. French, E. L. Harman, H. Dunaway, C. L. Ockerson, O. N. Axtell, C. G. Weyl, C. A. Bonfils, T. C. Hockridge, Geo. H. Wolbrecht, K. A. Widen, C. R. Appiano, H. A. H. d'Ailly, and A. W. Swanitz.

Size of each: 58.6 x 93.9 cm.

(630: [1893]: P181msr Extra-Oversize)

1337. 1897

District map of the lower Mississippi River published by the Mississippi River Commission. U.S. Corps of Engineers. H. E. Waterman . . . Secretary. [New York: J. Bien & Co.]: 1897.

Scale: 1 in. = 5 mi.

Cartographer: H. A. H. d'Ailly.

Lithographer: J. Bien.

Size: 92.6 x 76.7 cm.

Colored.

On left side of map is: Table of midstream distances from Cairo to Head of Passes.

(634: 1897: Un38mis Large)

1338. 1898

Survey of the Mississippi River made under the direction of the Mississippi River Commission. 1898. New York: Julius Bien & Co.: 1898. Charts nos. 201-242.

Scale: 1:10,000.

Cartographers: H. E. Waterman, Mason M. Patrick, et al.

Size of each: ca. 58.3 x 94 cm.

(630: 1898: W311msr Extra-Oversize)

1339. 1899

Map of the alluvial valley of the upper Mississippi River from the Falls of St. Anthony to the mouth of the Ohio River. Showing lands subject to overflow, bluff lines defining the limit of the alluvial valley, location of levees and trans-valley profiles. U.S. Corps of Engineers. Published by the Mississippi River Commission. New York: Julius Bien: 1899. 4 sheets.

Scale: 1:316,800.

Cartographers: Mason M. Patrick, J. A. Ockerman, C. W. Clark, H. A. H. d'Ailly, and A. A. Aguirre.

Lithographer: Julius Bien.

Size of each: 70.3 x 125.5 cm.

Colored.

(670: 1899: P278msr Extra-Oversize)

1340. [1900]

Survey of the Mississippi River. Made under the direction of the Mississippi River Commission. U.S. Corps of Engineers. Lake Itasca basin, Minnesota, within Itasca State Park. Projected from a trigonometrical survey made in 1900. New York: Bien: [1900].

Scale: 1:15,000.

Cartographers: C. W. Clark, K. A. Widen, and H. A. H. d'Ailly.

Lithographer: Julius Bien.

Size: 58.6 x 93.7 cm.

Colored.

(677: [1900]: Un38ibm Large)

1341. 1907

Map of the alluvial valley of the Mississippi River from the head of the St. Francis Basin to the Gulf of Mexico, showing lands subject to overflow, location of levees and trans-alluvial profiles. Published by the Mississippi River Commission. 1887. 3rd ed. 1907. N.p.: 1907. 8 sheets.

Scale: 1:316,800.

Cartographers: C. B. Comstock, Henry M. Adams, A. A. Humphreys, H. L. Abbot, Amos Stickney, C. B. Sears, J. G. D. Knight, Smith S. Leach, T. G. Dabney, William Starling, H. B. Richardson, C. W. Babbitt, Thomas Turtle, Edward Molitar, and C. W. Clark.

Size of each sheet: ca. 86.5 x 64.5 cm.

Colored.

(670: 1907: C738msr Extra-Oversize)

1342. 1907

Map of the Saint Francis Basin in two parts. Mississippi River Commission. 1903. New York: J. Bien: 1907.

Scale: ½ in. = 1 mi.

Cartographers: Mason M. Patrick, George P. Howell, and C. W. Clark.

Lithographer: J. Bien.

Size: 129.5 x 69.9 cm.

Colored.

(670: 1907: P278msr Large)

1343. [1909]

Atlas illustrating report of March 20, 1909, of Board on examination and survey of Mississippi River. Created by Act of Congress approved March 2, 1907. N.p.: [1909]. 52 plates.

Size: 63.3 x 101.8 cm.

See: U.S. 61st Congress. House of Representatives. Documents, no. 50. U.S. Engineer Department . . . *Report of a special board of engineers on survey of the Mississippi River.* Washington: 1909.

(630: [1909]: M695msr Large)

MICHIGAN Nos. 1344-1353

1344. [1856]

Black Lake Harbor, Michigan. U.S. Topographical Engineers. From survey made between the 12th. & 29th. September 1856 by Assistant John R. Mayer under the direction of Brevet Lt. Col. J. D. Graham. Map G. No. 48. Chicago: Ed. Mendel: [1856].

Scale: 1 in. = 200 ft., or 1:2400.

Cartographers: John R. Mayer and J. D. Graham.

Lithographer: Ed. Mendel.

Size: 65 x 92.3 cm.

Provenance: Presented by J. D. Graham, 10 March 1858.

This accompanies J. D. Graham's annual report, no. 161, to the Chief Topographical Engineer from Chicago, 15 November 1856.

(675: [1856]: Un38bLm Large)

1345. [1856]

Grand River Harbor, including part of the town of Grand Haven, Michigan. U.S. Topographical Engineers. From survey made in October 1856 by Assistant John R. Mayer under the direction of Brevet Lieut. Col. J. D. Graham. Map G. No. 49. Chicago: Ed. Mendel: [1856].

Scale: 1 in. = 200 ft., or 1:2400.

Cartographers: John R. Mayer and J. D. Graham.

Lithographer: Ed. Mendel.

Size: 64.5 x 99.9 cm.

Provenance: Presented by J. D. Graham, 10 March 1858.

This accompanies J. D. Graham's annual report, no. 161, to the Chief Topographical Engineer, 15 November 1856.

(675: [1856]: Un38grh Large)

1346. [1856]

Mouth of Kalamazoo River, Michigan. U.S. Topographical Engineers. From survey made between the 2d. & 10th. September 1856 by Assistant John R. Mayer under the direction of Brevet Lieut. Col. J. D. Graham. Map G. No. 51. Chicago: Ed. Mendel, lith.: [1856].

Scale: 1 in. = 200 ft., or 1:2400.

Cartographers: J. D. Graham and John R. Mayer.

Lithographer: Ed. Mendel.

Size: 60.8 x 91.9 cm.

Provenance: Presented by J. D. Graham, 10 March 1858.

This accompanies J. D. Graham's annual report, no. 200, to the Chief Topographical Engineer, Chicago, 30 September 1857.

(675: [1856]: Un38mkr Large)

1347. [1856]

St. Joseph Harbor, Michigan. U.S. Topographical Engineers. From survey made between the 14th. & 27th. August 1856 by Assistant John R. Mayer under the direction of Brevet Lieut. Col. J. D. Graham. Map G. No. 47. Chicago: Ed. Mendel: [1856].

Scale: 1 in. = 200 ft., or 1:2,400.

Cartographers: John R. Mayer and J. D. Graham.

Lithographer: Ed. Mendel.

Size: 66.5 x 96.5 cm.

Provenance: Presented by J. D. Graham, 10 March 1858.

Accompanies J. D. Graham's annual report, no. 161, to the Chief Topographical Engineer, 15 November 1856.

(675: [1856]: Un38sjh Large)

1348. 1857

Mouth of South Black River, Michigan. U.S. Topographical Engineers. From survey made in August 1856 by Assistant John R. Mayer under the direction of Brevet Lieut. Col. J. D. Graham. Map G. No. 50. Chicago: Ed. Mendel: [1857].

Scale: 1 in. = 200 ft., or 1:2,400.

Cartographers: John R. Mayer and J. D. Graham.

Lithographer: Ed. Mendel.

Size: 51.7 x 70.7 cm.

Provenance: Presented by J. D. Graham, 10 March 1858.

Accompanies J. D. Graham's annual report, no. 200, to the Chief Topographical Engineer, Chicago, 30 September 1857.

(675: 1857: Un38msb Large)

1349. [1857]

New Buffalo, and the marshes of Lake Pottowottomee at the mouth of the Galien River, Michigan. U.S. Topographical Engineers. Shewing the facilities of constructing a harbor of refuge. From survey made in September 1857, by Assistant John R. Mayer, under the direction of Brevet Lieut. Col. J. D. Graham. Map G. No. 54. Chicago: Ed. Mendel: [1857].

Scale: 1 in. = 200 ft., or 1:2,400.

Cartographers: John R. Mayer and J. D. Graham.

Lithographer: Ed. Mendel.

Size: 65.8 x 101.2 cm.

Provenance: Presented by J. D. Graham, 10 March 1858.

Accompanies J. D. Graham's annual report, no. 200, to the Chief Topographical Engineer, Chicago, 30 September 1857.

(675: [1857]: Un38bpg Large)

1350. 1904

State of Michigan. U.S. Department of the Interior. Compiled from the official records of the General Land Office and other sources under the direction of Frank Bond. [Washington: 1904].

Scale: 1 in. = 16 mi.

Cartographers: Frank Bond and Charles J. Helm.

Size: 77 x 65 cm.

Colored.

(675: 1904: Un38mic Large)

1351. 1916

Geological map of Michigan. Geological and biological survey of Michigan. Compiled by R. C. Allen, R. A. Smith, and L. P. Barrett, from surveys by the Michigan Geological Survey and the U.S. Geological Survey. 1916. Publication XXIII. Baltimore: A. Hoen, Lith.: 1916.

Scale: 1:750,000.

Cartographers: R. C. Allen, R. A. Smith, and L. P. Barrett.

Lithographer: A. Hoen, lithographer.

Size: 104 x 86.1 cm.

Colored.

(675: 1916: M585geo Large)

1352. 1936

The centennial geological map of the northern peninsula of Michigan. Michigan Department of Conservation. Geological Survey Division, R. A. Smith, State Geologist. Compiled by Helen M. Martin 1936. N.p.: [1936].

Scale: 1:500,000.

Cartographers: R. A. Smith and Helen M. Martin.

Size: 79 x 108.4 cm. *Size of paper:* 86.6 x 137.4 cm.

Colored.

Along left side is a legend describing the map.

Publication 39, Geological series 33. A part of the annual report for 1936.

(675: 1936: Sm51nop Large)

1353. 1936

The centennial geological map of the southern peninsula of Michigan. Michigan Department of Conservation. Geological Survey Division, R. A. Smith, State Geologist. Compiled by Helen M. Martin 1936. N.p.: [1936].

Scale: 1¼ in. = 10 mi.

Cartographers: R. A. Smith and Helen M. Martin.

Size: 101.8 x 75.2 cm. *Size of paper:* 117 x 104.5 cm.

Colored.

Along the left side is a legend describing the map; at the bottom is a chart of rocks of different periods.

(675: 1936: Sm51sop Small)

MINNESOTA Nos. 1354-1357

1354. [1835]

A reconnaissance of the Minnay Sotor Watapāh; or St. Peter's River to its sources: made in the year 1835. [By G. W. Featherstonhaugh].

Scale: 1 in. = 8 mi.

Cartographer: G. W. Featherstonhaugh.

Size: 59.3 x 107.4 cm.

Accompanies: "Report of a geological reconnaissance made in 1835 from the seat of government, by way of the Green Bay and the Wisconsin territory, to the Coteau de Prairie, an elevated ridge dividing the Missouri from the St. Peter's River," in U.S. 24th Congress. 1st Session. Senate. Documents, no. 333.

(677: [1835]: Un38msw Large)

1355. 1905

State of Minnesota. U.S. Department of the Interior. Compiled from the official records of the General Land Office and other sources under the direction of Frank Bond. 1905. New York: Brett Lithographic Co.: 1905.

Scale: 1 in. = 12 mi.

Cartographers: Frank Bond, A. F. Dinsmore, and Charles J. Helm.

Lithographer: Brett Lithographic Co.

Size: 94.5 x 80.4 cm.

Colored.

Inset: Minneapolis and vicinity and St. Paul and vicinity.

(677: 1905: Un38mgL Large)

1356. 1908

Minnesota. Rand McNally & Co's New Business Atlas Map of Minnesota. 1908.

Scale: 1 in. = 16 mi.

Size: 65.8 x 47.9 cm.

Colored.

Insets: North east corner of Minnesota showing Cook County; Minneapolis, St. Paul and vicinity.

Provenance: Presented by Mrs. A. I. Hallowell, November 1983.

(677: 1908: M666cms Large)

1357. [1910-1911]

State of Minnesota. U.S. Geological Survey. George Otis Smith, Director. Compiled in 1910-1911. [Washington]: 1910-1911.

Scale: 1:500,000.

Cartographers: George Otis Smith, R. B. Marshall, and A. F. Hassan.

Size: 133.8 x 118.5 cm.

Advance sheet subject to correction.

(677: [1910-1911]: Un38int Large)

WISCONSIN Nos. 1358-1373

1358. [ca. 1850]

Madison—The capital of Wisconsin. Map of Madison and the four lake country, Dane County, Wisconsin. N.p.: [ca. 1850].

Size: 41.3 x 57.2 cm. *Size of paper:* 47.1 x 72.5 cm.

The map shows projected railroads to Watertown, Fond du Lac, Portage and Lake Superior and the existing railroads. The four lakes are: Mendota (Great Lake), Memoma (Fairy Lake), Wanbesa (Swan Lake), and Kegonsa (Fish Lake).

On each side of the map are comments about Madison.

(676:951: [ca. 1850]: M265dcw Large)

1359. [1855]

A geological map of Wisconsin. By I. A. Lapham. [New York: J. H. Colton: 1855].

Scale: 1½ in. = 40 mi.

Cartographer: I. A. Lapham.

Size: 35.9 x 28.2 cm. *Size of paper:* 46.6 x 36.8 cm.

Colored.

Provenance: Presented and inscribed by I. A. Lapham, 17 August 1855.

Based on the geographical map of J. H. Colton.

(676: [1855]: L315gmw Small)

1360. [1856]

Manitowoc Harbor, Wisconsin. U.S. Topographical Engineers. From survey made between the 14th. & 19th. of August 1856 by assistant John O'Donoghue, under the direction of Brevet Lieut. Col. J. D. Graham. Map G. No. 46. N.p.: [1856].

Scale: 1 in. = 200 ft.

Cartographers: J. D. Graham and John O'Donoghue.

Size: 50.8 x 61.7 cm.

Provenance: Presented by James D. Graham, 10 March 1858.

Accompanies J. D. Graham's annual report, no. 161, to the Chief Topographical Engineer, Chicago, 15 November 1856.

(676: [1856]: Un38mhw Large)

1361. [1856]

Map of Milwaukee. By I. A. Lapham. N.p.: [1856].

Scale: 1 in. = 660 ft. or 10 chains.

Cartographer: I. A. Lapham.

Engravers: Sherman & Smith.

Size: 78.2 x 62 cm.

(676.953: [1856]: L315mmw Large)

1362. [1856]

Sheboygan Harbor, Wisconsin. U.S. Corps of Topographical Engineers. From survey made between the 19th & 24th of August 1856 by John Donoghue, under the direction of Brevet Lieut. Col. J. D. Graham. Map G. No. 45. Chicago: Ed. Mendel, lith.: [1856].

Scale: 1:2,400.

Cartographers: J. D. Graham and John Donoghue.

Lithographer: Ed. Mendel.

Size: 59.8 x 85.8 cm.

Provenance: Presented by J. D. Graham, 10 March 1858.

Accompanying J. D. Graham's annual report, no. 161, to the Chief Topographical Engineer, Chicago, 15 November 1856.

(676: [1856]: Un38shw Small)

1363. 1857

Colton's township map of the state of Wisconsin, showing the Milwaukee, Watertown & Madison R. R. and its connections. 1857. New York: J. H. Colton & Co.: 1857.

Scale: 9 cm. = 5 mi.

Cartographer: George W. Colton.

Engraver: J. M. Atwood.

Size: 63.2 x 55.3 cm. *Size of paper:* 76.7 x 66 cm.

Colored.

(676: 1857: C67ltmw Large)

1364. [1857]

Kenosha Harbor, Wisconsin. U.S. Topographical Engineers. From survey made in December 1855 by Assistant John R. Mayer under the direction of Brevet Lieut. Col. J. D. Graham . . . and shewing the state of the channel between the U.S. piers and of the bar from survey made, as above, between the 8th & 12th of May, 1857. Map G. No. 53. Chicago: Mendel: [1857].

Scale: 1 in. = 200 ft.

Cartographers: John R. Mayer and J. D. Graham.

Lithographer: Ed. Mendel.

Size: 61.1 x 82.2 cm.

Provenance: Presented by J. D. Graham, 10 March 1858.

(676: [1857]: Un38khw Large)

1365. [ca. 1857]

Madison, the capital of Wisconsin. N.p.: [ca. 1857].

Size of paper: 68.5 x 106.3 cm.

Beloit and Madison Rail Road connecting with Chicago; Milwaukee & Mississippi Rail Road; Madison and Watertown-Fond du Lac and Lake Superior Rail Roads are shown on the map.

(676.951: [ca. 1857]: M265mcw Large)

1366. 1893

Map of Madison, Wisconsin. A.A.A.S. XLII meeting, August 17 to 24, 1893. Madison: Tracy, Gibbs & Company: 1893.

Size of paper: 24.8 x 55.8 cm.

(676.951: 1893: M265pfm Book map)

1367. 1896

Map of the state of Wisconsin. U.S. Department of the Interior. Compiled from the official records of the General Land Office and other sources under the direction of Harry King. Baltimore: Friedenwald, lith.: 1896.

Scale: 1 in. = 12 mi.

Cartographers: Harry King and M. Hendges.

Lithographer: Friedenwald Co.

Size: 65.9 x 73.5 cm.

Colored.

(676: 1896: Un38diw Large)

1368. 1906

Map showing the location of creameries and cheese factories in Wisconsin in 1906. Issued by Wis. Expt. Station and Wis. Dairy & Food Com. N.p.: 1906.

Scale: 3 cm. = 10 mi.

Cartographer: E. Boynton.

Size: 111 x 87.4 cm.

Colored.

Amounts and values of dairy products in 1905 listed on map.

Accompanying *Bulletin,* no. 140.

(676: 1906: B718ccf Large)

1369. 1912

State of Wisconsin. U.S. Department of the Interior. Compiled from the official records of the General Land Office and other sources under the direction of I. P. Berthrong. Washington: Eckert, lith.: 1912.

Scale: 1 in. = 12 mi.

Cartographer: I. P. Berthrong.

Lithographer: Eckert Lithographing Co.

Size: 75 x 66.4 cm.

Colored.

(676: 1912: Un38wdi Large)

1370. 1918

General map of the soils of northern Wisconsin, by A. R. Whitson, assisted by Carl Thompson and T. J. Dunnewald. Geological and Natural History Survey of Wisconsin. 1918. Baltimore: A. Hoen & Company, lith.: 1918.

Scale: 1 in. = 6 mi.

Cartographers: A. R. Whitson, Carl Thompson, and T. J. Dunnewald.

Lithographer: A. Hoen & Company.

Size: 76.2 x 127 cm.

Colored.

(676: 1918: W759soi Small)

1371. 1939

Lac des Flambeau Quadrangle. U.S. Department of the International Survey. 1939. Advance sheet subject to correction. Preliminary edition, 1939.

Scale: 1:48,000.

Cartographer: Glenn S. Smith.

Size of paper: 68.5 x 51.7 cm.

Colored.

Provenance: Presented by Mrs. A. I. Hallowell, November 1983.

(676: 1939: Un42smi Large)

1372. 1945

1945 official highway map of Wisconsin. Prepared by the State Highway Commission of Wisconsin. Madison. Racine, Wisconsin: 1945.

Scale: 1 in. = 13 mi.

Lithographer: Western Printing & Lithographing Company.

Size: 69.2 x 59.5 cm.

Colored.

Insets: Racine; Kenosha; La Crosse; Beliot; Milwaukee; Madison; Eau Claire; Waukesha; Wausau; Sheboygan; Stevens Point; Fond du Lac; Janesville; Manitonoc; Oshkosh; Appleton; Green Bay.

Provenance: Presented by Mrs. A. I. Hallowell, November 1983.

On reverse is a regional map of Wisconsin, photographs, etc.

(676: 1945: W755sch Large)

1373. n.d.

Land of Lakes, northern Wisconsin. Map no. 1. Minocqua, Wisconsin: Land of Lakes maps: n.d.

Scale: 1½ in. = 2 mi.

Cartographer: M. B. Shaeffer.

Lithographer: Standard Map Company.

Size of paper: 103.3 x 98.3 cm.

Colored.

Provenance: Presented by Mrs. A. I. Hallowell, November 1983.

(676: nd.: Sh11LoL Large)

CENTRAL STATES Nos. 1374-1380

1374. [ca. 1720]

Carte de la Louisiane et du cours du Mississippi dressée sur un grand nombre de mémoires, entrau tres [sic] sur ceux de M. de le Maire, par Guillaume de l'Isle de l'Académie R[oya]le des Sciences. [Paris?: ca. 1720].

Cartographer: Guillaume de l'Isle.

Size: 35.5 x 41.2 cm.

Colored.

Provenance: Presented by Peter S. Du Ponceau, 15 January 1841.

(636: [ca. 1720]: D372Lt Small)

1375. [1836]

Map of a portion of the Indian country lying east and west of the Mississippi River to the forty sixth degree of north latitude from personal observation made in the autumn of 1835 and recent authentic documents. Constructed for the U.S. War Department Topographical Bureau, by G. W. F[eatherstonhaugh]. Washington: 1836.

Scale: 1 in. = 16 mi.

Cartographer: G. W. Featherstonhaugh.

Size: 66.7 x 97.8 cm.

Accompanies "Report of a geological reconnaissance made in 1835 . . ." by G. W. Featherstonhaugh. In U.S. Congress. 1st Session. Senate Documents, No. 353.

(634: [1835]: Un38mr Large)

1376. [1836]

Map showing the lands assigned to emigrant Indians west of Arkansas and Missouri. U.S. War Department. Topographical Bureau. [Washington: 1836].

Scale: 1 in. = 40 mi.

Size: 47.6 x 45.4 cm.

Colored.

Provenance: Presented by Joseph R. Ingersoll, 15 July 1836.

In *American state papers.* Washington: Gates and Seaton: 1861. Military affairs, vol. 6, p. 130.

(634: [1836]: Un38iL Small)

1377. 1880

General geologic map of the area explored and mapped by F. V. Hayden and the surveys under his charge 1869 to 1880. U.S. Geological Survey. New York: J. Bien & Co.: 1880.

Scale: 1 in. = 41.03 mi., or 1:2,6000,000.

Cartographer: F. V. Hayden.

Lithographer: J. Bien.

Size: 60.7 x 88.8 cm.

Colored.

(634: 1880: Un38axm Large)

1378. n.d.

[Iowa, Kansas, and Missouri, showing movement of the Sac and Fox Indians. n.d.].

Size: 45.7 x 33.9 cm.

Provenance: Presented by Mrs. A. I. Hallowell, November 1983.

Photostat.

(634.4: n.d.: Io91sfi Small)

1379. n.d.

[Minnesota, Iowa and Missouri, showing Indian claims. n.d.]. 2 sheets.

Size: 44 x 57 cm.; 45 x 56 cm.

Provenance: Presented by Mrs. A. I. Hallowell, November 1983.

Photostat.

(634.4: n.d.: M669inm Small)

1380. n.d.

[Minnesota, Iowa and Missouri showing Indian movements. n.d.].

Size: 59.8 x 46 cm.

Provenance: Presented by Mrs. A. I. Hallowell, November 1983.

Photostat.

(634.4: n.d.: M669inc Small)

IOWA Nos. 1381-1382

1381. 1885

State of Iowa. U.S. Department of the Interior. Compiled from the official records of the General Land Office and other sources under supervision of G. P. Strum. Washington: A. B. Graham, Lithographer: 1885.

Scale: 1 in. = 12 mi.

Cartographer: G. P. Strum.

Lithographer: A. B. Graham.

Size: 50.9 x 83.4 cm.

Colored.

(678: 1885: Un38gLo Large)

1382. 1911-1912

State of Iowa. U.S. Geological Survey. George Otis Smith, Director. [Washington]: U.S. Geological Survey: 1911-1912.

Scale: 1:500,000.

Cartographers: George Otis Smith, R. B. Marshall, and A. F. Hassan.

Engraver: U.S. Geological Survey.

Size: 71.6 x 109.9 cm.

(678: 1911-1912: Un38int Large)

MISSOURI Nos. 1383-1388

1383. [1837]

Map of the harbor of St. Louis, Mississippi River. U.S. Corps of Engineers. October 1837. Surveyed by Lt. R. E. Lee, M. C. Meigs, J. S. Morehead, H. Kayser and M. C. Ewing. Drawn by Lt. Meigs. Copied from the original by M. C. Ewing. No. 3. [Washington: 1837].

Scale: 5 in. = 1 mi.

Cartographers: R. E. Lee, M. C. Meigs, J. S. Morehead, H. Kayser, and M. C. Ewing.

Engraver: W. J. Stone.

Size of paper: 49.2 x 112.1 cm.

See: U.S. 25th Congress. 2nd Session. Senate Documents, No. 139. *A collection of maps etc., published by order of Congress.* Washington, 1843.

(679: [1837]: Un38hsL Large)

1384. 1893-1896

Geological survey of Missouri. Arthur Winslow, State Geologist of the Geological Survey of Missouri. Sheets nos. 2-4. Jefferson City: 1893-1896.

Scale: 1 in. = 8 mi.

Cartographer: Arthur Winslow.

Size: ca. 54 x 43 cm.

Colored.

See: Missouri Geological Survey, *Reports.* vol. i-xiii. 1891-1900.

Original wrappers.

(679: 1893-1896: M695geo Small)

1385. 1904

Ground plan [of the] Louisiana Purchase Exposition, St. Louis, Missouri. Worlds Fair. 1904. Issued by Brown Brothers & Company: New York, Philadelphia and Boston: Brown, Shipley & Company: London. Brooklyn: 1904.

Scale: 4 cm. = 1,000 ft.

Size: 27.9 x 45.1 cm.

Colored.

(679: 1904: H993Lou Small)

1386. [1904]

Map of the city of St. Louis, Mo. Issued by Brown Brothers & Company, New York, Philadelphia and Boston, and Brown, Shipley & Company, London. Brooklyn: E. Belcher Hyde: [1904].

Size: 45.7 x 44.5 cm.

Colored.

(679.971: [1904]: Sa27mstL Small)

1387. 1904

World's Fair, St. Louis. [St. Louis]: 1904.

Scale: 1¼ in. = 1,000 ft.

Cartographers: John Carson and W. J. Brown.

Size: 20.9 x 30.4 cm.

(679.971: 1904: C232wfs Small)

1388. 1911

State of Missouri. U.S. Department of the Interior. Compiled from the official records of the General Land Office and other sources under the direction of I. P. Berthrong. 1911. Washington: Eckert Lithographing Co.: 1911.

Scale: 1 in. = 12 mi.

Cartographers: I. P. Berthrong, M. Hendges, and Thos. O. Wansleben.

Lithographer: Eckert Lithographing Co.

Size: 72.3 x 83.6 cm.

Colored.

(679: 1911: Un38mLo Large)

NORTH DAKOTA Nos. 1389-1390

1389. 1910

State of North Dakota. U.S. Department of the Interior. Compiled from the official records of the General Land Office and other sources under the direction of I. P. Berthrong. 1910. Washington: Eckert Lith. Co.: 1910.

Scale: 1 in. = 12 mi.

Cartographers: I. P. Berthrong, M. Hendges, and Wm. Bauman, Jr.

Lithographer: Eckert Lith. Co.

Size: 66.1 x 85.8 cm.

Colored.

(681: 1910: Un38snd Large)

1390. 1918

State of North Dakota. U.S. Department of the Interior. Compiled from official records of the General Land Office and other sources under the direction of I. P. Berthrong. 1918. Washington: Eckert Lith. Co.: 1918.

Scale: 1 in. = 12 mi.

Cartographers: I. P. Berthrong, M. Hendges, and Wm. Bauman, Jr.

Lithographer: Eckert Lith. Co.

Size: 84.4 x 66 cm.

Colored.

(681: 1918: Un38sne Large)

SOUTH DAKOTA Nos. 1391-1397

1391. [ca. 1874]

Map of a reconnaissance of the Black Hills, July and August 1874, with troops under command of Lt. Col. G. A. Custer, 7th cavalry, by Capt. Wm. Ludlow, Corps of Engineers. U.S. Army Corps of Engineers. N.p.: [ca. 1874].

Scale: 1 in. = 12 mi.

Cartographers: William Ludlow and G. A. Custer.

Size: 64.8 x 54.3 cm.

(682: 1874: Un38bhL Extra-Oversize)

1392. [ca. 1874]

Map of a reconnaissance of the Black Hills, July and August 1874, with troops under command of Lt. Col. G. A. Custer . . . [and] Capt. Wm. Ludlow. U.S. War Department. [ca. 1874].

Scale: 1 in. = 12 mi.

Cartographers: G. A. Custer and Wm. Ludlow.

Size: 64 x 54.2 cm.

(628: [ca. 1874]: Un38bhLc Extra-Oversize)

1393. [ca. 1874]

Map of the Black Hills. From a reconnaissance by Capt. William Ludlow 1874 and maps of Warren and Raynolds. U.S. Army Corps of Engineers. N.p.: [ca. 1874].

Scale: 1 in. = 3 mi.

Cartographers: William Ludlow, Warren, Raynolds, and C. Becker.

Size: 96.5 x 60 cm.

(682: 1874: Un38bhL Extra-Oversize)

1394. 1879

Topographical and geological atlas of the Black Hills of Dakota. U.S. Department of the Interior. Geological and Geographical Survey of the Rocky Mountain region. J. W. Powell, in charge. New York: J. Bien: 1879. 2 maps and original wrappers.

Scale: 1 in. = 4 mi.

Cartographers: J. W. Powell, W. P. Jenny, H. Newton, V. T. M'Gillycuddy, H. P. Tuttle, and Emil Mahlo.

Lithographer: J. Bien.

Size of paper: 82 x 56.5 cm.

Colored.

(682: 1879: Un38bhd Large)

1395. 1883

Map of Dakota. [Offi]cial plats of public surveys, and published in the interests of immigration by the [Chi]cago, Milwaukee & St. Paul Railway Company. Chicago: Rand, McNally & Co.: 1883.

Scale: 1 in. = 22 statute mi.

Size: 64.2 x 49.2 cm.

Colored.

Upper left-hand corner of map is torn off.

(634.4: 1883: R157dak Small)

1396. 1901

State of South Dakota. U.S. Department of the Interior. Compiled from the official records of the General Land Office and other sources under the direction of Harry King. Washington: A. B. Graham Lith. Co.: 1901.

Scale: 1 in. = 12 mi.

Cartographers: Harry King and M. Hendges.

Lithographer: A. B. Graham Lithographic Co.

Size: 60 x 88.6 cm.

Colored.

(682: 1901: Un38ssd Large)

1397. 1910

State of South Dakota. U.S. Department of the Interior. Compiled from the official records of the General Land Office and other sources under the direction of I. P. Berthrong. 1910. Washington: Eckert Lithographing Co.: 1910.

Scale: 1 in. = 12 mi.

Cartographers: I. P. Berthrong, M. Hendges, A. F. Dinsmore, and Wm. Bauman, Jr.

Lithographer: Eckert Lithographing Co.

Size: 59.7 x 87.9 cm.

Colored.

(682: 1910: Un38gLo Large)

NEBRASKA No. 1398

1398. 1908

State of Nebraska. U.S. Department of the Interior. Compiled from the official records of the General Land Office and other sources under the direction of I. P. Berthrong. 1908. Washington: A. B. Graham Co.: 1908.

Scale: 1 in. = 12 mi.

Cartographers: I. P. Berthrong, A. F. Dinsmore, and Wm. Bauman, Jr.

Lithographer: A. B. Graham Co.

Size: 68 x 100 cm.

Colored.

Inset: City of Omaha.

(683: 1908: Un38gLo Large)

KANSAS Nos. 1399-1401

1399. 1856

Map of eastern Kansas [townships] by E. B. Whitman and A. D. Searl, general land agents, Lawrence, Kansas. 1856. Boston: J. P. Jewett & Co.: 1856.

Scale: 1 in. = 8 mi.

Cartographers: E. B. Whitman and A. D. Searl.

Lithographer: L. H. Bradford.

Size: 53.2 x 68.2 cm.

Colored.

Contains depictions of Constitution Hall, Topeka: Eldridge House, Lawrence, Kansas; and Ruins of Eldridge House, Lawrence, Kansas—all destroyed 21 May 1856.

(684: 1856: W599kan Book map)

1400. 1912

State of Kansas. U.S. Department of the Interior. Compiled from the official records of the General Land Office and other sources under the direction of I. P. Berthrong. 1912. [Washington]: 1912.

Scale: 1 in. = 12 mi.

Cartographer: I. P. Berthrong.

Size: 61.3 x 91.5 cm.

Colored.

(684: 1912: Un38gLk Large)

1401. 1930

Geological map of Kansas by Raymond C. Moore. Geological Survey of Kansas. [Kansas City]: 1930.

Scale: 1 in. = 20 mi.

Cartographer: Raymond C. Moore.

Size of paper: 59.7 x 65.1 cm.

Colored.

Bottom half of paper contains explanation of map.

(684: 1930: K134sur Large)

MOUNTAIN STATES Nos. 1402-1407

1402. 1859-1860

Map of the Yellowstone and Missouri Rivers and their tributaries. U.S. War Department. Explored by Capt. W. F. Raynolds and 1st. Lieut. H. E. Maynadier, 1859-1860. [Washington: 1860].

Scale: 1:1,200,000.

Cartographers: W. F. Raynolds and H. E. Maynadier.

Engraver: U.S. Engineering Bureau.

Size: 69.4 x 104.2 cm.

To accompany a report to the Bureau of Topographical Engineers. Lt. Col. Hartman Bache in charge.

(685: 1859-1860: Un38ymt Large)

1403. [1872]

Parts of Idaho, Montana and Wyoming territories from surveys made under the direction of F. V. Hayden, U.S. Geologist, and other authorities, 1871. Preliminary map for field use by F. V. Hayden, 1872. Compiled and drawn by E. Hergesheimer. U.S. Department of the Interior. Geological Survey of the Territories. [Washington: 1872].

Scale: 1 in. = 10 mi.

Cartographers: F. V. Hayden and E. Hergesheimer.

Size: 55.5 x 63 cm.

(632.3: [1872]: Un38gst Large)

1404. [ca. 1873]

Map of the sources of the Snake River with its tributaries, together with portions of the headwaters of the Madison and Yellowstone. U.S. Department of the Interior. Geological Survey of the Territories. From surveys and observations . . . by Gustavus R. Bechler . . . and Frank H. Bradley. [New York: Bien: ca. 1873].

Scale: 1:316,800.

Cartographers: Gustavus R. Bechler and Frank H. Bradley.

Size: 65.8 x 66.8 cm.

Colored.

Inset: Four profiles of mountain ranges.

Provenance: Presented by the U.S. Department of the Interior, 20 February 1874.

(623.3: [ca. 1873]: Un38geo Large)

1405. [ca. 1873]

Map of the sources of the Snake River with its tributaries, together with portions of the headwaters of the Madison and Yellowstone. U.S. Department of the Interior. Geological Survey of the Territories. From surveys and observations . . . by Gustavus R. Bechler. New York: Bien: [ca. 1873].

Scale: 1:316,800.

Cartographer: Gustavus R. Bechler.

Lithographer: J. Bien.

Size: 66.6 x 66.5 cm.

Colored.

Inset: Four profiles of mountain ranges.

Provenance: Presented by the U.S. Department of the Interior, 20 February 1874.

(623.3: [ca. 1873]: Un38ssr Large)

1406. 1877

Parts of western Wyoming, southeastern Idaho and northeastern Utah. Surveyed in 1877. U.S. Geological and Geographical Survey of the Territories. F. V. Hayden, U.S. Geologist in charge. New York: J. Bien: 1877.

Scale: 1:253,440.

Cartographers: Henry Gannett and F. V. Hayden.

Size: 57.5 x 84 cm.

(632.3: 1877: Un38wiu Large)

1407. 1879

Drainage map showing portions of Wyoming, Idaho and Utah. Primary triangulation of A. D. Wilson. Topography by Henry Gannett, G. B. Chittenden, G. R. Bechler and F. A. Clark. April 1879. U.S. Department of the Interior. Geological and Geographical Survey of the Territories. F. V. Hayden, U.S. Geologist in charge. New York: J. Bien: 1879.

Scale: 1 in. = 8 mi.

Cartographers: Henry Gannett, G. B. Chittenden, G. R. Bechler, F. A. Clark, A. D. Wilson, and F. V. Hayden.

Size: 52.1 x 92.1 cm.

(632.3: 1879: Un38ggs Large)

COLORADO Nos. 1408-1421

1408. 1873

[Colorado]. U.S. Department of the Interior. Geological and Geographical Survey of the Territories. Sketch showing the primary and secondary triangulation of 1873. Primary triangulation by Jas. T. Gardner. Secondary triangulation by A. D. Wilson, Henry Gannett and G. R. Bechler. N.p.: 1873.

Scale: 1 in. = 8 mi.

Cartographers: James T. Gardner, A. D. Wilson, Henry Gannett, and G. R. Bechler.

Size: 57 x 42.8 cm.

(687: 1873: Un38pst Large)

1409. [1873]

Preliminary map of central Colorado, showing the region surveyed in 1873. U.S. Geological and Geographical Survey of the Territories. F. V. Hayden . . . in charge. [Washington: 1873].

Scale: 1:506,880.

Cartographers: F. V. Hayden, J. T. Gardner, G. R. Bechler, Henry Gannett, A. D. Wilson, and Geo. B. Chittenden.

Size: 58 x 43.6 cm.

(687: [1873]: Un38stc Small)

1410. [1874]

Preliminary map of central Colorado. Showing the region surveyed in 1873 and 1874. U.S. Department of the Interior. Geological and Geographical Survey of the Territories. Preliminary triangulation by J. T. Gardner. Topography by G. R. Bechler, Henry Gannett, A. D. Wilson and S. B. Ladd. [Washington: 1874].

Scale: 1 in. = 10 mi.

Cartographers: J. T. Gardner, G. R. Bechler, Henry Gannett, A. D. Wilson, and S. B. Ladd.

Size: 63.4 x 57.4 cm.

(687: [1874]: Un38pcc Large)

1411. 1877

Geological and geographical atlas of Colorado and portions of adjacent territory, by F. V. Hayden. U.S. Department of the Interior. Geological and Geographical Survey of the Territories. [New York]: J. Bien, litho.: 1877. 20 maps, etc., with title page and original wrappers.

Scale: 1 in. = 4 mi.

Cartographer: F. V. Hayden.

Lithographer: Julius Bien.

Size of paper: ca. 69 x 97 cm.

Colored.

Provenance: Presented by the U.S. Geological Survey, 3 May 1878.

Contains panoramic views of Pike's Peake, Sawatch Range, Elk, La Plata, etc.

(687: 1877: Un38ter Large)

1412. 1905

State of Colorado. U.S. Department of the Interior. Compiled from the official records of the General Land Office and other sources under the direction of Frank Bond. 1905. [Washington]: Brett Litho. Co.: 1905.

Scale: 1 in. = 12 mi.

Cartographers: Frank Bond, A. F. Dinsmore, and Wm. Bauman, Jr.

Lithographer: Brett Litho. Co.

Size: 73.2 x 86.1 cm.

Colored.

Insets: Leadville; City of Denver; Cripple Creek.

(687: 1905: Un38dib Large)

1413. 1910

State of Colorado. U.S. Department of the Interior. Compiled from the official records of the General Land Office and other sources under the direction of I. P. Berthrong. 1910. Washington: Eckert Lithographers: 1910.

Scale: 1 in. = 12 mi.

Cartographers: I. P. Berthrong, A. F. Dinsmore, and Wm. Bauman, Jr.

Lithographer: Eckert Lithographers.

Size: 73 x 86 cm.

Colored.

Insets: Leadville; City of Denver; Cripple Creek.

(687: 1910: Un38soc Small)

1414. 1911

State of Colorado. U.S. Department of the Interior. Lands designated by the Secretary of the Interior as subject to entry under the provisions of the enlarged Homestead Act of February 19, 1909. N.p.: 1911 [1 April 1911].

Scale: 1 in. = 24 mi.

Cartographers: A. F. Dinsmore and Wm. Bauman, Jr.

Size: 36.4 x 43 cm.

Colored.

Insets: Leadville; City of Denver; Cripple Creek.

(687: 1911: Un38eha Large)

1415. [ca. 1914]

Panoramic view of the Mesa Verde National Park, Colorado. U.S. Department of the Interior. Prepared by John Renshawe. [Washington]: U.S. Geological Survey: [ca. 1914].

Scale: 1:45,000.

Cartographer: John Renshawe.

Engraver: U.S. Geological Survey.

Size: 48.2 x 57.6 cm.

Colored.

(687: [ca. 1914]: Un38pmv Large)

1416. [ca. 1914]

Panoramic view of the Rocky Mountain National Park, Colorado. U.S. Department of the Interior. Prepared by John Renshawe. [Washington]: U.S. Geological Survey: [ca. 1914].

Scale: 1:125,000.

Cartographer: John Renshawe.

Engraver: U.S. Geological Survey.

Size: 44.5 x 35.4 cm.

Colored.

(687: [ca. 1914]: Un38prm Small)

1417. 1921

State of Colorado. U.S. Department of the Interior. Compiled chiefly from the official records of the General Land Office with supplemental data from other map making agencies under the direction of I. P. Berthrong. 1921. Washington: Columbia Planograph Co.: 1921.

Scale: 1 in. = 12 mi.

Cartographers: I. P. Berthrong, A. F. Dinsmore, and Wm. Bauman, Jr.

Size: 72.8 x 86.3 cm.

Colored.

(687: 1921: Un38dic Large)

1418. [1941]

Geologic cross-sections of the Red Mountain, Sneffels, and Telluride districts [Colorado]. Plate 2. N.p.: [1941].

Scale: 1 in. = 1,000 ft.

Size of paper: 86.5 x 96.5 cm.

To accompany Colorado Scientific Society, *Proceedings*, vol. 14, no. 5.

(687: [1941]: G294rst Large)

1419. 1941

Preliminary geologic map of the Red Mountain, Sneffels, and Telluride districts of the Silverton, Caldera, Ouray and San Miguel Counties. Colorado. U.S. Department of the Interior. Geological Survey. Plate 1. 1941. N.p.: 1941.

Scale: 1 in. = 1,000 ft.

Cartographers: W. S. Burbank, C. F. Park, Jr., E. B. Eckel, V. C. Kelley, M. G. Barclay, M. G. Dings, R. S. Duce, J. E. Spurr, and R. S. Moehlman.

Size: 103.2 x 136.4 cm.

Colored.

To accompany Colorado Scientific Society, *Proceedings*, vol. 14, no. 5.

(687: 1941: B899rst Large)

1420. [ca. 1941?]

State of Colorado. Ironton and vicinity. Department of the Interior. Geological Survey. N.p.: [ca. 1941?].

Scale: 1:12,000.

Cartographers: Glenn S. Smith, C. A. Ecklund, and R. O. Davis.

Size of paper: 112.2 x 89.8 cm.

Colored.

Advance sheet subject to correction.

Polygonic projection. North American datum.

To accompany Colorado Scientific Society, *Proceedings*, vol. 14, no. 5.

(687: [ca. 1941?]: Sm51irt Large)

1421. [ca. 1941?]

State of Colorado. Telluride mining district. Department of the Interior. Geological Survey. N.p.: [ca. 1941?].

Scale: 1 in. = 1,000 ft.

Cartographers: H. H. Hodgeson and C. A. Ecklund.

Size of paper: 52.9 x 89.1 cm.

Colored.

Advance sheet subject to correction.

To accompany Colorado Scientific Society, *Proceedings*, vol. 14, no. 5.

(687: [ca. 1941?]: H667tmd Large)

NEW MEXICO Nos. 1422-1424

1422. 1860

Map of explorations and surveys in New Mexico and Utah made under the direction of the Secretary of War, by J. N. Macomb assisted by C. H. Dimmock. U.S. Army. Corps of Topographical Engineers. 1860. New York: Geographical Institute: 1864.

Scale: 1:760,320.

Cartographers: J. N. Macomb, C. H. Dimmock, John L. Hazzard, Samuel Sartain, and F. W. Egloffstein.

Size: 71.3 x 87.5 cm.

Provenance: Presented by the U.S. Government, 20 July 1877.

See: *Report of exploring expedition from Santa Fe to the juncture of the Grand and Green Rivers.*

(680: 1860: Un38nmu Large)

1423. 1908

Reduced reproduction of map of territory of New Mexico. U.S. Department of the Interior. Compiled from the official records of the General Land Office and other sources under the direction of I. P. Berthrong. 1908. Washington: Eckert Litho. Co.: 1908.

Scale: 1 in. = 20 mi.

Cartographers: I. P. Berthrong, A. F. Dinsmore, and Wm. Bauman.

Lithographer: Eckert Lithographing Co.

Size: 53.6 x 45.3 cm.

Colored.

(688: 1908: Un38tnm Large)

1424. 1909

Territory of New Mexico. U.S. Department of the Interior. Lands designated by the Secretary of the Interior as subject to entry under the provisions of the enlarged Homestead Act of February 19, 1909. [Washington]: 1911 [1 April 1911].

Scale: 1 in. = 24 mi.

Cartographers: A. F. Dinsmore and Wm. Bauman.

Size: 47.4 x 39.9 cm.

Colored.

(688: 1909: Un38eha Large)

WESTERN STATES Nos. 1425-1429

1425. [ca. 1873]

Montana and Wyoming territories, embracing most of the country drained by the Madison, Gallatin and upper Yellowstone Rivers. U.S. Department of the Interior. Geological and Geographical Survey of the Territories. Geology by F. V. Hayden and A. C. Peale. Drawn by Henry Gannett from notes and sketches by Adolf Burch. [Washington: ca. 1873].

Scale: 1 in. = 4 mi.

Cartographers: F. V. Hayden, A. C. Peale, Henry Gannett, and Adolf Burch.

Size: 73.1 x 53.2 cm.

Colored.

(685: [ca. 1873]: Un38myt Large)

1426. 1846

Topographical map of the road from Missouri to Oregon commencing at the mouth of the Kansas in the Missouri River and ending at the mouth of the Wallah-Wallah in the Columbia. From the field notes and journal of Capt. J. C. Frémont, and from sketches and notes made on the ground by his assistant Charles Preuss. Baltimore: E. Weber: 1846. Bound in 7 sections.

Scale: 1 in. = 10 mi.

Cartographers: J. C. Frémont and Charles Preuss.

Lithographer: E. Weber & Co.

Size of paper: 40.6 x 66.7 cm.

Provenance: Presented by Thomas H. Benton, 16 April 1847.

(633: 1846: Un38mok Large)

1427. 1883

United States geographical explorations and surveys west of the 100th meridian topographical atlas. Expeditions of 1869-1879 under the command of Lt. Col. George M. Wheeler. U.S. Geographical Survey. 1883. [Washington: 1883]. Atlas sheets nos. 32-84 in 53 pieces.

Scales: 1:126,720; 1:235,440; 1:506,880.

Size of paper: 48.5 x 61 cm.

Colored.

Provenance: Presented by George M. Wheeler.

Surveys cover California, Colorado, Idaho, Nevada, New Mexico, Oregon, and Utah. Table of contents with maps.

(633: 1883: Un38esw Large)

1428. 1913

Part of the United States west of the Mississippi River. U.S. Department of the Interior General Land Office. Prepared under the direction of I. P. Berthrong. Washington: Andrew B. Graham, lith.: 1913.

Cartographers: Thomas O. Wansleben and George O. Daidy.

Lithographer: Andrew B. Graham.

Size: 45.6 x 66.9 cm.

Colored.

(633: 1913: Un38diw Large)

1429. 1916

Part of the United States west of the Mississippi River showing activities of bureaus of the U.S. Department of the Interior. Prepared under the direction of I. P. Berthrong. [Washington]: U.S. Geological Survey: 1916.

Scale: 1 in. = ca. 150 mi.

Cartographers: M. Hendges and I. P. Berthrong.

Engravers: G. A. Daidy and U.S. Geological Survey.

Size: 45.6 x 67 cm.

Colored.

(633: 1916: Un38wmr Large)

MONTANA Nos. 1430-1432

1430. 1911

State of Montana. U.S. Department of the Interior. Compiled from the official records of the General Land Office and other sources under the direction of I. P. Berthrong. 1911. Washington: Eckert Lith. Co.: 1911.

Scale: 1 in. = 12 mi.

Cartographers: I. P. Berthrong and Daniel O'Hare.

Engraver: Eckert Lithographing Co.

Size: 85.6 x 125.2 cm.

Colored.

Insets: Great Falls; Helena; Butte; Missoula.

(685: 1911: Un38som Large)

1431. 1911

State of Montana. U.S. Department of the Interior. Lands designated by the Secretary of the Interior as subject to entry under the provisions of the enlarged Homestead Act of February 19, 1909. Edition of April 1, 1911. N.p.: 1911.

Scale: 1 in. = 24 mi.

Cartographer: Daniel O'Hare.

Size: 42.8 x 62.6 cm.

Colored.

Insets: Great Falls; Helena; Butte; Missoula.

(685: 1911: Un38eha Large)

1432. [ca. 1914]

Panoramic view of the Glacier National Park, Montana. U.S. Department of the Interior. Geological survey. Prepared by John H. Renshawe. [Washington]: U.S. Geological Survey: [ca. 1914].

Scale: 1:187,500.

Cartographer: John H. Renshawe.

Engraver: U.S. Geological Survey.

Size: 47.5 x 53 cm.

Colored.

(685: [ca. 1914]: Un38pvm Large)

WYOMING Nos. 1433-1446

1433. 1871

Lower Geyser Basin, Fire Hole River, Wyoming Territory. U.S. Department of the Interior. Geological and Geographic Survey of the Territories. Surveyed by the party in charge of F. V. Hayden. 1871. Compiled and drawn from field notes and sketches of A. Schönborn by E. Hergesheimer. [Washington]: 1871.

Scale: 1 in. = 500 yd.

Cartographers: F. V. Hayden, A. Schönborn, and E. Hergesheimer.

Size: 58.2 x 53.9 cm.

(686: 1871: Un38Lgb Large)

1434. 1871

Upper Geyser Basin, Fire Hole River, Wyoming Territory. U.S. Department of the Interior. Geological and Geographical Survey of the Territories. Surveyed by the party in charge of F. V. Hayden. 1871. Compiled and drawn from field notes and sketches of A. Schönborn by E. Hergesheimer. [Washington]: 1871.

Scale: 1 in. = 550 ft.

Cartographers: F. V. Hayden, A. Schönborn, and E. Hergesheimer.

Size: 46.9 x 58.6 cm.

Provenance: Presented by F. V. Hayden, 5 January 1872.

(686: 1871: Un38ugb Large)

1435. 1871

Yellowstone Lake. Showing the watershed of the Snake, Madison and Yellowstone Rivers. U.S. Department of the Interior. Geological Survey of the Territories. Dr. F. V. Hayden in charge. Surveyed and drawn by Henry W. Elliott. N.p.: 1871.

Scale: 1.1 cm. = 1 mi.

Cartographers: F. V. Hayden and Henry W. Elliott.

Size: 35.6 x 38.1 cm.

Provenance: Presented by F. V. Hayden, 6 January 1872.

(686: 1871: Un38Lake Small)

1436. 1871

Yellowstone Lake, Wyoming Territory. U.S. Department of the Interior. Surveyed by the party in charge of F. V. Hayden. 1871. N.p.: 1871.

Scale: 1 in. = 2 mi.

Cartographers: F. V. Hayden, A. Schönborn, E. Hergesheimer, and H. W. Elliott.

Size: 37 x 37.8 cm.

(686: 1871: Un38Lak Small)

1437. 1871

Yellowstone National Park, from surveys made under the direction of F. V. Hayden. U.S. Department of the Interior. Geological Survey of the Territories. 1871. [New York]: J. Bien, lithog.: 1871.

Scale: 1 in. = 10 mi.

Cartographers: F. V. Hayden and E. Hergesheimer.

Lithographer: Julius Bien.

Size: 30.2 x 28.3 cm.

Provenance: Presented by F. V. Hayden, 15 March 1872.

See: *American Journal of Science and Arts,* Series 3, Vol. 3, 1872.

(686: 1871: Un38par Small)

1438. [ca. 1876]

Map of the Lower Geyser Basin on the upper Madison River, after a reconnaissance by Gustavus R. Bechler. U.S. Department of the Interior. Geological and Geographical Survey of the Territories. N.p.: [ca. 1876].

Scale: 6 in. = 1 mi.

Cartographer: Gustavus R. Bechler.

Size: 54.7 x 89.6 cm.

Colored.

(686: [ca. 1876]: Un38fhr Large)

1439. [ca. 1876]

Map of the Upper Geyser Basin on the upper Madison River, Montana territory, after a reconnaissance by Gustavus R. Bechler. U.S. Department of the Interior. Geological and Geographical Survey of the Territories. N.p.: [ca. 1876].

Scale: 6 in. = 1 mi.

Cartographer: Gustavus R. Bechler.

Size: 45.4 x 50.3 cm.

(686: [ca. 1876]: Un38gbm Large)

1440. 1877

Part of central Wyoming. U.S. Department of the Interior. Geological and Geographical Survey of the Territories. F. V. Hayden, U.S. Geologist in charge. Surveyed in 1877. New York: Julius Bien: 1877.

Scale: 1:253,440.

Cartographers: F. V. Hayden and George B. Chittenden.

Lithographer: Julius Bien.

Size: 58 x 84.9 cm.

(686: 1877: Un38pcw Large)

1441. 1877-1878

Parts of western Wyoming and southeastern Idaho. U.S. Department of the Interior. Geological and Geographical Survey of the Territories. Surveyed in 1877 and 1878. F. V. Hayden, in charge. New York: Julius Bien: [1878].

Scale: 1:253,440.

Cartographers: G. R. Bechler, Fred A. Clark, and F. V. Hayden.

Lithographer: Julius Bien.

Size: 57.6 x 83.3 cm.

Provenance: Presented by the U.S. Geological and Geographical Survey of the Territories, 7 May 1880.

(686: 1877-1878: Un38wyi Large)

1442. 1878

Yellowstone National Park. U.S. Department of the Interior. Geological and Geographical Surveys of the Territories. Primary triangulation by A. D. Wilson. Secondary triangulation and topography by Henry Gannett. New York: Julius Bien: 1878.

Scale: 1:126,720.

Cartographers: A. D. Wilson and Henry Gannett.

Lithographer: Julius Bien.

Size: 81.1 x 72.2 cm.

(686: 1878: Un38ynp Large)

1443. 1907

State of Wyoming. U.S. Department of the Interior. Compiled from the official records of the General Land Office and other sources under the direction of I. P. Berthrong. 1907. Washington: A. B. Graham Co., Litho.: 1907.

Scale: 1 in. = 12 mi.

Cartographer: I. P. Berthrong.

Lithographer: A. B. Graham Co.

Size: 72.4 x 84.6 cm.

Colored.

Insets: Cheyenne and vicinity; Sheridan and vicinity; Evanston; Laramie.

(686: 1907: Un38gLo Large)

1444. 1911

State of Wyoming. U.S. Department of the Interior. Lands designated by the Secretary of the Interior as subject to entry under the provisions of the enlarged Homestead Act of February 19, 1909. N.p.: 1911 [1 April 1911].

Scale: 1 in. = 24 mi.

Cartographer: I. P. Berthrong.

Size: 36.2 x 42.2 cm.

Colored.

Insets: Sheridan and vicinity; Cheyenne and vicinity; Evanston; Laramie.

(686: 1911: Un38eha Large)

1445. [ca. 1914]

Panoramic view of the Yellowstone National Park, Wyoming-Montana-Idaho. U.S. Department of the Interior. Prepared by John R. Renshawe. Washington: U.S. Geological Survey: [ca. 1914].

Scale: 1:187,500.

Cartographer: John H. Renshawe.

Engraver: U.S. Geological Survey.

Size: 52.7 x 46.2 cm.

Colored.

(686: [ca. 1914]: Un38pvy Large)

1446. 1923

State of Wyoming. U.S. Department of the Interior. Compiled chiefly from the official records of the General Land Office with supplemental data from other map making agencies under the direction of I. P. Berthrong. Washington: Columbia Planograph Co.: 1923.

Scale: 1 in. = 12 mi.

Cartographers: I. P. Berthrong and J. J. Black.

Size: 73.8 x 83.6 cm.

Colored.

(686: 1923: Un38diw Large)

IDAHO Nos. 1447-1448

1447. 1905

State of Idaho. U.S. Department of the Interior. Compiled from the official records of the General Land Office and other sources under the direction of Frank Bond. 1905. Washington: Andrew B. Graham Co.: 1905.

Scale: 1 in. = 12 mi.

Cartographers: Frank Bond, A. F. Dinsmore, C. J. Helm, and I. P. Berthrong.

Lithographer: Andrew B. Graham Co.

Size: 114.2 x 76.2 cm.

Colored.

Insets: Pocatello and vicinity; Wallace and vicinity; Lewiston and vicinity; Boise and vicinity.

(692: 1905: Un38gLo Large)

1448. 1911

State of Idaho. U.S. Department of the Interior. Lands designated by the Secretary of the Interior as subject to entry under the provisions of the enlarged Homestead Act of June 17, 1910. [Washington]: 1911 [1 April 1911].

Scale: 1 in. = 24 mi.

Cartographer: Charles J. Helm.

Size: 55.5 x 36.4 cm.

Colored.

Insets: Pocatello and vicinity; Wallace and vicinity; Lewiston and vicinity; Boise and vicinity.

(692: 1911: Un38eha Large)

UTAH Nos. 1449-1453

1449. 1876

Atlas accompanying the report on the geology of a portion of the Uinta Mountains and a region of country adjacent thereto. U.S. Department of the Interior. Geological and Geographical Survey of the Territories. Second division. By J. W. Powell. 1876. [New York]: J. Bien, lith.: 1876. 2 maps and 6 profiles.

Scale: 4 in. = 1 mi.

Cartographers: W. H. Graves, J. W. Powell, and G. K. Gilbert.

Engraver: S. J. Kübul.

Size of paper: ca. 64.3 x 46 cm.

Colored.

Original wrappers.

(693: 1876: Un38g Large)

1450 1879-1880

Topographical and geological atlas of the district of the high plateaus of Utah. To accompany the report of Capt. C. E. Dutton. U.S. Department of the Interior. Geographical and Geological Survey of the Rocky Mountain Region. J. W. Powell in charge. New York: Julius Bien: 1879-1880.

Cartographers: John W. Powell and C. E. Dutton.

Lithographer: Julius Bien.

Size: 96.2 x 73.4 cm.

See: U.S. Geographical and Geological Survey, *Report* of Clarence Edward Dutton (557.3: Un2ro).

(693: 1879-1880: P876hpu Extra-Oversize)

1451. 1908

State of Utah. U.S. Department of the Interior. Compiled from the official records of the General Land Office and other sources under the direction of I. P. Berthrong. 1908. Washington: A. B. Andrews Co.: 1908.

Scale: 1 in. = 12 mi.

Cartographers: I. P. Berthrong, A. F. Dinsmore, and Wm. Bauman, Jr.

Lithographer: A. B. Andrews Co.

Size: 85.5 x 66.4 cm.

Colored.

(693: 1908: Un38gLo Large)

1452. 1911

State of Utah. U.S. Department of the Interior. Lands designated by the Secretary of the Interior as subject to entry under the provisions of the enlarged Homestead Act of February 19, 1909. [Washington]: 1911 [1 April 1911].

Scale: 1 in. = 24 mi.

Cartographers: A. F. Dinsmore and Wm. Bauman, Jr.

Size: 42.8 x 33.2 cm.

Colored.

(693: 1911: Un38eha Large)

1453. 1915

State of Utah. U.S. Department of the Interior. Compiled from the official records of the General Land Office and other sources under the direction of I. P. Berthrong. 1915. Washington: Eckert Lithographers: 1915.

Scale: 1 in. = 12 statute mi.

Cartographers: I. P. Berthrong, A. F. Dinsmore, Wm. Bauman, Jr., and Thos. O. Wansleben.

Lithographer: Eckert Lithographers.

Size: 85.1 x 66.8 cm.

Colored.

(693: 1915: Un38aso Small)

ARIZONA Nos. 1454-1456

1454. 1908

Territory of Arizona. U.S. Department of the Interior. Compiled from the official records of the General Land Office and other sources under the direction of I. P. Berthrong. 1908. Washington: Norris Peters Company: 1908.

Scale: 1:760,320.

Cartographers: I. P. Berthrong and Daniel O'Hare.

Size: 51.6 x 43 cm.

Colored.

Inset: Southern Arizona and Sonora, Mexico.

(694: 1908: Un38toa Small)

1455. 1911

Territory of Arizona. U.S. Department of the Interior. Lands designated by the Secretary of the Interior as subject to entry under the provisions of the enlarged Homestead Act of February 19, 1909. [Washington]: 1911 [1 April 1911].

Scale: 1 in. = 24 mi.

Cartographers: Daniel O'Hare and Wm. Bauman, Jr.

Size: 47.6 x 39 cm.

Colored.

Insets: Tucson; Phoenix; connections to railroads in Arizona.

(694: 1911: Un38eha Large)

1456. 1912

State of Arizona. U.S. Department of the Interior. Compiled from the official records of the General Land Office and the other sources under the direction of I. P. Berthrong. 1912. Washington: Eckert Litho. Co.: 1912.

Scale: 1 in. = 12 mi.

Cartographers: I. P. Berthrong, Daniel O'Hare, and Wm. Bauman, Jr.

Lithographer: Eckert Litho. Co.

Size: 95.4 x 78.3 cm.

Colored.

Insets: Connections to railroads in Arizona; Tucson; Phoenix.

(694: 1912: Un38gLo Large)

NEVADA Nos. 1457-1460

1457. [1903]

Topographical map of Washoe mining district, and Outline map of Washoe district, Nevada, showing Comstock Lode. . . . 1879. New York: J. Bien: [1903].

Scale: 1:18,000; 1:24,000.

Size: 112.8 x 86 cm.; 87.5 x 64.5 cm.

Colored.

Provenance: Presented by George M. Wheeler.

(633: 1883: Un38esw Large)

1458. 1908

State of Nevada. Department of the Interior. Compiled from the official records of the General Land Office and other sources under the direction of I. P. Berthrong. 1908. Washington: Andrew B. Graham Co. lithog.: 1908.

Scale: 1 in. = 12 mi.

Cartographers: I. P. Berthrong, Daniel O'Hare, and Chas. J. Helm.

Lithographer: Andrew B. Graham Co.

Size: 109 x 77.5 cm.

Colored.

Insets: City of Reno; Goldfield and North Goldfield.

(695: 1908: Un38gLo Large)

1459. 1911

State of Nevada. U.S. Department of the Interior. Lands designated by the Secretary of the Interior as subject to entry under the provisions of the enlarged Homestead Act of February 19, 1909. [Washington]: 1911 [1 April 1911].

Scale: 1 in. = 24 mi.

Size: 54.3 x 38.6 cm.

Colored.

Insets: City of Reno; Goldfield and North Goldfield.

(695: 1911: Un38eha Large)

1460. 1914

State of Nevada. U.S. Department of the Interior. Compiled from the official records of the General Land Office and other sources under the direction of I. P. Berthrong. 1914. Washington: Eckert Litho. Co.: 1914.

Scale: 1 in. = 12 mi.

Cartographers: I. P. Berthrong and Chas. J. Helm.

Lithographer: Eckert Litho. Co.

Size: 108.9 x 77.5 cm.

Colored.

Insets: Reno; Winnemucca; Elko; Ely; Goldfield and Columbia; Tonopah; Carson City.

(695: 1914: Un38gLo Large)

THE NORTHWEST No. 1461

1461. 1866

United States north west boundary survey. From notes by John G. Parke, U.S. Engineer, Chief Astronomer and Surveyor . . . exhibiting also surveys and reconnaissances by the British Boundary Commission, U.S. Coast Survey, British Admiralty, U.S. Engineer Bureau; Pacific Rail Road Explorations; U.S. Land Office; Capt. John Paliser; U.S. Exploring Expedition and Lewis and Clark. U.S. Boundary Survey Office. Archibald Campbell U.S. Commissioner. Washington: 1866. 2 pieces.

Scale: 1:1,000,000.

Cartographers: Lemuel D. Williams, Theodor Kolecki, and Edward Freyhold.

Photographer: Alexander Gardner.

Sizes: 40.9 x 61 cm. and 41 x 61 cm.

Provenance: Presented by Archibald Campbell, 20 September 1867.

(632.3: 1866: Un38bs Large)

PACIFIC COAST Nos. 1462-1465

1462. 1853

Reconnaissance of the western coast of the United States from Gray's Harbor to the entrance of Admiralty Inlet. U.S. Coast Survey. A. D. Bache, Superintendent. By the hydrographic party under the command of Lieut. James Alden. [Washington]: 1853.

Scale: 1:600,000.

Cartographers: A. D. Bache, James Alden, and E. Hergesheimer.

Engravers: G. McCoy, E. F. Woodward, and W. Smith.

Size: 36.9 x 44.7 cm.

Provenance: Presented through A. D. Bache, 10 October 1854; received 5 January 1855.

Contains four scenes of rocks and points.

(697: 1853: Un38gLa Small)

1463. 1854

Reconnaissance of the western coast of the United States (middle sheet) from San Francisco to Umpquah River. U.S. Coast Survey. A. D. Bache, Superintendent. By the hydrographic party under the command of Lieut. James Alden. Geographical positions by G. Davidson. 1854. [Washington: 1854].

Scale: 1:1,200,000.

Cartographers: A. D. Bache, James Alden, G. Davidson, W. B. McMurtrie, and A. Lindenkohl.

Engravers: G. McCoy, J. Knight, J. L. Hazzard, and G. B. Metzeroth.

Size: 62 x 56.6 cm.

Provenance: Presented through A. D. Bache, 6 March 1857.

Contains twelve engravings of views along the coast.

(696: 1854: Un38sfu Large)

1464. 1855

Reconnaissance of the western coast of the United States (northern sheet) from Umpquah River to the boundary. U.S. Coast Survey. A. D. Bache, Superintendent. By the hydrographic party under the command of Lieut. James Alden. Geographical positions by G. Davidson. 1855. [Washington: 1855].

Scale: 1:1,200,000.

Cartographers: A. D. Bache, James Alden, G. Davidson, W. B. McMurtrie, and A. Lindenkohl.

Engravers: G. McCoy, J. V. N. Throop, and J. J. Knight.

Size: 61.9 x 56.4 cm.

Provenance: Presented through A. D. Bache, 6 March 1857.

Contains eight engravings of headlands, capes, island, rocks and entrance to the Columbia River.

(696: 1855: Un38urb Large)

1465. 1861

Charts of the lights on the Pacific coast of the United States. Published by the order and under the direction of the Light House Board. The coast line prepared from the archives of the U.S. Coast Survey Office. New York: J. Bien: 1861. Series 3 (6 sheets).

Scale: 1:600,000.

Lithographer: J. Bien.

Size of paper: 116.5 x 67 cm.

Colored.

(696: 1861: Un48Lhb Large)

WASHINGTON Nos. 1466-1473

1466. 1853

Cape Flattery and Neé-ah Harbor, Washington. U.S. Coast Survey. Topography by G. Davidson. Hydrography by the party under the command of Lieut. James Alden. [Washington]: 1853.

Scale: 1:40,000.

Cartographers: G. Davidson, James Alden, and J. Lambert.

Engravers: S. Siebert and J. Young.

Size: 34.2 x 43.2 cm.

Provenance: Presented through A. D. Bache, 10 October 1854; received 5 January 1855.

Contains engraving of Cape Flattery.

(697: 1853: Un38cfn Small)

1467. [1853]

Preliminary survey of Shoalwater Bay, Washington. U.S. Coast Survey. A. D. Bache, Superintendent. By the hydrographic party under the command of Lieut. James Alden. 1853. [Washington: 1853].

Scale: 1:80,000.

Cartographers: A. D. Bache, James Alden, and A. Boschke.

Engravers: J. Young and S. E. Stull.

Size: 34.3 x 43.6 cm.

Provenance: Presented through A. D. Bache, 10 October 1854; received 5 January 1855.

697: [1853]: Un38swb Small)

1468. 1853

Reconnaissance of False Dungeness Harbor, Washington. U.S. Coast Survey. A. D. Bache, Superintendent. By the hydrographic party under the command of Lieut. James Alden. 1853. [Washington]: 1853.

Scale: 1:30,000

Cartographers: A. D. Bache, James Alden, and J. R. Key.

Engravers: W. Smith and J. S. Pettit.

Size: 24.6 x 31.1 cm.

Provenance: Presented through A. D. Bache, 10 October 1854; received 5 January 1855.

(697: 1853: Un38fdh Small)

1469. 1858

Reconnaissance of Canal de Haro and Strait of Rosario and approaches, Washington territory. U.S. Coast Survey. A. D. Bache, Supdt. Hydrography by the parties under the command of Comdr. J. Alden and R. M. Cuyler. [Washington]. 1858.

Scale: 1:200,000.

Cartographers: A. D. Bache, J. Alden, R. M. Cuyler, G. Davidson, J. S. Lawson, A. Boschke, J. J. Ricketts, and L. D. Williams.

Engravers: A. Petersen, R. T. Knight, and E. H. Sipe.

Size: 64.4 x 68 cm.

Inset: Sub sketch of Admiralty Inlet and Puget Sound.

Provenance: Presented by Hartman Bache, 21 October 1859.

(697: 1858: Un38hsr Large)

1470. 1885

Puget Sound, Washington territory. U.S. Coast and Geodetic Survey. 1867. [Washington]: 1885.

Scale: 1:200,000.

Cartographers: James Alden, George Davidson, Charles Wilkes, A. Lindenkohl, and J. S. Lawson.

Size: 70.7 x 44.6 cm.

Provenance: Presented by U.S. Coast and Geodetic Survey through Spencer C. McCorkle, 14 December 1885.

(697: 1885: Un38psw Large)

1471. 1909

State of Washington. U.S. Department of the Interior. Compiled from the official records of the General Land Office and other sources under the direction of I. P. Berthrong. 1909. Washington: Andrew B. Graham Co., litho.: 1909.

Scale: 1 in. = 12 mi.

Cartographers: I. P. Berthrong, Daniel O'Hare, and Wm. Bauman, Jr.

Lithographer: Andrew B. Graham Co., litho.

Size: 58.7 x 88.2 cm.

Colored.

Insets: Tacoma; Seattle.

(697: 1909: Un38gLo Large)

1472. 1911

State of Washington. U.S. Department of the Interior. Lands designated by the Secretary of the Interior as subject to entry under the provisions of the enlarged Homestead Act of February 19, 1909. [Washington]: 1911 [1 April 1911].

Scale: 1 in. = 24 mi.

Cartographer: Wm. Bauman, Jr.

Size: 29.3 x 44.1 cm.

Colored.

(697: 1911: Un38eha Large)

1473. [ca. 1915]

Panoramic view of the Mount Rainier National Park, Washington. U.S. Department of the Interior. Geological Survey. Prepared by John H. Renshawe from topographic sheet of the U.S. Geological Survey. [Washington]: U.S. Geological Survey: [ca. 1915].

Scale: 1:62,500.

Cartographer: John H. Renshawe.

Engraver: U.S. Geological Survey.

Size: 48.6 x 50.7 cm.

Colored.

(697: [ca. 1915]: Un38rnp Large)

OREGON Nos. 1474-1478

1474. [1841]

Map of the Oregon Territory. By the United States Ex[ploring] Ex[pedition]. Charles Wilkes, Esqr. Commander. 1841. New York: J. H. Young & Sherman & Smith [1841].

Scale: 1 in. = 51 mi.

Cartographer: Charles Wilkes.

Lithographers: J. H. Young and Sherman & Smith.

Size: 58.8 x 89.2 cm.

Inset: Columbia River, reduced from a survey made by the U. S. Exploring Expedition.

Provenance: Presented by J. R. Ingersoll, 3 April 1846.

(696.2: [1841]: Un38otw Large)

1475. 1910

State of Oregon. U.S. Department of the Interior. Compiled from the official records of General Land Office and other sources under the direction of I. P. Berthrong. 1910. Washington: Eckert Litho. Co.: 1910.

Scale: 1 in. = 12 mi.

Cartographers: I. P. Berthrong, M. Hendges, A. F. Dinsmore, and Wm. Bauman, Jr.

Lithographer: Eckert Litho. Co.

Size: 72.5 x 92.8 cm.

Colored.

Inset: City of Portland.

(698: 1910: Un38gLo Large)

1476. 1911

State of Oregon. U.S. Department of the Interior. Lands designated by the Secretary of the Interior as subject to entry under the provisions of the enlarged Homestead Act of June 17, 1910. [Washington]: 1911 [1 April 1911].

Scale: 1 in. = 24 mi.

Cartographers: M. Hendges, A. F. Dinsmore, and Wm. Bauman, Jr.

Size: 35.8 x 46 cm.

Colored.

Inset: City of Portland.

(698: 1911: Un38eha Large)

1477. [ca. 1913]

Panoramic view of the Crater Lake National Park, Oregon. U.S. Department of the Interior. Geological Survey. Prepared by John H. Renshawe from topographic sheet of the U.S. Geological Survey. Washington: U.S. Geological Survey: [ca. 1913].

Scale: 1:62,500.

Cartographer: John H. Renshawe.

Engraver: U.S. Geological Survey.

Size: 47.5 x 42.4 cm.

Colored.

(698: [ca. 1913]: Un38cra Large)

1478. 1914

Potash from kelp. By Frank Cameron. U.S. Department of Agriculture. Report No. 100, maps. Contribution from the Bureau of Soils. Kelp groves of the Pacific coast and islands of the United States and lower California. Washington: Government Printing Office: 1914.

Scale: 1:200,000 and 1:100,000.

Size: Various.

Colored.

Contains maps Nos. A-G, 1-2, 4-8, 11-14, 20-21, 23-27, 29-32, 34-47, 49-58, 60-61.

(696: 1914: P844pcc Large)

CALIFORNIA Nos. 1479-1494

1479. 1852

Monterey Harbor, California. U.S. Coast Survey. From a trigonometrical survey under the direction of A. D. Bache. [Washington]: 1852.

Scale: 1:40,000.

Cartographers: A. D. Bache, R. D. Cutts, A. M. Harrison, James Alden, W. B. McMurtrie, and J. Lambert.

Engravers: S. Siebert, S. V. Hunt, and E. F. Woodward.

Size: 33.5 x 29.7 cm.

Provenance: Presented through A. D. Bache, 10 October 1854; received 5 January 1855. Presented by the U.S. Coast and Geodetic Survey through Spencer C. McCorkle, 14 December 1885.

Contains view of Point Pinos.

(699: 1852: Un38mhc Small)

1480. 1853

Preliminary sketch of Santa Barbara, California. U.S. Coast Survey. A. D. Bache Superintendent. [Washington]: 1853.

Scale: 1:20,000.

Cartographers: A. D. Bache, A. M. Harrison, James Alden, and J. Lambert.

Engravers: M. F. O. Strobel, J. Jung, and W. Smith.

Size: 34.4 x 43 cm.

Provenance: Presented through A. D. Bache, 10 October 1854; received 5 January 1855.

Contains: View of the Town and Mission of Santa Barbara.

(699.971: 1853: Un38ssb Small)

1481. 1853

Reconnaissance of the western coast of the United States from San Francisco to San Diego by the hydrographic party under the command of Lieut. James Alden. U.S. Coast Survey. A. D. Bache, Superintendent. [Washington]: 1853.

Scale: 1:1,200,000.

Cartographers: James Alden, W. B. McMurtrie, and W. M. C. Fairfax.

Engravers: G. McCoy, S. V. Hunt, and W. Smith.

Size: 55.8 x 58 cm.

Provenance: Presented by Alexander Dallas Bache, 7 March 1857.

Contains 17 views of the coast, sailing instructions, tides, etc.

(633: 1853: Un38wss Large)

1482. 1853

San Diego entrance and approaches, California. U.S. Coast Survey. From a trigonometrical survey under the direction of A. D. Bache. [Washington]: 1853.

Scale: 1:25,000.

Cartographers: A. D. Bache, R. D. Cutts, Geo. Davidson, A. M. Harrison, and F. Herbst.

Engravers: S. Siebert, G. McCoy, and E. Yeager.

Size: 29.6 x 33.1 cm.

Inset: General sketch of San Diego bay and Los Coronados.

Provenance: Presented through A. D. Bache, 10 October 1854; received 5 January 1855.

(699: 1853: Un38sde Small)

1483. [1869]

San Francisco Peninsula. U.S. Coast Survey. Map is reduced from surveys made under the direction of A. D. Bache . . . by R. D. Cutts. [Washington: 1869].

Scale: 1:40,000.

Cartographers: A. D. Bache, R. D. Cutts, A. M. Harrison, A. F. Rodgers, and C. Rockwell.

Size: 45.7 x 70.6 cm.

(699: [1869]: Un38sfp Large)

1484. 1877

Map showing the distribution of Indian tribes of California to illustrate report of Stephen Powers. U.S. Department of the Interior. Geographical and Geological Survey of the Rocky Mountain region: J. W. Powell in charge. [New York]: Julius Bien: 1877.

Scale: 4.5 cm. = 40 mi.

Cartographers: H. Lindenkohl, Stephen Powers, and J. W. Powell.

Lithographer: J. Bien.

Size: 65.1 x 56.8 cm.

Colored.

(699: 1877: L646rsp Large)

1485. 1885

Entrance to San Francisco Bay, California. U.S. Coast Survey. From a trigonometrical survey made under the direction of A. D. Bache. 1859. [Washington]: 1885.

Scale: 1:50,000.

Cartographers: A. D. Bache, R. D. Cutts, A. F. Rodgers, A. M. Harrison, James Alden, W. M. C. Fairfax, J. Lambert, and J. J. Ricketts.

Engravers: J. Knight, A. Blondeau, and G. B. Metzeroth.

Size: 59.6 x 99 cm.

Inset: Sub sketch of entrance to San Francisco Bay.

Provenance: Presented by the U. S. Coast and Geodetic Survey through Spencer C. McCorkle, 14 December 1885.

Contains two views of the entrance to San Francisco Bay.

(699: 1885: Un38sfb Large)

1486. [ca. 1900]

Geological map of Shasta county. Prepared for the 11th. report of the State Mineralogist . . . By H. W. Fairbanks. San Francisco: Britton & Rey: [ca. 1910].

Scale: 1⅜ in. = 6 mi.

Cartographer: H. W. Fairbanks.

Lithographers: Britton & Rey.

Size: 38.7 x 55.2 cm.

Colored.

(699.774: [ca. 1900]: F153gms Small)

1487. 1903

Pacific coast from Point Pinos to Bodega Head, California. U.S. Coast and Geodetic Survey. Washington: 1903.

Scale: 1:200,000.

Size: 98.5 x 70.6 cm.

(699: 1903: Un38pbh Large)

1488. 1903

San Francisco entrance, California. U.S. Coast and Geodetic Survey. Washington: 1903.

Scale: 1:40,000.

Size: 85.5 x 105.6 cm.

Colored.

(699: 1903: Un38sfe Large)

1489. 1906

Geological map of the San Francisco Peninsula by Roderic Crandall of the U.S. Coast Survey. Plate no. 1179. 1869. May 1906. [Washington]: 1906.

Scale: 1:40,000.

Cartographer: Roderic Crandall.

Size: 70.5 x 45.1 cm.

Colored and emended by hand.

(699: 1906: C852usc Large)

1490. 1907

State of California. U.S. Department of the Interior. Compiled from the official records of the General Land Office and other sources under the direction of I. P. Berthrong. 1907. Washington: Andrew B. Graham Co., Litho.: 1907.

Scale: 1 in. = 12 mi.

Cartographers: I. P. Berthrong, A. F. Dinsmore, and Wm. Bauman, Jr.

Lithographer: Andrew B. Graham Co., Litho.

Size: 146.7 x 121.2 cm.

Colored.

Insets: City of Los Angeles; City of San Diego and vicinity; City and county of San Francisco.

(699: 1907: Un38gLo Large)

1491. 1914

San Francisco Bay. U.S. Geological Survey. Compiled and engraved and published by the U.S Geological Survey, George Otis Smith, Director. Washington: U.S. Geological Survey: 1914.

Scale: 1:1,000,000.

Cartographer: George Otis Smith.

Engraver: U.S. Geological Survey.

Size: 46.5 x 56.1 cm.

Colored.

(699: 1914: Un38int Large)

1492. [ca. 1915]

Panoramic view of the Yosemite National Park, California. Prepared by John H. Renshawe from the topographic sheet of the U.S. Geological Survey. [Washington]: U.S. Geological Survey: [ca. 1915].

Scale: 1:187,500.

Cartographer: John H. Renshawe.

Engraver: U.S. Geological Survey.

Size: 45.6 x 47.3 cm.

Colored.

(699: [ca. 1915]: Un38yos Large)

1493. [1922]

Fault map of the state of California, compiled from data assembled by the Seismological Society of America in cooperation with the Hydrographic Office, Navy Department, the U.S. Geological Survey, the Carnegie Institution of Washington, the University of California and Stanford University. N.p.: [1922]. 4 sheets.

Scale: 1:506,880.

Cartographers: Fred B. Scobey, Lot Bowen, John H. Renshawe, Bailey Willis, and H. O. Wood.

Size of paper: each ca. 132 x 101.5 cm.

Colored.

(699: [1922]: Sco39ftm Large)

1494. 1929

Geological map of the State of California. Issued by State Mining Bureau. Compiled under the direction of Fletcher Hamilton. 1916. San Francisco: Union Lithograph Co.: 1929.

Scale: 1:823,680.

Cartographers: Fletcher Hamilton and James Perrin Smith.

Lithographer: Union Lithograph Co.

Size: 147 x 119.3 cm.

Colored.

(699: 1929: C132gmc Small)

HAWAII No. 1495

1495. 1901

Map of the territory of Hawaii compiled from data on file in the U.S. Coast and Geodetic Survey, Hydrographic Office, Hawaiian Government Surveys, and other authentic sources under the direction of Harry King. U.S. Department of the Interior. General Land Office. Washington: A. B. Graham: 1901.

Scale: 1 in. = 12 mi.

Cartographers: Harry King and I. P. Berthrong.

Lithographer: Andrew B. Graham.

Size: 56 x 84 cm.

Colored.

Inset: Hawaiian Archipelago.

(960: 1901: Un38dih Large)

ALASKA Nos. 1496-1500

1496. 1766

Nouvelle carte des découvertes faites par des vaisseaux Russiens aux côtes inconnues de l'Amérique Septentrionale avec le pais adiacents. Dressée sur des mémoires authentiques de ceux qui ont assisté à ces découvertes, et sur d'autres connoissances dont on rend raison dans un mémoire separé. A St. Petersbourg à l'Académie Impériale de Sciences; 1758. Amsterdam: Marc Michel Rey: 1766.

Engraver: L. Schenk Jansz, 1765.

Size: 45.1 x 63.3 cm.

Shows routes of Captains Tschirikow and Bering.

(603: 1766: N756nca Large)

1497. North western America showing the territory ceded by Russia to the United States. Compiled for the Department of State at the United States Coast Survey Office. B. Peirce, Sup[erintenden]t 1867. [Washington]: Krebs, lith.: 1867. 2nd ed.

Scale: 1:5,000,000.

Cartographer: A. Lindenkohl.

Lithographer: Charles G. Krebs.

Size: 56.1 x 91.7 cm.

Colored.

Insets: Sitka and its approaches; North Pacific Ocean.

(605.7: 1867: Un38nwa Large)

1498. [1868]

Isothermal lines of Alaska. By Loren Blodget. N.p.: [1868].

Cartographer: Loren Blodget.

Lithographer: F. Bourquin.

Size: 20.5 x 25.5 cm.

Colored.

(605: [1868]: B621miL Small)

1499. [1903]

Southeastern Alaska and part of British Columbia. Shewing award of Alaska Boundary Tribunal, Oct. 20th. 1903. Ottawa: Mortimer: [1903].

Scale: 1:960,000.

Size: 93 x 49.2 cm.

Colored.

(605.6: [1903]: C162abc Large)

1500. 1909

Alaska compiled from the official records of the U.S. Department of the Interior. General Land Office; Coast and Geodetic Survey Geological Survey; Canadian and other sources under the direction of I. P. Berthrong. Washington: Eckert Lith. Co.: 1909.

Scale: 1 in. = 60 statute mi.

Cartographers: Daniel O'Hare and I. P. Berthrong.

Size: 77.9 x 106.2 cm.

Colored.

(605.7: 1909: Un38gLo Large)

CENTRAL AMERICA Nos. 1501-1502

1501. 1768

Nuevo mapa geographico de la America septentrional, perteneciente al virreynato de Mexico . . . por . . . Don Joseph Antonio de Alzate y Ramirez. 1768. Paris: Dezauche: 1768.

Scale: 105 leugas Castillanas de 17½ en grado.

Cartographer: Joseph Antonio de Alzate y Ramirez.

Size: 53.8 x 66.9 cm.

Colored.

(702: 1768: AL91nmm Large)

1502: 1851

A map of Central America. New York: Colton: 1851.

Scale: 6.7 cm. = 200 mi.

Size: 31.3 x 45.1 cm.

Inset: Isthmus of Panama.

Map given originally to Maine Historical Society by the estate of A. W. Longfellow, 9 September 1908.

(710: 1851: C672cen Book map)

MEXICO Nos. 1503-1507

1503. [ca. 1771]

Plan of Vera Crux, lying in the Gulph of Mexico, in lattitude 19° 10" north. [London: ca. 1771].

Scale: 3.5 cm. = 3 mi.

Engraver: J. Gibson.

Size: 22.3 x 36.5 cm.

See: Joseph Smith Speer, *West-India pilot.* London: Hooper: 1771.

(703: [ca. 1771]: G353vcm Small)

1504. 1819

A map of Mexico, Louisiana and the Missouri territory, including also the state of Mississippi, Alabama territory, east and west Florida, Georgia, South Carolina & part of the island of Cuba, by John H. Robinson, M.D. Philadelphia: John L. Narstin: 1819.

Scale: 69½ American mi. = 1 degree.

Cartographer: John H. Robinson.

Engraver: H. Anderson.

Size: 171 x 164.8 cm.

Colored.

Provenance: Presented by Thomas L. Plowman, 6 October 1820.

(702: 1819: R567mLm Large)

1505. 1847

Map of Mexico, including Yucatan and upper California, exhibiting the chief cities and towns, the principal travelling routes etc. Philadelphia: S. Augustus Mitchell: 1847.

Scale: 1 in. = 120 mi.

Cartographer: Samuel Augustus Mitchell.

Size: 44.3 x 64.3 cm.

Colored.

Inset: "The late battlefield" [of Monterey].

Provenance: Presented and autographed by Henry Phillips, Jr., 5 November 1880.

(702: 1847: M695mex Book map)

1506. 1848

Battles of Mexico. Survey of the line of operations of the U.S. Army, under command of Major General Winfield Scott, on the 19th & 20th August & on the 8th, 12th & 13th September, 1847 . . . U.S. Topographical Engineers. Washington: C. B. Graham: 1848.

Scale: 1 in. = 1 mi.

Cartographers: McClellan, Turnbull, and Hardcastle.

Lithographer: Curtis B. Graham.

Size: 60.4 x 87.9 cm.

Colored.

(707.6: 1848: Un38wsm Large)

1507. 1929

Carta de los ferrocarriles de la republica Mexicana. Published under the Direccion de Estudios Geográficos Climatológicos.

Scale: 1:3,000,000.

Size: 64.5 x 66.5 cm.

Colored.

Insets: Monterrey; Torreon; Yucatan; Baja California; Mexico, Puebla y Pachuca.

In same folder is: Plano de la cuidad de México. 1929; *Scale:* 1:20,000; *Size:* 66 x 87.5 cm.

(702: 1929: M575rea Book map)

GUATEMALA No. 1508

1508. 1902

Guatemala. From official and other sources. Prepared in the Bureau of the American Republics, William Woodville Rockhill, Director. Washington: Graham: 1902. 2 maps.

Scale: 1:792,000.

Cartographer: M. Hendges.

Lithographer: Andrew B. Graham.

Size: 63.2 x 61.2 cm.

Colored.

One map is topographical and the other is concerned with agriculture and elevations.

(711: 1902: Un38arg Large)

HONDURAS Nos. 1509-1510

1509. [ca. 1771]

A plan of Port San Fernando de Omoa near Honduras. To face p. 26. N.p.: [ca. 1771].

Scale: 600 fathoms = 6.5 cm.

Engraver: Prinald.

Size: 20 x 24.7 cm.

See: Joseph Smith Speer, *West-India pilot.* London: Hooper: 1771.

(713: [ca. 1771]: P936hom Small)

1510. [1771]

Plan of the fortification now finishing in the port of St. Fernando de Omoa in lattd. 15.50 n. N.p.: [1771].

Scale: 8.5 cm. = 100 toises.

Engraver: Bayly.

Size: 21.7 x 26.3 cm.

See: Joseph Smith Speer, *West-India pilot.* London: Hooper: 1771.

(713: [1771]: B341sfo Small)

BELIZE (British Honduras) Nos. 1511-1512

1511. [ca. 1771]

Plan of Blewfields Harbor, on the Mosquito shore. To face p. 31. [London: ca. 1771].

Engraver: Prinald.

Size: 24.1 x 36.5 cm.

See: Joseph Smith Speer, *West-India pilot.* London: Hooper: 1771.

(715: [ca. 1771]: P936nbm Small)

1512. 1773

A plan of Black River on the Mosquito Shore. [London]: 1773 [6 May 1773].

Cartographer: Captain Speer.

Engraver: Thos. Bowen.

Size: 32.2 x 49.9 cm.

See: Joseph Smith Speer, *West-India pilot.* London: Hooper: 1771.

(713: 1773: B671cam Small)

PANAMA Nos. 1513-1516

1513. [ca. 1771]

Plan of the town, and harbour of Puerto Vello [Portobello]. N.p.: [ca. 1771].

Scale: 1 in. = ¾ mi.

Engraver: J. Gibson.

Size: 23.3 x 32.9 cm.

See: Joseph Smith Speer, *West-India pilot.* London: Hooper: 1771.

(717: [ca. 1771]: G353ppv Small)

1514. 1865

Carta corográfica del estado de Panamá construida con los datos de la Comision Corográfica . . . por Manuel Ponce de Leon, injeniero, y Manuel Maria Paz, Bogotá, 1864. Estados unitas de Colombia. Paris: Monrocq: 1864. Washington: Norris Peters Lith. Co.: n.d.

Scale: 1:810,000.

Cartographers: Manuel Ponce de Leon and Manuel Maria Paz.

Engraver: Erhard Schieble, 1865.

Size: 52.6 x 82.3 cm.

Inset: Golfo de Urabá o del Darien del Norte.

To accompany *Notes on Panama* by H. C. Hale.

(717: 1865: P786ccp Large)

1515. [ca. 1904]

General plan of the Panama Canal from the plans of the Panama Canal Company, [ca. 1880]. Washington: Norris Peters Lith. Co.: [ca. 1904].

Scale: 1:100,000.

Lithographer: Norris Peters Lith. Co.

Size of paper: 58.1 x 90 cm.

Colored.

To accompany *Notes on Panama* by H. C. Hale.

(718: [ca. 1904]: P196gpp Large)

1516. 1904

Map of the Republic of Panama prepared in the War Department, Office of Chief of Staff, Second (military information) Division. January, 1904. Washington: Norris Peters Lith. Co.: 1904.

Scale: 8 cm. = 25 mi.

Cartographers: Charles H. Ourand and D. M. Blakelock.

Size: 57.3 x 112 cm.

Colored.

(717: 1904: Un38rpm Large)

1516a. 1905

Lock canal project. Map in the vicinity of Gatun, showing location of proposed dam and locks. Board of Consulting Engineers. Panama Canal. 1905. Blueprint. Pl.XI, 1905.

Scale: 1:5,000.

Size: 57.5 x 55.3 cm.

Provenance: Presented by William Gerig, division engineer.

(717: 1905: Un38pan Extra-Oversize)

1516b. 1905

Profile of Gatun Dam site, showing completed borings to indurated clay: also various artesian flows encountered in the strata. Made in the office of the division engineer of Cristobal division from the original field notes by F. B. Maltby. 1905. Blueprint.

Scale: Horizontal: 1 in. = 200 ft. Vertical: 1 in. = 20 ft.

Cartographer: F. B. Maltby.

Size: 50 x 118 cm.

Provenance: Presented by William Gerig, division engineer.

This is to accompany the surface map of Gatun Dam site.

(717: 1905: Un38pan Extra-Oversize)

WEST INDIES Nos. 1517-1553

1517. 1748/9

A new and exact map of the island of Antigua in America according to an actual and accurate survey made in the years 1746, 1747 & 1748. Describing the limits & boundaries of the several parishes, with the churches, also the divisions, with their respective boundaries; high roads, jury paths, the situation of every gentleman's plantation, mills, works and houses, with the harbors, bays, creeks, islands, rocks, shoals and soundings that surround the whole. By Robert Baker. [London]: Carington Bowles and Robert Wilkinson: 1748/9 [2 February 1748/9]. Map in four pieces.

Scale: 2 cm. = 5 furlongs.

Cartographer: Robert Baker.

Engraver: J. Mynde.

Size: 114 x 143.4 cm.

Colored.

Inset: English Harbour.

Contains: An alphabetical list of the subscribers' names, with references to the divisions in which they have land.

(732.44: 1748/9: B171aab Large)

1518. [1753]

A new and exact map of the island of St. Christopher in America, according to an actual and accurate survey made in the year 1753. Describing the several parishes, with their respective limits, contents, & churches; also the high ways, the situation of every gentleman's plantation, mills, & houses; with the rivers and gutts. Likewise the bays, roads, rocks, shoals and soundings that surround the whole. By Sam Baker. London: Carington Bowles & Robt. Wilkinson: [1753].

Scale: 1 in. = ½ mi.

Cartographer: Samuel Baker.

Engraver: James Mynde.

Size: 114.2 x 142.1 cm.

Colored.

Inset: The Leeward Caribbee Islands.

Contains list of subscribers' names on borders.

(732.24: [1753]: B171sca Large)

1519. [ca. 1763]

A new & exact map of the island of Barbadoes in America according to an actual & accurate survey made by William Mayo. N.p.: [ca. 1763].

Scale: 2 in. = 1 mi.

Cartographer: William Mayo.

Size: 94.2 x 111.7 cm.

Colored.

Insets: Prospect of Codrington College, and a plan of Bridge Town.

(736: [ca. 1763]: M455bia Large)

1520. [ca. 1771]

Aiguiade de Port o Rico, situated on the west end of that island, in 18° 23" north latitude. N.p.: [ca. 1771].

Scale: 3 mi. divided into fathoms or toises.

Engraver: J. Gibson.

Size: 22.9 x 36.8 cm.

See: Joseph Smith Speer, *West-India pilot.* London: Hooper: 1771.

(728: [ca. 1771]: G353apr Small)

1521. [ca. 1771]

The bay of Ocoa, on the south side of the Island of St. Domingo, in the latitude 18° 12" north. N.p.: [ca. 1771].

Scale of 3 mi.

Engraver: J. Gibson.

Size: 23 x 36 cm.

See: Joseph Smith Speer, *West-India pilot.* London: Hooper: 1771.

(725: [ca. 1771]: G353boc Small)

1522. [ca. 1771]

Plan de la rade du Port Paix, à la côte septentrionale de Saint Domingue. To face p. 48. [London: ca. 1771].

Scale: 2⅜ in. = 400 toises.

Engraver: [Prinald].

Size: 20.1 x 33.3 cm.

See: Joseph Smith Speer, *West-India pilot.* London: Hooper: 1771.

(725: [ca. 1771]: P936ssd Small)

1523. [ca. 1771]

Plan du Cap François et de ses environs. To face p. 12. [London: ca. 1771].

Scale: 1½ in. = 400 toises.

Engraver: [Prinald].

Size: 22.3 x 33.5 cm.

See: Joseph Smith Speer, *West-India pilot.* London: Hooper: 1771.

(725: [ca. 1771]: P936pcf Small)

1524. [ca. 1771]

Plan du cul de sac de Leogane, ou le Port au Prince. To face p. 46. [London: ca. 1771].

Scale: 4.8 cm. = 3 leagues.

Engraver: Prinald.

Size: 23 x 31 cm.

See: Joseph Smith Speer, *West-India pilot.* London: Hooper: 1771.

(725: [ca. 1771]: P936cds Small)

1525. [ca. 1771]

Plan of Blewfield's Harbour, in the island of Jamaica. To face p. 7. [London: ca. 1771].

Scale: 1 in. = 1 mi.

Engraver: Prinald.

Size: 24 x 31.2 cm.

See: Joseph Smith Speer, *West-India pilot.* London: Hooper: 1771.

(727: [ca. 1771]: P936bhj Small)

1526. [ca. 1771]

A plan of Cape Nichola mole, at the n. w. end of the island of Hispaniola. To face p. 10. [London: ca. 1771].

Scale: 1½ in. = 1 mi.

Engraver: Prinald.

Size: 25.1 x 24.9 cm.

See: Joseph Smith Speer, *West-India pilot.* London: Hooper: 1771.
(724: [ca. 1771]: P936nmh Small)

1527. [ca. 1771]

Plan of Lucia Harbour on the north side of Jamaica in lat[itu]d[e] 18° 23". To face p. 8. [London: ca. 1771].

Scale: 1 in. = 1 mi.

Engraver: [Prinald].

Size: 22.2 x 31.5 cm.

See: Joseph Smith Speer, *West-India pilot.* London: Hooper: 1771.
(727: [ca. 1771]: L965Lhj Small)

1528. [ca. 1771]

Plan of Mantica Bay on the north side of Jamaica. To face p. 9. [London: ca. 1771].

Scale: 1 in. = 1⅓ mi.

Engraver: Prinald.

Size: 23.2 x 37 cm.

See: Joseph Smith Speer, *West-India pilot.* London: Hooper: 1771.
(727: [ca. 1771]: M745nsj Small)

1529. [ca. 1771]

A plan of Port Antonio, on the northeast side of Jamaica. [London: ca. 1771].

Scale: 20 cm. = 8 furlongs, or 1 English mi.

Engraver: Prinald.

Size: 27.7 x 40.4 cm.

See: Joseph Smith Speer, *West-India pilot.* London: Hooper: 1771.
(727: [ca. 1771]: P936paj Small)

1530. [ca. 1771]

A plan of Port Royal harbour rattan. [London: ca. 1771].

Scale: 1 in. = ½ mi.

Engraver: Prinald.

Size: 16.8 x 30.2 cm.

See: Joseph Smith Speer, *West-India pilot.* London: Hooper: 1771.
(727: [ca. 1771]: P936har Small)

1531. [ca. 1771]

Plan of the bay of Matanzas, lying on the north side of Cuba, in the latt[itu]de 23° 10" north. [London: ca. 1771].

Scale: 60 mi. = 1 degree.

Engraver: J. Gibson.

Size: 22.8 x 37.

See: Joseph Smith Speer, *West-India pilot.* London: Hooper: 1771.
(723: [ca. 1771]: C353mbc Small)

1532. 1771

Plan of the harbour, and city, of Havannah, on the north side of the island of Cuba. London: S. Hooper: 1771 [1 October 1771].

Scale: 2 in. = 500 fathoms.

Engraver: I. Bayly.

Size: 23.2 x 37 cm.

See: Joseph Smith Speer, *West-India pilot.* London: Hooper: 1771.
(723.925: 1771: B341hch Small)

1533. [ca. 1771]

A plan of the harbour of Port Royal in the island of Jamaica. To face p. 1. [London: ca. 1771].

Scale: 1 in. = 800 fathoms.

Engraver: Prinald.

Size: 27 x 34 cm.

See: Joseph Smith Speer, *West-India pilot.* London: Hooper: 1771.
(727: [ca. 1771]: P936prj Small)

1534. [ca. 1771]

Plan of the harbour of St. Juan de Port o Rico. [London: ca. 1771].

Scale: "Half a mile of 60 to a degree."

Engraver: J. Gibson.

Size: 29.6 x 43 cm.

See: Joseph Smith Speer, *West-India pilot.* London: Hooper: 1771.
(728: [ca. 1771]: G353prh Small)

1535. 1774

. . . Chart of the West Indies. By Joseph Smith Speer. [London]: Capt. Speer: 1774 [20 May 1771].

Scale: 69½ British statute mi. = 1 degree.

Cartographer: Joseph Smith Speer.

Engraver: Thos. Bowen.

Size: 72.2 x ca. 118 cm.

Colored.

Provenance: Purchased from the library of William Priestman, 15 July 1831.

Contains engraving of the gold medal awarded to Speer by the Society of Arts and Commerce for his surveys and charts of the West Indies.
(720: 1774: Sp37cha Large)

1536. 1774

A compleat map of the West Indies, containing the coasts of Florida, Louisiana, New Spain, and Terra Firma: with all the islands. By Samuel Dunn, mathematician. London: Robert Sayer: 1774 [10 January 1774].

Scale: 69½ English mi. = 1 degree.

Cartographer: Samuel Dunn.

Size: 30.5 x 45 cm.

Colored.

This was removed from: *The American military pocket atlas.* R. Sayer and J. Bennett: [1776].

(720: 1774: D912cwi Small)

1537. 1779

Plan of the island of S[an]ta Lucia, in the West Indies. [Also] Plan of the Careenage Harbour. London: Speer: 1779 [16 March 1779]. 2 maps on one sheet.

Scale: 9.5 cm. = 6 mi.; 23 cm. = 2 mi.

Cartographer: Captain Speer.

Engraver: T. Bowen.

Size: 35.1 x 46.5 cm.

(735: 1779: B671stL Small)

1538. [1792]

Charta öfver canalerna och utloppen emellan Öarna fran St. Barthelemy till Dog och Prickle Pear efter de sednaste observationer utlagd 1772. Af Samuel Fahlberg. [Stockholm: 1792].

Scale: 6 cm. = 2,000 franska fot.

Cartographer: Samuel Fahlberg.

Engraver: B. Akerland.

Size: 54.5 x 64.4 cm.

Inset: Speciale-Charta öfver Inloppet och Hamnen Carenage samt Staden Gustavia.

Provenance: Presented and endorsed by Samuel Fahlberg, 4 April 1800.

It is dedicated to the "Kongl. Vetenskaps Academien." Fahlberg was the physician of the Swedish government in the island of St. Barthelemy.

(730: [1792]: F143bsa Large)

1539. [1793]

West Indies. [1793].

Scale: 60 mi. = 1 degree.

Engraver: S. Allardice.

Size: 17 x 28 cm.

Provenance: Presented by Thomas Dobson, 1798.

In Thomas Dobson, publisher, *Encyclopaedia,* vol. 9, following p. 118.

Wheat: 684

(032: En2)

1540. [ca. 1793]

West Indies, according to the best authorities. Boston: Thomas & Andrews: [ca. 1793].

Engraver: A. Doolittle.

Size: 19.8 x 31 cm.

Engraved for Morse's *American geography.*

Wheat: 686

(720: [ca. 1793]: T368wit Small)

1541. 1794

Plan of the island of St. Vincent laid down by actual survey . . . by John Byres chief surveyor 1776. London: Robert Wilkinson: 1794 [republished 14 July 1794].

Scale: Nearly 69½ British statute mi. = 1 degree.

Cartographer: John Byres.

Engraver: J [or I] Bayly.

Size: 94.2 x 61.5 cm.

Colored.

(735: 1794: B992stv Large)

1542. [ca. 1795]

West Indies according to the best authorities. N.p.: [ca. 1795].

Engraver: Gonne.

Size: 28.9 x 31.6 cm.

Insets: Harbour of Kingston and Port Royal in Jamaica; harbour of Port Royal in Martinico; and harbour of Havana in Cuba.

Provenance: Presented by Mathew Carey, 18 October 1805.

The *Transactions* (vol. 6) state that Carey presented the "Materials from which Guthrie's Geography were compiled." This is one of those maps. See: William Guthrie, *The general atlas for Carey's edition of his geography improved.* Philadelphia: Carey: 1795 [1 May 1795].

Engraved for Chamber's edition of Guthrie's new system of geography.

(720: [ca. 1795]: G583cwi Small)

1543. 1796

Carte de l'isle St. Domingue dressée pour l'ouvrage de M. L. E. Moreau de St. Méry. 1796.

Scale: 2.5 cm. = 9 lieues of 2,000 toises each.

Cartographer: I. Sonis.

Engraver: Vallance.

Size: 36.5 x 62 cm.

Provenance: Presented by Médéric L. E. Moreau de Saint Méry, 20 May 1796.

Published in Médéric L. E. Moreau de St. Méry, *Description topographique et politique de la partie Espagnole de l'isle Saint-Domingue,* opposite p. 1. Also English version, 1796.

Wheat: 693,694

(917.293: M81, and 917.293: M81c)

1544. [1797]

West Indies from the best authorities. [1797].

Scale: 2.5 cm. = 225 British statute mi.

Engraver: Hill.

Size: 18.6 x 33 cm.

In: Jedidiah Morse, *The American gazetteer,* opposite "West Indies."

Wheat: 699

(917.3: M83am)

1545. [ca. 1798]

Karte des Mexicanischen Meerbusens und der anstossenden Inseln und Laender zu Campens Entdekkung von America. [Hamburg: ca. 1798].

Scale: 69 English mi. = 1 degree.

Engraver: T. A. Pingeling.

Size: 20.3 x 31.9 cm.

(720: [ca. 1798]: P656kma Small)

1546. [1799]

West Indies. [1799].

Size: 18.5 x 32.8 cm.

In: John Payne, *A new and complete system of universal geography,* vol. 4, opposite p. 486.

Wheat: 700

(910: P29)

1547. [ca. 1800]

West Indies. N.p.: [ca. 1800].

Scale: British statute miles 69 to a degree.

Cartographer: Joseph Smith Speer (?).

Size: 27.8 x 46.5 cm.

(720: [ca. 1800]: W529bwi Small)

1548. [ca. 1806]

Carte de la partie françoise de St. Domingue faite par Bellin . . . et par P. C. Varlé et autres ingrs. [Philadelphia: ca. 1806].

Scale: 25 French common leagues of 2282 toises each.

Cartographers: Bellin and P. C. Varlé.

Engraver: J. T. Scott, Philadelphia.

Size: 36.9 x 48.3 cm.

Colored.

(725: [ca. 1806]: B411fph Small)

1549. 1807

A new chart of the Caribbean Sea and Isles, including the coast of Guayana; accurately drawn from authentic documents by R. Blanchford. London: Author: 1807 [14 April 1807].

Cartographer: R. Blanchford.

Size: 72.8 x 161.4 cm.

Insets: The island of Curazao; Berbice River; the river Surinam; Essequebo.

(701.6: 1807: B561csg Large)

1550. 1871

Mapa de la Isla de Santo Domingo . . . Philadelphia: Charles A. Poizat: 1871.

Scale: 60 geographical or sea mi. = 1 degree.

Cartographer: Robert H. Schomburck.

Size: 21.4 x 41 cm.

Photograph by F. Gutekunst, Philadelphia.

(724: 1871: Sch67pp Small)

1551. 1872

Geological map of the Republic of Santo Domingo by William M. Gabb, A. Pennell, C. Runnebaum and L. Pennell . . . Part of the province of Axua and all of Hayti copied from the map of Robert Schomburck. N.p.: 1872.

Scale: 1 in. = 12 English mi.

Cartographers: W. A. Gabb, J. A. Reade, J. de la C. Martinez, A. Pennell, C. Runnebaum, L. Pennell, and Robert Schomburck.

Size: 39.5 x 62.5 cm.

(725: 1872: G123sdg Large)

1552. 1911

Map of Cuba prepared in the War College Division, General Staff War Department. Based on military reconnaissances by the United States Army during the first and second occupation of Cuba, 1898-1902 and 1906-1909. Washington: 1911. 2 pieces.

Scale: 1 in. = 10 mi.

Cartographers: Frank B. Essex and A. B. Williams.

Size: each, 76 x 98.5 cm.

Colored.

(723: 1911: Un38rca Large)

1553. 1915

Pilot chart of the Central American waters. January 1915. Prepared from data furnished by the Hydrographic Office of the Navy Department and by the Weather Bureau of the Department of Agriculture. N.p.: 1915.

Size: 59.5 x 89.3 cm.

Colored.

Inset: Gulf of Mexico and Caribbean Sea, with surrounding areas.

(701.2: 1915: P643caw Large)

SOUTH AMERICA Nos. 1553a-1562

1553a. 1775

Mappa geográfico de America Meridional, dispueto y Gravado por D. Juan de la Cruz Cano y Olmedilla. 1775.

Scale: 17½ Spanish leagues = 1 degree.

Cartographer: Juan de la Cruz Cano y Olmedilla.

Engravers: Juan de la Cruz Cano y Olmedilla and Ricarte.

Size: 184 x 130.5 cm.

Provenance: Presented by Palaullo.

(740: 1775: C882sam Large)

1554. 1785

A map of South America according to the best authorities. London: C. Dilly & G. Robinson: 1785 [1 January 1785].

Scale: 69 British statute mi. = 1 degree.

Size: 34.2 x 37 cm.

Colored.

Provenance: Presented by Mathew Carey, 18 October 1805.

The *Transactions* (vol. 6) state that Carey presented the "Materials from which Guthrie's Geography were compiled." This is one of those maps. See: William Guthrie, *The general atlas for Carey's edition of his geography improved.* Philadelphia: Carey: 1795 [1 May 1795].

Engraved for William Guthrie, *Atlas to his system of geography* . . . London: Dilly: 1785.

(740: 1785: D582gga Small)

1555. [ca. 1790]

A map of South America according to the best authorities. N.p.: [ca. 1790].

Scale: 69 British statue mi. = 1 degree.

Size: 33.6 x 36.2 cm.

Provenance: Presented by Mathew Carey, 18 October 1805.

The *Transactions* (vol. 6) state that Carey presented the "Materials from which Guthrie's Geography were compiled." This is one of those maps. See: William Guthrie, *The general atlas for Carey's edition of his geography improved.* Philadelphia: Carey: 1795 [1 May 1795].

Engraved for Chambers's edition of Guthrie's *New system of geography.*

(740: [ca. 1790]: So87ces Small)

1556. [1790]

South America from the best authorities. [1790].

Engraver: Scot.

Size: 17.8 x 21.4 cm.

Provenance: Presented by Thomas Dobson, 1798.

In: Thomas Dobson, publisher, *Encyclopaedia,* vol. 1, following p. 538, after the map of North America.

Wheat: 701

(032: En2)

1557. [ca. 1793]

South America. Boston: Thomas & Andrews: [ca. 1793].

Engraver: A. Doolittle.

Size: 16.1 x 20.7 cm.

Engraved for Morse's *American geography.* [*The American universal geography*]. I, opposite p. 642.

Wheat: 706

(740: [ca. 1793]: D722sua Small)

1558. 1797

A map of South America and the adjacent islands. 1797.

Engraver: Callender.

Size: 16.1 x 11.2 cm.

In: Jedidiah Morse, *The American gazetteer,* opposite "America."

Wheat: 714

(917.3: M83am)

1559. 1799

South America from the best surveys. 1799.

Engraver: Scoles.

Size: 18.6 x 21.8 cm.

In: John Payne, *A new and complete system of universal geography,* vol. 4, opposite p. 461.

Wheat: 719

(910: P29)

1560. 1815

Carte encyprotype, de l'Amérique Méridionale, dressée par H. Brué. Paris: Desray: 1815.

Scale: 69½ mi. = 1 degree.

Cartographer: Adrien Hubert Brué.

Size: 109.5 x 134 cm.

Colored.

Inset: Chile, Argentina, and the Falkland Islands [the southern part of the continent].

See: Adrien Hubert Brué, *Atlas universel.* Paris: Desray: 1816.

(740: 1815: B831cam Large)

1561. 1839

South America from original documents, including the survey by the officers of H.M. Ships Adventure and Beagle. Dedicated to Captain R. Fitz Roy, R.N. by John Arrowsmith. London: John Arrowsmith: 1839.

Scale: 69 British mi. = 1 degree.

Size: ca. 61 x 50 cm.

Insets: Galapagos Islands; Port San Carlos; Falkland Islands; Southern tip of South America.

Removed from *Narrative of the surveying voyages of His Majesty's ships Adventure and Beagle* . . . London: 1839.

(700: 1839: So88abe Small)

1562. 1927-1935

[Maps of South America, Central America, West Indies and Mexico] compiled and drawn by the American Geographical Society of New York. Baltimore: A. Hoen: 1927-1935. 10 maps.

Scales: 1:1,000,000; 1:400,000.

Sizes: 46 x 60 cm.; 46 x 71 cm.

Contents: 1. Atacama, Chile; 2. Panama; 3. Barranquilla, Colombia; 4. Santo Domingo-San Juan; 5. Para, Brazil; 6. Cuyaba, Brazil; 7. Paranapanema, Brazil; 8. Chihuahua, Mexico; 9. Tierra del Fuego; 10. Cordoba-Santa Fe, Argentina.

(700: 1927-1935: Am31his Large)

CENTRAL SOUTH AMERICA Nos. 1563-1564

1563. [1855]

[Track surveys of the tributaries of the Rio la Plata]. By Commander Thomas J. Page, *U.S.S. Water Witch.* 1855. Washington: 1855. 19 maps.

Scale: 1:100,000; 1:2,000,000.

Cartographer: Thomas J. Page.

Engraver and lithographer: G. Stern and J. Bien.

Sizes: ca. 95.6 x 67.5 cm.; and no. 19: 100.5 x 31.5 cm.

Provenance: Presented by T. J. Page on 16 October and 18 December 1857, and on 2 April and 15-16 October 1858.

Contents: 1. Track survey of the River Uruguay with portions of the Rios Negros and Gualeguaychu. Sheet no. 1; 2. . . . Sheet no. 2. Laguna and Sausal Islands to Salto Grande; 3. Track survey of the Rivers Salado, Parana and Colastiné; 4. Mouths of the Parana and Uruguay. Sheet no. 1. Martin Garcia and Martin Chico Channels; 5. Track survey of the River Parana. Sheet no. 2. (Buenos Aires to Paloma Island); 6. . . . Sheet no. 3. (Paloma Island to Curumbé Island); 7. . . . Sheet no. 4. (Curumbé Island to Arroyo Soldado); 8. . . . Sheet no. 5. (Arroya Soldado to Acollardo Island); 9. . . . Sheet no. 6. (Acollardo Island to mouth of Colastiné); 10. . . . Sheet no. 7. (Mouth of Colastiné to lat. 29° 40"); 11. . . . Sheet no. 8. (Lat. 29° 55" to lat. 28° 40"); 12. . . . Sheet no. 9. (Lat. 28° 40" to Guardia Cerrito); 13. Track survey of the River Paraguay. Sheet no. 10. (From Guardia Cerrito to lat. 26° 5" near Oliva); 14. . . . Sheet no. 11. (Oliva to Rio Paraguaymy); 15. . . . Sheet no. 12. (Rio Paraguaymy to Concepcion); 16. . . . Sheet no. 13. (From Concepcion to Lat. 21° 40"); 17. . . . Sheet no. 14. (From Lat. 21° 40" to 20° 10"); 18. . . . Sheet no. 15. (From Lat. 20° 10" to Curumba); 19. Reference chart to the track survey of the tributaries of Rio la Plata.

(744: [1855]: Un36tLp Large)

1564. [1856]

Map of the basin of La Plata, based upon the results of the expedition under the command of Thomas J. Page, U.S. Navy, in the years 1853, '54, '55 & 56. And of the adjacent countries. Compiled from the best authorities. U.S. Navy Department. [New York: 1856].

Scale: 1:3,000,000.

Cartographer: Thomas J. Page.

Engraver: J. Bien.

Size: 90.8 x 65.6 cm.

Colored.

Provenance: Presented by Thomas J. Page, 17 September 1858.

See: Track surveys . . . map no. 1563.

(744: [1856]: Un36bLp Book map)

COLOMBIA Nos. 1565-1568

1565. [ca. 1771]

Island of old Providence. N.p.: [ca. 1771].

Engraver: Prinald.

Size: 17.7 x 14.5 cm.

See: Joseph Smith Speer, *West-India pilot.* London: Hooper: 1771.

(752.74: [ca. 1771]: P936iop Small)

1566. 1771

Plan of Carthagena Harbour and city in lattd. 0/10:26. to 0/10:15 north. London: S. Hooper: 1771 [1 October 1771].

Engraver: Bayly.

Size: 47.3 x 34.1 cm.

See: Joseph Smith Speer, *West-India pilot.* London: Hooper: 1771.

(752.923: 1771: B341chc Small)

1567. [ca. 1771]

[Plan of the city and suburbs of Carthagena]. N.p.: [ca. 1771].

Scale: 9 in. = 400 toises.

Engraver: J. Gibson.

Size: 25 x 32.8 cm.

Along left side is list of references for the plan of the city.

See: Joseph Smith Speer, *West-India pilot.* London: Hooper: 1771.

(752.923: [ca. 1771]: G353pcs Small)

1568. [ca. 1860-1861?]

Interoceanic ship canal via the Atrato and Truando rivers. N.p.: [ca. 1860-1861?]

Size: 56 x 114 cm.

To accompany U.S. 36th Congress. 2nd Session Senate Executive Documents, No. 9.
(752: [ca. 1860-1861?]: In83cat Small)

VENEZUELA No. 1569

1569. 1950

Geologic-tectonic map of the United States of Venezuela (except the territory of Amazonas and part of the state of Bolivar), compiled by Walter H. Bucher, Columbia University. Published by the Geological Society of America. Washington: Williams & Heintz Co.: 1950.

Scale: 1 in. = 25 mi.

Cartographer: Walter H. Bucher.

Size: 81.5 x 147 cm.

See: W. H. Bucher, *Geologic structure and orogenic history of Venezuela.* [New York]: Geological Society of America: 1952.

(753: 1950: V558geo Roller)

PERU No. 1570

1570. 1861-1913

[Collection of maps of the rivers, cities, etc., of Peru]. Sociedad Geográfica de Lima, Peru. Lima: 1861-1913. 31 maps.

Scale: Various

Cartographer: Fitz-Carrald.

Size: Various.

Colored.

Contents: 1. Mapa popular del Peru; 2-3. Mapa del departamento de Loreto; 4. Plano del rio Ucayali; 5. Vias de comunicacion entre Paita, Eten y parte navegable del rio Alto Marañon; 6. Mapa del Alto Yurua y Alto Purus; 7. Mapa de la provincia de Pacasmayo; 8. Plano del Rio Tigre; 9. Plano del Rio Napo; 10. Plano del Rios Pachitea y Pichis; 11. Lago Titicaca . . .; 12. . . . los Rios Tigre, Pastaza y Morona; 13. Mapa que comprende las ultimas exploraciones y estudios verificados desde 1900 hasta 1906; 14. Plano del Rio Putumayo; 15. Plano del Rio Bajo Marañon; 16. Plano del Rio Serjali; 17. Plano del Cuzco; 18. Plano del Rio Huallago; 19. Plano del Rio Tambopata; 20. Mapa histórico geográfico de los valles de Paucartambo; 21. Plano del Rio Pastaza. 22. Plano del Rio Manú. 23. . . . Casma á Chacas; 24. Region oriental del Peru; 25. Plano del Rio "Ucayali" . . .; 26. Geologia de los inmediaciones de Morococha; 27. Croquis de los Rios Napo y Putumayo; 28. . . . Navegacion del Alto Marañon; 29. Plano del Rio Amazonas Peruano; 30. Croquis de los rios Alto Ucayali y Bajo Urubamba; 31. Croquis general de los rios estudiados por la comision exploradora del Istmo Fitz-Carrald.

(764.2: 1861-1913: P436geo Large)

BOLIVIA No. 1571

1571. 1901

Mappa de la República de Bolivia mandado organizar y publicar por el Presidente Constitucional, General José Manuel Pando. Formado por Eduardo Idiaquez. Reprinted by the International Bureau of the American Republics. Washington: Graham: 1901.

Scale: 1:2,000,000.

Cartographers: Eduardo Idiaquez and Victor Puig.

Engraver: Erhard Hermanos, Paris.

Size: 104.6 x 89.8 cm.

Colored.

(766: 1901: Id14mrb Large)

BRAZIL Nos. 1572-1574

1572. 1905-1911

Folhas topographicas. Commissão Geographica e Geologica de S. Paulo. São Paulo: 1905-1911. 9 sheets.

Scale: 1:100,000.

Size: Various.

Colored.

Contents: 1. Folha de Bragança. Preliminary edition, 1909; 2. Folha de Casa Branca. Preliminary edition, 1905; 3. Folha de S. Bento. Preliminary edition, 1909; 4. Folha de Pindamonhangaba. Preliminary edition, 1905; 5. Folha de Pirassununga. Preliminary edition, 1905; 6. Folha de São Paulo. Preliminary edition, 1905; 7. Folha de Franca. Preliminary edition, 1911; 8. Folha de S. Sebastião do Paraizo. Preliminary edition, 1911; 9. Folha de Mococa. Preliminary edition, 1911.

(774.8: 1905-1911: Sa67top Large)

1573. 1908

Carta general do Estado de S. Paulo. Organisada pela Commissão Geographica e Geologica. São Paulo: 1908.

Scale: 1:1,000,000.

Size: 66.2 x 106.2 cm.; 1910 map, 33 x 52.4 cm.

Colored.

Inset: Planta da Ciudade de S. Paulo.

Contains also editions of 1910 and 1912.

(774.8: 1912: Sa67geo Large)

1574. [ca. 1913]

Geological sketch map of the region about Natal, Rio Grande do Norte, Brazil. By Olaf P. Jenkins. N.p.: [ca. 1913].

Scale: 2.5 cm. = 2 mi.

Cartographer: Olaf P. Jenkins.

Size: 49.2 x 48.5 cm.

Inset: Index map.

(770: [ca. 1913]: J425nat Large)

CHILE Nos. 1575-1577

1575. 1839

The Strait of Magalhaens, commonly called Magellan. Surveyed by the officers of His Majesty's Ships, Adventure and Beagle. Under the direction of Captains Phillip Parker King, F.R.S., Pringle Stokes & Robert Fitz Roy. 1826-34. London: Henry Colburn: 1839.

Cartographers: Phillip Parker King, Pringle Stokes, and Robert Fitz Roy.

Engraver: J. Gardner.

Size: 31.4 x 46.9 cm.

Removed from *Narrative of the surveying voyages of His Majesty's ships Adventure and Beagle* . . . London: 1839.

(782.669: 1839: K588abe Small)

1576. 1839

Part of Tierra del Fuego from H.M.S. Beagle. 1834. London: Henry Colburn: 1839.

Engraver: J. & C. Walker.

Size: 31.4 x 47.1 cm.

Removed from *Narrative of the surveying voyages of His Majesty's ships Adventure and Beagle* . . . London: 1839.

(782: 1839: P255bea Small)

1577. 1839

Chiloe and parts of the adjacent coasts from H.M.S. Beagle. 1835. London: Henry Colburn: 1839.

Engraver: J. & C. Walker.

Size: 46.6 x 31.8 cm.

Removed from *Narrative of the surveying voyages of His Majesty's ships Adventure and Beagle* . . . London: 1839.

(782: 1839: C435bea Small)

ARGENTINA Nos. 1578-1580

1578. 1839

Southern portion of South America. London: Henry Colburn: 1839.

Scale: 1 in. = 180 geographic mi.

Engraver: J. Dower.

Size: 33.1 x 35.6 cm.

Removed from *Narrative of the surveying voyages of His Majesty's ships Adventure and Beagle* . . . London: 1839.

(780: 1839: So88sop Small)

1579. 1900

Mapa de los ferrocarriles de la republica Argentina. Ministerio de Obras Publicas. N.p.: 1900.

Size: 25.5 x 18.8 cm.

Colored.

(784: 1900: Ar4lrra Small)

1580. 1909

Maps of the Argentine Republic. N.p.: 1909. 4 maps.

Scale: 1:8,500,000.

Size: 46.4 x 29 cm.; the hyposometric card is 26.9 x 21.1 cm.

Colored.

Contents: map of the Argentine Republic; map of the meteorological stations; map of the hydrometrical stations: hypsometric card.

(744: 1909: Ar11rep Small)

AUSTRALIA Nos. 1581-1610

SOUTH AUSTRALIA Nos. 1581-1583

1581. [1892]

Map showing the explorations and discoveries in south Australia and Western Australia made by the Elder Scientific Exploring Expedition originated and equipped by Sir Thomas Elder . . . commanded by David Lindsay; 1891-1892. Royal Geographical Society of Australiasia. Adelaide: A. Vaughan, Litho.: [1892].

Scale: 1 in. = 15 mi.

Cartographers: D. Lindsay and L. A. Wells.

Size: 69.2 x 151 cm.

(810: [1892]: R817esx Large)

1582. 1898

Geological map of the Northern Territory of South Australia, prepared . . . by H. Y. L. Brown . . . Physical geography compiled by C. Winnecke . . . from private and official records. Surveyor General's Office, South Australia. Adelaide: A. Vaughan: 1898.

Scale: 1 in. = 20 mi.

Cartographer: H. Y. L. Brown.

Size: 132.5 x 73 cm.

Colored.

(814: 1898: So87ter Large)

1583. 1899

Geological map of South Australia, prepared . . . by H. Y. L. Brown. Surveyor General's Office, South Australia. Adelaide: 1899.

Scale: 1 in. = 16 mi.

Cartographer: H. Y. L. Brown.

Size: 142.5 x 128 cm.

Colored.

(814: 1899: So87ade Large)

WESTERN AUSTRALIA Nos. 1584-1585

1584. 1906-1907

[Maps belonging to *Bulletins* nos. 21, 24, 28, 29 of the Geological Survey of Western Australia]. Perth: 1906-1907.

Scales: Various.

Sizes: Various.

Contents *Bulletin,* no. 21: 1. Princess Royal Gold Mine; 2. Norseman Gold Mines; 3. Cumberland Gold Mine; 4. Lady Mary & Valkyrie Gold Mine; 5. Geological & topographical map of Norseman Gold Mines.

Contents *Bulletin,* no. 24: 6. Geological sketch map of Laverton; 7. Geological sketch map of Lancefield; 8. Geological sketch map of Heaphy's Find; 9. Geological sketch map of Burtville.

Contents *Bulletin,* no. 28: 10. Geological sketch map of Lawlers; 11. Geological sketch map of Sir Samuel; 12. Great Eastern Gold Mine; 13. Never-Can-Tell Gold Mine; 14. Waroonga Gold Mine; 15. Vivien Gold Mine; 16. Bellevue Gold Mine.

Contents *Bulletin,* no. 29: 17. Kangaroo Mine; 18. Leviathan Mine; 19. Jubilee Mine; 20. Salisbury Mine; 21. Cue No. 1 group; 22. Princess Ada Mine; 23. Cue Victory Mine; 24. Light of Asia; 25. Queen of the May Mine; 26. Agamemnon Mine; 27. Princess Royal Mine; 28. Lady Forrest Mine; 29. Polar Star Mine; 30. Golden Gate Mine; 31. Victory United Mine; 32 and 33. Great Fingall Mine, nos. 2, 3; 34. Great Fingal Associated Mine, nos. 1 & 2; 35 and 36. Rubicon Mine; 37. Croesus Mine; 38. Great Fingal Consolidated . . . 39. Trenton Mine; 40. East Fingall Mine; 41. Kinsella Mine.

(813: 1906-1907: W529gms Large)

1585. 1911

Topographical map of Meekatharra, by H. W. B. Taylor . . . [Geological Survey of Western Australia]. Perth: H. J. Pether: 1911.

Scale: 1 in. = 40 chains.

Cartographer: H. W. B. Taylor.

Lithographer: H. J. Pether.

Size: 98.4 x 47.5 cm.

Inset: Locality plan.

Provenance: Presented by H. W. B. Taylor.

(813: 1911: W529top Large)

QUEENSLAND Nos. 1586-1605

1586. 1898

Geological map of Charters Towers Goldfield, Queensland. Geological lines by R. L. Jack, W. H. Rands and A. Gibb Maitland. Topography by William Thompson . . . Second edition shewing underground workings. Geological Survey of Queensland. Brisbane: Government Engraving & Lithographic Office, Queensland: 1898. 6 sheets.

Scale: 1 in. = 4 chains.

Cartographers: R. L. Jack, W.H. Rands, A. Gibb Maitland, and William Thompson.

Lithographers: A.A. Wright and H. W. Fox.

Size: 169.9 x 196 cm.

Colored.

Inset: Millett's Wellington.

(815: 1898: Q37geo Large)

1587. 1898

Map of the Etheridge Goldfield, Queensland. Geological Survey of Queensland. Surveyed by T. R. Geraghty . . . No. 137 of Geological Society Publications. Georgetown: 1898. 4 sheets.

Scale: 1 in. = 1 mi.

Cartographer: T. R. Geraghty.

Size: 146.8 x 103.9 cm.

(815: 1898: Q37eth Large)

1588. 1899

Geological map of Queensland compiled by Robert Jack . . . government geologist, from the latest information available, including surveys of Robert L. Jack, William H. Rands, A. Gibb Mitland, S. B. J. Skerchly, B. Dunstan, W. E. Cameron and L. C. Green . . . Issued by authority of the . . . Minister for Mines. Geological Survey of Queensland. London: Wyman & Sons: 1899. 6 sheets.

Scale: 16 statute mi.

Cartographers: Robert L. Jack, William H. Rands, A. Gibb Maitland, S. B. J. Skerchly, B. Dunstan, W. E. Cameron, and L. C. Green.

Size: 189.4 x 174.2 cm.

Colored.

Insets: Map of islands annexed to & forming part of the Colony of Queensland; Cape York.

(815: 1899: Q37mine Large)

1589. 1899

Geological Survey of Queensland. Geological map of part of Gympie Goldfield by William H. Rands. 1899. Brisbane: Edmund Gregory: 1899. 3 sheets.

Scale: 1 in. = 4 chains

Cartographer: William H. Rands.

Lithographers: H. W. Fox and A. McLaren.

Size: 95 x 189 cm.

Colored.

(815: 1899: Q37gymp Large)

1590. 1902

Geological sketch map of Queensland prepared under the supervision of B. Dunstan . . . compiled by H. W. Fox. No. 182 of the Geological Survey of Queensland Publications. Brisbane: George Arthur Vaughan: 1902.

Scale: 1 in. = 20 mi.

Cartographers: B. Dunstan and H. W. Fox.

Lithographer: H. W. Fox.

Size: 87.2 x 66.7 cm.

Colored.

(815: 1902: Q37rept Large)

1591. 1904

Sketch map of the Walsh & Tinaroo mineral field, prepared at the Geological Survey Office, from the latest surveys of the lands, railways and mines departments, under the supervision of Walter E. Cameron. No. 187 of the Geological Survey of Queensland publications. Brisbane: G. A. Vaughan: 1904. 2 sheets.

Scale: 1 in. = 2 mi.

Cartographers and lithographers: H. W. Fox and W. E. Cameron.

Size: 77.7 x 124.8 cm.

Inset: Locality map.

(815: 1904: Q37tina Large)

1592. 1905

Geological sketch map of Queensland showing mineral localities prepared under the supervision of B. Dunstan . . . 1905 . . . and compiled by H. W. Fox. No. 206 of Geological Survey of Queensland Publications. Brisbane: Government Printing Office: 1906.

Scale: 1 in. = 20 mi.

Cartographers: B. Dunstan and H. W. Fox.

Lithographer: H. W. Fox.

Size: 87.2 x 66.4 cm.

Colored.

(815: 1905: Q37mine Large)

1593. 1905

Geological sketch map of Queensland showing mineral localities prepared under the supervision of B. Dunstan . . . and compiled by H. W. Fox. Brisbane: 1905. Revised to December 1907. No. 217 of Geological Survey of Queensland Publications. Brisbane: 1908. 3rd ed.

Scale: 1 in. = 20 mi.

Cartographers: B. Dunstan and H. W. Fox.

Size: 87 x 66.7 cm.

Colored.

(815: 1905: Q37Loca Large)

1594. 1905

Plan 2 showing underground workings on the Golden Gate line of reef. Illustrating the report on "Some Croydon gold mines" by B. Dunstan. Geological Survey of Queensland. Brisbane: Government Printing Office: 1905. 4 sheets.

Scale: 1 in. = 100 ft.

Cartographers: H. W. Fox and B. Dunstan.

Size: 63.4 x 339.6 cm.

Colored.

See: Queensland Geological Survey, *Publication.* Brisbane: 1905, No. 202.

(815: 1905: Q37goLg Large)

1595. 1907

Copper mining district of Cloncurry north western Queensland based on Queensland 4 mile map issued by Department of Public Lands and showing freeholds & mineral leases . . . under the supervision of L. C. Ball . . . No. 213 of the Geological Survey of Queensland Publications. [Brisbane]: 1907.

Scale: 1 in. = 6 mi.

Cartographer: L. C. Ball.

Lithographer: A. McLaren.

Size: 93.5 x 57.3 cm.

Colored.

Index, etc., on left side and bottom of map.

(815: 1907: Q37gLoy Large)

1596. 1908

Queensland showing principal mining centres and railways. B. Dunstan . . . geologist. Geological Survey of Queensland. Brisbane: 1908.

Scale: 1 in. = 100 mi.

Cartographers: B. Dunstan and W. H. Greenfield.

Size: 41 x 29.3 cm.

Colored.

(815: 1908: Au71aqs Small)

1597. 1908

Sketch map of the Etheridge Goldfield prepared at the Geological Survey Office, Department of Mines. By W. H. Greenfield. Geological Survey Publications no. 213. Brisbane: Government Printing Office: 1908.

Scale: 1 in. = 6 mi.

Cartographer: W. H. Greenfield.

Size: 66.8 x 95.2 cm.

Colored.

Inset: The Knobs Oaks Rush.

On right side is Index of mineral leases, etc.

(815: 1908: Q37sur Large)

1598. 1909

Sketch map of the Herberton & Chillagoe gold and mineral fields compiled from official and other sources under the supervision of W. H. Greenfield, at the Geological Survey Office, Department of Mines. No. 220 of Geological Survey Publications. [Brisbane]: 1909.

Scale: 1 in. = 6 mi.

Cartographers: E. F. Eberhardt and W. H. Greenfield.

Size: 96.3 x 73 cm.

Colored.

(815: 1909: Q38herb Large)

1599. 1910

Sketch map of east central Queensland gold, mineral and coal fields between Rockhampton, Gladstone and the Dawson & Mackenzie Rivers . . . by W. H. Greenfield. No. 226 of Geological Survey Publications. Brisbane: Government Printing Office: 1910.

Scale: 1 in. = 4 mi.

Cartographers: W. H. Greenfield and E. F. Eberhardt.

Size: 97.4 x 73.4 cm.

Colored.

Inset: Showing the Dawson and Mackenzie coalfield area.

(815: 1909: Q37coaL Large)

1600. 1911

Sketch map of Cape York peninsula gold and mineral fields. Palmer, Cooktown, Starcke, Alice River, Hamilton, Coen, Rocky River, Bowden, Potallah, Lochinvar and Hayes Creek. Prepared at the Geological Survey Office, Department of Mines . . . by W. H. Greenfield. Publication no. 233. Brisbane: Anthony J. Cumming: 1911.

Scale: 1 in. = 6 mi.

Cartographers: W. H. Greenfield and D. Fox.

Size: 96.8 x 73.6 cm.

Colored.

Inset: Bowden M.F.

(815: 1911: Q37york Large)

1601. 1911

Sketch map of goldfields in the vicinity of Charters Towers. Charters Towers, Ravenswood, Windsor, Carrington, Strathalbyn, Piccadilly and part of Star River mineral field. Prepared at the Geological Survey Office, Department of Mines . . . Publication no. 236. Brisbane: Government Printing Office: 1911.

Scale: 1 in. = 4 mi.

Size: 56.2 x 83.6 cm.

Colored.

(815: 1911: Q37goLd Large)

1602. 1911

Sketch map of the Croydon and Etheridge goldfields prepared at the Geological Survey Office, Department of Mines, by W. H. Greenfield . . . Publication no. 230 [Queensland Geological Survey]. Brisbane: A. J. Cumming: 1911.

Scale: 1 in. = 6 mi.

Cartographer: W. H. Greenfield.

Size: 66.5 x 86 cm.

Colored.

On right side is index, etc.

(815: 1911: Q37cret Large)

1603. 1919

Geological map of Charters Towers and environs to accompany report on the Charters Towers goldfield. Geological Survey of Queensland. Publication no. 256. By John H. Reid. N.p.: 1919. 6 sheets.

Scale: 1:6,336.

Cartographers: John H. Reid and Cecil C. Morton.

Size: Each 61.4 x 69.9 cm.

Colored.

See publication 244.

(815: 1919: R274ctg Small)

1604. n.d

Ipswich coalfield map. No. 1. Geological Survey of Queensland. N.p.: n.d.

Scale: 1 in. = 1 mi.

Cartographers: W. E. Cameron and T. H. C. Bath.

Size: 53.5 x 59 cm.

Colored.

Report 271. Contains also below ground drawings of strata.

(815: n.d.: Ip65cLd.1 Small)

1605. n.d.

Ipswich coalfield. Map. no. 2. Geological Survey of Queensland. N.p.: n.d. 6 sheets.

Scale: 1 in. = 20 chains.

Cartographers: W. E. Cameron, T. C. Naylor, A. C. Reid, and T. H. C. Bath.

Size: Each 43.6 x 36.3 cm.

Colored.

See report 271.

(815: n.d.: Ip65cLd.2 Small)

VICTORIA No. 1606

1606. 1875

Victoria. Compiled & engraved at the Department of Lands and Survey, Melbourne, under the direction of A. J. Skene. Melbourne: 1875.

Scale: 3.1 cm. = 10 mi.

Cartographers: A. J. Skene and S. B. Bonney.

Engraver: James Slight.

Size: 56.8 x 83.2 cm.

Inset: Australia and adjacent islands.

(816: 1875: Au71dsv Large)

NEW SOUTH WALES Nos. 1607-1610

1607. 1903-1905

Geological map of the Gerringong District (with sections), [and] geological map of Little Forest and Conjola, with sections . . . Records. Geological Survey of New South Wales. Sydney: W. A. Gullick: 1903-1905. 3 maps.

Scales: Various.

Sizes: Various.

Colored.

See: New South Wales Geological Survey, *Records* . . . Sydney: 1905-1909. Vol. 8, pt. 2.

(817: 1903-1905: N426geo Large)

1608. 1903

Geological sketch map of the country in the vicinity of Sydney. [Geological Survey of New South Wales]. Prepared under the direction of E. F. Pittman based on the work of the Revd. W. B. Clarke. Sydney: W. A. Gullick: 1903.

Scale: 4 cm. = 5 mi.

Cartographers: E. F. Pittman and W. B. Clarke.

Lithographer: W. A. Gullick.

Size: 73.3 x 71.8 cm.

Colored.

(817: 1903: N426syd Large)

1609. 1914

Geological map of New South Wales . . . Prepared under the direction of E. F. Pittman . . . Department of Mines, Sydney, New South Wales, 1914. Sydney: 1914. 2 sheets.

Scale: 1 in. = 16 mi.

Cartographer: E. F. Pittman.

Size: 109.5 x 133.1 cm.

Colored.

Inset: Australia.

(817: 1914: N426min Large)

1610. [ca. 1922]

[Maps, plans and sections to accompany geological memoir no. 8: "The geology of the Broken Hill district."] New South Wales Geological Survey. Sydney: [ca. 1922]. 21 maps.

Cartographer: E. C. Andrews, et al.

Sizes: Various.

Colored.

Geological maps: 1. That portion of the Broken Hill District which contains the principal silver, lead, and zinc deposits, 2. Northerly termination of the Broken Hill Basin at Poole Hills, 3. Broken Hill and immediate neighbourhood (3 sheets), 4. The Broken Hill lode and the country in the immediate vicinity. (2 sheets), 5. The country in the vicinity of the de Bavay Fault, Broken Hill, 6. The Pinnacles, 7. "The Sisters" lode, 17 miles north-easterly from Broken Hill, 8. The Broken Hill lode, including: (a) Longitudinal section showing profile of Hill surface. (b) Plan of floor levels. (c) Selected sections across the Broken Hill lode (2 sheets), 9. Geological plans of nos. 6 and 7 levels, south Blocks Mine, Zinc Corporation Ltd., 10. Geological plan of 1,400 ft. level, North Broken Hill Ltd., 11. Broken Hill south mine. Plans of main levels, showing distribution of the principal minerals in the ore body.

Sections: 12. Geological section across the Barrier Range, Broken Hill District, 13. North Broken Hill mine, 14. Junction North Mine, main shaft, and British Mine, marsh shaft, 15. British Mine, Thompson shaft, 16. British Mine, Blackwood shaft, 17. Block 14 mine, 18. Broken Hill proprietary mine: (a) Delprat and Stewart shafts. (b) McBryde shaft, 19. Central Mine, Sulphide Corporation, Ltd., 20. Broken Hill south mine, 21. South Blocks mine, Zinc Corporation Ltd.

(817: [c. 1922]: An21bhd Large)

OCEANS—World Nos. 1611-1615

1611. 1849-1860

Maury's wind and current charts. 3rd ed. 1849. No. 2. Series A. N.p.: 1849-1860.

Cartographer: M. F. Maury.

Size: Ca. 40 x 26 cm.

(551:47: m44 Oversize vertical and horizontal)

1612. [1889?]

Mean barometrical pressure for the year; and Range of mean barometrical pressure for the year. London: [1899?].

Size of paper: 66.6 x 49.8 cm.

This is "Supplement to charts showing the mean barometrical pressure over the Atlantic, Indian and Pacific Oceans. (Official No. 76.) Issued by the authority of the Meteorological Council."

(980: [1889?]: M568mbp Large)

1613. [ca. 1910-1920]

Carte générale bathymétrique des océans dressée par ordre de S.A.S. le Prince de Monaco d'après le mémoire de M. le Professeur Thoulet . . . sous le direction de M. Charles Sauerwein, enseigne de vaisseau. Par M. Tollemer. Paris: Erhard Frères: [ca. 1910-1920]. 25 maps and title page.

Scale: 1:10,000,000.

Cartographers: Thoulet, Charles Sauerwein, Tollemer, Battaille, Bolzé, Lebas, Lévéque, Morell, and Normand.

Engraver: Erhard Frères.

Sizes: Various.

Colored.

(900: [ca. 1910-1920]:T393gmo Large)

1614. 1935

Carte générale bathymétrique des océans. Bureau Hydrographique International, Monaco. Monaco: 1935. 4 sections.

Size: 58.5 x 100 cm.

"La lère et la 2ème édition ont été publiées par ordre de S.A.S. le prince Albert ler de Monaco."

(900: 1935: M745hyd Roller)

1615. v.d.

U.S. Hydrographic Office. Pilot charts. Washington: various dates.

Provenance: Presented by the U.S. Navy, Hydrographic Office, as they were pulled.

(900: Un3p Extra-oversize)

PACIFIC OCEAN Nos. 1616-1618

1616. 1799

A chart shewing the tract of Capt. Cook's last voyage. 1799.

Scale: (2.5 cm. = ca. 19° latitude).

Size: 20.4 x 25.7 cm.

In: John Payne, *A new and complete system of universal geography,* vol. 4, opposite p. 521.

Wheat: 914

(910: P29)

1617. 1934

The voyages of Captain Cook. London: A. & C. Black: 1934.

Cartographer and engraver: Emery Walker, Ltd.

Size: 35.5 x 55.7 cm.

Provenance: Presented by Thomas D. Cope, 9 December 1949.

Photostat, taken from *Exploration of the Pacific,* by J. C. Beaglehole [London: A. C. Black: 1934].

(910: 1934: B35lvcc Small)

1618. n.d.

Maury's wind and current chart. Whale sheet No. 2. Series F. U.S. Hydrographical Office. N.p.: n.d.

Cartographer: M. F. Maury.

Size: 59.8 x 90.6 cm.

Colored.

Contains part of the book plate of J. Peter Lesley.

(900: n.d.: M444wcc Small)

POLYNESIA Nos. 1619-1620

1619. 1780

Carte de la Polynesie ou la cinquième partie de la terre [title also in Swedish]. Par Daniel Djurberg. Stockholm: 1780.

Cartographer: Daniel Djurberg.

Size: 47.5 x 71.8 cm.

Colored.

Inset: Easter and S. Daniel Islands.

Shows tracks of Captain Cook, de Surville, Tasman, Roggewin, Wallis, Bougainville, Biron, Mendonna, Mendarina, La Maire, and Carteret.

(940: 1780: D642ppt Large)

1620. 1839

Keeling Islands. H.M.S. Beagle. London: Henry Colburn: 1839.

Engravers: J. & C. Walker.

Size: 31.5 x 46.6 cm.

Insets: South Keeling Islands; North Keeling Islands.

Removed from *Narrative of the surveying voyages of His Majesty's Ships Adventure and Beagle* . . . London: 1839.

(995.3: 1839: K257son Small)

ATLANTIC OCEAN Nos. 1621-1633

1621. [1771]

The present state of Europe & America or the Man in the Moon taking a view of the English Armade. [1771].

Size: 6.1 x 10.5 cm.

Provenance: One copy was presented by the Estate of Richard Gimbel, 1974.

In: *The plea of the colonies on the charges brought against them by Lord Mansfield* . . . [1771].

Wheat: 2 [or 3]

(973.3: P71)

1622. [ca. 1785]

Remarques sur la navigation de Terre-Neuve à New-York afin d'éviter les Courrants et les bas-fonds au sud de Nantuckett et du Banc de George. [Franklin's second chart of the Gulf Stream]. Paris: Le Rouge: [ca. 1785].

Cartographer: Benjamin Franklin.

Size: 31.7 x 36.2 cm.

See: Franklin Bache, "Where is Franklin's first chart of the Gulf Stream?" American Philosophical Society, *Proceedings,* vol. 76, p. 731, 1936; and Benjamin Franklin, "A letter from Dr. Benjamin Franklin to Mr. Alphonsus Le Roy . . . August 1785," American Philosophical Society, *Transactions,* vol. 2. Philadelphia: 1786.

(980: [ca. 1785]: F843sLn Small)

1623. [1786]

A chart of the Gulf Stream. [1786].

Scale: [2.5 cm. = ca. 4° latitude].

Engraver: James Poupard.

Size: 20.3 x 25.5 cm.

Inset: [Annual passage of herrings].

In: Benjamin Franklin, "A letter from Benjamin Franklin . . . " with "Remarks upon the Navigation . . . " as an appendix to Franklin's "A letter . . . " The "Annual passage of herrings" was published for: John Gilpin, "Observations on the annual passage of herrings." Both articles in: American Philosophical Society. *Transactions,* vol. 3.

(506:73: Am4)

1624. [1789]

Chart of the Gulf Stream. [1789].

Size: 18.5 x 21.7 cm.

In: Benjamin Franklin, "Remarks upon the navigation . . . Gulf Stream," *American Museum,* vol. 5, opposite p. 213.

Wheat: 723

(050: Am33)

1625. [1792]

[Chart of the North Atlantic Ocean. 1792].

Scale: [2.5 cm. = ca. 4° latitude].

Size: 19.9 x 41.2 cm.

Provenance: Presented by Jonathan Williams, Jr., 20 November 1795.

See: Jonathan Williams, Jr., "Memoir on the use of the thermometer in discovering banks, soundings, &c.," American Philosophical Society, *Transactions,* vol. 3, copy 4, opposite p. 84.

Wheat: 724

(506.73: Am4)

1626. [1792]

[Chart of the North Atlantic Ocean. 1792].

Scale: [2.5 cm. = ca. 4° latitude].

Size: 20.1 x 42.1 cm.

Provenance: One copy presented by Jonathan Williams, Jr., 19 November 1799.

See: Jonathan Williams, Jr., "Memoir on the use of the thermometer in discovering banks, soundings, &c.," American Philosophical Society, *Transactions,* vol. 3, copy 1, opposite p. 84.

Wheat: 725

(506.73: am4t; Pam., vol. 316, no. 1; and, 656: W67)

1627. 1815

Chart of North Atlantic Ocean with tracks of the shipping to West Indies, North America, etc. [Edinburgh]: 1815.

Engraver: J. Moffat.

Size: 50.1 x 63.3 cm.

Colored.

Drawn and engraved for *Thomson's new general atlas,* 1 May 1815.

(983: 1815: M735cna Small)

1628. [1816]

America to the British Channel, 1816. Navigation Chart no. 1. [London]: Davis & Dickson: [1816].

Scale: 1 in. = 300 mi.

Cartographer: I. Garnett.

Engraver: H. Frost.

Size: 11.2 x 36 cm. *Size of paper:* 26.5 x 37.9 cm.

Provenance: Presented by John Garnett prior to 1818.

On 20 November 1807, Garnett's nautical chart was ordered to be printed in the *Transactions.* This chart had been received 16 October 1807 as an essay for the Extra Magellanic Premium and the members "determined by ballot [it] to be worthy, and awarded $45.00, or a medal of the value thereof" to Garnett. He took the chart and had it published in New Brunswick, New Jersey, in 1807.

(983: [1816]: G163abc Small)

1629. 1837

Chart of Georges Shoal & Bank, surveyed by Charles Wilkes, Lieut. Commandant . . . in U.S. brig *Porpoise,* schooners *Maria* and *Hadassah* . . . Published under the direction of the Navy Commissioners, 1837. N.p.: 1837.

Cartographers: C. Wilkes, J. A. Alden, and W. May.

Engravers: S. Stiles, Sherman, Smith.

Size: 99.1 x 110.8 cm.

Provenance: Presented by the Athenaeum of Philadelphia, 4 January 1839; and J. Forsyth for Capt. Wilkes, 16 August 1839.

Original in U.S. Congress, *Collection of maps* etc., published by order of Congress, Washington, 1843, no. 110.

(983: 1837: Un36gs Small)

1630. 1893

Wreck chart of the North Atlantic coast of America. U.S. Navy Department. Hydrographic Office. Bureau of Navigation. Richardson Clover, Lt. Commander, U.S. Navy, Hydrographer. Washington: 1893.

Cartographer: Richardson Clover.

Size: 62.1 x 58 cm. *Size of paper:* 87 x 61.5 cm.

Colored.

Gives list of wrecks and derelicts, etc., from 1887 to 1891.

(983: 1893: Un36nho Large)

1631. 1901-1902.

Monthly pilot charts of the North Atlantic and Mediterranean. London: E. Weller & Grahams Ltd. Litho.: 1901-1902. 16 maps.

Lithographers: E. Weller & Grahams Ltd.

Size: 46 x 65.7 cm.

Colored.

Charts for January, April-December 1901, and January-May 1902.

(983: 1901-1902: M765nam Large)

1632. 1957

Physiographic diagram Atlantic Ocean (Sheet 1) by Bruce C. Heezen and Marie Tharp. Geological Society of America, Special Paper 65. Reprinted by permission of the Lamont Geological Observatory (Columbia University). [Washington]: 1957.

Scale: Vertical exaggeration about 20:1.

Cartographers: Bruce C. Heezen and Marie Tharp.

Size: 71.7 x 141.2 cm.

Colored.

See: Geological Society of America, *Special paper,* no. 65. New York: Geological Society of America: 1957.

(983: 1957: H363ao Large)

1633. 1961

Physiographic diagram of the South Atlantic Ocean, the Caribbean Sea, the Scotia Sea and the eastern margin of the South Pacific Ocean. By Bruce C. Heezen and Marie Tharp. Lamont Geological Observatory, Columbia University. Published by the Geological Society of America. Washington: Williams & Heintz Map Corp.: 1961.

Cartographers: Bruce C. Heezen and Marie Tharp.

Size: 142.2 x 118.7 cm.

Colored.

(987: 1961: G293pdsa Large)

INDIAN OCEAN Nos. 1634-1635

1634. [ca. 1860]

Indischer Ocean. Commodore B. V. Wüllerstorf-Urbair, S. M. Fregatte Novara. 1857. [Vienna]: K.-K. Hof- u. Staats-Druckerei: [ca. 1860]. 7 maps.

Scales: Various.

Cartographer: B. V. Wüllerstorf-Urbair.

Size of paper: Ca. 80 x 56.5 cm.

(990: [ca. 1860]: W957fno Large)

1635. 1964

Physiographic diagram of the Indian Ocean the Red Sea, the South China Sea, the Sulu Sea and the Celebes Sea by Bruce C. Heezen and Marie Tharp. Lamont Geological Observatory, Department of Geology, Columbia University. Published by the Geological Society of America. Washington: Williams & Heintz Map Corp.: 1964.

Cartographers: Bruce C. Heezen and Marie Tharp.

Size: 155.5 x 120 cm.

Colored.

Descriptive sheet accompanies physiographic diagram. It has five figures, of which three are maps: 3. Control chart. All sounding lines used in the preparation; 4. Physiographic provinces; and 5. Bathymetric sketch of the Indian Ocean. Size of these three maps: ca. 51 x 39 cm.

(990: 1964: G293pdio Large)

PART III

Atlases

1636. [ca. 1569-1570]

Theatrum orbis terrarum. A. Ortelius. Antwerp: 1569-1570.

Cartographer: Abraham Ortelius.

Size: 41 cm.

The title page is in manuscript.

(912: Or8 Oversize)

1637. [ca. 1670?]

Atlas. By Frederick de Wit. Amsterdam: Frederick de Wit: [ca. 1670?]. 100 plates.

Cartographer: Frederick de Wit.

Size: 55.5 cm.

Provenance: Presented by John Vaughan, 19 January 1816.

Original binding in vellum.

(912: W77 Oversize)

1638. 1680

La primera parte del monte de la turba, ardiente Alumbrando con la claridad de su fuego toda la India-Occidental, empeçando desde el Rio Amazonas. Y fenesciendo al norte de Tierra Nueva. Descrita por Arnoldo Roggeveen. Amsterdam: Pedro Goos: 1680.

Cartographer: Arnold Roggeveen.

Size: 45 cm.

Original vellum binding.

(912.6: R63p Oversize)

1639. [1693]

Atlas maritime (The French Neptune). [Amsterdam: Pierre Mortier: 1693]. Volume 1.

Cartographer: Romein de Hooge.

Engraver: Romein de Hooge.

Size: 62 cm.

Provenance: Presented by John G. Watmough, 17 February 1837.

Title page is wanting. For volume 2, bound in same volume, see: *Zee Atlas tot het gebruik van de Vlouten des Konings van Groot Britanje* (no. 1640).

(912: H76 Oversize)

1640. 1693-1694

Zee Atlas tot het gebruik van de Vlouten des Konings van Groot Britanje, gemaakt volgens de nieuwste Memorien der ervarenste Ingenieurs en Stuurlieden, en verrykt met Profilen van de Vermaardste Zeehavenen en Zeesteden van Europa . . . Gegraveerd en verzameld door den Heer Romein de Hooge. Amsterdam: Pieter Mortier: 1694. Volume 2.

Cartographer: Romein de Hooge.

Engraver: Romein de Hooge.

Size: 62 cm.

Provenance: Presented by John G. Watmough, 17 February 1837.

For vol. 1, bound in same volume, see: *Atlas maritime (The French Neptune).*

(912: H76 Oversize)

1641. 1695-1753.

Maps of Holland, Germany, Bohemia, Austria & Italy. 1695-1753. 2 volumes.

Size: 54 cm.

Provenance: Presented by Joseph P. Engles, October 1839.

Binder's title: *German atlas.*

(912.4: M32 Oversize)

1642. [1712]

[*Grande carte des Pays Bas catholiques et des contrées qui les bordent* en XXIV feuilles par E. H. Friex. Amsterdam: Covens: 1712].

Cartographer: E. H. Friex [Fricx].

Size: 50.5 cm.

(912.4: F91g Oversize)

1643. [1739]

[*Plan de Paris,* commencé l'année 1734, dessiné et gravé sous les ordres de Messire Michel Etienne Turgot . . . Levé et dessiné par Louis Bretez; gravé par Claude Lucas; écrit par Aubin. N.p.: 1739].

Cartographer: Louis Bretez.

Engraver: Claude Lucas.

Size: 55.9 cm.

Provenance: Presented by Peter Stephen Du Ponceau.

(912.44: B75 Oversize)

1644. 1742

Atlas novus coelestis in quo mundus spectabilis, et in eodem tam errantium quam inerrantium stellarum phoenomena notablilia, circa ipsarum lumen, figuram, faciem, motum, eclipses, occultationes, transitus magnitudines, distantias, aliaque. Secundum Nic. Copernici, et ex parte Tychonis de Brahe, hypothesin. Nürnberg: Sumptibus Heredum Homannianorum: 1742.

Cartographers: Copernicus and Tycho Brahe.

Size: 23 cm.

(524: D72a Oversize)

1645. [ca. 1742?]

A new map of part of North America from the latitude of 40 to 68 degrees. Including the late discoveries made on the Furnace bomb ketch in 1742 and the western rivers & lakes falling into Nelson River in Hudson's Bay, as described by Joseph La France, a French Canadese Indian who traveled thro those countries and lakes for 3 years from 1739 to 1742. N.p.: [ca. 1742?].

Cartographer: Joseph La France.

Size: 20 cm.

Provenance: Presented by J. J. Stevenson, 7 December 1883.

(912.7: L13)

1646. [ca. 1745-1750]

[Uranographia Britannica, or exact view of the heavens.] By John Bevis. N.p.: ca. 1745-1750.

Cartographer: John Bevis.

Size: 37 cm.

Contains a manuscript note by John Williams, assistant secretary of the Royal Astronomical Society, giving an explanation of the work and partial provenance of this copy which contains the bookplate of Sir George Shuckburgh.

(523.89: B465 Oversize)

1647. 1752

Carte des nouvelles découvertes au nord de la Mer du Sud, tant à l'est de la Siberie et du Kamtchatka, qu'à l'ouest de la Nouvelle France . . . Par M. De l'Isle. Paris: 1752.

Scale: 25 French common leagues = 1 degree.

Cartographer: De l'Isle.

Size: 26 cm.

With this is: "Explication de la carte . . ."

(912.7: D37)

1648. 1753

Atlas Germaniae specialis seu systema tabularum geographicarum, in quibus imperium Romano-Germanicum . . . Opus inceptum a Ioh. Bapt. Homanno. Nürnberg: Homanniana: 1753.

Cartographer: Joh. Bapt. Homann.

Size: 55 cm.

Provenance: Presented by John Vaughan, 17 January 1806.

(912.43: H75 Oversize)

1649. [1761]

A new, general, and universal atlas, containing forty-five maps. Engraved by Mr. Kitchin, & others. [London]: A[ndrew] Dury: [1761].

Engraver: Thomas Kitchin.

Size: 12 cm.

Provenance: Presented by Murphy D. Smith, 27 March 1986.

(912: D93n)

1650. 1764

A compleat set of new charts on thirty-eight large plates, containing an accurate survey of the coast of Portugal and the Mediterranean Sea, in which are included seventy-five charts, of the principal harbours in the Straits, shewing the rocks, shoals, soundings, &c.: anchoring places, with their true latitude. The whole being accurately surveyed by J. Giacomo Alagna of Messina. London: J. Mount and T. Page: 1764.

Cartographer: J. Giacomo Alagna.

Size: 60.5 cm.

(912:4: AL12 Oversize)

1651. 1764

Le petit atlas maritime. Recueil de cartes et plans des quatre parties du monde. Paris: 1764. 5 volumes.

Cartographer: Jacques Nicolas Bellin.

Size: 32.6 cm.

Provenance: This title was purchased at the sale of the library of Benjamin Franklin, 18 March 1803.

(912: B41 Oversize)

1652. 1774

Map of the city and liberties of Philadelphia. John Reed. Philadelphia: J. Reed: 1774. 3 sheets.

Cartographer: John Reed.

Engraver: James Smither.

Size of each sheet: ca. 75 x 49.8 cm.

Provenance: Presented by Joseph Reed.

Contains pictures of the House of Employment and Alms House; State House.

(912.748: R25)

1653. 1774

New atlas of the mundane system; or, of geography and cosmography: describing the heavens and the earth, the distances, motions and magnitudes, of the celestial bodies; the various empires, kingdoms, states, republics, and islands, throughout the known world. By Samuel Dunn. London: Robert Sayer: 1774.

Cartographer: Samuel Dunn.

Size: 48 cm.

Provenance: Presented by Samuel Dunn through the hands of Benjamin Franklin, 17 December 1774.

(912: D92 Oversize)

1654. 1775

A complete body of ancient geography. By Monsr. d'Anville . . . to which are added Britannia Romana by Mr. Horsley; Graeciae . . . by Mons. Charles de l'Isle; Germany, France, Italy, Spain, and the British Isles, in an intermediate state between ancient and modern geography; by Mons. d'Anville. And the whole improved by inserting the modern names of places under the ancient. London: R. Sayer and J. Bennett: 1775.

Cartographers: D'Anville, Horsley, and de l'Isle.

Size: 56.4 cm.

Provenance: Presented by Mr. and Mrs. Charles Chauncey, 13 June 1813.

(912.3: An9c Oversize)

1655. [1776]

The American military pocket atlas; being an approved collection of correct maps, both general and particular, of the British colonies; especially those which now are, or probably may be the theatre of war: taken principally from the actual survey and judicious observations of engineers de Brahm and Romans, Cook, Jackson, and Collet: Major Holland, and other officers, employed in his Majesty's fleets and armies. London: R. Sayer and J. Bennett: [1776].

Cartographers: De Brahm, Romans, Cook, Jackson, Collet, and Samuel Holland.

Size: 21.5 cm.

Provenance: Presented by John Wilson.

Maps have been removed. See nos. 454, 627, 784, 1081, and 1536.

(912.7: Am3)

1656. [ca. 1776]

Charts of the coast and harbours of New England. Composed and engraved by Joseph Frederick Wallet Des Barres, esq; in consequence of an application, of the right honourable Lord Viscount Howe, Commander-in-Chief of his majesty's ships in North America. From the surveys taken under the direction of the Lords of Trade by Samuel Holland, Esq. N.p.: [ca. 1776].

Cartographers: Samuel Holland and J. F. W. Des Barres.

Engraver: J. F. W. Des Barres.

Size: 58 cm.

(912.74: D45 Oversize)

1657. 1776

A maritime survey of Ireland and the west of Great Britain; taken by the order of the right honourable the Lords Commissioners of Admiralty; in two volumes: accompanied with a book of nautical descriptions and directions to each volume: by Murdoch Mackenzie. London: Murdoch Mackenzie: 1776. 2 volumes.

Cartographer: Murdoch Mackenzie.

Size: 56.7 cm.

(912.42: M19 Oversize)

1658. [ca. 1777]

The sea coast of Nova Scotia; exhibiting the diversities of the coast, and face of the country near it; the banks, rocks, shoals, soundings, &c. together with remarks and directions for the conveniency of navigation and pilotage. Surveyed by order of the right honourable the Lords Commissioners of the Admiralty by Joseph Frederick Wallet Des Barres esqr. [London: ca. 1777].

Cartographer: J. F. W. Des Barres.

Size: 58 cm.

(912.71: D45 Oversize)

1659. 1780 [reprint of 1966]

The Atlantic Neptune, published for the use of the Royal Navy of Great Britain, by Joseph F. W. Des Barres, Esq. under the directions of the Right Hon[oura]ble the Lords Commissioners of the Admiralty. London: 1780. Reprinted by Barre Publishers: 1966.

Two of the original plates of the atlas were presented to the Society by the British Admiralty through Harold Braham, British consul general in Philadelphia, 6 February 1948. They are: Series III, plates 39 [Delaware Bay] and 38 [Plan of the environs of Philadelphia (1 of 2)].

(912.7: D45a.r Extra-oversize)

1660. 1780

A Bengal atlas: containing maps of the theatre of war and commerce on that side of Hindoostan. Compiled from the original surveys; and published by order of the honourable the court of directors for the affairs of the East India Company. By James Rennell. N.p.: 1780.

Cartographer: James Rennell.

Size: 49 cm.

Provenance: Presented by J. P. Engles, 18 October 1839.

(912.54: R29 Oversize)

1661. 1780

The English pilot. The fourth book. Describing the West-India navigation. From Hudson's Bay to the river Amazones. Particularly delineating the coasts, capes, headlands, rivers, bays, roads, havens, harbours, streights, rocks, sands, shoals, banks, depths or water, and anchorages, with all the islands therein; as Jamaica, Cuba, Hispaniola, Barbadoes, Antigua, Bermudas, Porto Rico, and the rest of the Caribbee and Bahama islands. Also, a new description of Newfoundland, New England, New York, East and West New Jersey, Dellawar Bay, Virginia, Maryland, Carolina, &c . . . The whole being much enlarged and corrected, with the additions of several new charts and descriptions. By the information of divers able navigators of our own and other nations. London: J. Mount and T. Page, W. Mount and T. Page: 1780.

Size: 48.3 cm.

(912.6: En3)

1662. 1781

Great-Britain's coasting pilot: being a new and exact survey of the sea-coast of England and Scotland from the river of the Thames to the westward and northward; with the islands of Scilly, and from thence to Carlisle; likewise the islands of Orkney and Shetland, describing all the harbours, rivers, bays, roads, rocks, sounds, buoys, beacons, sea-marks, depths of water, latitude, bearings and distances from place to place the setting and flowing of the tides; with directions for the knowing of any place, and how to harbour a ship in the same with safety. With directions for coming into the channel between England and France. By Captain Greenville Collins. London: Mount and Page: 1781. 2 volumes.

Cartographer: Greenville Collins.

Size: 53.2 cm.

Provenance: Presented by Mathew Carey, 17 August 1832.

(912.42: C69 Oversize)

1663. 1782

Neptune occidental. A compleat pilot for the West-Indies, including the British Channel, Bay of Biscay, & all the Atlantic Islands; with their bays, harbours, keys, rocks, land-marks, depths of water, latitudes, longitudes, &c. &c. Done from actual surveys, by the late Thomas Jefferys Geographer to the King. London: R. Sayer and J. Bennett: 1782.

Cartographer: Thomas Jefferys.

Size: 54 cm.

(912.729: J36n Oversize)

1664. 1783

The West-India atlas; or, a compendious description of the West Indies: illustrated with forty-one correct charts and maps, taken from actual surveys. Together with an historical account of the several countries and islands which compose that part of the world. Their discovery, situation, extent, boundaries, product, trade, inhabitants, strength, government, religion, &c. By the late Thomas Jefferys, geographer to the King. London: Robert Sayer and John Bennett: 1783.

Cartographer: Thomas Jefferys.

Size: 54.5 cm.

Provenance: Presented by Jacob Snider from the library of John Vaughan, 17 March 1842.

An engraved title page reads: The West Indian atlas; or, a compendious description of the West Indies; taken from actual surveys and observations, by Thomas Jefferys, geographer to the King. London: Robert Sayer and John Bennett: 1783.

(919.729: J36w Oversize)

1665. 1784

Atlas antiquus, Danvillianus, conspectus tabularum geographicarum. Nürnberg: Chr. Weigelio-Schneideriana: 1784.

Cartographer: J. B. Bourguignon d'Anville.

Size: 56 cm.

Provenance: Presented by Samuel Moore, 21 November 1806.

(912.3: An9a Oversize)

1666. 1787-1788

Atlas encyclopédique, contenant la géographie ancienne, et quelques cartes sur la géographie du moyen âge, la géographie moderne, et les cartes relatives à la géographie physique. Rigobert Bonne. Paris: 1787-1788. 2 volumes.

Cartographer: Rigobert Bonne.

Size: 31.5 cm.

(912: B64 Oversize)

1667. [1788-1812]

[*Atlante geografico del regno di Napoli.* G. A. R. Zannoni. Naples: Guerra: 1788-1812].

Cartographer: G. A. R. Zannoni.

Size: 83.5 cm.

Provenance: Presented by Chevalier Morelli, 21 August 1835.

Title page wanting.

(912.45: Z13 Extra-oversize)

1668. [1788-1790]

An hydrographical collection of forty accurate plans on a large scale, of the principal ports, bays, roads, and harbours in the West Indies, viz. On the Spanish Main and Florida, the islands of Jamaica, Hispaniola or St. Domingo, Cuba and Porto Rico; wherein the anchoring and watering-places, soundings, rocks, shoals, &c. are minutely laid down: accompanied with a general chart of the West-Indies. London: Robert Sayer: [1788-1789].

Size: 52.3 cm.

Provenance: Presented by Jacob Snider, Jr., from the library of John Vaughan, 18 March 1842.

(912.729: Sa9 Oversize)

1669. 1792

D'Anville's atlas, containing a map of the world, the world in twelve maps; & twelve maps of the most interesting parts of the world upon a large scale. Also eleven maps of ancient geography for the study of ancient history. London: Engraved and printed for John Harrison: 1792.

Cartographer: Bourguignon d'Anville.

Size: 53.8 cm.

(912: An9 Oversize)

1670. 1792

Atlas national et général de la France, divisée en ses quatre-vingt-trois départemens et en cinq cens quarante-trois districts, conformément aux décrets de l'Assemblée Nationale, sanctionnés par le Roi, dressé sur les meilleurs cartes qui ont paru jusqu' à présent, et sur la carte des triangles levée géométriquement par ordre du Roi, et mis au jour par M. Cassini de Thury . . . revu, vérifié au Comité de Constitution, par Sr. Desnos. [Paris]: Veuve Hérissont: 1792 [16 Juin 1792].

Cartographers: Cassini de Thury and Sr. Desnos.

Size: 56 cm.

(912.44: C27 Oversize)

1671. 1795

The general atlas for Carey's edition of Guthrie's geography improved. Philadelphia: Mathew Carey: 1795.

Size: 42.6 cm.

Provenance: Presented by Jacob Snider from the library of John Vaughan, 18 March 1842.

(912: G98c Oversize)

1672. 1796

New atlas of the mundane system; or of geography and cosmography: describing the heavens and the earth, the distances motions and magnitudes, of the celestial bodies; the various empires, kingdoms, states and republics throughout the known world. By Samuel Dunn. London: Robert Laurie & James Whittle: 1796.

Cartographer: Samuel Dunn.

Size: 48 cm.

Provenance: Presented by Peter Stephen Du Ponceau, 20 March 1840.

(912: D923)

1673. 1798

Chart atlas. A voyage around the world, performed in the years 1785, 1786, 1787, and 1788 by the Boussole and Astrolabe, under the command of J. F. G. de La Perouse: published by order of the

National Assembly under the superintendence of L. A. Milet-Mureau. 2nd ed. London: G. G. and J. Robinson: 1798.

Cartographers: J. F. G. de La Perouse and L. A. Milet-Mureau.

Size: 58 cm.

(910.4: L31.2 Oversize)

1674. [ca. 1800?]

A compleat pilot for the West-Indies, including the British Channel, Bay of Biscay, and all the Atlantic islands: done from actual surveys and observations of the most experienced navigators in his majesty's and the merchant's service. By the late Thomas Jefferys. A new edition, corrected to the present time. London: Robert Laurie and James Whittle: [ca. 1800?].

Cartographer: Thomas Jefferys.

Size: 54.1 cm.

(912.729: J361n Oversize)

1675. [ca. 1802?]

[Maps of the coast of Africa]. By Aaron Arrowsmith. N.p.: [ca. 1802?]. 5 maps.

Cartographer: Aaron Arrowsmith.

Size: 31 cm.

Title page wanting.

(912.6: Ar6 Oversize)

1676. [1803]

[*Atlas historique, généalogique, chronologique, et géographique.* Par E. A. D. Las Casas]. N.p.: [1803].

Cartographer: E. A. D. Las Casas.

Size: 56 cm.

Title page wanting. Stamped on the cover is the name: A. Le Sage.

(911: L33 Oversize)

1677. [ca. 1804]

Allgemeiner Hand-Atlas der ganzen Erde nach den besten astronomischen Bestimmungen neuesten Entdeckungen und kritischen Untersuchungen entworfen und zu A. C. Gaspari. Weimar: [ca. 1804].

Cartographer: A. C. Gaspari.

Size: 56.5 x 67.6 cm.

(912: G21 Extra-oversize)

1678. 1804

A new and elegant general atlas, comprising all the new discoveries, to the present time, containing sixty-three maps drawn by Arrowsmith and Lewis. Philadelphia: John Conrad & Co.: 1804.

Cartographers: Arrowsmith and Lewis.

Size: 28.4 cm.

(912: Ar6)

1679. [ca. 1810]

Wallis's new pocket edition of the English counties, or travellers companion in which are carefully laid down all the direct & cross-roads, cities, townes, villages, parks, seats and rivers. With a general map of England & Wales. London: James Wallis: [ca. 1810].

Size: 15 cm.

(912.42: W15)

1680. 1812

A new and elegant general atlas. Comprising all the new discoveries, to the present time. Containing sixty three maps, drawn by Arrowsmith and Lewis. Intended to accompany the new improved edition of Morse's geography, but equally well calculated to be used with his gazetteer, or any other geographical work. Boston: Thomas & Andrews: 1812.

Cartographers: Arrowsmith and Lewis.

Size: 27.5 cm.

Provenance: Presented by the Estate of Richard Gimbel, 1974.

(Paine 71: Ar6n)

1681. 1814

A map of Wayne & Pike counties, Pennsylvania; shewing the situations & forms of the warrantee tracts with the numbers by which the respective tracts are designated on the maps & books in the office of the Commissioners of Rates for Wayne county, the townships, boundaries, roads, waters and principal places. By Jason Torrey. Philadelphia: 1814 [7 April 1814].

Scale: 1 cm. = 1 mi.

Cartographer: Jason Torrey.

Engraver: H . S. Tanner.

Size: 65.5 x 47.4 cm.

Provenance: Presented by John Melish, 16 April 1819.

Bound with: Jason Torrey. *An index to the map of Wayne and Pike counties, Pennsylvania.* Philadelphia: Joseph Rakestraw: 1814.

(912.748: T63)

1682. [1816]

[*Grand atlas universel, ou collection des cartes encyprotypes, générales et détaillés des cinq parties du monde.* Paris: Desray: 1816.]

Cartographer: H. Brué.

Size: 62.6 cm.

Provenance: Presented by Peter S. Du Ponceau, 15 May 1818.

Title page lacking.

(912: B83 Oversize)

1683. 1818

A modern atlas, from the latest and best authorities, exhibiting the various divisions of the world, with its chief empires, kingdoms, and states; in sixty maps, carefully reduced from the largest and most authentic sources. Directed and superintended by John Pinkerton. Philadelphia: Thomas Dobson & Son: 1818.

Cartographer: John Pinkerton.

Size: 58.8 cm.

Provenance: Presented by Thomas Astley, 5 April 1816–17 July 1818. As the maps were pulled, Astley presented copies which were bound into the present volume.

See: John Pinkerton, *General collection of voyages and travels.*

(912: P65 Oversize)

1684. 1820

A complete genealogical, historical, chronological, and geographical atlas; being a general guide to history, both ancient and modern . . . the whole forming a complete system of history and geography. By Lavoisne. From the last London edition, improved by C. Gros, of the University of Paris, and J. Aspin. 2nd American ed. Philadelphia: M. Carey & Son: 1820.

Cartographers: Lavoisne, C. Gros, and J. Aspin.

Size: 45.3 cm.

Provenance: Presented by M. Carey & Son, 16 December 1820.

(911: L39 Oversize)

1685. 1822

A complete historical, chronological, and geographical American atlas, being a guide to the history of North and South America, and the West Indies . . . to the year 1822. Philadelphia: H. C. Carey and I. Lea: 1822.

Size: 45.5 cm.

Provenance: Presented by Henry C. Carey and Isaac Lea.

(911: C18 Oversize)

1686. 1823

Atlas des Marais Pontins. Paris: Firmin Didot: 1823.

Size: 42.3 cm.

Provenance: Presented by Mr. and Mrs. Charles Chauncey.

(912.45: Ar6m Oversize)

1687. 1823

A new American atlas containing maps of the several states of the North American Union. Projected and drawn on a uniform scale from documents found in the public offices of the United States and state governments, and other original and authentic information. By Henry S. Tanner. Philadelphia: H. S. Tanner: 1823.

Cartographer: Henry S. Tanner.

Size: 60.2 cm.

Provenance: Presented by Henry S. Tanner, 17 July 1818–15 July 1825. As the maps were pulled, Tanner presented copies to the Society. Later, they were bound into this volume.

(912.73: T15n Oversize)

1688. 1824

Atlas de l'Océan Pacifique dressé par M. de Krusenstern commandant de la Marine Impériale de Russie. St. Petersburg: 1824.

Cartographer: M. de Krusenstern.

Size: 66.5 cm.

Provenance: Presented by Admiral Krusenstern, 16 September 1825.

(656: K94 Oversize)

1689. [1825]

Atlas of the state of South Carolina made under the authority of the legislature, prefaced with a geographical, statistical and historical map of the state. By Robert Mills. N.p.: [1825].

Cartographer: Robert Mills.

Size: 54 cm.

Provenance: Presented by Robert Mills, 6 October 1826. Later, a copy was presented by Samuel Francis DuPont on 13 November 1861, which had been "captured in Nov. 61 in Beaufort the Head Quarters of the Rebel Commanding General."

S. F. DuPont understood that only one library, Cambridge, had a copy and he thought the atlas to

> possess some interest as giving an idea of the inland waters, where we have been in part operating, with their remarkable interconnection. The Map of the Beaufort District very striking in this particular, should be mended [Archives. S. F. Du Pont to B. Gerhard, 13 November 1861].

There is also a facsimile copy, published in 1938: "This edition is limited to 350 copies of which this is book No. 66."

(912.757: M62 Oversize)

1690. 1827-1828

[General atlas.] New edition. Philadelphia: H. S. Tanner: 1827-1828. 11 maps.

Cartographer: Henry S. Tanner.

Size: 53.8 cm.

(912: T16g Oversize)

1691. 1832

Atlante corografico storico e statistico del regno delle Due Sicilie eseguito litograficamente, compilato, e dedicato a S. M. il Re Ferdinando II. Dal suo umilissimo e fidelissimo suddito Benedetto Marzolla. Naples: Reale Litografia Milint[ ]e: 1832.

Cartographer: Benedetto Marzolla.

Size: 47.5 x 63 cm.

"Presented by Chevalier Morrelli to Daniel J. Desmond esq." is stamped on the cover.

(912.45: M36a Oversize)

1692. 1833

Atlas de l'Europe. Établissement géographique de Bruxelles. 165 feuilles; Échelle 1:600,000; projection modifiée de Flamsteed, d'après les meilleurs matériaux. Dessiné par Henry Perkin, Gravé sur pierre . . . P. Doms, F. Charles, L. Bulens. Brussels: 1833. 2 volumes.

Cartographers: Henry Perkin.

Lithographers: P. Doms, F. Charles, and L. Bulens.

Size: 57.5 cm.

Provenance: Presented by H. S. Tanner, John Vaughan, John Kintzing Kane, William Gibson, Mathew Carey, Nathaniel Chapman, Peter S. Du Ponceau, J. Francis Fisher, Job R. Tyson, M. Robeson, George B. Wood, William Drayton, Thomas Harris, Nathan Dunn, Wm. Hembel, John P. Wetherell, Nicholas Biddle, James Mease, Robert M. Patterson, T. Sargeant, Charles N. Bancker, and Henry C. Carey, 2 March 1838.

(912.4: Et12 Oversize)

1693. 1833

Tanner's universal atlas. No. 2. Philadelphia: Tanner: 1833. 5 maps.

Cartographer: Henry S. Tanner.

Size: 40 cm.

Provenance: Tanner presented copies of these maps as they were pulled, 20 September 1833–5 December 1834.

(912: T16u Oversize)

1694. 1834-1838

Atlas Classica. Philadelphia: H. S. Tanner: 1834-1838. Nos. 1-9.

Size: 40 cm.

Provenance: Presented by H. S. Tanner as the maps were produced, 4 May, 15 June, 17 August 1838, etc.

(911: T15 Oversize)

1695. 1835

A comprehensive atlas: geographic, historical & commercial. By T. G. Bradford. Boston: William D. Tichnor: New York: Wiley & Long: 1835.

Cartographer: T. C. Bradford.

Size: 32.8 cm.

Provenance: Presented by Robert Cushman Murphy, May 1958.

(911: B73 Oversize)

1696. [1840?]

[*Histoire et description des voies de communication aux Etats-Unis et des travaux d'art qui en dépendent.* Par Michel Chevalier]. N.p.: [1840?].

Cartographer: Michel Chevalier.

Size: 55 cm.

Provenance: Presented by Mr. and Mrs. Charles Chauncey.

This is the atlas volume.

(912.73: C42 Oversize)

1697. 1849-1860

Wind, current and thermal charts by M. F. Maury. Ser. A-F. Washington: 1849-1860.

Cartographer: M. F. Maury.

Size: 33.7 cm.

Provenance: Presented, in part, by M. F. Maury, November 1852.

(551.47: M44 Oversize)

1698. 1850

The atlas of physical geography: constructed by Augustus Petermann, F. R. G. S. . . . with descriptive letter-press, embracing a general view of the physical phenomena of the globe. By Thomas Milner. London: Wm. S. Orr & Co.: 1850.

Cartographer: Thomas Milner.

Size: 37.7 cm.

(551: P44 Oversize)

1699. 1851

Atlas von Land- und Seekarten vom Japanischen Reiche Dai-Nip-Pon und dessen neben- und Schutzländern Jezo mit den Südlichen Kurilen, Krafto, Kôrai und den Liu-Kiu-Inseln nach Originalkarten und astronomischen Beobachtungen der Japaner mit Hinweisung auf die Entdeckungen in älterer und neuerer Zeit. Nebst einer Seekarte von der Küste von China und der Insel Formosa nach unausgegebenen Holländischen Seekarten vom 17 Jahrhundert bearbeitet und herausgegeben von Philipp Franz von Siebold. Berlin: Simon Schropp & Co.: 1851.

Cartographer: Philipp Franz von Siebold.

Size: 71 cm.

(912.52: Si15 Oversize)

1700. 1855-1860

Report of explorations and surveys . . . for a railroad from the Mississippi River to the Pacific Ocean. U.S. Army. [Washington]: 1855-1860.

Size: 27 x 48 cm.

(508.73: Un3 Extra-oversize)

1701. 1865

Atlas of the oil region of Pennsylvania from actual surveys under the direction of F. W. Beers. Assisted by Beach Nichols, J. M. Beers, A. Leavenworth, C. S. Peck, C. A. Curtis & Geo. Stewart. With a few facts relating to petroleum, historically, scientifically, and commercially, reviewed by Ivan C. Michels, Editor of the Philadelphia Coal Oil Circular and Petroleum Price Current. New York: F. W. Beers, A. D. Ellis and G. G. Soule: 1865. 34 maps.

Cartographers: F. W. Beers, Beach Nichols, J. M. Beers, A. Leavenworth, C. S. Peck, C. A. Curtis, and Geo. Stewart.

Lithographer: Ferd. Mayer & Co.

Size: 40 cm.

(553.28: B391a)

1702. 1869

Mitchell's new general atlas, containing maps of the various countries of the world, plans of cities, etc., embraced in sixty-three quarto maps, forming a series of one hundred maps and plans, together with valuable statistical tables. Philadelphia: S. Augustus Mitchell, Jr.: 1869.

Size: 40 cm.

Provenance: Presented by Mr. and Mrs. Charles Chauncey, June 1913.

(912: M69 Oversize)

1703. 1872

Rand, McNally & Co.'s railway guide. The travelers' hand book to all railway and steamboat lines of North America. By Robert A. Bower. Chicago: Rand, McNally & Co.: 1872.

Cartographer: Robert A. Bower.

Size: 16.5 x 12.7 cm.

(656.6: R15r)

1704. 1874

Statistical atlas of the United States based on the results of the ninth census 1870, with contributions from many eminent men of science and several departments of the government. U.S. Census Office. 9th census, 1870. Compiled under the authority of Congress by Francis A. Walker. New York: J. Bien, lith.: 1874.

Cartographer: Francis A. Walker.

Lithographer: Julius Bien.

Size: 55.7 cm.

(917.3: W15 Oversize)

1705. [ca. 1876-]

Detail map of the Upper Mississippi River from the mouth of the Ohio River to Minneapolis, Minnesota. In eighty-nine sheets. Published by the Mississippi River Commission. N.p.: [ca. 1876-].

Size: 63.5 cm.

(912.73: Un3m Oversize)

1706. 1876

Geological and topographical atlas accompanying the report of the geological exploration of the fortieth parallel, made by authority of the honorable Secretary of War under the direction of Brig. and Brvt. Major General A. A. Humphreys, chief of Engineers U.S. By Clarence King, U. S. Geologist in charge. 1876. [New York]: Julius Bien, lith.: 1876.

Cartographers: Clarence King and A. A. Humphreys.

Lithographer: Julius Bien.

Size: 83.9 cm.

(620: Un3pr and 620: Un3py Oversize)

1707. [1877]

[Atlas to accompany] *Annual report upon the geographical survey west of the 100th meridian.* New York: Graphic Co. photo-lith.: [1877].

Cartographers: Geo. M. Wheeler; Weyss, Herman, Maulo, Louis Nell, and Lang.

Lithographer: Graphic Co. photo-lith.

Size: 19¼ in.

To accompany the copies of Appendix N N.

(557.3: Un15ra Oversize)

1708. 1879

Topographical and geological atlas of the district of the High Plateau of Utah, to accompany the report of Cap. C. E. Dutton. U. S. Dept. of the Interior. U.S. Geographical and Geological Survey of the Rocky Mountain region. J. W. Powell in charge. New York: Julius Bien: 1879.

Cartographers: J. W. Powell and C. E. Dutton.

Size: 31½ in.

(557.3: Un2ro Oversize)

1709. 1880

Atlas of Delaware County, Pennsylvania, containing nineteen maps exhibiting the early grants & patents, compiled from official records; together with a history of the land titles in the county. By Benjamin H. Smith. Philadelphia: Henry B. Ashmead: 1880.

Cartographer: Benjamin H. Smith.

Size: 39.9 cm.

(912.748: Sm5 Oversize)

1710. [ca. 1880-]

Detail map of the Lower Mississippi River from the mouth of the Ohio River to Head of the Passes. Published by the Mississippi River Commission. N.p.: [ca. 1880-]. 83 sheets.

Size: 63.5 cm.

(912.73: Un3m, vol. 5. Oversize)

1711. 1881-1897

Map of the Lower Mississippi River from the mouth of the Ohio River to the Head of the Passes. Scale, one inch to one mile. Published in 1881-1897 by the Mississippi River Commission. N.p.: 1881-1897. 32 sheets.

Scale: 1 in. = 1 mi.

Size: 38.4 cm.

(912.73: Un3m, vol. 3 Oversize)

1712. 1884-1885

Grand atlas. Divisions 1-5. Second Geological Survey of Pennsylvania. J. P. Lesley, state geologist. Harrisburg: Lane S. Hart, state printer: 1884-1885.

Cartographer: J. P. Lesley.

Size: 32½ in.

Incomplete.

1713. 1887

The royal atlas of modern geography exhibiting, in a series of entirely original and authentic maps the present condition of geographical discovery and research in the several countries, empires, and states of the world. By the late Alexander Keith Johnston. With additions and corrections to the present date by T. B. Johnston. A new edition. Edinburgh and London: W. & A. K. Johnston: 1887.

Cartographers: Alexander Keith Johnston and T. B. Johnston.

Size: 51 cm.

1714. 1888

Bradley's atlas of the world for commercial and library reference. A complete American and foreign atlas, compiled from official state, national and international surveys, supplemented by information furnished by the postal and interior departments of the United States, and from many reliable private sources. Officially approved and adopted by the departments of the United States government. With isometric index to each map. Philadelphia: Wm. M. Bradley & Bro.: 1888.

Size: 48.7 cm.

1715. 1888

Handy reference atlas of the world. By John Bartholomew. With complete index and geographical statistics. New York: George Routledge: 1888.

Cartographer: John Bartholomew.

Size: 19 cm.

1716. 1889

Facsimile-atlas to the early history of cartography with reproductions of the most important maps printed in the XV & XVI centuries. Nils Adolf Erik Nordenskiöld. Translated from the Swedish original by J. A. Ekelöf and Clements R. Markham. Stockholm: 1889.

Cartographer: Nils Adolf Erik Nordenskiöld.

Size: 51.9 cm.

1717. 1891

Rand, McNally & Company's neuer Familien-Atlas der Welt. Chicago: Rand-McNally: 1891.

Size: 36.6 cm.

1718. 1892-1895

Map of the Missouri River from its mouth to Three Forks, Montana. In eighty-four sheets. Scale, one inch to one mile [1:63,360], and nine index sheets. Published in 1892-95 by the Missouri River Commission, N.p.: 1892-1895.

Scale: 1 in. = 1 mi.

Size: 38.4 cm.

(912.73: Un3mi Oversize)

1719. 1897

The Century atlas of the world. Prepared under the superintendence of Benjamin E. Smith. New York: The Century Co.: 1897.

Cartographer: Benjamin E. Smith.

Size: 33.5 cm.

(912: C33)

1720. [1898]

Adolf Steiler's Handatlas über alle Theile Erde und über das Weltgebäude. Gotha: Justus Perthes: [1898].

Cartographer: Adolf Karl Stieler.

Size: 39.4 cm.

Provenance: Presented by William Welch Flexner, James Thomas Flexner, and the Rockefeller University, January 1964.

Contains inscription: Simon Flexner, Johns Hopkins Hospital, 1899. From Dr. W. H. Welch.

(912: St5h.n Oversize)

1721. 1898

Atlas of ports, cities, and localities of the island of Cuba. Containing reproductions of maps, charts, and plans obtained from the United States Congressional Library, Coast and Geodetic Survey, Hydrographic Office, Engineer Department U.S. Army; also Pichardo's map of the island of Cuba, and other sources. Washington, D.C.: 1898.

Cartographer: Pichardo.

Size: 55.5 x 70.8 cm.

(912.7291: Un3 Extra-oversize)

1722. 1899-1911

Physical atlas. J. G. Bartholomew, editor. Edinburgh: John Bartholomew & Co.: 1899-1911. Vol. 3. *Meteorology.* By J. G. Bartholomew, A. J. Herbertson and Alexander Buchan. Vol. 5. *Zoogeography.* By J. G. Bartholomew, W. Eagle Clarke and Percy H. Grimshaw.

Cartographers: J. G. Bartholomew, W. Eagle Clarke, Percy H. Grimshaw, A. J. Herbertson, and Alexander Buchan.

Size: 46.3 cm.

(500: B28 Oversize)

1723. 1900

Atlas of the Philippine Islands. U.S. Treasury Department. Coast and Geodetic Survey. Henry S. Prichett, Superintendent. Special publication no. 3. Washington: Government Printing Office: 1900. 30 maps.

Cartographer: Henry S. Prichett.

Size: 38 x 33.4 cm.

There is another title page: *Atlas de Filipinas. Colección de 30 mapas.* . . . 1899.

(526: Un3sp: no. 3 Oversize)

1724. 1906

Atlas of Canada, prepared under the direction of James White. Department of the Interior of Canada. [Toronto]: Toronto Lithographic Co., Ltd.: 1906.

Cartographer: James White.

Size: 45 cm.

Provenance: Presented by James White, 1906.

(912.71: C16 Oversize)

1725. [ca. 1906?]

Research in China. 1903-1904. Geographical and geological maps. Bailey Willis, geologist in charge. Carnegie Institution of Washington. [Washington, D.C.: ca. 1906?].

Cartographer: Bailey Willis.

Size: 55.1 x 46.7 cm.

In: Carnegie Institution of Washington, *Publication,* no. 54.

(506.73: C21p Oversize)

1726. 1911

Cram's modern new census atlas of the United States and world. Thirteenth census edition. A complete series of maps, of the United States and possessions, foreign countries, world, polar regions and hemispheres . . . Carefully edited, compiled and published by George F. Cram. New York, Chicago: G. F. Cram: 1911.

Cartographer: George F. Cram.

Size: 39.4 cm.

(912: C84 Oversize)

1727. 1912

Rand McNally & Co.'s unrivaled atlas of the world, containing colored maps of every country and civil division upon the face of the globe. With marginal index to maps of foreign countries together with historical, descriptive, and statistical matter pertaining to each, and city maps. A new complete ready reference index to the United States showing full returns from the census of 1910. Chicago: Rand, McNally & Co.: 1912.

Size: 36.5 cm.

Provenance: Presented by the Estate of Colonel Richard Gimbel, 1974.

(Paine 71: R15u)

1728. 1914

Statistical atlas of the United States. U.S. Department of Commerce. Bureau of the Census. Wm. J. Harris, director. 13th census. Prepared under the supervision of Charles S. Sloane, geographer of the census. Washington: Government Printing Office: 1914.

Cartographers: Wm. J. Harris and Charles S. Sloane.

Size: 29.5 cm.

(317.3: Un3ms)

1729. 1915

Johnston's royal atlas of modern geography exhibiting, in a series of sixty entirely original and authentic maps the present condition of geographical discovery and research in the several countries, empires and states of the world. With a special index to each map. Edinburgh and London: W. & A. K. Johnston: 1915.

Cartographer: Alexander Keith Johnston.

Size: 50.8 cm.

(912: J64r Oversize)

1730. 1921

World atlas of commercial geology. U.S. Department of the Interior. Geological Survey. George Otis Smith, director. Washington: U.S. Geological Survey: 1921. Part I. Distribution of mineral production. Part II. Water power of the world.

Cartographer: George Otis Smith.

Size: 35 cm.

(553: Un3w Oversize)

1731. 1924

An historical atlas of modern Europe from 1789 to 1922 with an historical and explanatory text by C. Grant Robertson . . . and J. C. Bartholomew. 2nd ed., revised and enlarged. London: Oxford University Press: 1924.

Size: 35.8 cm.

(912.4: R542h.2 Oversize)

1732. 1925

Statistical atlas of the United States. U.S. Department of Commerce. Bureau of the Census. W. M. Steuart, director. [14th census.] Prepared under the supervision of Charles S. Sloane, geographer of the census. Washington: Government Printing Office: 1925.

Cartographers: W. M. Steuart and Charles S. Sloane.

Size: 30 cm.

(317.3: Un3ns)

1733. 1925-1926

Stielers Hand-atlas. 254 Haupt- und Nebenkarten in Kupferstich. Zehnte Auflage, Hundertjahr-Ausgabe. Von grund auf neubearbeitet unter Leitung von Prof. Dr. H. Haack. Gotha: Justus Perthes: 1925-1926.

Cartographers: H. Haack and Adolf Karl Steiler.

Size: 40 cm.

(912: St5h.10 Oversize)

1734. 1926

A book of old maps delineating American history, from the earliest days down to the close of the Revolutionary War. Compiled and edited by Emerson D. Fite & Archibald Freeman. Cambridge: Harvard University Press: 1926.

Size: 42 cm.

(912: F55 Oversize)

1735. 1932

Atlas of the historical geography of the United States. By Charles Oscar Paullin. Edited by John K. Wright. Washington, New York: Carnegie Institution of Washington and the American Geographical Society of New York: 1932. 688 maps (partly colored) on 166 plates.

Cartographers: Charles Oscar Paullin and John K. Wright.

Size: 36.5 cm.

In: Carnegie Institution of Washington, *Publication,* no. 401.

(506.73: C21p Oversize)

1736. 1932

The George Washington atlas. A collection of eighty-five maps, including twenty-eight made by George Washington; seven used and annotated by him; eight made under his direction, or for his use or otherwise associated with him; and forty-two new maps concerning his activities in peace and war and his place in history. Edited by Lawrence Martin. U.S. George Washington Bicentennial Commission. Washington, D.C.: 1932.

Size: 46.2 cm.

Provenance: Presented by the George Washington Bicentennial Commission, 28 October 1932.

(912.73: M36g Oversize)

1737. 1933-1939

Atlante internazionale del touring club Italiano centosettantasei tavoli principali centotrentanove carte parziali di sviluppo. Opera redatta ed eseguita nell' ufficio cartografico del T.C.I. sotto la direzione di L. V. Bertarelli, O. Marinelli, P. Corbellini. Quarta edizione. Milan: Touring Club Italiano: 1933-1939.

Cartographers: L. V. Bertarelli, O. Marinelli, and P. Corbellini.

Size: 49.5 cm.

(035: Enl.a)

1738. 1933

Piri Reis Haritası [Piri Reis's map]. Istanbul: 1933.

Cartographer: Piri Reis.

Size: 38 cm.

Provenance: Presented by Halet Çambel, 1981.

With this is a pamphlet describing the map. Also Piri Reis Kitabı Bahriye. Istanbul: Devlet Basimevi: 1935.

This facsimile is supposed to be based on Christopher Columbus's map of 1498.

(912: P66h.f Oversize)

1739. 1934-1938

Steiler's atlas of modern geography. 263 maps on 114 sheets. Engraved on copper. 10th ed. International edition: Published by Prof. Dr. Hermann Haack with the cooperation of Dr. Berthold Carlberg & Rudolf Schleifer. Gotha: Justus Perthes: 1934-1938.

Cartographers: Adolf Karl Stieler, Hermann Haack, Berthold Carlberg, and Rudolf Schleifer.

Size: 40.1 cm.

(912: St5a.10i Oversize)

1740. 1935

Historical and commercial atlas of China. By Albert Herrmann. Harvard Yenching Institute. Monograph series. Vol. 1. Cambridge: Harvard University Press: 1935.

Cartographer: Albert Herrmann.

Size: 33.4 cm.

(912.51: H433 Oversize)

1741. 1935

The march of man; a chronological record of peoples and events from prehistoric times to the present day: comprising a comparative time chart of universal history in seven sections, an historical atlas of 96 pages and 64 plates of illustration. General editor: L. H. Dawson. Historical atlas prepared by George Philip and Ramsay Muir with the collaboration in the American section of Robert McElroy. [London]: The Encyclopaedia Britannica Co., Ltd.: 1935.

Cartographers: L. H. Dawson, George Philip, Ramsay Muir, and Robert McElroy.

Size: 30.2 cm.

(902: M33 Oversize)

1742. 1935

New revised atlas of the world, latest maps, new indexes, up-to-date gazetteer. Cleveland: World Syndicate Publishing Company: 1935.

Size: 33.7 cm.

(912: W89)

1743. 1936

Commercial atlas of the world. New census by L. E. Smith . . . edited by Frederick K. Branom. Chicago: Geographical Publishing Co.: 1936.

Cartographers: Lloyd Edwin Smith and Frederick K. Branom.

Size: 55.9 cm.

At head of title: The Geographical Publishing Company's Commercial Atlas.
(912: Sm5c Oversize)

1744. 1942

Encyclopaedia Britannica world atlas; with physical and political maps, geographical comparisons, a glossary of geographical terms, a gazetteer index, and with geographical summaries, world spheres of influence, prepared by G. Donald Hudson under the editorial direction of Walter Yust. Chicago: Encyclopaedia Britannica, Inc.: 1942.

Cartographers: C. Donald Hudson and Walter Yust.

Size: 42.2 cm.

Laid in is "Your post-World War II supplement."
(912: En12 Oversize)

1745. 1955-

The Times atlas of the world. Edited by John Bartholomew. Mid-century edition. London: Times Publishing Co.: 1955- .

Cartographer: John Bartholomew.

Size: 50 cm.

Each volume includes index-gazetteer.
(912: T485)

1746. [1955]

World atlas and gazetteer, presenting the world in its geographical, physical, and commercial aspects. New York: P. F. Collier & Son Corp.: [1955].

Size: 36.8 cm.
(912: C69w)

1747. 1960

Photographic lunar atlas based on photographs taken at the Mount Wilson, Lick, Pic du Midi, McDonald and Yerkes Observatories. Edited by G. P. Kuiper with the collaboration of D. W. G. Arthur, E. Moore, J. W. Tapscot, and E. A. Whitaker. Chicago: The University of Chicago Press: 1960. Photographs.

Size: 54 cm.

Provenance: Presented by Roger Butterfield, March 1974.
(523.39: K95p Oversize)

1748. 1961

Early maps of North America. Robert M. Lunny. Newark, New Jersey: The New Jersey Historical Society: 1961.

Size: 24.8 cm.
(912.7: L97e)

1749. 1965

A collection of charts of the coasts of Newfoundland and Labrador, &c . . . Drawn from original surveys taken by James Cook and Michael Lane, surveyors, Joseph Gilbert, and other officers in the king's service. London. San Francisco: 1965.

Cartographers: James Cook, Michael Lane, and Joseph Gilbert.

Engraver: Thomas Jefferys.

Size: 46 x 31 cm.

(917.18: C77c.s Oversize)

1750. 1966

A map of the inhabited part of Virginia containing the whole province of Maryland with part of Pennsilvania, New Jersey and North Carolina. Drawn by Joshua Fry & Peter Jefferson in 1751. London: Robert Sayer: [1794]. Charlottesville, Virginia: University of Virginia Press: 1966.

Cartographers: Joshua Fry and Peter Jefferson.

Engraver: Thos. Jefferys.

Size of each sheet: 40.5 x 63.3 cm.

There is an index with this map.

Facsimile.

(912.75: F94f.2)

1751. 1969

Atlas géographique et physique du royaume de la Nouvelle-Espagne vom verfasser auch kurz Benannt: Mexico-Atlas. By Alexander von Humboldt. Paris: 1811. Stuttgart: 1969.

Cartographer: Alexander von Humboldt.

Size: 47 cm.

Provenance: Presented by the Federal Republic of Germany, 1970.

(917.2: H88.ar Oversize)

1752. [ca. 1969]

Benjamin Franklin's campaign on the Lehigh. 1756. N.p.: [ca. 1969].

Size: 44.3 cm.

(912.748: F85 Oversize)

1753. 1970

The national atlas of the United States of America. Washington, D.C.: U.S. Department of the Interior. Geological Survey: 1970.

Size: 49 cm.

(912.73: Un4n)

1754. 1970

The water masses of the North Atlantic Ocean. A volumetric census of temperature and salinity. W. R. Wright and L. V. Worthington. Serial atlas of the marine environment, fol. 19. New York: American Geographical Society: 1970.

Cartographers: W. R. Wright and L. V. Worthington.

Size: 41 cm.

Provenance: Presented by the Woods Hole Oceanographic Institute, 1970.

(912.7: W93w Oversize)

1755. 1974

[World ocean atlas]. Sergei G. Gorshkov, ed. vol. 1. Pacific Ocean. New York: Pergamon Press: 1974.

Cartographer: Sergei G. Gorshkov.

Size: 46.8 cm.

In Russian, with an introduction and index in English.

(912: G68w Oversize)

1756. 1976

America in maps, dating from 1500 to 1856. Compiled and edited by Egon Klemp. New York & London: Holmes & Meier publishers: 1976.

Size: 51.7 cm.

(912: K67a Oversize)

1757. 1976

Atlas of early American history. The Revolutionary era. 1760-1790. Lester J. Cappon, editor-in-chief. Princeton, N.J.: Princeton University Press: 1976.

Size: 37.5 cm.

(911.7: C17a)

1758. 1976

Hydrographical basin of the Upper Mississippi River from astronomical and barometrical observations, surveys and information by J. N. Nicollet. In the years 1836, 37, 38, 39 and 40; assisted in 1838, 39, & 40, by Lieut. J. C. Fremont, of the Corps of Topographical Engineers under the supervision of the Bureau of the Corps of Topographical Engineers and authorized by the War Department. Reduced and compiled under the direction of Col. J. J. Abert and W. H. Emory. N.p.: 1843. Printed from an original copy, 1976.

Cartographers: J. N. Nicollet, J. C. Fremont, J. J. Abert, and W. H. Emory.

Engraver: E. F. Woodward.

Size: 24 cm.

Provenance: Presented by Edmund C. and Martha C. Bray, 30 December 1976.

(912.77: N54h)

1759. 1976

Topografiske atlas Danmark 82 Kortudsmit med beskrivelse. Atlas over Danmark. Ser. II, Bd. 1-2. Det. K. Danske Geografiske Selskab. Copenhagen: C. A. Reitzels Forlag: 1976.

Cartographers: N. Kingo Jacobsen, Ruth Helkiaer Jensen, and Kr. Marius Jensen.

Size: 34 cm.

(910.6489: C79a Oversize)

1760. [ca. 1977]

Atlas of the oceans. New York: Rand McNally & Co.: [ca. 1977].

Size: 38 cm.

(551.46: R15a)

1761. 1978

A historical atlas of colonial Virginia. John S. Hale, cartographer. Verona, Virginia: Old Dominion Publications: 1978.

Cartographer: John S. Hale.

Size: 40.7 cm.

Provenance: Presented by John S. Hale, 1978.

(912.755: H13h Oversize)

1762. [ca. 1978]

The Rand McNally new concise atlas of the universe. New York: Rand McNally & Co.: [ca. 1978].

Size: 38 cm.

(523.1: M78r.rev)

1763. 1981

Early maps. By Tony Campbell. New York: Abbeville Press: 1981.

Size: 38.6 cm.

(912: C14e Oversize)

1764. 1982

The Hammond-Harwood House atlas of historical maps of Maryland, 1608-1908. By Edward C. Papenfuse and Joseph M. Cole III. Baltimore and London: The Johns Hopkins Press: 1982.

Size: 28.5 x 36.4 cm.

(911.7: P19h Oversize)

1765. 1983

Atlas of the Lewis and Clark expedition. Gary E. Moulton, editor. Lincoln, Nebraska, and London: University of Nebraska Press: 1983.

Cartographers: Lewis and Clark.

Size: 40.5 cm.

Provenance: Presented by the University of Nebraska Press, September 1983.

This is volume 1 of *The Journals of the Lewis & Clark Expedition.* Gary E. Moulton, editor.

(917.3: L58j.m)

1766. 1983

Early sea charts by Robert Putnam. New York: Abbeville Press Publishers: 1983.

Size: 38.8 cm.

(912: P98e Oversize)

1767. 1983

Washington College presents: On the Map. An exhibit and catalogue of maps relating to Maryland and the Chesapeake Bay honoring George Washington at the beginning of the third century of Washington College at Chestertown, Maryland. February 21-March 6, 1983. By Russell Morrison, Edward C. Papenfuse, Nancy M. Bramucci and Robert J. H. Janson-La Palme. Chestertown, Maryland: Washington College: 1983.

Size: 28.5 cm.

Provenance: Presented and inscribed by Russell Morrison, March 1983.

(912.73: M83o)

1768. 1984

United States coastal charts 1783-1861. By Peter J. Guthorn. Exton, Pa.: Schiffer Publishing Co.: 1984.

Size: 31 cm.

(912.73: G98u)

1769. n.d.

Atlas accompanying volume III on mining industry. U.S. Geological Expedition of the Fortieth Parallel. Clarence King, geologist in charge. New York: Julius Bien engraver and printer: n.d.

Cartographers: Clarence King, R. H. Stretch, I. E. James, Marlette, Hunt, and T. D. Parkinson.

Engraver: Julius Bien.

Size: 19½ in.

(557.31: Un3k, vol. 3 Oversize)

1770. n.d.

Atlas of magnetic declination of Europe for epoch, 1944-5. Washington: Army Map Service: n.d.

Size: 66.2 cm.

(538.7: B62 Oversize)

1771. n.d.

Sectional maps, showing 2,500,000 acres farm and wood lands, of the Illinois Central Rail Road Company, in all parts of the state of Illinois, with the line of their rail road, and other interesting rail roads. The shaded sections show the lands of the company. N.p.: n.d.

Size: 23.5 cm.

(912.773: I16)

PART IV

Globes and Model

1772. 1798

Celestial globe, inscribed: To the Rev. Nevil Maskelyne, D.D., F.R.S., Astronomer Royal, this new British Celestial Globe. Containing the positions of nearly 6000 Stars, Clusters, Nebulae, Planetary Nebulae, etc. correctly computed and laid down for the year 1800 from the latest observations and discoveries by Dr. Maskelyne, Dr. Herschel, the Rev. Mr. Wollaston, etc. etc. is respectfully dedicated by his most obedient h[um]ble servents [sic] Wm. & T. M. Bardin. Label pasted on: Sold by W. & S. Jones, Holborn, London. 1798.

Cartographers: William Bardin and T. M. Bardin.

Size: 18 in. diameter.

Colored.

Provenance: Presented by William Jones, 24 June 1799. For terrestrial globe, see below, No. 1773.

The globe is contained by a 15¼ in. high mahogany four-legged ringstand with crossbars and horizon circle and a brass meridian circle support for turning the globe.

William Jones, a London optician and member of the Society, presented this celestial globe, and a terrestrial one [see no. 1773], to the Society on 24 June 1799. He wrote on 3 November 1795 that he was "chiefly engaged with others, in completing a new pair of 18 in. globes more accurate than any hitherto constructed" which he expected to be ready "about next spring." He promised to "present them with a [sic] folios when they are ready." He wrote again on 24 June 1799:

> Enclosed you have a bill of Lading for a pair of Globes, which are shipt on board the Active, and if they have a safe arrival I will trouble you to forward; and further, to favour me with the presentation of them to the Philosophical [Society]. . . . I deferred sending them until the Geography had received the best additions and discoveries, and I hope they will be considered an useful addition to the collections of the society [Archives. W. Jones to J. Vaughan and the APS, 3 November 1795, and 24 June 1799].

Jones did forward some folios with the globes but the one he hoped to include and which was not published until 1803 was: George Adams, *Astronomical and geographical essays: containing a full and comprehensive view, on a new plan, on the general principles of astronomy; the use of the celestial and terrestrial globes* . . . 5th ed. Corrected and enlarged by William Jones, London. Printed for W. and S. Jones. 1803.

See also: APS *Memoirs*, vol. 53. Robert P. Multhauf, *A catalogue of instruments and models in the possession of the American Philosophical Society* (Philadelphia: American Philosophical Society: 1961) for additional information pertaining to the globes and model.

(Philosophical Hall)

1773. [1798]

Terrestrial globe, inscribed: To the Rt. Honorable Sir Joseph Banks, Bart., President of the Royal Society, this new British terrestrial globe containing all the latest discoveries and communications from the most correct and authentic observations and surveys to the year 1799 by Capt[ai]n Cook and more recent navigators. Engraved from an accurate drawing by Mr. Arrowsmith, Geographer, is respectfully dedicated by His most obedient h[um]ble servent [sic] Wm. & T. M. Bardin. Label pasted on: Sold by W. & S. Jones, Holborn, London. 1798.

Cartographers: William Bardin, T. M. Bardin, and Aaron Arrowsmith.

Size: 18 in. diameter.

Colored.

Provenance: Presented by William Jones, 24 June 1799. For comments on the presentation, see celestial globe notes, no. 1772, above.

The globe is contained by a 15¼ in. high mahogany four-legged ringstand with crossbars and horizon circle and a brass meridian circle support for turning the globe. The printing is difficult to read: this title follows that of the globe in the Essex Institute, Salem, Mass., as described in *A catalogue of early globes made prior to 1850 and conserved in the United States. A preliminary listing* by Ena L. Yonge. New York: American Geographical Society: 1968.

(Philosophical Hall)

1774. n.d.

Model in plaster of "Mont Blanc en relief," Switzerland. N.d.

Size: 4 in. x 6½ in. x 2 in. high.

Colored.

Provenance: Presented by Ferdinand Rudolph Hassler, 1805.

(Philosophical Hall)

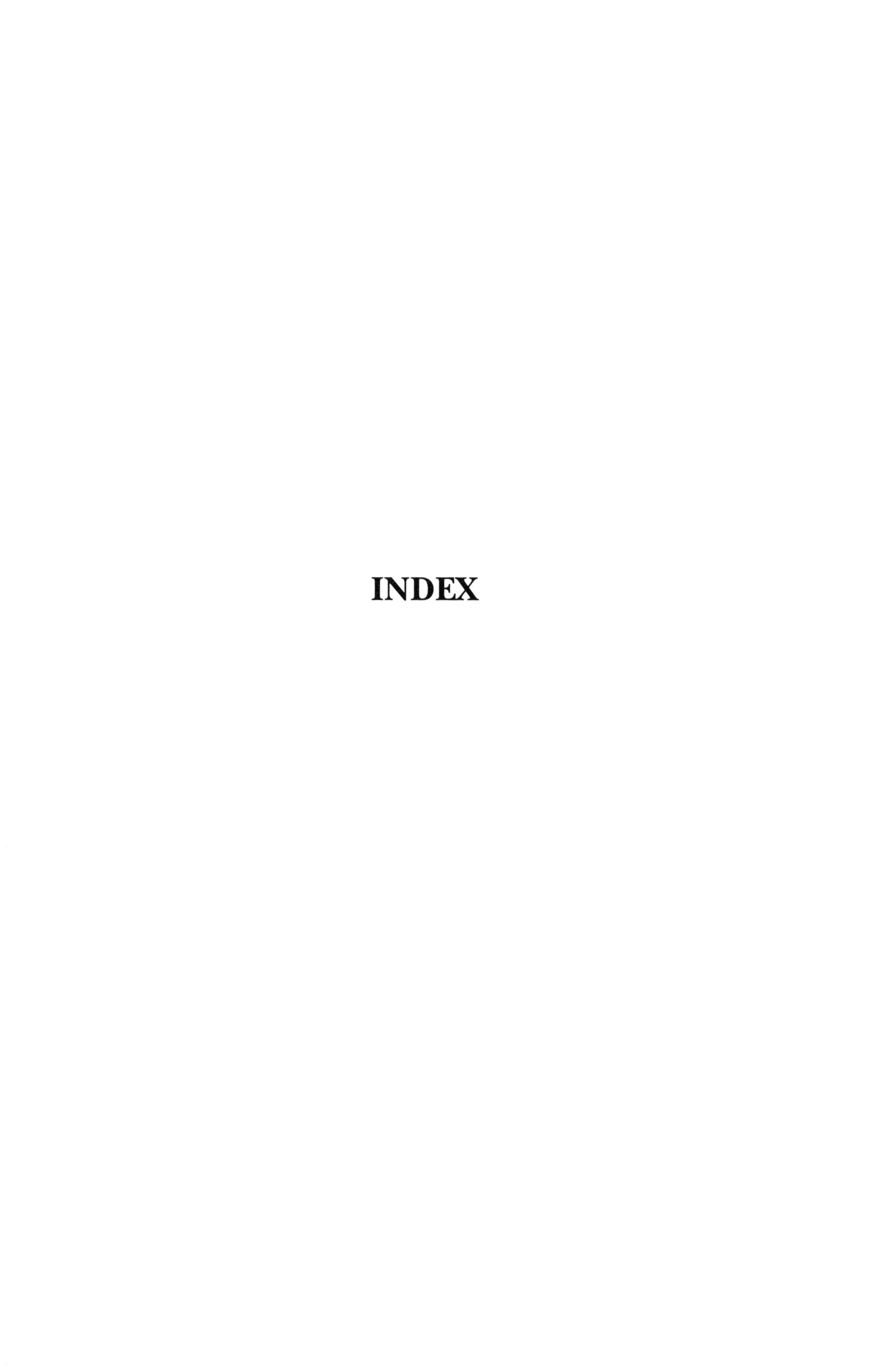

INDEX

A——, D., cartographer, 845
Aakier, Denmark, 174
Aalborghuus, Denmark, 176, 178, 180
Aargau, Brugge, Switzerland, geology, 237
Abacou Point to Grande Baye du Mesle, Vache island, St. Domingue, 32 (53)
Abbeville Press, publisher, 1763, 1766
Abbot, Henry Larcom (1831-1927, soldier), cartographer, 1124-28, 1131, 1141; donor 1123, 1126-28, 1131
Abella y Casariego, Enrique (fl. 1884-1898, Spanish geologist), cartographer, 413; *Descripción física, geológica y mineria en la isla de Panay,* 413
Aberdeen, Scotland, 148
Abert, John James (1788-1863, geologist, soldier), donor, 1332
Abert, S. T., cartographer, 1080, 1229, 1310, 1758
Absecon Inlet, NJ, 882
"An abstract of sundry papers and proposals for improving the inland navigation of Pennsylvania and Maryland . . . " APS *Transactions,* 767
Académie royale des sciences, Paris. *See:* Institut de France.
Acadie or Acadia. *See:* Nova Scotia, Canada.
Account of a voyage up the Mississippi River by Z. M. Pike, 1330
An account of the improvements carried on by Sir John Sinclair on his country estate in Scotland, 156
Account of expeditions to the source of the Mississippi, by Z. M. Pike, 1330
Acheson, H., lithographer, 1325; publisher, 1325
Ackworth, GA, 1181
Acul Bay, St. Domingue, 32 (33); to Petit Anse, St. Domingue, 32 (52)
Adair Co., KY, 1249
Adams, George (1750-1795, English instrument maker), *Astronomical and geographical Essays* . . . 5th ed. Corrected and enlarged by William Jones, 1803., 1772
Adams, H., cartographer, 853, 1153
Adams, Henry Martin, cartographer, 1341
Adams, I. H., cartographer, 1195
Adams, John Alexander (1803-1880, engraver), 878
Adams, J. W., cartographer, 740
Adams County, MS, Washington, 1 (3)
Adams County, PA, 927, 955
Address delivered on the dedication of the cemetery, Joseph Story, 704
Adirondack Mountains, NY, 32 (84), 842, 1298
Adlum, John (1759-1836, pioneer in viticulture), cartographer, 913
Admiralty Inlet, Canada, to Gray's harbor, Washington, 1462
Admiralty Inlet, Washington, 1469
Adriatic Sea, 358
Advenier de Breuilly, A. P., cartographer, 32 (34), (36), (48), (53)
Adventure (ship), 1561, 1575
Aerøe, Denmark, 173
Africa, 74a, 87a, 87b, 112, 421-26, 443-44, 446; Central Africa, 443-46; Gold Coast, 422, 335; Guinea, 182; North Africa, 117-18, 325; West Africa, 434-42; West Africa, Cape Blanco to Cape Virga, 434; West Africa, coast line, 439; West Africa, coast line, Gulf of Guinea, 441
Agate harbor, Lake Superior, 1288
Agra, India, 399
Agriculture, Guatemala, 1508; Jackson Purchase, KY, 1248
Aguirre, A. A., cartographer, 1339
Ailly, H. A. H. d', cartographer, 1336-37, 1339-40
Aire and Calder Navigation, England, 139
Airy, Sir George Biddle (1801-92, astronomer), 63
Aitken, Robert (1734-1802, engraver, publisher), 697; engraver, 497-98, 626, 1082, 1095; publisher, 785
Akerland, B., engraver, 1538
Akin, James (1773-1846, engraver), engraver, 1161-63
Akudnirn, North West Territory, Canada, 487
Alabama, 1134, 1179, 1200-208; 1504; Bessemer, 1208; Birmingham, 1208; Black Warrior River, 1202; Bon Secour Bay, 23 (6); French emigrants, 1202; Mobile, 1208; Mobile bar, 1200; Mobile Bay, 23 (10), (11), 644, 1190; Tuscaloosa River, 1202
Alachua Savana, in East Florida, 4 (1)
Alagna, J. Giacomo, *A complete set of new charts . . . of the coast of Portugal and the Mediterranean Sea,* 1650
Alaska, 74b, 456, 546, 587, 598-99, 608, 610, 616, 621-22, 1496-1500; Boundary tribunal, 1499; geology, 623; loess, 623; meteorology, 1498; Sitka, 1497
Albany, NY, 576-77, 826, 838; to Newborough, NY, 787-794; to Poughkeepsie, NY, 804-809

Albemarle Co., VA, 32 (54)
Albemarle Sound, NC, 32 (91), 759
Albert I (prince of Monaco, 1848-1922), 1613-14
Alberta, Canada, 482, 516-19, 531-36, 543; Calgary, 543; elevators (grain), 519; Jasper, 543-44; Lake Louise, Rocky Mountains, 533; Lesser Slave Lake, 531; Peace River, 534, 536; Rocky Mountains, Banff, 532
Albright, Chester E., cartographer, 1022
Alden, James (1810-1877, U.S. naval officer, surveyor), cartographer, 1204-205, 1462-64, 1466-68, 1470, 1479-81, 1485, 1629
Alden, William C., cartographer, 465
Aleutian Islands, U.S., 463, 608
Alexander, James, publisher, 919
Alexandria, MD, to Annapolis, MD, 1054-55
Alger, Francis (1807-1863, Canadian mineralogist), cartographer, 494
Algiers, 436
Algoma, Ontario, Canada, 509
Algonkian Indians, 48
Allahabad, India, 399
Allardice, Robert Barclay (1779-1854, engraver), engraver, 147, 154, 315, 352, 405, 1068, 1539
Allegany Co., PA, 32 (31)
Allegany Mountains, 32 (3), 759
Allen, John, Jr.(?) (d. 1831, English publisher), publisher, 761, 1084
Allen, Rolland Craten (1881-1948, geologist), cartographer, 1351
Allen, William Frederick (1846-1915, railroad expert), cartographer, 589
Allentown, PA, to N.Y.C., NY 870
Allgemeiner Hand-Atlas der Ganzen Erde . . . A. C. Gaspari, 1677
Almon, John (1737-1805, British bookseller), publisher, 757, 1321
Almy, J. J., cartographer, 1138
Alps, geology, 281
Alsek River, Yukon Territory, Canada, 547
Altamaha River, GA, 32 (12)
Altimetry-U.S., 591
Alto Marañon River, Paita to Eten, Peru, 1570
Alto Purvus, Peru, 1570
Alto Yurua, Peru, 1570
Alton, IL, 1324
Alzate y Ramirez, Joseph Antonio de (1738-1799, Spanish meteorologist), cartographer, 1501
Amazon River 1638, 1666; Peru, 1570
Amazonas Territory, Venezuela, 1569
Amboy, NJ, 890
Amenta, ____, cartographer, 1040
America in maps, Egon Klemp, 1756
American Anthropological Association, 37
American Association of Petroleum Geologists, 466
The American atlas, T. Jefferys, 550
The American atlas, John Reid, 917, 1159, 1175, 1240
The American Bank Note Co., engraver 589; publisher 594
American Colonization Society, 440
American Ethnological Society, 464, 468
The American gazetteer, Jedidiah Morse, 406, 458, 629, 1085, 1544, 1558
American Geographical Society of New York, 100, 102-103, 1562; publisher, 1735, 1754
The American geography, Jedidiah Morse, 628, 1083, 1540, 1557
American harbor to Kingawa, Canada, 6 (4)
American Indians. *See:* Indians
American Indian Life, E. C. Parsons, 37
American Institute of Architecture, Philadelphia chapter, 1033
American Journal of Science and Arts, 1437
American military atlas, 756
American military pocket atlas, 454, 627, 784, 1081, 1536, 1655
The American museum, 74, 1624; "A description of the surprising cataract, in the great river Connecticut," 728
American Museum of Natural History, N.Y.C., 35
American Philosophical Society, 49, 466, 551, 631, 1042, 1077, 1122; archives 1 (1-3), manuscripts communicated to the APS 2 (1-5); Magellanic Premium 1 (1); *Memoirs,* 1772; president, Thomas Jefferson 1 (1); *Proceedings* 1 (3), (87), 1053, 1227, 1622-23; *Transactions* 1 (2), 2 (1), 32 (18a), (19), (23-4), (84), 66-67, 73, 92, 107, 128, 153, 160, 192, 195, 233, 308, 350, 361, 367, 400, 424, 567, 570, 755, 760, 1166, 1298, 1542, 1554-55, 1625-26, 1628; "An abstract of sundry papers and proposals for improving the inland navigation of Pennsylvania and Maryland," 767
American Photo-Lithographic Co., lithographer, 462; publisher, 1181, 1261
American Revolution, 32 (6), (9), 38 (1), 469, 551, 626, 669, 766, 785, 1081, 1089, 1093, 1095, 1171, 1656, 1757; Boston, MA,

700; Charlestown, MA, 702; Philadelphia, PA, 995-96
American State Papers, 1376
American universal geography, Jedidiah Morse, 76, 368
Ames, Charles Lesley, donor, 26
Amherst, Jeffery, baron Amherst (1717-1797, British field-marshal), 784
Amstelland, The Netherlands, 213
Amsterdam, The Netherlands, 213
Amur River, 378
Anderson, ——, 760
Anderson, Alexander (1775-1870, U.S. engraver), cartographer, 656, 665, 674, 739, 1069; engraver, 262, 817, 1119, 1149, 1240, 1245
Anderson, Hugh (1811-1824, engraver), 1504
Anderson, John Alexander (1834-1892, U.S. Presbyterian clergyman, congressman), cartographer, 779
Andrade, ——, cartographer, 1040
André, ——, engraver, 1237
Andrée, Saloman August (1854-1897), *Handatlas,* 100
Andrews, A. B., Co., lithographer, 1451; publisher, 1451
Andrews, Ebenezer T., publisher, 368, 1540, 1557, 1680
Andrews, Emma B., collection, 3
Andromeda (constellation), 50 (1)
Angelo, G. N. (fl. 1776-1806, Italian engraver in Denmark), engraver, 174, 176, 178-79, 180-81, 183, 185, 300
Ani mines, Japan, 30
Ankor, Cambodia, archaeology, 402
Anna, czarina of Russia, 29
Annapolis, MD, harbor, 1073; to Alexandria, MD, 1054-55; to Bladensburg, MD, 1056-57; to Bowling-green Ordinary, VA, 1096; to Dumfries, VA, 1097; to Fredericksburg, VA, 1098-99; to Hanover and Newcastle, VA, 1100; to Hanover Court-House, VA, 1101; to Head Lynchs Ordinary, VA, 1102; to New Kent Court-house, VA, 1103-104; to Philadelphia, PA, 908-11, 1058-63; to Todd's Ordinary, VA, 1105; to Williamsburg, VA, 1106-107; to York, VA, 1108-109
Annin, William B. (fl. 1813-1839, engraver), engraver, 641, 678
"Annual passage of herrings," 1623
Annual report, 1900, NJ Geological Survey, Forests of NJ, 891
Antarctic Ocean, bathymetry, 102
Antarctic regions, 101-104
Antarctica, 32 (93)
Antietam, MD, 774
Antigua, 1517, 1661; English harbor, 1517
Antigua, Mexico, 32 (18)
Antinous (constellation), 50 (1-2)
Antler Creek, British Columbia, Canada, 537
Antwerp, Belgium, 216
Anville, Jean Baptiste Bourguinon d' (1697-1782, French cartographer), 32 (7), 70, 88, 90, 347, 363, 365, 370, 422-23, 435, 451, 1665, 1669; *Atlas antiquus,* 1665, 1669; *A complete body of ancient geography,* 1654
Apenrade, Denmark, 300
Apfel, E. T., cartographer, 465
Appenzell, Switzerland, 228
Appiano, C. R., cartographer, 1336
Appleton, WI, 1372
Appomattox Court House, VA, 775; battlefield, 774
Aquanishuonîgy, 752-53
Arabia, 371
Archaeology, 134; Ankor, Cambodia, 402; Egypt, 3 (1-3)
Archives of the American Philosophical Society, 1 (1-3)
Arctic exploration, 6 (23), 32 (79); regions, 32 (88), 92-100
Arctic Exploring Expedition of 1860 and 1861, 96
Arctic Ocean, 93, 95
Arctic regions, 32 (88), 92-100
Argentina, 1560, 1578-80; Cordoba-Santa Fe, 1562; hydrometry, 1580; La Plata tributaries, track survey, 1563-64; meteorology, 1580; Ministerio de obras publicas, 1579; railroads, 1579; San Carlos, 1561; Strait of Magellan, 1575; Tierra del Fuego, 1576
Arizona, 1454-56; Phoenix, 1455-56; railroads, 1455-56; Tucson, 1455-56
Arkansas, 1263-64, 1376; Hot Springs, 1263; Little Rock, 1263
Arkansas River, 39
Arkton (?), 46 (3)
Arlett, William, cartographer (fl. 1844-1851), 336-37
Armaschevsky, P., cartographer, 319
Armstrong, John, cartographer, 33 (4)
Armstrong, Marcus, cartographer, 152
Armstrong County, PA, Kittanning, 919
Arndt, H. H., cartographer, 984
Arras, France, 261
Arrowsmith, Aaron (1750-1823, English car-

tographer), cartographer 91, 460, 1675, 1678, 1680, 1773; and Lewis, *A new and elegant general atlas,* 1678, 1680
Arrowsmith, John, publisher, 1561
Arsebémiling to Ipiakdiuak, Canada, 6 (3)
Art du tunnelier, Duhamel du Monceau, 14
Arthur, D. W. G., 1747
Artibonite bay to Baradaires bay, Saint Domingue, 32 (27)
Artois, France, 248, 251
Ashburner, Charles A., cartographer, 958, 960
Ashford, ——, publisher, 469
Ashford, Kent: Harry Margery, Lympne Castle, publishers, 469
Ashman, J., cartographer, 440
Ashmead, Henry B., publisher, 1709
Asia, 74a, 87a, 112, 363-70, 389; Asia Minor, 117; North 318
Aspin, Jehoshophat (fl. 1814-1840), cartographer, 1684
Assiniboia, Saskatchewan, Canada, 516
Association internationale du Congo, 443
Assyria, 374; Nimrúd, 374; Ninevah, 374; Selamiyeh, 374
Astley, Thomas, donor, 1683
Astrolabe (ship), 1673
Astronomical and geographical essays . . . by George Adams. 5th ed. Corrected and enlarged by William Jones, 1803, 1772
Astrup, Denmark, 179
Atacama, Chile, 1562
The Athenaeum of Philadelphia, donor, 1629
Atkinson, W. G., cartographer, 1123
Atlanta, GA, military campaign, 1180-81
Atlante corografico storico . . . Due Sicilie . . . Benedetto Marzolla, 1691
Atlante geografico del regno di Napoli. G.A.R. Zannoni, 1667
Atlante internazionale del touring clvb Italiano . . . L.V. Bertarelli, *et al.*, 1737
Atlantic City, NJ, 890
The Atlantic Neptune, J. F. W. Des Barres, 1659
Atlantic Ocean, 1621-33; bathymetry, 1612; chart of a voyage from Philadelphia, PA, to Cork, Ireland, l; coast (U.S.), 32 (86); coast (U.S.) lighthouses, 624; coast (U.S.) to Rocky Mountains, glaciers, 625; coast (U.S.) Savannah River, GA, to St. Marys River, FL, 1178; Delaware breakwater, 1047; North Atlantic, 553, 1754; north FL coast, 1178; North Atlantic navigation chart, 1628; physiographic diagram, 1632-33; pilot charts, 1631; wreck chart 1630: South Atlantic, 102; bathymetry, 102; physiographic diagram, 1633
Atlas . . . , Samuel Holland, 634
Atlas, Frederick de Wit, 1637
Atlas accompanying volume III on mining industry, U.S. Geological Expedition to the 40th Parallel, 1769
Atlas Américain septentrionale, G. L. Le Rouge, 904
Atlas Antiquus, D'Anville, 1665
Atlas classica, H. S. Tanner, 1694
Atlas de Filipinas, 1723
Atlas de l'Europe, Henry Perkin, 1692
Atlas de l'océan pacifique, A. J. von Krusenstern, 1688
Atlas des marais pontins, 1686
Atlas encyclopédique . . . , Rigobert Bonne, 1666
Atlas géographique et physique du royaume de la Nouvelle Espagne, A. von Humboldt, 1751
Atlas germaniae, J. B. Homann, 1648
Atlas historique, généalogique, chronologique, et géographique, E. A. D. Las Casas, 1676
Atlas maritime [The French Neptune], 1639
Atlas national et général de la France, Cassini de Thury, 1670
Atlas novus coelestis in quo mundus spectabilis . . . , Copernicus and Tycho Brahe, 1644
Atlas of Canada, James White, 1724
Atlas of Delaware County, PA, Benjamin H. Smith, 1709
Atlas of early American history, Lester J. Cappon, ed.-in-chief, 1757
Atlas of magnetic declination of Europe for epoch 1944-45, 1770
Atlas of physical geography, Thomas Milner, 1698
Atlas of ports, cities, and localities of the island of Cuba, 1721
Atlas of the historical geography of the U.S., Charles Oscar Paullin, 1735
Atlas of the Lewis and Clark expedition, Gary E. Moulton, ed., 1765
Atlas of the oceans, Rand, McNally & Co., 1760
Atlas of the oil region of Pennsylvania, F. W. Beers, 1701
Atlas of the Philippine Islands, U.S. Coast and Geodetic Survey, 1723
Atlas of the state of South Carolina, Robert Mills, 1689
[Atlas to accompany] *Annual report upon the geographical survey west of the 100th meridien,* 1707

Atlas to [William Guthrie's) system of geography, 92, 107, 1554 (*see also:* M. Carey)
Atlas universel, Adrien Hubert Brué, 1560
Atlas von Land- und Seekarten vom Japanischen Reich, Philippe Franz Siebold, 1699
Atlases, 1636-1771
Atranto River, Colombia, 1568
Attahvich, Sam, cartographer, 621
Attaslopi (?), Labrador, Canada, 48 (8)
Atwood, J. M., engraver, 1363
Aub, ——, lithographer, 941, 1017
Aubin, ——, 1643
Audendried, William, 32 (75-76)
Augustin, J. J. Inc., publisher, 464
Austria, 308, 1641, 1741; geology, 278; Krems, geology, 278; Manhardsberge, geology, 278; roads, 280
Australia, 1581-1610; Dawson River, 1599; Mackenzie, River 1599; New South Wales, 1607-10; N.S.W. Broken Hills district, geology, 1610; N.S.W. Conjola, 1607; N.S.W. Department of Mines, 1609; N.S.W. Gerringong district, 1607; N.S.W. Geological Survey, 1607-08, 1610; N.S.W. geology, 1607-10; N.S.W. Little Forest 1607; N.S.W. Sydney, geology, 1608; Queensland, 1586-1605: Queensland, Bowden, 1600; Charter's Tower gold field, 1586; Queensland, coal, 1599, 1604-105; Geological Survey, 1586; Queensland, geology, 1586-1605; Queensland, Gladstone, 1599; Queensland, gold fields, 1586-87, 1589, 1594-1603; Queensland, Knobs Oaks Rush, 1597; Queensland, mining centers, 1596; Queensland, railroads, 1596; Queensland, Rockhampton, 1599; Queensland, Cape York, 1588, 1600; South Australia, 1581; South Australia, geology, 1582-83; Victoria, 1606; Victoria, Department of Lands and Survey, 1606; Western Australia, 1584-85; Western Australia, Geological Survey, 1584; Western Australia, geology, 1585; Western Australia, gold mines, 1584; Western Australia, Meekatharra, 1585
Austro-Turkish wars, 29
Avon (ship), 441
Axtell, O. N., cartographer, 1336
Axua province, Santo Domingo, 1551
Aylett's Warehouse, VA, to Williamsburg, VA, 1110-11
Ayres, H. R., cartographer, 1249
Azores Islands, Portugal, 422
Babbitt, C. W., cartographer, 1341
Bach, J. G., lithographer, 304
Bache, Alexander Dallas (1806-1867, physicist), 684, 686, 690, 692-93, 743-46, 773, 832-33, 847-51, 1073-74, 1227; cartographer, 581-82, 642-47, 684-88, 690-93, 705, 746, 773, 849-51, 853, 1049, 1073-75, 1130, 1138, 1153, 1167-68, 1178, 1182, 1195, 1203-205, 1209, 1230-31, 1462-64, 1467-68, 1479-83, 1485; donor, 56, 581-82, 642-43, 688, 691, 741, 745, 881, 1049, 1130, 1153, 1167, 1195, 1203-204, 1209, 1230, 1462-64, 1466-68, 1479, 1481-82
Bache, Charles M., cartographer, 882-83, 885, 1137
Bache, Franklin (d. 1936), "Where is Franklin's first chart of the Gulf Stream?" APS, *Proceedings,* 1622
Bache, George Mifflin (1811-1846, hydrographer), 881, 885, 1049, 1073, 1075
Bache, Hartman (1798-1872), 1402: cartographer, 880, 1166; donor, 740, 880, 1166, 1191, 1279, 1310, 1469
Bache, Richard Meade (1830-1907, engineer, APS 1884), cartographer, 851 957, 1020-21, 1046
Bachelder, John Badger (1825-1894, map publisher, author), publisher, 956
Baffin Bay, Canada, 6 (2), 23 (8), 88, 90, 93, 99; Cumberland Sound, 32 (85); Disco Bay, 93; Niantilac harbor, 32 (85); Omenak fiord, 93
Baffin Island, Canada, 6 (5); Cumberland sound, 6
Baffinland, Canada, 6 (6)
Bagby, F. H., cartographer, 1249
Bahama Banks, 1187, 1194
Bahama Islands, 32 (29a), 1085-87, 1661
Bahar, India, 399
Bailey, Loring Woart (Canadian geologist), 485
Baine, H. E., engraver, 531
Baja California, Mexico, 1507
Bajo Marañon, Peru, 1570
Baker, Charles, cartographer, 671
Baker, John, 33 (4)
Baker, Newton Diehl (1871-1937, soldier), 970
Baker, Robert (surveyor), cartographer, 1517
Baker, Samuel (British naval officer), cartographer, 1518
Balbach, A., cartographer, 647, 853, 1230
Baldwin, Aug. S. (d. 1876, naval officer), car-

tographer, 851, 1230
Baldwin, Robert (1749-1810, British publisher), publisher, 786, 995
Ball, Lionel Clive (1878-, Australian geologist), cartographer, 1595
Baltimore, MD, 576-77, 766, 769, 770, 772; harbor, 644, 1075
Bancker, Charles Nicoll (1778?-1869, financier), donor, 1692
Bancker, Gerald (British engraver, cartographer), cartographer, 861
Banff, Alberta, Canada, Rocky Mountains, 532
Banjalucka, Yugoslavia, 29 (50)
Banks, Sir Joseph, bart., (1743-1820, English, president of the Royal Society of London, APS 1787), 1773
Banks (chartered), Ontario, Canada, 508; Quebec, Canada, 508
Bankson estate, 46 (12)
Bar Harbour, ME, 636
Baradaires Bay to Artibonite Bay, Saint Domingue, 32 (27)
Barbados, 1519, 1661; Bridgetown, 1519; Codrington College, 1519
Barber, John Warner (1798-1885, engraver and publisher), 12
Barclay, M. G., cartographer, 1419
Bardin, T. M. (fl. 1782-1820, English cartographer), 1772-73
Bardin, William (fl. 1782-1820, English cartographer), 1772-73
Bargeveen, The Netherlands, 197
Barker, J., engraver, 877
Barker, John R., cartographer, 642
Barker, William (fl. 1794-1803, engraver), 74c, 295, 556, 673, 714-15, 738, 814, 916, 1067, 1118, 1147, 1174, 1241, 1255, 1272
Barlow, John Whitney (1838-1914, geologist, soldier), cartographer, 1206
Barnard, H. S., engraver, 96, 1138, 1182
Barnard, John Gross (fl. 1851-1862, cartographer), cartographer, 1131
Barnard College, Columbia University, N.Y.C., NY, 37
Barnegat Bay, NJ, 886
Barnegat City, Long Beach, NJ, 894
Barnegat Inlet, NJ, 882, 888
Barnegat lighthouse, NJ, 894; to Sandy Hook, NJ, 644
Barnes, R. L. (fl. 1862-1874, publisher), publisher, 949, 1017, 1139
Barney, Joshua (d. 1867, soldier), cartographer, 1290, 1294, 1296-97
Barney, William Chase (d. 1892, cartographer), cartographer, 745
Barnstable, MA, Bay, 677; shore to N.Y.C., NY, 635
Baron, Bernard (1696-1766, French engraver), engraver, 449
Barosh, Patrick J., cartographer, 633
Barranquilla, Colombia, 1562
Barre, ——, publisher, 1659
Barrett, Leslie Park (1887-, geologist), cartographer, 1351
Barry, ——, donor, 276-80
Bartholomew, J. C., 1731
Bartholomew, J. G., cartographer, 1722; *Physical atlas,* 1722
Bartholomew, John (1831-1893, Scottish publisher), cartographer, 1715, 1745; *Handy reference atlas of the world,* 1715
Bartholomew, John, editor, *The Times atlas of the world,* 1745
Bartholomew, John, and Co., cartographer, 117, 275, 389, 1722
Bartle, R. F., company, engravers, 607, 611, 615, 854, 1196
Barton, Benjamin Smith (1766-1815, physician, naturalist, APS 1789), 32 (67), 126, 1177; cartographer, 33 (6), 1314; collection, 4
Barton, William (1754-1817, APS 1787), donor, 719
Bartram, John (1699-1777, botanist, naturalist, APS 1743), cartographer, 32 (3)
Bartram, William (1739-1823, traveller and naturalist, APS 1791), 4 (1), 1186; donor, 1186; *Travels through North & South Carolina, Georgia, East and West Florida,* 1186
Base map, U.S., 600, 613
Basel, Switzerland, geology, 238
Basire, James (1769-1822, engraver), engraver, 129
Bass Island, Lake Erie, 1278
The Bass, Scotland, 148
Basse Terre anchorage, St. Domingue, 32 (64)
Batan, Panay Island, Philippines, 409
Bates, Waldron, cartographer, 649
Bath, T. H. C., cartographer, 1604-605
Bathymetry, Antarctic Ocean, 102; Atlantic Ocean, 1612; Atlantic Ocean, south, 102; Indian Ocean, 1612; Indian Ocean, south, 102; Japan, 385; maps, 102; Norway, 165; oceans (world) 1613-14; Pacific Ocean, 1612; Pacific Ocean, south, 102
Battaille, ——, cartographer, 1613

Battle formations, 29 (22-24)
Battle of Montmorency, Quebec, Canada, 44 (1)
Battle of New Orleans, La., 1220
Bauman, Sebastian (1739-1803, soldier), cartographer, 32 (9)
Bauman, William, cartographer, 1423-24
Bauman, William, Jr., cartographer, 1198, 1224, 1389-90, 1397-98; 1412-14, 1417, 1451-6, 1471-72, 1475-76, 1490
Bayamo, Cuba, 32 (26)
Bayfield, Henry Wolsey (1795-1885, British Admiralty surveyor), cartographer, 1277
Baylor, J. B., cartographer, 1020-21, 1196
Bayly, J. [or I.] (fl. 1765-1772, British engraver), engraver, 1510, 1532, 1541, 1566
Bays, Lake of, Ontario, Canada, 490
Beach, Lansing Hoskins [b. 1860, cartographer], cartographer, 1223
Beach, R., cartographer, 394
Beagle (ship), 1561, 1575, 1577, 1620
Beaglehole, John Cawte, author, 1617; *The voyages of Capt. Cook,* 1617
Bear Island, Norway, 100
Beauchamp, William Martin (1830-1925, anthropologist), 32 (67)
Beaver Creeks, Big and Little, 33 (6)
Beaver Island, MI, 1285
Bechler, Gustave R. (fl. 1856-1876, geographer), cartographer, 1405-405, 1407-10, 1438-39, 1441
Beck, Richard, collector and cartographer, 5 (1-2)
Becker, C., cartographer, 1393
Bedford County, PA, geology, 954, 966
Beer, Wilhelm, cartographer, 56
Beers, Frederick W., cartographer, 1701; publisher, 1701
Beers, F. W., & Co., publisher, 951; *Atlas of the oil region of Pennsylvania,* 1701
Beers, J. M., cartographer, 1701
Beijing, China, 389, 396
Beiträge zur geologischen Karte der Schweis, Schweizerische Naturforschende Gesellschaft, 238
Belgium, 187, 192, 214-16, 275, 283, 444-45; Antwerp, 216; Brabant, 188, 190; Cambray, 215; Flanders, 188, 190, 214, 251; Hainaut, 215, 251; Liège, 189; Namur, 215, 251; Scheldt River, 188; Stavelo, 189
Belgrade, Yugoslavia, 29, 46, 57, 58-59
Belize (British Honduras), 1511-12; Black River, 1512; Blewfields Harbor, 1511
Belknap, Jeremy (1744-1798, cartographer), cartographer, 653; donor, 638, 663; *History of New Hampshire,* 653
Bell, Robert (1841-1917, director of the Canadian Geological Survey), cartographer, 488, 548
Bell Telephone Co. lines, map, 609
Bellanger, Jean Achille (French engraver), engraver, 294
Bellin, Jacques Nicolas (1703-1772, cartographer), cartographer, 1548, 1651
Bellin, Samuel (1799-1894, British engraver), cartographer, 360; engraver, 372
Bellows (or Great Falls), Connecticut River, 728
Bells Point Shoal, NC, 1151
Beloit and Madison railroad, 1365, 1372
Bender, ——, 29 (55)
Benedetti, Ignazio (fl. 1773, Italian engraver), engraver, 343
Bengal, India, 182, 399, 1660
A Bengal atlas, James Rennell, 1660
Benjamin Franklin Parkway, Phila., PA, 1028
Benjamin Franklin's last campaign on the Lehigh, 1752
Benner, ——, engraver, 744
Benner, H., publisher, 744
Bennett, John (fl. 1775-1784. British publisher), publisher, 152, 453, 550, 627, 756, 784, 1081, 1093, 1536, 1654-55, 1663-64
Bennett, Richard (fl. 1758-1780, British engraver, draughtsman, engineer), cartographer and engraver, 106
Benton, Thomas Hart [1782-1858, Senator], donor, 1426
Berbice River, Guyana, 1549
Bergen County, NJ, 886
Bergsten, J.M., cartographer, 984
Bering, Vitus (1680-1741, Danish navigator), 1496
Bering Strait, 100
Berkeley County, VA, 1120
Berks County, PA, Birdsboro quadrangle, 977; geology, 959
Berlin, K. Akademie der Wissenschaften, 289, 291
Berliner Lithog. Institut, lithographer, 115, 303
Bermuda, 35 (18), 1661
Bermuda Hundred, VA, 774
Bern, Switzerland, 219
Bernard, Simon (1779-1839, cartographer), cartographer, 769
Berniere, Henry de, cartographer, 32 (6)
Bertarelli, L. V., cartographer, 1737; *et al.,*

Atlante internazionale del touring Club Italiano, 1737
Berthrong, I. P., cartographer, 607, 611, 615, 1198, 1208, 1213, 1224, 1263, 1267, 1316, 1320, 1327, 1369, 1388-90, 1397-98, 1400, 1413, 1417, 1423, 1428-30, 1443-44, 1446-47, 1451, 1453-54, 1456, 1458, 1460, 1471, 1475, 1490, 1495, 1500
Bertre, ——, cartographer, 432
Bessemer, AL, 1208
Bethlehem, PA, 32 (1)
Béthune, France, 271
Betuwe, The Netherlands, 198
Beverly, MA, 675
Bevis [or Bevans], John (1693-1771, British astronomer), cartographer, 1646; *Uranographia Britannica,* 1646
Beyrich, Heinrich Ernst (1815-1896, German cartographer), cartographer, 115
Bianchi, Giulio Cesare (fl. 1800, Italian engraver), cartographer and engraver, 356
Bibert, ——, cartographer, 273-74
Bicharieh Arabs, 442
Bicon, ——, 29 (4)
Biddle, Charles John (1819-1873), donor, 8, 27-28
Biddle, Mary (publisher), cartographer, 1037
Biddle, Nicholas (1786-1844, banker), donor, 28, 1692
Bien, Julius (1826-1909, publisher, engraver, lithographer), publisher, 417, 624, 659, 774, 836, 884, 958, 1179, 1207, 1248, 1250, 1313-14, 1337, 1339-40, 1342, 1404-406, 1411, 1437, 1440-42, 1449-50, 1457, 1465, 1484, 1706, 1708, 1769; engraver, 956, 1563-64, 1769; lithographer, 94, 415-17, 580, 587, 624, 884, 886, 955, 958, 1070, 1079, 1207, 1248-50, 1313-14, 1337-40, 1342, 1377, 1394, 1405, 1411, 1437, 1440, 1442, 1450, 1465, 1484, 1704, 1706
Bien & Sterner, lithographer, 580
Biesbosch, The Netherlands, 199
Big Sandy Valley, KY, 1251
Big Walker Mountain, VA, 32 (80)
Biles, Langhorne, 15 (4)
A bill in the chancery of N.J. board of general proprietors of the eastern division of N.J., 859-60
Billin, C.E., cartographer, 960
Bingen, Germany, 29 (8)
Birdsboro, Berks Co., PA, quadrangle, 977
Birmingham, AL, 1208
Biron, ——, 1619
Biscay, bay, France, 1663, 1673-74
Biscoe, ——, 82
Bizard, ——, cartographer, 273-74
Bjelaja Gora, U.S.S.R., 317
Black, A., & Co., publisher, 1617
Black, Barbury, & Allen, publishers, 79
Black, John Charles (b. 1839, bureaucrat), cartographer, 592
Black, J. J., cartographer, 1446
Black Hills, SD, 1391-93; geology, 1394
Black Lake Harbor, MI, 1344
Black Rock Harbor, CT, 743
Black River, Belize, 1512
Black Sea, 276, 312, 360
Black Warrior River, AL, 1202
Blackburn, George, 1165
Blackburn, J. E., cartographer, 1235
Blackhorse Pike, NJ, to N.Y.C., NY, 871
Blackwell, Thomas E., cartographer, 32 (84), 1298
Blackwood, William, & Sons, publishers, 87
Bladensburg, MD, to Annapolis, MD, 1056-57
Blaeu, Willem Janszoon, (1571-1638, Dutch astronomer, geographer), cartographer, 245
Blake, George S., cartographer, 684-85, 689, 716-17, 741-45, 832-33, 849, 1049
Blake, Jacob Edmund (d. 1846, soldier), cartographer, 1193, 1227
Blakeley, J., 42 (1)
Blakelock, D. M., cartographer, 1516
Blanchford, Rufus, cartographer and publisher, 1549
Blanco, Cape, to Cape Virga, West Africa, 434
Blenheim Palace, England, 31
Blewfields Harbor, Belize, 1511
Bloch, Arthur, donor, 9, 974
Blodget, Lorin (1823-1901, statistician, climatologist, publisher, APS 1872), cartographer, 1312, 1498
Blondeau, Alexandre (1799-1838, French engraver), engraver, 182, 357, 428, 1485
Blue Hill Bay, ME, 645
Bluefield Harbor, Jamaica, 1525
Blunt, Edmund (1770-1862, hydrographer, publisher), cartographer, 717, 741-42, 744, 746, 832-33, 847-51, 853-54, 885, 1049, 1194; donor, 1194
Blunt, F., cartographer, 1138
Blunt, George William (1802-1878, hydrographer, publisher), donor, 1194
Board of general proprietors of the eastern division of NJ, *A bill in the chancery of N.J.,*

859-60
Boas, Franz (1858-1942, anthropologist, APS 1903), collection, 6; cartographer, 486-487; "The Central Eskimo," U.S. Bureau of American Ethnology, Sixth annual report, 486-87
Bode, Johann Elert (1747-1826, German astronomer of Hamburg), 75
Bodega Head, to Point Pinos, CA, 1487
Boeckh, Lt., cartographer, 1259, 1262
Boer War, 447
Børglum, Denmark, 179
Bøvling, Denmark, 181, 183
Boggs, Samuel Whittemore (1889-1954, cartographer), cartographer, 394
Bohemia. *See:* Czechoslovakia
Bohol, Island, Philippines, 417
Bois St. Lys, G. J., cartographer, 32 (25-29), (32-33), (35), (37-39), (40-47), (49-52), (55-59), (60-64). Some of these maps may be by Philippeaux.
Boise, ID, 1447-48
Bolivia, 23 (5), 1571
Bolivar (state), Venezuela, 1569
Bolles, C. P., cartographer, 1182
Bolzé, ——, cartographer, 1613
Bombay Hook to Cherry Island Flats, Delaware River, 1046
Bon Secour Bay, AL, 23 (6)
Bond, Frank, cartographer, 1350, 1355, 1412, 1447
Bonfils, C. A., cartographer, 1336
Bonne, Rigobert (1727-1794, author), *Atlas encyclopédique,* 1666
Bonney, S. B., cartographer, 1606
Bonsall, Joseph B. (civil engineer and publisher), 1037
Bontz, George F., cartographer, 284
A book of old maps delineating American history, Emerson D. Fite and Archibald Freeman, 1734
Boone County, WV, 1140
Boonton, NJ, 892
Boothia Isthmus, N.W.T., Canada, 487
Boqueta de Woiseri, J. L., cartographer, 1216
Bordentown sheet, NJ, geology, 888
Bornholm, Denmark, 185
Boschke, A. (fl. 1857-1861, surveyor and publisher), cartographer, 705, 746, 854, 1079, 1167, 1283, 1467, 1469
Bostock, Hugh Samuel (b. 1901, geologist), cartographer, 465
Boston and Providence railroad, 678
Boston, Mass., 89, 576-77, 633, 636, 695, 697-710, 718; American revolution, 700; fortifications on Boston neck, 25 (1); harbor, 644, 698, 705; to Cape Hatteras, NC, 750; Mount Auburn cemetery, 704; public library, trustees, 709
Boston Marine Insurance Co., 682
Boston Society of Natural History, 462
Bouchette, Josiah (1744-1841, British general), 631
Bougainville, Louis Antoine de (1729-1811, French navigator), 1619
Bouguer gravity anomaly map, NY, 841; gravity map, northeastern U.S., 633; southeastern Canada, 633
Boulton, S., cartographer, 423
Bouquet, Col. Henry (1719-1765, soldier), 901, 1302, 1765
Bourne, A. (fl. 1815-1820, cartographer), cartographer and publisher, 1309
Bourquin, Frederick (fl. 1868, lithographer), lithographer, 887, 1498; publisher, 887
Boutelle, Charles Addison (1839-1901, naval officer, journalist, Congressman), cartographer, 644, 647, 691, 693, 747, 1167-68, 1182
Bouvé, Elisha W. (fl. 1838-1849, lithographer), cartographer, lithographer, and publisher, 681
Bowden, Queensland, Australia, 1600
Bowditch, Nathaniel (1773-1838, seaman, mathematician, and astronomer), cartographer, 675; donor, 675
Bowen, Emanuel (d. 1767, British geographer), cartographer, 121; engraver, 122
Bowen, J.M., publisher, 575
Bowen, John T. (b. 1801, lithographer), lithographer, 942
Bowen, Lot, cartographer, 1493
Bowen, Nicholas (d. 1871, cartographer), cartographer, 1127, 1131
Bowen, Thomas (d. 1790, British engraver), engraver, 1512, 1535, 1537
Bower, John (fl. 1810-1815, engraver), engraver, 768, 926
Bower, Robert A., cartographer, 1703; *Rand, McNally & Co.'s railway guide,* 1703
Bowes, Joseph (fl. 1796 in Phila.), engraver, 1000, 1798
Bowles, Carington (1724-1793, British publisher), publisher, 289, 1517; and Robert Wilkinson, pubs., 289, 1517
Bowles, John (1701-1779, publisher), publisher, 120, 145

Bowles, Thomas (fl. 1700-1763, publisher), publisher, 120; and Robert Sayer and John Bowles, publishers, 120
Bowling-green ordinary, VA, to Annapolis, MD, 1096
Bowman, Amos (d. 1894, Canadian mining engineer), cartographer, 537-38
Bowman, Isaiah (1878-1905, geographer, APS 1923), donor, 103
Boyce, William M. (d. 1855, cartographer), cartographer, 685, 688, 741, 1020-21, 1049, 1153, 1166
Boyd, W. H. cartographer, 542, 1333, 1336
Boyé, Martin Hans (1812-1909, chemist), donor by heirs, 928, 1122, 1124
Boyertown, PA, quadrangle, 982
Boyle County, KY, 1249
Boynes, —— de, 360
Boyneton, Edward Carlisle (d. 1893), cartographer, 1368
Braam Houckgeest, A. E. van (Dutch diplomat), *Voyage de l'Ambassade* . . . , 386a-387
Brabant, Belgium, 188, 190
Bradford, L. H., lithographer, 1399
Bradford, Samuel (1803-188?), 949
Bradford, Thomas Gamaliel (1802-1887, cartographer), cartographer, 1695; *A comprehensive atlas,* 1695
Bradford, William (1721/2-1791, publisher), publisher, 566
Bradford & Inskip, publishers, 566
Bradley, Abraham (1731-1824), cartographer, 555, 559, 629, 1085, 1093
Bradley, Frank Howe (1838-1879, geologist), 1404
Bradley, William M., & brother, publishers, 1714; *Bradley's atlas of the world,* 1714
Bradley's atlas of the world, W.M. Bradley & bro., 1714
Bragança, Brazil, 1572
Braham, Harold, 1659
Brahe, Tycho (1546-1601, astronomer), 1644; and Copernicus, *Atlas novus coelestis in quo mundus spectabilis* . . ., 1644
Braid, A., cartographer, 1197
Bramucci, Nancy M., 1767
Branom, Frederick Kenneth (b. 1891, geographer), cartographer, 1743
Bransfield, Edward, cartographer, 101
Brassier, William, cartographer, 784
Bray, Edmund, donor, 1758
Bray, Martha, donor, 1758
Brazil, 1572-74; Bragança, 1572; Casa Branca, 1572; Cuyaba, 1562; Franca, 1572; geology, 1572, 1574; Mococa, 1572; Para, 1562; Paranapanema, 1562; Pindamonhangaba, 1752; Paraçununga, 1572; Rio Grande do Norte, Natal, geology, 1574; São Bento, 1752; São Paulo (state), 1572-73; São Paulo (city), 1573; São Sebastião do Paraizo, 1572
Breckenridge County, KY, 1249
Bredsted, Denmark, 300
Breeze, Sidney, 578
Bregante, F., cartographer, 335, 408, 410-12, 441, 689; engraver, 268a, 335-38, 410-11, 441
Bressanini, Rinaldo (?) (1803-1864, Swiss engraver), engraver, 239
Bretez, Louis, cartographer, 1643; (*Plan de Paris* . . .), 1643
Brett lithographic co., lithographer, 1355, 1412; publisher, 1412
Bretzenheim, ——, 29 (10)
Brewster County, TX, 1234
Bridesburg to Cherry Island Flats, Delaware River, 644, 957
Bridgeport harbor, CT, 743
Bridgeport quadrangle, NJ-PA, 782
Bridges, William, cartographer, 826, 844
Bridgeton, NJ, 886
Bridgetown, Barbados, 1519
Bridgman, E. C., publisher, 1169
Brie, France, 245
Bristol, England, 123, 125
Bristol, PA, to N.Y.C., NY, 868
British Columbia, Canada, 482, 488, 517, 537-44, 1499; Antler Creek, 537; Caribou mining district; 537; Cunningham Creek, 537; Dragon Creek, 537; geology, 537-38; Grouse Creek, 537; Hardscrabble Creek, 537; Harvey Creek, 537; Jasper, 543-44; Keithley, 537; Lightning Creek, 537; Little Snowshoe Creek, 537; Nelson Creek, 537; railroads, 540-41; Rossland, 542; Slough Creek, 537; Sugar Creek, 537; Vancouver Island, 541; Williams Creek, 537; Willow Run, 537
British empire in America. *See:* U.S. and Canada
British, French, and Spanish empires in America. *See:* Canada, Mexico, Florida, and the U.S.
British Honduras. *See:* Belize
British Isles, 23 (7), 144, 1654
British Museum, 63
Britton, John, publisher, 1486
Britton & Rey, publishers, 1486

Broad, Wallace, cartographer, 485
Broad Top coal field, PA, 966
Brobst. *See:* Bropst
Brockhaus, Friedrich Arnold (1772-1823, German geologist), cartographer, 305
Broken Hill district, N.S.W., Australia, geology, 1610
Brongniart, Alexandre (1770-1847, French mineralogist, geologist), cartographer, 113
Brooke, Robert, cartographer, 32 (68)
Brookline, MA, 699
Brooks, Vincent, publisher, 63
Bropst (Brobst), Christian (fl. 1835), cartographer, 32 (77), 986
Broughton, John Gerald (b. 1914, geologist), cartographer, 839
Brown, ____, 288, 438; publisher 1312
Brown, Amos Peaslee (1864-1917, geologist), cartographer, 961
Brown, David J., collection, 7; cartographer, 7 (4-6)
Brown, Eman, cartographer, 32 (7)
Brown, Henry Yorke Lyell (b. 1844, Australian cartographer), cartographer, 1582-83
Brown, John Carter, Library, 467
Brown, Peter, 1150
Brown, Samuel (1769-1830, physician), 2 (2), cartographer, 2 (1); donor, 2 (1), 32 (66)
Brown, Shipley & Co., 1385-86
Brown, W. J., cartographer, 1387
Brown Brothers & Co., 1385-86
Brué, Adrien Herbert (1786-1832, French geographer), cartographer, 266, 1560, 1682; *Atlas universel,* 1560
Brugge (Aargau), Switzerland, geology, 237
Brunhouse, Robert Levere (1908-, historian), "David Ramsay, selections from his writings," APS *Transactions,* vol. 55, part 2; 567
Brunswick, Luneburg, Germany, 293
Brunswick, NJ, to N.Y.C., NY, 864-65
Buchan, earl of. *See:* Erskine, David Steuart
Buchan, Alexander (1829-1907, oceanographer), cartographer, 1722
Bucher, Walter Herman (1888-1965, geologist), cartographer, 1569; *Geologic structure and orogenic history of Venezuela,* 1569
Buckholz, L.V., cartographer, 1124
Buckingham, publisher, 704
Bucks County, PA, 15 (4), 897
Büler, Joh. Jacob (German cartographer), cartographer, 232
Buffalo and Philadelphia railroad, 776
Buffalo and Western NY railroad, 952
Buffalo harbor, NY, 834
Buffalo hunt, 579
Buffalo, Pittsburgh & Western railroad, 778
Bugge, Thomas (1740-1815, Danish traveller), cartographer, 184-85, 300
Buisson, ____, publisher, 1237
Bulens, L., lithographer, 1692
The Bullard Company, publisher, 636, 695
Bulletin de la classe d'agriculture, Société des arts de Genève, 239
Bulletin, Geographical Society of Philadelphia, 99
Bulletin, Geological and Natural History Survey of CT, 749
Bulletin, Geological Survey of PA, 985
Bulletin, Philippine Islands, Division of Geology and Mines, 414
Bulletin, U.S. Geological Survey, 1093
Bunce, ____, publisher, 1265
Burbank, C. F., cartographer, 1419
Burch, Adolf, cartographer, 1425
Burchmore, George (surveyor), cartographer, 675
Burd, Edward Shippen (b. 1779), 45, 46 (20)
Bureau de longitude, France, 116
Bureau of the American Republics, 1508
Burgess, R. W., cartographer, 1278-79
Burgoyne, John (1722-1792, English soldier), 499
Burias Island, Philippine Islands, Busainga, 410
Burland Lithographer Co. (Canadian publisher), publisher, 471, 538
Burlingame, E. W., 32 (97)
Burma, Schwemo district, 32 (97)
Burr, David H. (1803-1875, geographer, cartographer, topographer), cartographer, 1275, 1311
Burt, William A., cartographer, 1277
Busainga, Burias Island, Philippine Islands, 410
Bushy Run, PA, battle, 901
Bussole (ship), 1673
Bute, earl of. *See:* Stuart, John, 3rd earl of Bute
Butler, Lt., cartographer, 1191
Butte, MT, 1430-31
Butterfield, Roger Place (b. 1907), donor, 1747
Buttermilk Sound, GA, 32 (12)
Butts, J. R., cartographer, 1178
Buzzards Bay, MA, 677

Byrd, William (1674-1744, Virginia planter), cartographer, 1094
Byres, John, cartographer, 1541

Cadiz, Spain, 329, 332
Caesar, Gaius Julius [100 or 102 B.C.-44 B.C., Roman emperor], 129
Cailliaud, Frédéric [1787-1869, French traveller, author], 427-28, 438; *Voyage à l'Est . . .*, 428; *Voyage à l'Ouest*, 427
Cairo, Egypt, 432
Cairo, IL, 1337
Caldera County, CO, 1419
Caldwell, J., engraver, 286
Calgary, Alberta, Canada, 543
California, 40, 74d, 88, 90, 456, 1427, 1479-94, 1505; fault map, 1493; geology, 1493-94; Indians, 1484; Los Angeles, 1490; Los Coronados, 1482; Monterey, 644; Monterey harbor, 1479; Point Pinos to Bodega Head, 1487; San Diego, 1490; San Diego entrance, 1482; San Francisco, 85, 1490; San Francisco bay, 1491; San Francisco entrance, 644, 1488; San Francisco peninsula, geology, 1489; to San Diego, 1481; to Umpquah River, Oregon, 1463; Santa Barbara, 1480; Shasta County, geology, 1486; Yosemite National Park, 1492
Callahan, Dennis (fl. 1857-1863, cartographer), 1132, 1179
Callendar [Callender], Benjamin (1773-1856, engraver), engraver, 406, 629, 661, 1558, 1797
Çambel, Halet, donor, 1738
Cambray, Belgium, 215
Cambreleng, Churchill C., 1688
Cambria County, PA, 937
Cambrian rocks, NY, 839
Cambridge, MA, 699; Harvard University, 696; Radcliffe College, 696
Camden, NJ, 886, 890, 1014, 1029
Camden, Ontario, Canada, 504-514
Cameron, Frank, cartographer, 1478
Cameron, Walter Evan (b. 1871, Australian mining engineer), cartographer, 1588, 1604-605
Camp Gray, 1277
Campanius Holm, Thomas (1624-1695, Swedish cartographer), *Kort Beskrifning om provincien Nya Swerige uti America*, 896, 898, 1048
Campbell, ——, publisher, 761, 1084
Campbell, Archibald, 1461; donor, 1461
Campbell, Col. ——, cartographer, 1178
Campbell, John, 4th earl of Loudon (1705-1782, Scottish general), 29
Campbell, Tony, *Early maps*, 1763
Campbell, William Martin, cartographer, 981, 1033
Campobello Island, New Brunswick, Canada, 493
Campus Martius, Marietta, OH, 1306
Canada, 32 (98), 449-51, 453-55, 471-482, 550, 552, 557, 569, 572, 577, 598, 626, 752-53, 757, 761, 780, 783, 785, 816, 819, 828, 965, 1724; Admiralty Inlet to Gray's Harbor, Washington, 1462; *Alberta*, 482, 516-19, 543, Calgary, 543, elevators (grain), 519, Jasper, 543-44, Lake Louise, Rocky Mountains, 533, Lesser Slave Lake, 531, Peace River, 534, 536; Rock Mountains, 532; American harbor to Kingawa, 6 (4); Arsebémiling to Ipiakdiuak, 6 (3), *Baffin Bay*, 6 (2), 23 (8), 93, 99; Disco bay, 93, Omenak fiord, 93; Baffin Island, 6 (5); Baffinland 6 (6); boundary, U.S., 631-32; *British Columbia*, 482, 488, 517, 537-4, 1499, Antler Creek, 537, Cunningham Creek, 537, Dragon Creek, 537, Geological and Natural History survey, 537, geology, 537-38, Grouse Creek, 537, Hardscrabble Creek, 537; Harvey Creek, 537, Keithley Creek, 537, Lightning Creek, 537, Little Snowshoe Creek, 537, Nelson Creek, 537, railroads, 540-41, Rossland, 542, Slough Creek, 537, Sugar Creek, 537, Vancouver Island, 541, Williams Creek, 537, Willow River, 537; Canals, 1298; Cumberland Sound, Baffin Island, 6; Cumberland Strait, 32 (81); *Department of the Interior*, 475-78, 502-503, 507-509, 510, 516-519, 520, 522, 524; 527-30, 532-34; 536, 538-39; 540-41, 543-47, 549, 1724; *Dictionary of altitudes*, 474; Geological and Natural History Survey, 537, Natural Resources Intelligence Service, 478-80; Department of Mines and Resources, 480, 482, 511-14, 542; east, roads, 490; Fort Churchill, 92: Frobisher Strait, 32 (81); Fullerton harbor, 489; Geological and Natural History Survey, 471; Geological Survey, 485, 488, 496, 531, 548; geology, 471, 481, 489, 623; Green Mountains, 32 (84), 1298; Hudson Bay, 6 (2), 48 (4), 452, 489, 548; hydrography, 472; *Labrador*, 452, 48 (8), Eskimo, 48 (11), Lake St. Francis, 780, Lake St. John, 48 (10),

(14), Lincoln Sea, 99; Loess, 623; *Manitoba* 515-26, 545, elevators (grain), 519, Lake Winnipeg, 525, Norway House, 522, Norway House, magnetic variations, 522, Oxford House, 523, Oxford House, magnetic variations, 523, Selkirk, 524, Selkirk, magnetic variations, 524, York Factory, 548; Melville Bay, 23 (1); Melville Sound, 95; Montreal, Quebec, 32 (84), 780, 1298; Niantilic harbor, Baffin Bay, Cumberland Sound, 32 (85); Nelson River, 1645; *New Brunswick,* 485, 488, 492-96, Campobello Island, 493, Deer Island, 493; Grand Manan Islands, 493, Passamaquoddy Bay, 493, St. Croix River, 493; Newfoundland, 1661, 1749; Nova Scotia, 455, 488, 492-96, 550, 761, 1658; *North West Territories,* 545, Akudnirn, 487, Eclipse Sound, 487, King William Land, 487, Oqo, 487, Repulse Bay, 487; *Ontario,* 488, Algoma, 509, banks (chartered), 508, Fort Severn, 548, Gatineau, 503, geology, 507, Georgian Bay, 32 (98), Kawartha Lakes, 490, Lake of Bays, 490, Mattagami, 509, Muskoga, 490, Rideau Lakes, 490, Sudbury, 509, Timiskaming, 509, Toroton, 39; orography, 473; Peace River, 531; *Prince Edward Island,* 485, 488; *Quebec,* 485, 488, 497-503, 756, banks (chartered), 508, Battle of Montmorency, 44 (1); Gaspe peninsula, 508, Magadelene Island,633, Montreal, 490, 498, 502, Quebec City, 44 (1), 490, 497, 501-2, 831, Ungava Bay, 48 (15); railroads, 476-77, 479-80, 583-84, 594; roads, 452, 780; Robeson Channel, 100; Rocky Mountains; 549, St. Charles River, 44 (1); St. Lawrence River, 48 (1), (14), St. Lawrence River gulf, 1298; *Saskatchewan,* 516-19, Assiniboia, 516, elevators (grain), 519 Grand Rapids, 520; Smith Sound, 100; southeast, Bonguer gravity map, 633, Sylva Mountain, 23 (2-3); Turner Valley, 491; to Umpquah River, Oregon, 1464; water power, 472; *Yukon Territory,* 539, 547, Alsek River, 547, Kluane River, 547

Canada Bank Note Co., publisher, 485

Canada Company, 484

Canadian lakes, 32 (7)

Canal navigation: Lake Superior to the gulf of the St. Lawrence, 32 (84)

Canals, Canada, 1298; Chesapeake and Delaware, 32 (73), 755, 760-60a, 1045; Chesapeake and Ohio Canal, 770; Delaware, 767; Delaware and Raritan canal, 771; Delaware Bay and Chesapeake Bay, 755, 760-60a; Erie canal, NY, 827; England, 136-37, 139; Florida, 1192; France, 252; Great Lakes, 1276; Lake Superior, 1298; Lockport route, Great Lakes, 1276, Louisiana, 1221, Maryland, 767, 1065, Muscle Shoals, Tennessee River, Alabama, 1206; New Jersey; 878, 935; New York, 829-31, 1300; Niagara Falls, NY, 828; Pennsylvania, 767, 913, 935, 945; Pennsylvania canal, 936; Scotland, 139; South Carolina, 1162; United States, 570, 573, 576, 587, 591, 598, 607, 611, 615, 1298; Wales, 139; world, 81

Canary Islands, Spanish, Fuertaventura, 336; Lanzarote Island, 337; Tenerife Island, 338

Canaveral, Cape, FL, 1186

Canton, China, 388-89, 601

Cap Français, St. Domingue, 32 (44), 1523; plain, 32 (52)

Cape Breton Island, 30

Cape Canaveral, FL. *See:* Canaveral, Cape, FL

Cape Cod, MA. *See:* Cod, Cape, MA

Cape May City, NJ, 644, 883

Cape May peninsula, NJ, 886

Cape May roads, NJ, 880

Capestang, France, 252

Cappon, Lester Jesse (b. 1900, historian), 1757; *Atlas of early American history,* 1757

Capps, S.R., cartographer, 465

Captain's Island harbor, east and west, CT, 745

Cardozo Casado Giraldes, Joaquim Pedro (d. 1845, Portuguese cartographer), cartographer, 333; donor, 333

Carey, Henry Charles (1793-1879, publisher, APS 1833), donor, 1685, 1692; publisher, 1685

Carey, H. C., and I. Lea, publishers, 1685

Carey, Mathew (1760-1839, publisher, APS 1821), 32 (18a), (19), (23-4), 66-7, 73, 92, 107, 128, 153, 160, 192, 195, 233, 254, 308, 350, 361, 367, 400, 424, 1542, 1554-55, 1662, 1692; *The general atlas for Carey's edition of [William Guthrie's] geography improved,* 32 (19), (23-4), 554, 1554-55; publisher, 32 (19), (23-4), 66-7, 73, 92, 107, 128, 153, 160, 192, 195, 233, 254, 308, 350, 361, 367, 400, 424, 554, 567, 1554-55, 1671; *A short account of Algiers,* 436

Carey's American atlas, 32 (18a), 556, 639, 655, 664, 673, 714, 738, 814, 876, 916, 1044, 1067, 1118, 1147, 1158, 1174, 1241,

1255, 1272
Carey, M., & Son, publishers, 316, 1684
Caribbean islands, 1661; Sea, 1549, 1553, physiographic design of the sea, 1633
Caribou mining district, B.C., Canada, 537
Carischa, 29 (39)
Carlberg, Berthold, cartographer, 1739
Carleton, Osgood (1742-1816, surveyor, publisher, teacher of mathematics), 32 (20), 553, 561, 638
Carletti, Nicolaus, cartographer, 342; publisher, 342
Carlisle, England, 1662
Carlsruh to Coblentz, Rhine River, Germany, 29 (1)
Carmelo y Bauermann, publisher, 413
Carnate, 182
Carnegie Institution of Washington, DC, 1493; publisher, 1735; *Publications,* 1725, 1735; *Research in China,* 1725
Carpenter, Edward Snow (1918-, archaeologist), 49
Carpenter, Samuel, 33 (2)
Carrigain, Philip (1772-1842, surveyor), cartographer, 657
Carrington, Col. ——, 22 (1)
Carrollton, MS, 1333
Carson, Sir John Wallace? [1864-1922. Cartographer], cartographer, 1387
Carson City, NV, 1460
Cartagena, Colombia, 1566-67
Carte des nouvelles découvertes au nord . . . Siberie, de l'Isle, 1647
Carter, Robert (1663-1732, Colonial official, landholder), 22 (1)
Carter & Wilkinson, 712
Carteret, Philip (d. 1796, English admiral and explorer), 1619
CARTOGRAPHERS: *See:* A——, D.; Abbot, Henry L.; Abella y Casariego, Enrique; Abert, John J.; Abert, S. T.; Adams, H.; Adams, Henry M.; Adams, I. H.; Adams, J. W.; Advenier De Breuilly, A. P.; Adlum, John; Aguirre, A. A.; D'Ailly, H. A. H.; Aitken, Robert; Albright, Chester E.; Alden, J. A.; Alden, James; Alden, James B.; Alden, William C.; Alexander, A.; Alger, Francis; Allen, R. C.; Allen, W. F.; Almy, J. J.; Alzante y Ramirez, Joseph Antonio; Amenta, ——; Andrade, ——; Andersen, A.; Anderson, J. A.; Andrews, Emma B.; Anville, J.B.B.d'; Apfel, E. T.; Appiano, C. R.; Arlett, ——; Armaschevsky, P.; Arndt, H. H.; Arrowsmith, Aaron; Armstrong, John; Armstrong, Marcus; Ashburner, Charles A.; Ashman, J.; Aspin, J.; Attahvich, Sam; Axtell, O. N.; Ayres, H. R.; Babbitt, C. W.; Bache, A. D.; Bache, Charles M.; Bache, George M.; Bache, Hartman; Bache, R.; Bache, R. M.; Bachelder, John D.; Bagby, F. H.; Bailey, Willis; Baker, Charles; Baker, Robert; Baker, Samuel; Balbach, A.; Baldwin, A. S.; Ball, L. C.; Banard, J.; Bancker, Gerard; Barclay, M. G.; Bardin, T. M.; Bardin, William; Barlow, J. W.; Barosh, Patrick J.; Barney, Joshua; Barney, W. C.; Barrett, L. P.; Bartholomew, J. G.; Bartholomew, John; Bartholomew, John and Co.; Barton, Benjamin Smith; Bartram, John; Bartram, William; Bates, Waldron; Bath, T. H. C.; Battaille, ——; Bauman, Sebastian; Bauman, William; Bauman, William, Jr.; Bayfield, ——; Baylor, J. B.; Beach, Lansing H.; Beach, R.; Bechler, Augustus R.; Beck, Richard; Becker, C.; Beer, Wilhelm; Beers, F. W.; Beers, J. M.; Belknap, Jeremy; Bell, Robert; Bellin, Jacques Nicolas; Bellin, S.; Bennett, Richard; Bergsten, J. M.; Bernard, S.; Berniere, Henry de; Bertarelli, L. V.; Berthrong, I. P.; Bertre, ——; Bevis, John; Beyrich, ——; Bianchi, G. C.; Bibert, ——; Biddle, Mary; Billin, C. E.; Bizard, ——; Black, John C.; Black, J. J.; Blackburn, J. E.; Beachwell, T. E.; Blaeu, W. J.; Blake, George S.; Blake, J. Edmond; Blakelock, D. M.; Blanchford, R.; Blodget, Loren; Blunt, E. C.; Blunt, Edmund; Blunt, F.; Boas, Franz; Boeckh, Lt.; Boggs, S. W.; Bois St. Lys, Georges Joseph; Bolles, C. P.; Bolze, ——; Bond, Frank; Bonfils, C. A.; Bonney, S. B.; Bontz, George F.; Boquetta De Woiseri, J. L.; Boschke, A.; Bostock, H. S.; Boulton, S.; Bourne, A.; Boutelle, C. O.; Bouvé, E. W.; Bowditch, Nathaniel; Bowen, Emanuel; Bowen, Lot; Bower, Robert A.; Bowman, Amos; Boyd, C. M.; Boyd, W. H.; Boyé, Herman; Boyer, C. M.; Boynton, E.; Bradford, T. G.; Bradley, Abraham, Jr.; Bradley, Frank H.; Brahe, Tycho; Braid, A.; Branom, Frederick K.; Bransfield, Edward; Brassier, William; Bregante, F.; Bretez, Louis; Bridges, William; Broad, Wallace; Brobst, Christian; Brockhous, F. A.; Brongniart, Alexandre; Brooke, Robert; Broughton, John G.; Brown, Amos P.; Brown, David B.; Brown, Eman; Brown, H. Y. L.; Brown, Samuel;

Brown, W. J.; Brué, Adrien Hubert; Buchan, Alexander; Bucher, Walter H.; Buckholz, L. V.; Buler, Jon. Jacob; Bugge, T.; Burbank, C. F.; Burch, Adolf; Burchmore, George; Burgess, R. W.; Burr, David H.; Burt, William A.; Butler, L. T.; Butts, J. R.; Byrd, William; Byres, John; Callahan, Dennis; Cameron, Frank; Cameron, W. E.; Campbell, William M.; Capps, S. R.; Cardozo Casado Giraldes, J. P.; Carlberg, Berthold; Carleton, Osgood; Carletti, Nicolaus; Carrigain, Philip; Carson, John; Cassini De Thury, ——; Cate, Addison S.; Catherwood, F.; Chalifour, J. E.; Chambers, Benjamin; Chantre, E.; Chapman, Isaac A.; Chittenden, G. B.; Choffat, Paul; Chastenet Peysegur, Comte de; Chauchard, ——; Chester, C. M.; Chevalier, Michel; Chewitt, James G.; Churchill, F. A.; Churchman, John; Churton, William; Clapp, F. G.; Clark, C. W.; Clark, E. B.; Clark, Fred A.; Clark, G. R.; Clark, William; Clark, William Bulloch; Clark, W. Eagle; Clarkson, Matthew; Cleveland, William C.; Clover, Richardson; Cochran, ——; Colahan, John B.; Colburn, Zerah; Cole, C. S.; Colles, Christopher; Collet, ——; Collins, Greenville; Colonna, B. A.; Colton, George W.; Colton, G. W., & C. B. Colton, & Co.; Colton, Roger B.; Comber, W. G.; Comstock, C. B.; Conlin, R. R.; Conway, T. R.; Coombs, D.; Cook, F. S.; Cook, George H.; Cook, James; Cooke, Charles E.; Cope, E. B.; Cope, Thomas D.; Copernicus, Nicolas; Coram, T.; Corbellini, P.; Coutans, G.; Cowden, James S.; Coyle, R.; Craig, Thomas; Cram, George F.; Cram, T. J.; Crandall, Roderich; Craven, T. A.; Craven, T. M.; Cret, Paul P.; Creux [Creuxius], François du; Crisp, Edward; Cruz Cano y Olmedilla, Juan De La; Cummin, R. D.; Curtis, C. A.; Custer, G. A.; Cutts, R. D.; Cuyler, R. M.; Czjzer, Joh.; Dabney, T. G.; DeBrahm, William G.; Dahl, Lt. ——; Daidy, George A.; Dale, J. B.; Dalrymple, ——; Dandridge, William; Darby, William; Daser, L.; Davidson, G.; Davis, C. H.; Davis, R. O.; Dawson, George M.; Dawson, L. H.; Day, ——; Day, Sherman; Day & Zimmerman; Dearborn, H; Dechen, H. Von; Dewes, J. H.; DeHaven, E. J.; Delgado, J. F. N.; Denham, H. M.; Dennis, W. H.; Desnos, ——; DesBarres, J. F. W.; Desray, ——; Detloff, C.; DeWees, H. M.; Dewes, J. H.; DeWitt, Sim.; Dezauche, ——; Dickens, C. L.; Dickison, A.; Digby, Captain; Diment, William D.; Dimmock, C. H.; Dings, M. G.; Dinsmore, A. F.; D'Invilliers, E. V.; Dixon, Jeremiah; Djurberg, Daniel; Dodson, James; Donn, J. W.; Donoghue, John; Doolittle, Amos; Dorr, F. W.; Dorsey, Clarence W.; Dougal, W. H.; Douglas, E. M.; Downes, D.; Drake, N. F.; Drayton, T. F.; Dreher, John; Duce, R. S.; DuFour, John James; Dufrenoy, ——; Dunaway, H.; Dunbar, Samuel O.; Duncan, Winifred; Dunham, A. B.; Dunn, Samuel; Dunnewald, T. J.; Dunstan, B.; Dury, Andrew; DuSimitière, P. S.; Dutton, C. E.; DuVal, H. L.; Eakin, C. M.; Eastburn, B.; Eberhardt, E. F.; Eckel, E. B.; Ecklund, C. A.; Edwards, W. S.; Egloffstein, F. W.; Elie de Beaumont, Léonce; Ellicott, Andrew; Ellicott, B.; Ellicott, Eugene; Ellicott, Joseph; Elliott, Henry W.; Ells, Robert Walpole; Elverson, J. S.; Ely, ——; Elyard, Hammond; Emory, W. H.; Erdmann, A.; Erichsen, P.; Escher von der Linth, Arnold; Eskimo; Espinalt y Garcia, Bernado; Essex, Frank B.; Estabrook, E. L.; Evans, A. W.; Evans, Lewis; Ewing, M. C.; Faden, William; Fahlberg, Samuel; Fairbanks, H. W.; Fairfax, Wilson McC.; Fales, J. C.; Falsan, A.; Farley, J.; Farquahar, C.; Faucou, Lucien; Fauntleroy, C. M.; Favre, Alphonse; Featherstonhaugh, G. W.; Federov, E.; Fendall, C.; Ferguson, James; Ferrer, J. J. de; Ferraris, Comte de; Fettke, C. R.; Filson, John; Finlayson, James; Fisher, Donald W.; Fisher, Joshua; Fisher, Thomas; Fisher, William; Fitch, Charles H.; Fitz-Carrald, ——; Fitz Roy, Robert; Fitzwilliams, Richard; Flint, Richard F.; Fohs, F. Julius; Fornaro, A.; Forshey, Caleb Goldsmith; Forster, J. H.; Fowler, ——; Fox, D.; Fox, H. W.; Fra(), Joseph; Franklin, Benjamin; Frazer, Persifor; Frehold, Edward; Fremont, J. C.; French, G. H.; Fréreault, P.; Friex (Fricx), E. H.; Fritsch, Karl von; Fry, Joshua; Fuller, M. L.; Gabb, William M.; Gannett, Henry; Garcias, Diego; Gardner, James H.; Garnett, John; Gaspari, A. C.; Gay, E. F.; Gedney, Thomas R.; Gerahty, T. R.; Gerdes, F. H.; Gerhard, Peter; Geyer, A. R.; Ghys, Martin; Gilbert, G. K.; Gilbert, Joseph; Gilbert, S. A.; Gilbert, W. W.; Gillman, H.; Giroux, N. J.; Glegg, J. G.; Glot, C. B.; Glück, John B.;

Goldsborough, J. R.; Goldthwait, J. W.; Goldthwaite, Richard P.; Goll, J. J.; Goodchild, J. G.; Goodfellow, E.; Gordon, Thomas; Gorman, Fred J.; Gorshkov, Sergei G.; Goujon, J.; Gould, L. M.; Gourov, A.; Graham, Campbell; Graham, J. D.; Graham, W. J.; Grant, ——; Gravelot, H.; Graves, W. H.; Gray, A. B.; Gray, Carlisle; Gray, George; Greeg, Robert; Green, L. C.; Greenfield, W. H.; Greene, F. B.; Greenleaf, Moses; Greenough, George Bellas; Greenwell, W. E.; Gregory, H. E.; Griffith, Dennis; Grimshaw, Percy H.; Gros, C.; Gross, Alexander; Groves, D. (?); Gruner, G. S.; Guedes, P.; Guillot, A. H.; Gunnison, J. W.; Gussefeld, F. L.; Guyot, Arnaud; Haack, Herman; Haggard, John L.; Hale, John S.; Hales, John G.; Hall, James; Hall, N. J.; Halley, Edmund; Hamilton, Fletcher; Hammond, S. A.; Hanus, G. C.; Harboe, P.; Hardcastle, Lt.; Harden, E. B.; Harden, O. B.; Harding, W. W.; Hare, C.; Hare, Horace Bimey; Harney, W. S.; Harris, Caleb; Harris, Harding; Harris, William J.; Harrison, A. M.; Hassan, A. F.; Hassler, F. R.; Hauchecorne, ——; Hauducoeur, C. P.; Hawkins, George T.; Hawley, J. M.; Hayden, F. V.; Hayward, James; Heap, G.; Hearding, W. H.; Heart, Jona.; Heckewelder, John G.; Hedges, M.; Heeren, Lillian; Heezen, Bruce C.; Heilprin, Angelo; Helm, Charles J.; Henderson, A. A.; Hendersen, John; Hennen, Ray V.; Henry, Alfred J.; Henry, D. F.; Henry, Mathew S.; Herald, Frank A.; Herbertson, A. J.; Herbst, F.; Hergesheimer, E.; Herisson, ——; Herman, ——; Heroguel, Robert; Herrmann, Albert; Heslop, ——; Hewitt, E. M.; Heymann, Knazio; Higginson, J. H.; Hildreth, C. T.; Hilgard, J. E.; Hill, Benjamin F.; Hill, Frank A.; Hill, John; Hill, Robert T.; Hills, John; "Hiram," ——; Hitchcock, C. H.; Hockridge, T. C.; Hodgeson, H. H.; Hoening, J. B.; Holland, Samuel; Holman, James D.; Holme, Thomas; Homann, Johann Baptist; Hood, Wash; Hooge, Romein de; Hooper, Robert Lettis, Jr.; Hoskins, D. M.; Hosmer, C.; Howell, C. W.; Howell, George W.; Howell, R. P., Jr.; Howell, Reading; Horsley, ——; Hough, B.; Houghton, D.; Hudson, G. Donald; Hughes, E. M.; Hulings, William E.; Humboldt, Alexander von; Humphreys, A. A.; Hutchins, Thomas; Hyslop, J. M.; Iardella, C. T.; Idiaquiz, Eduardo; Iljin, A. A.; Irving, Alex; Isachsen, Yngvar W.; Jackson, ——; Jacobsen, N. Kingo; Jacotin, ——; Jacques, Herbert; Jäck, R. L.; Jaillot, ——; James, I. E.; James, John A.; Jefferson, Peter; Jefferson, Thomas; Jefferys, Thomas; Jenkins, Olaf P.; Jenkins, T. A.; Jennings, J. H.; Jenny, W. L. B.; Jenny, W. P.; Jensen, Kr. Marius; Jensen, Ruth Helkiaer; Jobbins, J. R.; Johnson, Guy; Johnson, Willard J.; Johnston, Alexander Keith; Johnston, J.; Johnston, J. E.; Johnston, T. B.; Johnston, William; Jomard, Edmé François; Jones, Felix; Jones, John; Jones, N. F.; Jones, S. C.; Jones, Thomas H.; Junken, C.; Kane, Elisha Kent; Karpinsky, A.; Kay, George F.; Kayser, Henry; Kearny, James; Keen, Gregory S.; Kehn, T. M.; Keim, ——; Keiser, Frank; Keller, Charles; Kelly, V. C.; Kendall, E. O.; Key, J. R.; King, C. F.; King, Clarence; King, Harry; King, Henry; King, Philip Parker; Kitchin, Thomas; Kneass, William; Knight, J. G. D.; Knight, R. T.; Knorr, E. K.; Knott, W. T.; Köpp, Henry; Kolecki, Theodor; Konchin, A.; Kosak, William; Koversky, ——; Krusenstern, Adam Johann von; Kümmel, Henry B.; Kündig, Andreas; La Camp, J. de; Ladd, S. B.; La Fon, Benjamin; La France, Joseph; Lamb, George; Lambert, John; Lamson, A. G.; Lamson, G. W.; Lane, Michael; Lang, ——; Lapham, I. A.; La Perouse, J. F. G. de; Lapie, Pierre; Laporte, ——; Lapworth, Charles; Las Casas, E. A. D.; Latrobe, Benjamin F.; Lattré, ——; Laurie, Robert; Lavoisne, ——; Law, S. A.; Lawson, J. S.; Lay, Amos; Leach, Smith S.; Leavenworth, A.; Lebas, ——; Lecesne, ——; Le Conte, ——; Lee, Robert E.; Lee, Thomas J.; Lehman, Ambrose E.; Lehman, George M.; Leighton, M. M.; Ligowsky, Captain; Lemercier, ——; Lempriere, C.; Leon, Manuel Ponce de; Lesley, J. Peter; Lesley, Joseph, Jr.; Leth, Henry de; Leveque, ——; Leverett, Frank; Lewis, ——; Lewis, F.; Lewis, I. W. P.; Lewis, Meriwether; Lewis, Samuel; Li, C.; Linant de Bellefonds; Lindenkohl, A.; Lindenkohl, H.; Lindenkohl, W.; Lindsay, David; Lindström, Peter; Linney, W. M.; Lisle, G. de; Littrow, J. J.; Locke, John; Longfellow, A. L.; Longfellow, A. W.; Longshow, Robert; Lopez, Juan; Lorenz,

W.; Lotter, Matthieu Albrecht; Loughridge, R.; Lovell, W. H.; Low, Albert Peter; Lowthian, Joseph; Lucas, Claude; Lucas, F., Jr.; Luce, William; Ludlow, William; Lydecker, G. J.; Lyman, Benjamin Smith; Lynch, F. C. C.; Lyttle, W. S.; McAdoo, J. F.; McArthur, J. J.; McArthur, W. P.; McCalpin, William; McCaskill, Murray; McClery, M. I.; McClellan, General; Mac Clintock, Paul; McConnell, R. G.; McConomy, John A.; McCorkle, S. C.; McCoy, Isaac; McCoy, J. C.; McCoy, Lee H.; McCrackin, A.; McDowell, Captain; McElfatrick, Lt.; McElroy, Robert; McEvoy, James; M'Gillycuddy, V. T.; McInnes, William; McIntyre, Henry B.; Mackay, A. D.; McKendree, William G.; McKennon, I.; Mackenzie, Murdoch; McLaughlin, D. B.; McLaughlin, Fred; Maclure, William; Macomb, J. N.; McMurtrie, W. B.; Madler, Johann Heinrich; Maffit, J. N.; Mahlo, Emil; Mahon, Charles; Maitland, A. Gibb; Maltby, F. B.; Mandarin, H. L.; Mannert, C.; Mansfield, H. B.; Mansfield, J. M.; Marinden, H. L.; Marinelli, O.; Marion, Israel; Marlette, ——; Marshall, R. B.; Martin, G. C.; Martin, Helen M.; Martin, Richard A.; Martin, W. T.; Martinez, Juan de; Marzolla, Benedetto; Mason, Charles; Mathewson, A. J.; Maull, James L.; Maulo, ——; Maury, M. F.; Maverick, Peter; Maxwell, John; May, W.; Mayer, Christian; Mayer, John R.; Maynadier, H. E.; Mayo, George V.; Mayo, William; Meade, G. G.; Mechan, J.; Mechlin, J. R. P.; Meigs, M. C.; Meil, J. W.; Melish, John; Mendenhall, T. C.; Merrill, Frederick J. H.; Merrill, William E.; Messenger, John; Michler, N.; Mikhalsky, A.; Milet-Mureau, L. A.; Miller, A. M.; Miller, J. T.; Milliadge, T.; Mills, Edwin; Mills, Robert; Milne, Thomas; Milner, Thomas; Mintzer, William A.; Mitchell, A. M.; Mitchell, H.; Mitchell, John; Mitchell, Samuel Augustus; Mitchell, William; Moehlman, R. S.; Molitar, Edward; Moll, Herman; Montero y Gay, Claudio; Moore, J. H.; Moore, Raymond C.; Moore, Willis L.; Morehead, J. S.; Morgan, B.; Morgan, John H.; Morhardt, Major; Morrell, ——; Morris, E.; Morrow, A. T.; Morton, Cecil C.; Morton, Robert H.; Moser, J. F.; Mosman, A. T.; Mountaine, William; Mouschketov, J.; Mouzon, ——; Muller, Albr.; Mueller, John U.; Muldron, Robert; Muir, Ramsay; Muller, Ernest H.; Munroe, William H.; Munsell, Luke; Murphy, R. C.; Myers, Albert Cook; Na, J. F.; Nansen, Fridtjof; Narvaez, Jose Maria; Naumann, Carl F.; Nav, Anthony; Naylor, T. C.; Negrete, Pedro Celestino; Neilson, J. C.; Nell, Louis; Nes, F. F.; Newnan, John; Newton, H.; Nichols, Beach; Nicholas, D. A.; Nicholson, John P.; Nicholson, W. L.; Nicollet, J. N.; Nitikin, S.; Noetzel, G.; Noguero, J.; Nolin, I. B.; Nordeskiold, Nils Adolf Eric; Norie, J. W.; Norman, G. W. H.; Normand, ——; Norwood, Charles J.; Norwood, J. W.; Nyström, E. T.; Ockerman, J. A.; Ockerson, C. L.; O'Donoghue, John; Ogden, H. D.; Ogden, H. G.; O'Hare, Daniel; Ord, E. O. C.; Ortelius, Abraham; O'Sullivan, Owen; Ourand, Charles H.; Owen, F. W.; Pagenaud, Edward; Paige, James A.; Palfrey, Carl F.; Paliser, John; Palmer, W. R.; Park, C. F.; Parke, John G.; Parkinson, T. D.; Parrish, H. E.; Parsons, E. C.; Patrick, Mason M.; Patten, Richard; Patterson, C. P.; Paullin, Charles Oscar; Paxton, J. A.; Paz, Manuel Maria; Peale, A. C.; Peale, C. W.; Peck, C. S.; Peck, J. M.; Peirce, Benjamin; Peirce, John; Pelham, Henry; Penn, William; Pennell, A.; Pennell, L.; Perkin, Henry; Perkins, F. W.; Perrault, Major; Perthes, Justus; Peter, J. F.; Peters, W. J.; Pfyffer von Altishofen, F. X.; Phelps, T. S.; Philip, George; Phillips, William B.; Pichardo, ——; Pichon, ——; Picquet, Charles; Pike, Z. M.; Pilarski, Casimir von V; Pillans, P. J.; Pinkerton, John; Pittman, E. F.; Platt, R.; Poe, O. H.; Poggiali, Giuseppe; Pollock, William Wilson; Piri, Reis; Poole, C. H.; Popple, Henry; Porter, D. D.; Poussin, William Tell; Powell, John Wesley; Powers, Stephen; Pownall, Thomas; Pratt, Joseph Hyde; Preston, Captain; Preuss, Charles; Price, Jonathan; Price, Paul H.; Prime, F.; Pritchett, Henry S.; Proctor, John R.; Puig, Victor; Purcell, Joseph; Pursh, Frederick; Rabout, C. P.; Rameletti, Giuseppe; Ramsay, David; Ramsay, V.; Rand, Edward L.; Rands, H. W.; Rapkin, John; Ratzer, Bernard; Raynolds, W. F.; Rea, J. A.; Reed, John; Reed, J. G.; Reger, D. B.; Reichard, C. G. von; Reid, A. C.; Reid, John H.; Remington, T. J. L.; Renard, C.; Renevier, Eugène; Rennell, James; Renshawe, John H.; Renwick, J.; Revetta, Frank A.; Richard,

Albert; Richardson, H. B.; Ricketts, J. J.; Ridgely, Frederick; Rizzi Zannoni, G. A.; Robert, ——; Roberts, A. C.; Robertson, John; Robinson, H. H.; Robinson, John H.; Rockwell, C.; Rodgers, A. F.; Rodgers, John; Roeser, C.; Roggeveen, Arnold; Romans, Bernard; Roper, W. H.; Ropes, William III; Rose, Robert R.; Ross, J. W.; Rue, ——; Runnebaum, C.; Russell, Richard Joel; St. Barbot-De-Marny, N.; Sanders, R. H.; Sanderson, Ivan Terence; Sands, B. F.; Sanson, N.; Sartain, Samuel; Sauerwein, Charles; Sauthier, Claude Joseph; Sayer, Robert; Scammen, E. P.; Schade, Theodor; Schardt, Hans; Schleifer, Rudolf; Schmettau, Samuel von; Schmitz, M.; Schomburck, Robert H; Schönborn, A.; Schott, Charles A.; Schrivener, C. P.; Schruers, Edwin, Jr.; Schulz, Edward H.; Schull, John; Scott, Joshua; Scott, J. D.; Scobey, Fred; Scudder, Samuel H.; Scull, Nicholas; Scull, William; Seale, R. W.; Searl, A. D.; Searl, Frederick; Sears, C. B.; Sederholm, J. J.; Seib, J.; Selden, Henry; Sellers, John; Sellier, L. M.; Selwyn, Alfred R. C.; Senécal, C. O.; Senex, John; Sengteller, A.; Seutter, Matthew; Seybert, Adam; Sheafer, P. W.; Sheaff, J. A.; Shaeffer, M. B.; Shaffner, M. N.; Shanke, H.; Sheet, Charles Wilkins; Shepps; V. C.; Shippen, Joseph, Jr.; Shutting, Julius; Sieb, John; Siebold, Phillip Franz von; Siegen, P. M.; Simonel, ——; Simowitch, S.; Sitgreaves, L.; Skanke, H.; Skene, A. J.; Skertchly, S. B. J.; Skinner, ——; Skinner, Andrew; Slight, James; Sloane, Charles S.; Smith, Captain; Smith, Benjamin E.; Smith, Benjamin H.; Smith, David; Smith, E.; Smith, Erwin Frink; Smith, George B.; Smith, George Otis; Smith, Glenn; Smith, James Perrin; Smith, Lloyd Edwin; Smith, R. A.; Smith, W. F.; Smith, Warren D.; Smith, William; Smock, John C.; Smyth, David William; Sokolov, N.; Sokolow, A. A.; Sonis, I.; Sorokin, A.; Sotzmann, D. F.; Southack, Cyprian; Speer, Joseph Smith; Spofford, Elmer P.; Spurr, J. E.; Stanford, Edward; Stansbury, M. H.; Starling, W.; Steele, J. Dutton; Stellwagen, H. S.; Steuart, William; Stewart, George; Stickney, Amos; Stieler, Adolf Karl; Stiles, Arthur; Stockdale, John; Stokes, Pringle; Stone, C. H.; Stout, Peter F.; Strach, Cochran; Strauch, ——; Strausz, A.; Stretch, R. H.; Strickland, G.; Strickland, W.; Strother, D. H.; Strother, John; Strum, G. P.; Struther, ——; Sullivan, J. A.; Sutherland, A.; Sutton, Frank; Suziki, K.; Swanitz, A. W.; Swift, W. H.; Sword, James B.; Sydney, J. C.; Taitt, David; Talbot, A. D.; Talbot, William (?); Talcott, George; Tamura, E.; Tanner, H. S.; Tardieu, P.; Taylor, ——; Taylor, H. W. B.; Taylor, John; Temple, W. G.; Thackara, ——; Tharp, Marie; Thom, George; Thomas, B. E.; Thomas, C. M.; Thomas, E. B.; Thomas, E. J.; Thompson, ——; Thompson, C.; Thompson, Carl; Thompson, J. G.; Thompson, W. B.; Thompson, William; Thomson, A. R.; Thomson, J. Edgar; Thoulet, ——; Throop, O. H.; Thwaites, F. T.; Ting, V. K.; Tofino, Vicente; Togawa, T.; Tolleman, ——; Topley, William; Topper, James; Torrey, Jason; Trowbridge, W. P.; Trumbauer, Horace; Trumbull, John (?); Truxtun, Thomas; Tschernyschev, Th.; Tucker, R. C.; Turnbull, ——; Turtle, Thomas; Tuttle, H. P.; Tyrrell, J. B.; Udden, J. A.; Ughi, Ka. Lodovico; U.S. Army Military Intelligence; U.S. Geological Survey; U.S. Geological Survey, Topographic Branch; Hydrographic Office; Vance, David H.; Van Olden, A. E.; Vargas, Federico; Varlé, P. C.; Vermeuille, John Clarkson; Vermeule, C. C.; Vidal, A. T. E.; Viguier, Constant; Vinal, W. I.; Visscheri, Nikolaes; Voegelin, C. F.; Voegelin, E. W.; Voegelin, F. M.; Wadsworth, Alexander L.; Wagner, O. G.; Wainwright, R.; Walcott, Charles D.; Walker, Emery, Ltd.; Walker, Francis A.; Walker, J.; Wall, J. Sutton; Wallace, J.; Wallis, John; Walser, Gabriel; Wampler, J. M.; Wang, C. C.; Wansleben, Thomas O.; Warberg, O.; Wardle, J. M.; Warner, W. H.; Warren, ——; Warren, G. K.; Warren, J. G.; Waterman, H. E.; Wauters, A. J.; Webber, F. P.; Weber, E. A.; Wells, L. A.; Weldon, Daniel; Werner, T. W.; Wersebe, Col.; Wessel, Caspar; Weyl, C. G.; Weyss, John E.; Wheat, J. H.; Wheeler, George M.; Wheeler, J. B.; White, I. C.; White, James; Whitelaw, James; Whiting, H. L.; Whitman, E. B.; Whitney, Milton; Whitson, A. R.; Widen, K. A.; Wies, N.; Wilbur, D. C.; Wilkes, Charles; Wilkes, Charles, Jr.; Wilkins, Lt.; Wilkinson, J.; Wilkinson, Santiago; Willett, James R.; Williams, Arch B.; Williams, I.; Williams, J. S.; Williams, Lemuel D.; Williams, T. J.; Williams, W.; Williams, W. G.; Williams, W.

S.; Willis, Bailey; Willman, H. B.; Wilson, A. D.; Wilson, H. M.; Wilson, John; Wilster, F.; Wiltberger, C., Jr.; Winslow, Arthur; Wise, G. D.; Wit, Frederick de; Witzel, P.; Wolbrecht, George H.; Wood, H. O.; Wood, J. H., Jr.; Woodhull, M.; Woodruff, I. C.; Worthington, L. V.; Wragg, S.; Wright, John K.; Wright, W. R.; Wüllerstorf-Urbair, B. von; Young, G. A.; Young, J. H.; Young, R. E.; Yust, Donald; Zannoni, Giovanni Antonio Rizzi; Zantzinger, C. C.; Zimmerman, ——; Zimpel, Charles; Zinn, George A.
Cary, John (ca. 1754-1835, English cartographer, engraver, globe maker and publisher), 138
Casa Branca, Brazil, 1572
Case, Thomas Bristol, 33 (1)
Cash, Mr., donor, 1220
Cassini de Thury, César François (1714-1784, French scientist), 1670; *Atlas national et général de France,* 1670
Cassville, GA, 1181
Cat Island harbor, MS, 1209
A catalogue of early globes made prior to 1850 . . . by Ena L. Yonge, 1773
A catalogue of instruments and models in the possession of the APS, by Robert P. Multhauf, 1772
Catawba Indians, 48 (2)
Catawissa, PA, 32 (95), 986
Cate, Addison S., cartographer, 984-85
Catharine Street, Phila., PA, 46 (15)
Catherwood, Frederick (1799-1854, English artist, cartographer), 372
Catteau-Calleville, J. P. G., *Tableau de états danois . . .* , 182
Cawkin's Island harbor, CT, 744
Cayes, St. Domingue, 32 (39)
Cebu, Cebu Island, Philippine Islands, 412, 417
Cedar Street, Phila., PA, 46 (16-17)
Celebes Sea, physiographic diagram, 1635
Celestial globe, 1772
Center Square water works, Phila., PA, 1007
Central Africa, 443-46
Central America, 1501-502, 1562
Central Europe, 276-87
Central Park, N.Y.C., NY, 855
Central South America, 1563-64
Central U.S., 1374-1380; geology, 1377
Centre County, PA, 937
The Century atlas of the world, Benjamin E. Smith, 1719
The Century Co., publisher, 1719
Century Lithographic Co., lithographer, 1022
Chalco Lake, Mexico, 32 (11)
Chalifour, J. E. (Canadian geographer), cartographer, 447-79, 502-503, 509-10, 521, 529, 535, 543-44
Chalmandrier, Nicolas (fl. 1744-1760, French engraver and publisher), engraver, 252
Chambers, ——, publisher, 73, 128, 153, 160, 192, 195, 233, 306, 308, 350, 361, 367, 424, 1555
Chambers, Benjamin (fl. 1685, Pa. immigrant), cartographer, 973, 976
Chamouny valley, Switzerland, 244
Champagne, France, 249
Champlain, Lake, NY, 752-53, 784, 824, 1300; Rouse's point, 623
Champlain Society, 32 (98)
Chancellorsville, VA, 774
Channery town in Ross, Scotland, 148
Chantre, E., cartographer, 272
Chapin, William, engraver, 575
Chapman, Isaac A., cartographer, 2 (3)
Chapman, Nathaniel (1780-1853, physician, APS 1807), donor, 1692
Charles, prince of Lorraine, 189
Charles, F., lithographer, 1692
Charleston, SC, 32 (5) 576-77, 644, 1081, 1163, 1165; harbor, 1166-68
Charlestown, MA, 32 (6), 699; American Revolution, 702
Charpentier, René Jacques (1733-1770, French engraver), engraver, 65
Chart atlas, A voyage around the world. J. F. G. de la Peyrouse, 1673
Charters Tower gold fields, Queensland, Australia, 1586
Charts of the coast and harbours of New England, J. F. W. Des Barres, 1656
Chastenet Puysegur, comte de, cartographer, 32 (55)
Chatham, England, 125
Chattahoochee River, GA, 1181
Chattanooga, TN, 1208; battlefield, 1259, 1262
Chattanooga and Chickamauga National Park, 1208
Chauchard, M. (fl. 1790-1800, French cartographer), cartographer, 294, 351
Chauncey, Mr. and Mrs. Charles, donors, 1654, 1686, 1696, 1702
Chautauqua Lake, NY, 816, 819

Cheat River, PA, 45 (3)
Chelsea, MA, 699
Cheevers, John, engraver, 1304
Cheevers, Thomas, engraver, 758
Cherry Island flats to Bombay Hook, Delaware River, 1046; to Bridenberg, Delaware River, 644, 957
Chesapeake Bay, 20 (1), 644, 755, 1041-42, 1077, 1767; and Susquehanna River, MD, 1068; Hampton Roads, York River, 1138; to Delaware Bay, DE, canal, 760-60a
Chesapeake and Delaware Canal, 32 (73), 1045
Chesapeake and Delaware Canal Co., 760
Chesapeake and Ohio Canal, MD, 576, 770
Chester, C. M., cartographer, 1196-97
Chester County, PA, 897; Darby Creek, 43 (1)
Chester River, MD, 760, 1074
Chester, NJ, 892
Chevalier, Michel (1806-1879, French statesman, professor of political economy), *Histoire et description des voies de communication aux Etats Unis,* 1696
Chewett, James G. (fl. 1826-1833, Canadian cartographer), cartographer, 484
Cheyenne, WY, 1443-44
Chiapas interdency, Mexico, 32 (16)
Chicago, IL, 1365; harbor and bar, 1326
Chicago River and harbor, IL, 1325
Chicago, Milwaukee and St. Paul Railroad, 1395
Chickamauga and Chattanooga National Park, 1208
Chickasaw Indians, 32 (5)
Chidsey, Arnold D., Jr., "The Penn patents in the forks of the Delaware . . .," 32 (2)
Chihuahua, Mexico, 1562
Childs, Cephas Grier (1793-1871, lithographer, publisher), lithographer, 931, 940, 986, 1047, 1072; publisher, 931, 986, 1047
Childs and Inman, lithographers, 1047, 1072
Chile, 1560, 1575-78; Atacama, 1562; Chiloe Island, 1577; Easter Island, 1619; Strait of Magellan, 1575; Tierra del Fuego, 1576
Chiloe Island, Chile, 1577
China, 78, 386a-398, 1699, 1725, 1740; Beijing [Peking], 389, 396; coast, 389; Canton, 388-89; Geological Survey, 395; geology, 395; Hong Kong 388-89; Kwantung Peninsula, 390; Macao, 387; Manchuria, 379; railroads, 389, 393
China League, 391
Chincha Islands, Peru, 35 (1-2), (7)
Chippewa Indians, 1277
Chittenden, George B. (fl. 1873-1876, geologist), cartographer, 1407, 1409, 1440
Chocarne, Geoffroy Alphonse (b. 1797, French engraver), engraver, 427
Choco Island, Peru, 35 (12)
Chocolate River, 1277
Choctaw Indians, 32 (5)
Choffat, Paul (1849-1919, French geologist), cartographer, 339
Choiseul-Amboise, Étienne François, duc de (1719-1785, French minister of Louis XVI), 250
Chopunnish Indians' map of the Snake River, 28 (10)
Chouchou cove, St. Domingue, 32 (62)
Christian County, KY, 1249
Church, I. B., 821
Churchill, F. A., cartographer, 1127
Churchman, John (1753-1805, scientist), cartographer, 74, 1041-42, 1077; donor, 74; *An explanation of the magnetic atlas or variation chart,* 74
Churton, William (d. 1767), cartographer, 1094
Cincinnati, OH, 577, 942, 1311; quadrangle, OH-KY, 1315
City Island, NY, 849
Civil War, 1123-29; hospitals, Gettysburg battlefield, 969
Civitá Vecchia, Italy, 14 (1)
Clapp, F. G., cartographer, 395
Clarendon Press, publisher, 114
Clark, C. W., cartographer, 1335, 1339-42
Clark, E. B., cartographer, 1024, 1026
Clark, Fred A. (fl. 1870-1880, cartographer), 1407, 1441
Clark, G. R., cartographer, 33 (8)
Clark, William (1770-1838, explorer, soldier), 8 (27), cartographer, 8 (1-2), 28 (1), (3-7), (10-14), 560, 566, 1765; collection, 8; *History of the exploration under the command of Lewis and Clark,* 566
Clark, William, and Meriwether Lewis, Journals, 28
Clark, William Bullock, cartographer, 1076
Clark County, KY, 1249
Clark's River 28 (11); Yellowstone River fork, 560
Clarke, ——, 82
Clarke, S. J.; publishing co., publisher, 968
Clarke, Thomas (fl. 1798-1800), engraver, 296-97
Clarke, William Branwhite (fl. 1880-1893,

geologist), 1608
Clarke, W. Eagle, cartographer, 1722
Clarkson, Matthew (1735-1800, surveyor, entrepreneur, public official), cartographer, 1037
Clausner, (Claussner, Klaussner, Jakob), Joseph (fl. 1744-1797, Swiss surveyor and engraver), engraver, 234
Clearfield County, PA, 937
Clements, William L., Library, 32 (5)
Clerke, ——, 80
Cleveland, William C., cartographer, 462; donor 462
Cleveland, OH, 942
Clinton, De Witt (1769-1828, politician), donor, 827
Clinton County, KY, 1249
Clover, Richardson, cartographer, 747, 1630
Clymer, George, collection 9; Clymer and Meredith lands, 9 (2)
Coal, KY, 1247; OH (bituminous), 32 (83), Turnbull County, 1314; PA, 937, 948, 952, 961, 965, (anthracite), 945, 953, 958, (bituminous), 32 (83); Queensland, Australia, 1599, 1604-5; U.S., 965; WV, 1140-41, (bituminous) 32 (83)
Coatesville, PA-DE quadrangle, 973
Cobb Creek, PA, 43 (5)
Coblentz to Carlsruh, Rhine River, Germany, 29 (1)
Cochorn (?), Baron de, 29 (7)
Cochran, A. B. (fl. 1871-1875, surveyor), cartographer, 953
Cod, Cape, MA, 636, 679, 695; harbor, 679, 682-83; isthmus, 677; roads, 695
Codrington College, Barbados, 1519
Coe, William Robertson, collection of Western Americana, Yale University, 560
Coggins, E. H., publisher, 1016
Colahan, John B., cartographer, 1012
Colburn, Henry, publisher, 1575-78, 1620
Colburn, Zerah (1832-1870, American engineer), 580; *Railroad economy; the revised and extended report on European and American railroads,* 580
Cold Harbor battlefield, VA, 774
Colden, Cadwallader (1688-1766, loyalist, Lt. Gov. of NY), 1724; *Papers relating to and act of New York for encouragement of the Indian trade,* 783
Cole, C. S., cartographer, 504
Cole, Joseph M., III, 1764
Colinton, FL, 1189
A collection of maps . . . U.S. Congress, 1166, 1226-28, 1331, 1383, 1629
A collection of the charts of the coast of Newfoundland, 1749
Colles, Christopher (1738-1816, engineer and cartographer), cartographer, 71, 635, 721-27, 787-812, 863-74, 1054-64, 1096-116; *The geographical ledger and systematized atlas,* 71, 906-907; *A survey of the roads,* 635, 721-27, 787-812, 863-74, 906-11, 1054-64, 1096-116
Colles, Eliza, engraver, 71, 635
Collet, John Abraham (fl. 1767-1776, Swiss origin, engineer and surveyor), cartographer, 1081, 1093, 1655
Collier, P. F., & son, publisher, 1746; *World atlas and gazetteer,* 1746
Collin, Étienne (fl. 1798-1829, French engraver), engraver, 310, 333
Collin, Nicholas (1746-1831), 32 (65)
Collins, Greenville (fl. 1669-1698, hydrographer), cartographer, 1662; *Great Britain's coastal pilot,* 1662
Collins, Zaccheus (1764-1831, botanist, philanthropist, APS 1804), 987
Collinson, Peter (1694-1768, merchant, botanist, F.R.S.), 32 (3)
Collon, Jacques (French engraver), engraver, 113
Colombia, 1514, 1565-68; Atranto River, 1568; Barranquilla, 1562; Cartagena, 1566-67; Indians, 35 (13), Old Providence Island, 1565; Truando River, 1568
Colonna, B. A., cartographer, 1137
Colorado, 1408-21, 1427; Caldera County, 1419; Cripple Creek, 1412-14; Denver, 1412-14; Department of the Interior, Geological Survey, 1420-21; Elk, 1411; geology, 1420-21; Ironton, 1420; La Plata, 1411; Leadville, 1412-14; Mesa Verde National Park, 1415; Ouray County, 1419; Pike's Peak, 39, 1411; Pueblo, 39; Rocky Mountain National Park, 1416; San Miguel County, 1419; Sawatch Range, 1411; Silverton County, 1419; Telluride mining district, 1421
Colorado Scientific Society, *Proceedings,* 1418-21
Colton, C. B., publisher, 590, 1502
Colton, George Woolsworth (1827-1901, publisher), cartographer, 1018, 1363; *Colton's Atlas of the World,* 1018; lithographer, 590; publisher, 590
Colton, G. W. & C. B., & Co., cartographer, engraver, publisher, 590

Colton, Joseph Hutchins (1800-1893, publisher), 83, 846, 944, 1018, 1311, 1324, 1359, 1363
Colton, Roger B., cartographer, 625
Colton's atlas of the world, G. W. Colton, 1018
Columbia County, PA, 958
Columbia, NV, 1460
Columbia Planograph Co., publisher, 1417, 1446
Columbia Railroad, PA, 576, 1072
Columbia River, 28 (4), 560, 1426, 1474; confluence with Mult-no-mah, 28 (9), entrance, 1464; Fort Vancouver, 32 (82), Great Falls, 28 (3); Great Rapids, 28 (5), long and short narrows, 28 (6)
Columbia University, N.Y.C., NY, 844
Columbian Magazine, 700, 905, 1305-7
Columbus, Christopher (1451-1506, discoverer of the New World), 82, 1738
Commercial atlas of the world, L. E. Smith, 1743
Commoren, ——, 29 (51)
Communicações. Direccão dos trabalhos geologicos, Portugal, 340
Comorin, Cape, India, 401
Compagnie des Indes Hollandaises, 386a
A compleat pilot for the West-Indies, Thomas Jefferys, 1674
A complete body of ancient geography, d'Anville, 1654
A complete genealogical, historical, chronological, and geographical atlas, Lavoisne, 1684
A complete historical, chronological and geographical american atlas, 1685
A complete set of new charts . . . of the coast of Portugal and the Mediterranean Sea, J. Giacomo Alagna, 1650
A comprehensive atlas, T. G. Bradford, 1695
Comstock, C. B., cartographer, 1131, 1301, 1341
Comstock lode, Washoe mining district, NV, 1457
A concise natural history of East and West Florida, B. Romans, 1184-85, 1200
Condor, T., engraver, 1238, 1270, 1308
Conestoga navigation, PA, 939
Confederate States of America, army, 1123, 1181
Congo, Katanga. *See:* Zaire, Katanga
Congress Hall, Phila., PA, 1001
Conjola, N.S.W., Australia, 1607
Conlin, R. R., cartographer, 984
Connecticut, 455, 583, 719-49, 752-53, 761, 780, 835; Bridgeport harbor, 743; Captain's island harbor, east and west, 745; Cawkin's Island harbor, 744; Black Rock harbor, 743; Geological and Natural History Survey, *Bulletin,* 749; geology, 749; Guilford, 731; Hadley, 732-33; Judges Cave, 734; Litchfield Observatory, Hamilton College, 57; Milford, 737; New Haven, 729-30, 735-36; New Haven harbor, 741; New London harbor, 644, 742, 747; Norwich, 747; roads, 780; Sheffield Island harbor, 744; Stratford, 721-27; Stratford to N.Y.C., NY, 795-97
Connecticut River, Great Falls (or Bellows), 728; north, 740, 746
Connecticut, Thames River, 644, 747
Connelly, John Robert, donor, 32 (1), 899; "Independent confirmation of the magnetic declination, 1737-38," 32 (1), 899
Conover, Samuel F. (d. 1824), 920; donor, 920
Conrad, John, publisher, 921-22, 1274
Constance, Lake, Switzerland, 221
Constantinople, Turkey. *See:* Istanbul, Turkey
Constellations, ancient Egypt, 59; Andromeda, 50 (1); Antinous, 50 (1-2), Delphinus, 50 (1); Hercules, 50 (1); Sagitarius, 50 (2), Serpentarius, 50 (2)
Constitution Hall, Topeka, KN, 1399
Contour map, U.S., 614, 619
Conway, T. R., cartographer, 1227
Cook, F. S., cartographer, 776
Cook, George Hammell (1818-1889, geologist), cartographer, 886, 888
Cook, James (1728-1799, English explorer), 66-67, 69, 73, 74b-c, 80, 82-83, 312, 363, 370, 1616-17, 1619, 1655, 1749, 1773
Cook County, MN, 1356
Cooke, Charles E., cartographer, 1315
Coombs, D., cartographer, 46 (19)
Cope, E. B., cartographer, 969, 971
Cope, Thomas Darlington (1880-1964, scientist), cartographer, 973; donor, 973, 1037, 1617; "Degrees along the west line . . . ," 902, APS, *Proceedings,* vol. 93; "Maryland-Penna boundary survey," 1053
Copenhagen, Dansk Geographische Selskab, *Topografich Atlas Danmark,* 1759
Copenhagen, Denmark, royal palace, 166; K. Viderskabernes Societets, 300
Copernical system, 60
Copernicus, Nicolas (1473-1543, founder of modern astronomy), 1644; *Atlas Novus Coelestis in quo Mundus Spectabilis,* 1644
Copper mine, 33 (10)

Copper mining, Queensland, Australia, 1595
Copper plates, engraved: Delaware Bay, 1659, Lewis and Clark expedition, 566; Plan of the environs of Phila., 1659
Coral islands, 11 (1)
Coram, T., cartographer, 1611
Corbellini, P., cartographer, 1737
Cordoba-Santa Fe, Argentina, 1562
Cordoba, Mexico, 32 (18)
Core Sound, NC, 1151
Cork, Ireland, 145; voyage from Phila., 1 (1)
Corner, George Washington (1889-1981, anatomist, APS 1940), donor, 782, 977, 979-80, 982
Cornhill St., London, England, 119
Cornia, 29 (39)
Cornwall, England, 10 (1)
Cornwallis Charles, 1st marquis of Cornwallis (1738-1805, English soldier), 32 (9)
Corporation Commission, publisher, 1268
Corrientes, Cape, to Mazatlan, Mexico, 32 (15)
Corsica, France, 257, 269, 347
Cosamaluapan, Mexico, 32 (18)
Couch, Jonathan, collection, 10; *History of Polperro,* 10 (1)
Courland, U.S.S.R., 159
Courtenay, F., engraver, 645, 882, 885, 1091, 1196
Coutans, Guillaume (b. 1724, French cartographer), cartographer, 263
Couxsaxráge Indians, 752-53
Covens, publisher, 213, 287, 720
Cóvens, J., Jr., publisher, 213, 720
Cóvens & Mortier en J. Cóvens, Jr., publishers, 213, 287, 720
"Covington Map" in Hudson Bay Co., 32 (82)
Cowan, John, 1784
Cowden, James S., cartographer, 592
Cowperthwait, Thomas (fl. 1839-1895, publisher), 658, 1221, 1246, 1258
Cox, ——, 46 (2)
Cox, I. S., engraver, 484
Coyle, R., cartographer, 440
Craig, Thomas (fl. 1747-1778, surveyor, soldier), cartographer, 32 (1), 899
Cram, George F., cartographer, 1726; *Cram's modern new census atlas of the U.S. and the world,* 1726; engraver, 650; publisher, 1726
Cram's modern new census atlas of the U.S. and the world, 1726
Cram, T. D., cartographer, 643, 690
Cramoisy, ——, publisher, 32 (74)
Cranberry, NJ, to N.Y.C., NY, 869
Crandall, Roderic, cartographer, 1489
Crater Lake National Monument, Oregon, 1477
Craven, Tunis Augustus Macdonough (1813-1864, naval officer), cartographer, 853-54
Craven, T. M., cartographer, 1230-31
Cret, Paul Philippe (1876-1946, architect, APS 1928), cartographer, 1028
Cree Indians, 48 (14)
Creed, ——, engraver, 109
Creek Indians, 32 (5), 1177
Creux (Creuxius), François du, 32 (98)
Crevecoeur, Michel Guillaume St. Jean de (1735-1813, French agronomist, traveller), *Letters from an American farmer,* 670
Crimea, U.S.S.R., 29 (40)
Cripple Creek, CO, 1412-14
Crisp, Edward (fl. 1704-1711, surveyor and publisher), cartographer, 1163
Crittenden County, KY, 1249
Crooked Creek, KY, 2 (1, 2)
Crow Shoal, Delaware Bay, 880
Crown Point, NY, 784
Cruz Cano y Olmedilla, Juan de la (d. 1790, Spanish cartographer, traveller), cartographer, 1553
Cuajan Island. *See:* Guam, 608
Cuba, 32 (26), (29a), 1504, 1552, 1661, 1668, 1721; Bayamo, 32 (26); coast between Havana and Cape St. Antonio, 32 (72), Havana harbor, 1542; Matanzas, 1531; Santiago de Cuba, 32 (26), (38), (49)
Cucó, Pasqual (fl. 1778-1795, Spanish engraver), engraver, 330
Cumberland County, England, 120
Cumberland County, PA, 955
Cumberland, MD. *See:* Fort Cumberland, MD
Cumberland plain, England, 7 (1)
Cumberland River, KY, 1253
Cumberland Sound, Baffin Island, Canada, 6; Niantilac harbor, 32 (85)
Cumberland Strait, Canada, 32 (81)
Cummin, R. D., cartographer, 1024, 1026
Cumming, Anthony J. (Australian publisher), publisher, 1600, 1602
Cummings, Jacob Abbott (1773-1820, publisher, geographer), publisher, 641
Cummings & Hilliard, publisher, 641
Cundee, James (fl. 1807-1810, British publisher), publisher, 1002
Cunningham Creek, B.C., Canada, 537

Curaçao, The Netherlands, 1549
Currents, oceans, world, 1612
Curtis, C. A., cartographer, 1701
Custer, George Armstrong (1839-1876, soldier), 1391-92
Cutler, Manasseh (1742-1823, clergyman, botanist, colonizer, APS 1785), *Explanation of the map of the federal lands confirmed by the treaties of 1784,* 1303
Cutts, Richard Dominicus (d. 1883, soldier), cartographer, 1073-75, 1231, 1479, 1482-83, 1485
Cuyaba, Brazil, 1562
Cuyler, R. M., cartographer, 1469
Cuzco, Peru, 1570
Czechoslovakia, 287, 1641, 1741; railroads, 280; roads, 280
Czjzek, Joh. (fl. 1849-1860, German geologist), cartographer, 278

Dabney, T. G., cartographer, 1341
Dahl, Lt. ——, cartographer, 1259
Daidy, George A., cartographer, 1208; engraver, 1428
Dairy products, WI, 1049
Dale, J. B., cartographer, 1049
Dallas, GA, 1181
Dalrymple, Alexander (1737-1808, Scottish hydrographer and explorer of the Indian coast), cartographer, 460
Dame Marie Bay, St. Domingue, 32 (32)
Dandeleux, Nicholas (b. ca. 1749, French engraver), engraver, 432
Dandridge, William, cartographer, 1094
Dane County, WI, 1358
Dankworth, F., engraver, 577, 687, 693, 743, 848, 850-51, 1049, 1074, 1167, 1204-205
Danube River, 276
Danville, PA, 32 (95), 986
Danville and Pottsville Rail Road, PA, 932
Darby, William (1775-1854, writer), cartographer, 925, 1218; donor, 1218; publisher, 925, 1190; *A geographical description of Louisiana,* 1218
Darby, William, & B. Tanner, publishers, 1190
Darby, PA, 43 (4)
Darby Creek, Chester Co., PA, 43 (1): to Young's Ford on the Schuylkill, PA, 32 (68)
Darby Road, Phila., PA, 19 (1)
Darien del Norte, Gulfo de, Panama, 1514
Darton, William, Sr. (fl. 1785-1820, British engraver and publisher), engraver, 134
Darwin, Charles Robert (1809-1882, English scientist), cartographer, 11, collection, 11; *Origin of species,* 11
Daser, L., cartographer, 691
"David Ramsay, selection from his writings," Robert L. Brunhouse, 567. APS, *Transactions,* vol. 55
David, Theodore M., 3 (1-3)
Davidson, George (1825-1911, American astronomer and geographer), cartographer, 1463-64, 1466, 1469-70
Davies, ——, publisher, 670, 1628
Davis, Charles Henry (1807-1877, hydrographer), cartographer, 686, 705
Davis, John (fl. 1756-1759, publisher), publisher, 900
Davis, M. L., publisher, 900
Davis, M. L., & W. A., publishers, 32 (19)
Davis, R. O., cartographer, 1420
Davis & Dickson, publishers, 1628
Dawkins, Henry (fl. 1758-1780), engraver, 903-1302
Dawson, E. B. (fl. 1829-1836), engraver, 81
Dawson, George Mercer (1849-1901. Canadian geologist), cartographer, 473, 531, 538; "Note to accompany maps . . . ," Geological and Natural History Survey of Canada, *Annual Report,* 537
Dawson, L. H., cartographer, 1741
Dawson River, Australia, 1599
Day, Charles (1879-1931, architect, APS 1925), cartographer, 1032
Day, Sherman (1806-1884, merchant and engineer), cartographer, 12; collection, 12; *Historical collections of the state of Pennsylvania,* 12
Day & Son, publisher, 63
Day & Zimmerman, cartographers, 1032
Dearborn, Henry (d. 1829, cartographer), cartographer, 32 (6), 702
De Bernier, Henry de (?), 702
De Brahm, William Gerard (1717-ca. 1799, surveyor general of the Southern District, APS 1784), cartographer, 1081, 1093, 1655
De Bry, Theodor (1528-1598), *Zehender Theil der Orientalische Indien . . .,* 1645
Dechen, Heinrich von, *i.e.,* Ernst Heinrich Karl von (1800-1899, German geologist), cartographer, 303
Deel, Denmark, 172-73, 184
Deer Island, New Brunswick, Canada, 493
Defrénoy, ——, 271

"Degrees along the west line . . . ", Thomas D. Cope, 902
De Haven, Edwin Jesse (1816-1865, naval officer), cartographer, 1231
De la Camp, John (fl. 1857-1869, cartographer), cartographer, 775
De la Haye, Guillaume Nicolas (1727-1802, engraver and cartographer), engraver, 435
Delarochette, Louis Stanislas d'Arch (1731-1802, British cartographer), 289
Delaware, 455, 751-53, 755-56, 760-60a, 763, 767, 777, 782, 973, 1041-52, 1069, 1077; to Chesapeake Bay, MD, canal, 760; boundary, 902; Bay, 20 (2-4), 32 (92), 755, 762, 1041, 1659, 1661; Bay and River, 1049; Bay and River to Phila., PA, 1051; breakwater, 644, 1047, 1050, canals, 760-60a, 767, 1065, Cape Henlopen, lighthouse, 1003, Cape Henlopen to Gay Head, L.I., NY, 773; Chesapeake and Delaware canal, 32 (73), Coatesville quadrangle (DE-PA), 973; Crow Shoal, 880; Duck Creek, 760; Fort Christina, 1048; Georgetown, 1052; Marcus Hook quadrangle (DE-NJ-PA), 782; DE-NJ-PA quadrangle, 973; West Chester quadrangle (DE-PA), 973; roads, 767, 769, 777
Delaware and Raritan Canal, 771
Delaware County, PA, 1709; Garrett Road, 43 (2); Marshall Road, 43 (2); West Chester Road, 43 (2), (5)
Delaware River, 21 (1), 32 (92), 762, 895, 987, 1042, 1077; Bombay Hook to Cherry Island flats, 1046; canal, 760-60a; chart (Phila., PA), 644; Cherry Island flats, to Bridesberg, 644, 957; League Island, Phila., PA, 46 (6); lower part, 20 (7); Phila., PA, 46 (1), (6), (19), 1020
Delgado Joaquim Filippe Nery de Encarnação (1835-1908, Portuguese geologist), cartographer, 339
Delhi, India, 399
Delisle, Guillame. *See:* L'Isle, Guillame de
Delleker, George (1806-1830), engraver, 666
Delmarva peninsula, 1042, 1077
Delphinius (constellation), 50 (1)
Deming, Chauncey, 18 (1)
Denham, Henry Mangles (British naval officer), cartographer, 441
Denmark, 159-60, 166-85, 1759; Aakier, 174; Aalborghuus, 176, 178, 180; Aastrup, 179; Aeroe, 173; Apenrade, 300; Børglum, 179; Bövling, 181, 183; Bornholm, 185; Bredsted, 300; Copenhagen, royal palace, 166; Deel, 172-73, 184; Dronningburg, 176; Faroe Islands, 182; Flensborg, 300; Fyen, 171, 173; Haderslewhuus, 184, 300; Halds, 176; Haureballegards, 174, 176; Hertugdummet, 173; Jylland [Iylland], 172, 177, 184; Kaløe, 176; Kildinghuus, 174; Koldinghuus, 183-84; Laaland, 170-71; Langeland, 171, 173; Lugumcloster, 300; Lundenaes, 183; Mariager, 176; Möen, 171; Orum, 180; Riberhuus, 183-84; Samsöe, 171; Schleswig, 172-73, 300; Seigelstrup, 178; Siaeland, 167, 169, 171; Silkeborg, 176, 183; Skaane, 171; Skanderborg, 174, 176; Skivehuus, 181; Stiernholms, 174; Taasinge, 173; Tønder, 300; Thorsinge, 171; Vensyysel, 178-80; Vestervig, 180
Dennis, A., cartographer, 32 (8)
Dennis, N. H., cartographer, 644, 747, 1182
Denver, CO, 1412-14
Derréyeh, Egypt, 371
Des Barres, Joseph Frederick Wallet (1772-1824, British military engineer, cartographer), engraver, 1656; *The Atlantic Neptune,* 1659; *Charts of the coast and harbours of New England,* 1656; *The sea coast of Nova Scotia,* 1658
Desbarres, Lt. ——, cartographer, 501, 634, 852, 1656, 1658-59
Descripcion Fisica, Geologica y Minera en la Isla de Panay, E. Abella y Casariego, 413
Description Geologique de Canton Genève, Alphonse D. Favre, 239
Description of the Genessee Country, Charles Williamson, 764-65, 815
"A description of the surprising cataract, in the great Connecticut River," *American Museum,* 728
Description Topographique et Politique de la Partie Espagnole de l'Isle Saint-Domingue, M. L. E. Moreau de St. Méry, 1543
A descriptive list of maps and views of Philadelphia, P. L. Phillips, 998
De Silver, Charles, publisher, 946
Des Moines rapids, Mississippi River, 1331
Desmond, Daniel J., 1691
Desnos, Louis Charles (fl. 1750-1770, French publisher), cartographer, 1670
Desray, ——, (fl. 1807-1821), publisher, 266, 1560, 1682
Desray et J. Goujon, publishers, 266
Destarnick, ——, 29 (34)
Detail map of the lower Mississippi river, Missis-

sippi River Commission, 1710
Detail map of the upper Mississippi river, Mississippi River Commission, 1705
Detloff, C., cartographer, 238; publisher, 238
Deutsch, Morton (b. 1920, psychologist, editor), donor, 1171
Devlet Basımevi, publisher, 1738
Devon, England, 10 (1)
De Vorsey, Louis, Jr., 469
DeWees, H. M., cartographer, 882
Dewes, J. H., cartographer, 960
Dewing, Francis (fl. 1716-1723, engraver, publisher), engraver, 467; publisher, 467
De Witt, Simeon (1756-1834, Surveyor-General for NY), cartographer, 818, 820; donor, 818
Dezauche, J. A. (fl. 1768-1831, French editor and mapseller), cartographer, 258; publisher, 258, 294, 351, 1501
Dezertas Island, Portugal, 328
Dialing, 74a
Dickens, H. L., cartographer, 851
Dickison, A., cartographer, 542
Dickson, J. (British bookseller), publisher, 1628
Dictionary of altitudes . . . , Department of the Interior, Canada, 474
Didot, Firmin (1764-1836, French publisher), 301
Diessler, Oscar, lithographer, 302
Digby, Captain (fl. 1759, British surveyor, mathematician), cartographer, 501
Dilly, Charles (1739-1807, English bookseller), publisher, 66-67, 107, 254, 1554
Dilly, C., & G. Robinson, publishers, 66-67, 107, 254, 1554
Diment, William D., cartographer, 841
Dimmock, C. H., cartographer, 1422
Dings, M. G., cartographer, 1419
Dinsmore, A. F., cartographer, 598, 1210, 1213, 1224, 1267, 1320, 1355, 1397-98, 1412-14, 1417, 1423-24, 1447, 1451, 1453-54, 1475-76, 1490
D'Invilliers, Edward Vincent (b. 1857, geologist), cartographer, 959; *Geology of the South Mountain Belt of Berks Co.,* 959
Dirección de Hidrografia, Madrid, donor and publisher, 268a, 326, 333, 336-40, 408-12, 441, 689
Disappointment, Cape, 28 (7)
Disco Bay, Baffin Bay, Canada, 93
The discovery, settlement and present state of Kentucky, John Filson, 1236
District of Columbia, 82, 576-77, 644, 770, 1065, 1078-80, 1134; Washington, capitol, 579; and Georgetown harbors, 1080; to New Orleans, LA, 1090; to Williamsburg, VA, 1127; Washington Monument, 579; roads, 777
District of Columbia, Georgetown, 644
Dixon, Jeremiah (1733-1779, British cartographer, surveyor), cartographer, 32 (82), 754, 902
Djurberg, Daniel, 1744-1834 (Swedish cartographer), cartographer, 1619
Dneister River, U.S.S.R., 29 (42)
Dobson, Thomas (1751-1823, publisher), donor, 74a-74d, 87b, 130-31, 147, 154, 177, 196, 259, 261, 295-99, 309, 315, 331, 352, 354, 366, 405, 425, 457, 1539, 1556; *Encyclopaedia, see: Encyclopaedia;* publisher, 74a-74d, 87b, 130-31, 147, 154, 177, 196, 259, 261, 295-99, 309, 315, 331, 352, 354, 366, 405, 425, 457, 1273, 1329, 1539, 1556, 1683
Documents relative to the negotiations for peace between the U.S. and Great Britain [John Melish], 565
Dodsley, Robert (1703-1764, British poet, dramatist, bookseller), publisher, 752
Dodson, James (d. 1757, English mathematician), cartographer, 77
Dominican Republic, Santo Domingo, 32 (42)
Doms, P., lithographer, 1692
Doney, Richard, 18 (1)
Donn, J. W., cartographer, 1075, 1127, 1131, 1137-38, 1259, 1262
Donoghue, John (*see also:* O'Donoghue), cartographer, 1362
Donors: *See:* Abbot, H. L.; Albert, J. T.; Ames, Charles Lesley; Astley, Thomas; The Athenaeum of Philadelphia; Bache, A. D.; Bache, Hartman; Bancker, Charles W.; Barry, ——; Barton, William; Bartram, William; Belknap, Jeremy; Benton, Thomas Hart; Biddle, Charles J.; Biddle, Nicholas; Bloch, Arthur; Blunt, Edmund; Blunt, G. W.; Bowditch, Nathaniel; Bowman, Isaiah; Boyé, Martin H., heirs; Bray, Edmund; Bray, Martha; Brown, Samuel; Butterfield, Roger; Çambel, Halet; Campbell, Archibald; Carey, Henry C.; Carey, Mathew; Carey, M. & Son; Cardozo, Casado Geraldes, J. P.; Cash, Mr. Chapman, Isaac A.; Chapman, Nathaniel; Chauncey, Mr. & Mrs. Charles; Churchman, John; Cleveland, William C.; Clin-

ton, DeWitt; Connelly, John Robert; Conover, Samuel F.; Cope, Thomas D.; Corner, George W.; Darby, William; Deutsch, Morton; DeWitt, Simeon; Dirección de Hidrografia, Madrid; Dobson, Thomas; Dougal, W. H.; Drayton, William; Dunn, Nathan; Dunn, Samuel; DuPonceau, P. S.; DuPont, Samuel Francis; Durrett, R. T.; Dyer, Peter R.; Eastwick, Joseph L.; Eastwick, Suzanne Wister; Eckhard, F. J.; Edgell, J. A.; Engles, Joseph P.; Erskine, David Steuart, Earl of Buchan; Evanson, Howard N.; Fahlberg, Samuel; Federal Republic of Germany; Fisher, J. Francis; Fisher, Thomas; Flexner, James Thomas; Flexner, William Walsh; Flood, Margaret Toogood; Forshey, Caleb G.; Fox, Charles Pemberton; Fox, Mary; Friends of the Wissahickon, Inc.; Gardner, Robert M.; Garnett, John; Geological Society of London; George Washington Bicentennial Commission; Gerig, William; Gibbons, C.; Gibson, William; Gimbel, Richard, estate; Girard College; Gorman, Fred J.; Gould, George; Graham, James D.; Gray, Asa; Great Britain, Admiralty; Great Britain, Board of Ordnance; Greene, Robert; Greenwich Observatory; Hale, John S.; Hallowell, Mrs. A. I.; Harris, Thomas; Hassler, F. R.; Hayden, F. V.; Heazlitt, C. F.; Hembel, William; Henry, Mathew S.; Hill, Richard R.; Hinton, John; "Hiram"——; Hoopes, Alban; Horsfield, Timothy; Howell, Reading; Hudson, Edward; Ingersoll, C. J.; Ingersoll, Joseph R.; Ivanoff, ——; Jackson & Alger; Jefferson, Thomas; Jomard, E. F.; Jones, William; Kane, E. K.; Kane, John Kintzing; Karpinsky, A.; Kimber, Emmor; Komarzewsky, ——; Krusenstern, Adam Johann von; King of the Two Sicilies; Lafon, B.; Lanjuinais, J. D.; Lapham, I. A.; Lay, Amos; Lea, I.; Lee, Richard Henry, II; Lesoüef, A.; Lesueur, Charles A.; Long, Col; Lyman, B. S.; McCoy, Lee H.; McIlhenny, Francis S.; McMahon, B.; Martin, Lawrence; Martinez, Juan J.; Marvin, Mrs. Mary Vaughan; Maury, M. F.; Meade, George G.; Mease, James; Melish, John; Menzies, Donald; Merrick, George; Moreau de St. Méry, M. L. E.; Miller, C. William; Mills, Robert; Mitchell, S. Augustus; Montgomery Co. (PA) Planning Commission; Moore, Samuel; Moreledo, Severiano; Morelli, Chevalier; Morrison, Russell; Munsell, Luke; Murphy, R. C.; Myers, Albert Cook; Nagy, Károly; The Netherlands, Minister of the Interior; Neville, Rev. E.; New Hampshire; NY State Museum; Nikitin, S.; Nolan, James Bennett; Norris, Joseph Parke; Oakford, Joshua L.; Page, Thomas J.; Palaullo, ——; Palmer, William Ricketts; Parker, Daniel; Parsons family; Patterson, Robert; Patterson, Rev. Robert; Patterson, Robert M.; Pedersen, Peder; Phillips, Henry, Jr.; Plowman, Thomas L.; Poussin, William Tell; Proud, Robert; Reed, Joseph; Reps, John W.; Rhode Island Historical Society; Rivinus, Florens; Rockefeller University; Robbins, Mrs. Caroline; Robbins, W. J.; Robeson, M.; Rogers, Fred B.; Royal Asiatic Society of Calcutta; Sabin, Edward; Sargeant, T.; Short, William; Shyrock, R. H.; Sinclair, Thomas; Smith, M. D.; Smith, Thomas Peters; Snider, Jacob, Jr.; Stevenson, J. J.; Strickland, W.; Sveriges Geologiska Undersökning; Tanner, H. S.; Taylor, H. W. B.; Thornton, John; Trego, Charles B.; Truxtun, Thomas; Tyson, Job R.; U.S. Coast and Geodetic Survey; U.S. Dept. of the Interior; U.S. Geological and Geographical Survey of the Territories; U.S. Geological Survey; U.S. Government; U.S. Hydrographic Office; U.S. Treasury Department; U.S. War Dept.; University of Nebraska Press; Valltravers, Johann R.; Vance, David H.; Vaughan, John; Vaughan, Wm.; Virginia; Vogelin, C. F.; Watmough, John G.; Wetherell, John P.; Wheeler, George M.; White, James; Wilkes, Charles; Williams, Jonathan, Jr.; Williams, W. G.; Wilson, John; Winlock, Herbert E.; Wood, George B.; Woods Hole Oceanographic Institute; Yampolsky, Mrs. Helene Boas; Young, James H.

Doolittle, Amos (1754-1832, engraver), engraver, 76, 368, 628, 638-39, 663, 711, 728-37, 876, 1044, 1083, 1158, 1540; cartographer, 628; publisher, 663

Doral, Antonio (Spanish cartographer), 335

Dorchester, MA, 699

Dorn, Sebastian (d. ca. 1778, German engraver), engraver, 222

Dorr, F. W., cartographer, 854, 1127, 1131, 1259, 1262

Dorsey, Clarence W., cartographer, 1076

Dougal, William H. (1822-1895, engraver),

cartographer, 1285; engraver, 834, 1127-28, 1278, 1281-82, 1284, 1286-94, 1296-97
Douglas, E. M., cartographer, 1235
Dover-Stanhope, NJ, 892
Dower, John, engraver, 1578
Downes, D., cartographer, 1247
Dragon Creek, B. C., Canada, 537
Drake, N. F., cartographer, 32 (87)
Drayton, F., cartographer, 829-31, 1008, 1276
Drayton, John (1766-1822, governor of SC), *View of South Carolina, as respects her natural and civil concerns,* 1161-62
Drayton, Thomas Fenwick (d. 1891, soldier), cartographer, 828
Drayton, William, donor, 1692
Dreher, John, cartographer, 32 (75-76)
Drent, The Netherlands, 191
Dronningborg, Denmark, 176
Drovetti, 427
Drum Shoal, NC, 1151
Duane, William John (1780-1865, lawyer, politician), 32 (77), 986
Dublin, Ireland, 145
Duce, R. S., cartographer, 1419
Duche, ——, 46 (2)
Duche, Anthony, 46 (20)
Duche, John, 46 (11)
Duck Creek, DE, 760
Du Creux [Creusius], François (1596-1666, Jesuit), 32 (98)
Dürrenbach, 29 (11)
Du Four, John James, cartographer, 2 (2)
Dufrénoy, Pierre Armand (1792-1850, French geologist), cartographer, 271
Du Hamel du Monceau, Henri Louis (1700-1782, French botanist), *Art du Tunnelier; Mémoires sur la Formation des Os; Recherches sur les Ruins d'Herculanéum; Traité des Arbes et Arbustes,* 14; collection, 13
Dumfries, VA, to Annapolis, MD, 1097
Dunaway, H., cartographer, 1336
Dunbar, Samuel O., cartographer and publisher, 680
Dunbar, William (1749-1810, U.S. surveyor), 2 (4)
Duncan, Winifred, cartographer, 41 (1-2); *Your Mexico,* 41
Dunham, Albert B., cartographer, 32 (8)
Dunlap's American Daily Advertiser, 997
Dunmore, Lord. *See:* Murray, John, 4th earl, 1732-1809.
Dunn, Nathan, donor, 681, 1692
Dunn, Samuel (d. 1792, British mathematician, instrument maker, and inventor), cartographer, 32 (7), 53, 453-54, 1536, 1653, 1672; donor, 53; *New atlas of the mundane system,* 1653, 1672
Dunnewald, T. J., cartographer, 1370
Dunotyr Castle, Scotland, 148
Dunstan, Benjamin (b. 1865, Australian geologist), cartographer, 1588, 1590, 1592-94, 1596
Du Ponceau, Peter Stephen (1760-1844, lawyer, Indian scholar, linguist), donor, 294, 351, 1374, 1643, 1672, 1682, 1688, 1692
Du Pont, Samuel Francis (d. 1865), donor, 1689
Durrett, Reuben Thomas (1824-1913, author), donor, 1236
Dury, Andrew (fl. 1742-78, English publisher), publisher, 286, 1649
Du Simitière, Pierre Eugène (1736-1784, artist and museum owner, APS original member), cartographer, 1095
Du Tertre, Jean Baptiste (French publisher), 252
Dutton, Clarence Edward, cartographer, 1450, 1708; *Report,* U.S. Geographical and Geological Survey, 1450
Du Val, Henri Louis Nicolas (1783-1854, French cartographer and printseller), cartographer, 1182
Duval, Peter S. (fl. 1831-1879, lithographer), engraver, 939; lithographer, 939, 1014; publisher, 939, 1015
Dvina River, U.S.S.R., 313
Dybwad, Jacob, publisher, 165
Dyer, Peter R., donor, 12

Eagle harbor, Lake Superior, 1282
Eagle River, Lake Superior, 1289
Eakin, Constant Mathieu (d. 1869, soldier), cartographer, 684-88, 741, 743-45, 1020-21, 1204-205, 1166
Early maps, Tony Campbell, 1763
Early maps of North America, Robert M. Lunny, 1748
Early sea charts, Robert Putnam, 1766
The East India Company, 1660
East Indies. *See also:* Indonesia, 78, 363, 370
Eastburn, Benjamin (d. 1741, Pennsylvania surveyor-general), cartographer, 994
Easter Island, Chile, 1619
Eastern Canada, 483-491
Eastern hemisphere, 87a-87b
Eastern U.S., 624-25

Eastern Woodlands Indians, 48
Eastport Harbor, ME, 644
Eastwick, Joseph L., donor, 556, 639, 655, 664, 673, 714, 738, 814, 846, 876, 916, 944, 1044, 1067, 1118, 1147, 1158, 1174, 1241, 1255, 1272
Eastwick, Suzanne Wister, donor, 556, 639, 655, 664, 673, 714, 738, 814, 846, 876, 916, 944, 1044, 1067, 1118, 1147, 1158, 1174, 1241, 1255, 1272
Eastwood, E. M., engraver, 417
Eau Claire, WI, 1372
Eberhardt, E. F., cartographer, 1598-99
Eclipse Sound, N. W. T., Canada, 487
Eckard, F. J., donor, 398
Eckel, E. B., cartographer, 1419
Eckert Lithographic Co., lithographer, 284, 1198, 1213, 1224, 1316, 1320, 1327, 1369, 1388-90, 1397, 1413, 1430, 1453, 1460, 1475; publisher, 1224, 1263, 1327, 1397, 1413, 1423, 1430, 1453, 1456, 1460, 1475, 1500
Eckhard, F. J., donor, 398
Ecklund, C. A., cartographer, 1420-21
École Française d'Extreme-Orient. L'Aviation Militaire et Service Géographique de l'Indochine, 402-403
Ecuador, Galapagos Islands, 1561
Eden River, England, 71
Edenton, NC, 32 (91)
Edgartown harbor, MA, 686
Edgell, J. A., donor, 101
Edinburgh, Scotland, 148, 152
Edinburgh Castle, Scotland, 148-49
Edinburgh Geographical Institution, 275, 389
Edwards (James E.? Fl. 1791-1817, British cartographer), cartographer, 460
Edwards, W. S., cartographer, 885, 1168
Egg Harbor, NJ, 886
Egloffstein (Eggloffstein), F. W. (fl. 1855-64, topographer), cartographer, 1422
Egypt, 52, 325, 371, 427-33; (ancient) constellations, 59; archaeology, 3 (1-3); Cairo, 432; Esne, 427; geology, 433; Girgeh, 427; Nile River, 428; Syout, 427; Thebes, 427; Valley of the Kings, 3 (1-3);
Ehrgot, T., publisher, 1140
Ehrgott & Krebbs, publishers, 1140
Ekelöf, J. A., 1716
Elba Island, Italy, 344, 357; Porto-Ferrajo, 357
Elbe River, 276
Elder, Sir Thomas, 1581
Elder Scientific Exploring Expedition, 1581
Eldridge House, Lawrence, KS, 1399
Elevators (grain), Alberta, Canada 519; Manitoba, Canada, 519; Saskatchewan, Canada, 51
Elford, I. M. S., 1165
Elie de Beaumont, Léonce (*i.e.,* Jean Baptiste Armand Louis Léonce, 1798-1874, French geologist), cartographer, 271
Eliot, Jared (1685-1763, minister, doctor, scientist), 32 (3)
Elizabeth, NJ, 890
Elizabeth River, VA, 1137
Elizabethtown, NJ to N.Y.C., NY, 863
Elk, CO, 1411
Elk River, MD, 766
Elko, NV, 1460
Ellicott, Andrew (1754-1820, surveyor, mathematician, APS 1803), cartographer, 1273, 1329; *Journal,* 1273, 1329
Ellicott, Benjamin (1765-1827, cartographer), cartographer, 816, 819
Ellicott, Eugene, cartographer, 1137
Ellicott, Joseph (1760-1826, surveyor, cartographer), cartographer, 816, 819
Elliott, Henry Wood (1846-1930, geologist), cartographer, 1435-36
Ellis, A. D., publisher, 1701
Ells, Robert Wheelock (1845-1911, Canadian geologist), 485
El Paso County, TX, 1234
Elverson, J. S. (geologist), cartographer 961
Ely, ——, cartographer, 936, 1263; engraver, 936
Ely, NV, 1460
Ely and Hammond, cartographers, 936
Emory, William Hemsley (1811-1887, soldier), cartographer, 1229, 1758
Encyclopaedia (Dobson), 74a-d, 87b, 109, 130, 147, 154, 177, 196, 259, 261, 295-99, 309, 315, 331, 352, 354, 366, 405, 425, 457, 1539, 1556
Encyclopaedia Britannica, Inc., publisher, 1744
Encyclopaedia Britannica World Atlas, 1744
Endeavour (ship), 73
Endicott & Co., publishers, 956
Endicott, ——, lithographer, 772
Engelmann, Wilhelm (1808-1878, German publisher), publisher, 304
The engineering drawings of Benjamin Henry Latrobe, Darwin H. Stapleton, 33 (7)
England, 52, 119-143, 152, 1654, 1662, 1679; Bristol, 123, 125; canals, 136-37, 139; Car-

lisle, 1662; Chatham, 125; Cornwall, 10 (1); Cumberland County, 120; Cumberland plain, 7 (1); Devon, 10 (1); Eden River 7 (1); geology, 7 (1), 138-39, 142; Hull, 123, 125; Kent, Hollwood, 129; Liverpool, 123, 125; London, 123, 125-26, 129, 143; London, Cornhill Street, 119; London docks, 132; London, Royal Exchange, 131; Newcastle, 123, 125; Northamptonshire, 121; Northamptonshire, Grand Junction Canal, 136; Pennine fault, 7 (1); Plymouth, 123, 125; Portsmouth, 123, 125; railroads, 139, 141; roads, 124-25, 127-28, 133, 137; Sheerness, 125; Taunton, 135; Thames River, 132, 1662, Trent River, 136; Westmoreland County, 120; Yarmouth, 125

English armada, 1621

English Channel, 1663, 1674

English fort, Pittsburgh, PA, 45 (4)

English harbor, Antigua, 1517

English-Lenni Lenape and Lenni Lenape-English dictionary, 20

The English pilot . . . West-India navigation, 1661

Engles, Joseph P., donor, 55, 1641, 1660

Engravers: (See also: Lithographers): Adams, J. A.; Aitken, Robert; Akerland, B.; Akin, James; Allardice, S.; American Bank Note Co.; Anderson, A.; Anderson, H.; André, ——; Angelo, G. N.; Annin, W. B.; Atwood, J. M.; Baine, H. E.; Barker, W.; Barnard, H. S.; Baron, B.; Bartle, R. F., Co.; Basire, James; Bayly, I. [J.]; Bellanger, ——; Bellin, S.; Benedetti, ——; Benner, ——; Bennett, Richard; Bianchi, G. C.; Bien, Julius; Blondeau, Alexandre; Bowen, Emanuel; Bowen, Thomas; Bower, J.; Bowes, ——; Bregante, F.; Bressanini, ——; Caldwell, J.; Callender, B.; Cano y Olmedilla, Juan De la Cruz; Chalamandrier, Nicolas; Chapin, William; Charpentier, Jacques; Cheevers, John; Cheevers, T.; Chocarne, ——; Clarke, T.; Clausner, Joseph; Colles, Eliza; Collin, E.; Collon, J.; Colton, C. B.; Colton, G. W.; Colton, G. W. & C. B. Colton & Co.; Condor, T.; Courtenay, F.; Cox, I. S.; Cram, George F.; Creed, ——; Cuco, Pasqual; Daidy, G. A.; Dankworth, F.; Dandeleux, ——; Darton, W.; Dawkins, Henry; Dawson, E. B.; Delleker, ——; Des Barres, J. F. W.; Dewing, Francis; Doolittle, Amos; Dorn, Seb.; Dougal, W. H.; Dower, J.; DuVal, P. S.; Eastwood, E. M.; Ely, ——; Enthoffer, J.; Erhard Hermanos Also, Erhard frères; Evans, H. C.; Faden, William; Fairman, G.; Filbert, R. M.; Fisk, ——; Fisk & Rusell; Foot, T.; Fortif, ——; Franks, ——; Franks & Johnson; Frederich, ——; Fridrich, ——; Friend, ——; Frost, H.; Furrer, H.; Gardner, J.; Gaw, R. M.; Gibson, John; Gillingham, Edwin; Glot, C. B.; Goldthwaite, J. H.; Gonne, ——; Green, Clerkenwell; Gridley, ——; Groux, ——; Guerra, Gius; Guiter, ——; Haines, D.; Hall, ——; Hammond, ——; Harrison, C. P.; Harrison, D. R.; Harrison, Samuel; Harrison, W.; Harrison, William Jr.; Harrison, W. & S.; Hazzard, J. L.; Hergesheimer, E.; Herrlein, E.; Hewitt, ——; Hill, Samuel; Hooge, Romein De; Hooker, ——; Hunt, S. V.; Jäck, Carl; Jansz, L. Schenk; Jefferys, Thomas; Jobbins, J. R.; Jones, ——; Jones, Benjamin; Johnson, ——; Johnston, A. K.; Johnston, W.; Johnston, W. & A. K.; Jukes, Francis; Kearney & Co.; Kitchin, Thomas; Kleckhoff, K.; Knight, A.; Knight, H. M.; Knight, J.; Knight, J. J.; Knight, R. T.; Kneass, William; Kondrup, J. C.; Krebs, Charles G.; Kübul, S. J.; La Haye, Guill de; Lamb, F.; Lawson, Andrew; Lawson, O. A.; Leidenfrost, E.; Leth, Henry de; Lewis, S. M.; Lindenkohl, A.; Lindenkohl, H.; Lotter, I. C.; Lotter, M. A.; Lotter, Tob. Conr; Lownes, Caleb; McClellan, D.; McCoy, G.; Maedel, E. A.; Maeder, A.; Marchant, J.; Martin, D.; Matsudaira, N.; Maverick, Peter R.; Metzeroth, G. B.; Moffat, J.; Molkow, E.; Morghen, Philip; Moskaira, A.; Murray, G.; Mutlow, John; Mynde, James; Neele, S. J.; Neuman, Louis E.; Nicholls, Sutton; Noguera, J.; Norman, John(?); Ourand, C. H.; Pendleton, ——; Penner, G. F.; Petersen, A.; Pettit, J. S.; Pettitt, S. T.; Picquet, Charles; Pingeling, T. A.; Poggiali, Giuseppe; Poupard, James; Prinald, H. D.; Probst, J.; Purcell, ——; Queiroz, I.; Rausch, ——; Ricart, ——; Richomme, ——; Ridley, J.; Roads, W.; Rogers, J.; Rollé, A.; Rollinson, Charles(?) or William(?); Russell, ——; Schedler, J.; Schieble, Erhard; Schielde, Georges Erhard; Scoles, I.; Scot, ——; Scot & Allardice; Scott, J. T.; Seale, R. W.; Senex, John; Sengteller, A.; Service Géographique De l'Indochine; Seymour, ——; Shallus, Francis; Sherman, ——;

Shermith & Smith; Siebert, Selmar; Sim, W.; Sipe, E. H.; Smith, ——; Smith, B.; Smith, Elvino; Smith, W.; Smith, Jones & T. Foot; Smither, B.; Smither, James; Smither, John; Sonne, I.; Stern, ——; Stiles, S. & Co.; Stone, William J.; Stout, James D.; Strobel, M. F. O.; Stull, S. E.; Tanner, Benajmin; Tanner, H. S.; Tardieu, Ambroise; Tardieu, P. S. F. père; Thackara, ——; Thackara & Vallance; Thompson, J. G.; Thompson, W. A.; Throop, D. S.; Throop, J. V. N.; Throop, O. H.; Tiebout, Cornelius; Tilgmann, ——; Trenchard, James; Trummer, C.; Turner, James; U.S. Engineering Bureau U.S. Geological Survey; Vallance, F.; Vallance, J.; Vandergucht, Gerard; Verwest, A.; Vevelst, E.; Vicq, ——; Voysard, E.; Wagner, W.; Walker, Emery, Ltd.; Walker, C.; Walker, J.; Walker, J. & C.; Warr, John; Warr, John & William W.; Warr, William, W.; Wessel, C.; Williams, W.; Woodman, Richard; Woodman and Mutlow; Woodward, E. F.; Wright, C. C.; Wuhrer, L.; Yeager, E.; Young, J.; Young, J. H.; Young & Delleker; Zuliani, Felice
Enthofer, Joseph (1818-1901, Austrian geographer and engraver for the U.S. Coast Survey), 645
Erdmann, A., cartographer, 161
Erhard Hermanos, engraver, 1571
Erhard frères, engraver, 1613; publisher, 271, 1613
Erichsen, P., cartographer, 885
Erie Canal, NY, 573, 576, 827
Erie, lake, 759, 816, 819, 829-30, 834, 852-53, 758, 1275-76, 1279-80; Bass Island, 1278; Kelly's Island, 1278
Ering, 29 (3)
Erman, Axel Joakim (1814-1869, Swedish geologist), 317
Erskine, David Steuart (11th earl of Buchan, 1742-1829, Scottish peer), donor, 1078
Escher von der Linth, Arnold (1807-1872, Swiss geologist), cartographer, 240-41
Esk River Reservoir, Scotland, 7 (5)
Eskimo, cartographer, 32 (81), 486-87; Labrador, Canada, 48 (11)
Esnault, —— (fl. 1780-1807, French map seller), 265
Esne, Egypt, 427
Espinalt y Garcia, Bernardo (fl. 1778-1794, Spanish cartographer), cartographer, 330
Espiritu Santo, FL, 1190
Esplanade or Place of Arms, New Orleans, LA, 1216
Esequabo, Guyana, 1549
Essek(?), 29 (43)
Essex, Frank B., cartographer, 1552
Essex County, NJ, 886
Essex Institute, Salem, MA, 1773
Estabrook, E. L., cartographer, 395
Estaing, Charles Hector, comte d' (1729-1794, French admiral), 1171
Estonia, 313
Etang du Thau, France, 252
Etbaye, 442
L'Etbaye: pays inhabité par les Arabes Bicharieh, M. A. Linant de Bellefond, 442
Eten to Paita on the alto Marañon River, Peru, 1570
Ethnolinguistics, Indochina, 403
Eugene, Prince, properly François Eugène de Savoie Carignan (1663-1736, Austrian soldier), 29
Europe, 74a, 78, 87a-87b, 105-19, 456, 469, 1621, 1692, 1731; geology, 114-15; magnetism, 1770; solar eclipse, 116
Europe, Central, 276-287
Evans, Andrew Wallace, cartographer, 1182
Evans, Lewis (1700?-1755, cartographer), cartographer, 750-53, 757, 859-60
Evanson, Howard N., donor, 32 (99)
Evanston, WY, 1443-44
Evens, H. C., engraver, 692, 854, 882, 885, 1091, 1230
Everett, Edward (1794-1865, Secretary of State, APS 1841), 453
Ewing, Maskell C. (d. 1849, soldier), cartographer, 1383
An examination of the Connecticut claim to lands in Pennsylvania, William Smith, 719
An explanation of the magnetic atlas or variation chart, John Churchman, 74
An explanation of the map of the city, and liberties of Philadelphia, John Reed, 989-92, 1652
Explanation of the map of the federal lands confirmed by the treaties of 1784, M. Cutler, 1303

Facsimile-atlas to the early history of cartography . . . , Nils Adolf Erik Nordenskiöld, 1716
Faden, William (1750-1836, English map collector and engraver and publisher), cartographer, 175, 191, 861; engraver, 32 (7), 68, 291, 499, 552, 861, 966; publisher, 68, 124, 150, 175, 189-90, 244, 291, 293, 325, 328, 345, 399, 401, 499, 552, 652, 861, 996, 1013; *North American atlas,* 499, 1013

Fahlberg, Samuel (1758-1834, Swedish physician, APS 1801), cartographer, 32 (65), 1538; donor, 32 (65), 1538
Fairbanks, Harold Wellman (b. 1860, cartographer, geologist), cartographer, 1486
Fairfax, F., cartographer, 643
Fairfax, W. McC., cartographer, 631-32, 690, 1049, 1138, 1481, 1485
Fairman, Daniel (1782-1815, engraver), engraver, 675, 815
Fairman, Gideon (1744-1827, engraver), engraver, 818
Fairmount Park, Phila., PA 1022
Fairmount Park Association, Phila., PA, 1028; *Report of the Commission,* 1028
Fairmount Water Works, Phila., PA; *See also:* Phila. Waterworks, 1005
Fairmount Park, Wissahickon Valley, Phila., PA, roads, 1034
Fales, J. C., cartographer, 1249
Falkland Islands, Great Britain, 1560-61
Falls of St. Anthony, Mississippi River, 1339
Falsan, Albert (b. 1833, French geologist), cartographer, 272
False Dungeness Harbor, WA, 1468
False Presqu'ile Harbor, MI, Lake Huron, 1292
Far East, 378-380
Farden, John, autograph, 62
Farley, John (d. 1874, soldier), cartographer, 677, 741, 770, 882, 885, 1138
Farming, U.S., 620
Farmville, VA, 774; battlefield, 774
Faroe Islands, Denmark, 182
Farquahar, C., cartographer, 693
Farthest North, Fridtjof Nansen, 32 (88)
Faucou, Lucien (fl. 1887-1889, French geographer), cartographer, 273-74
Fault map, CA, 1493
Fauntleroy, C. M., cartographer, 1182
Fauntleroy, R. H., cartographer, 1203, 1230-31
Favre, Alphonse (1815-1890, Swiss paleontologist, geologist), cartographer, 239, 243, 279; *Description géologique du Canton de Genève, 239*
Fayette County, PA, 933
Fear, Cape, harbor, NC, 1195
Featherstonhaugh, George William (1780-1866, geologist), cartographer, 1354, 1375; *Report of a geological reconnaissance . . .*, 1375
Federal Republic of Germany, donor, 1751
Federov, E., cartographer, 319
Fell, PA, 947
Fendall, C., cartographer, 882, 885, 1182
Fenner, R., publisher, 135
Ferdinand II, king of the Two Sicilies, 1691
Ferguson, James, cartographer, 741, 743, 745, 847-48, 853-54, 885, 1049, 1073-75
Ferraris, comte de, cartographer, 190
Ferrer, J. J. de, 32 (29a)
Fettke, Charles Reinhard (b. 1888, geologist), cartographer, 985
Fez, Morocco, 436
Field, Henry, "Contributions to the anthropology of Iran," 377, "The anthropology of Iraq," 376
Field Museum of Natural History, Chicago, IL, 376-77
Filbert, R. M., engraver, 1008
Filson, John (1747-1788, cartographer), cartographer, 1236-37; *The discovery, settlement and present state of Kentucky,* 1236; *Histoire de Kentucke,* 1237
Finger Lakes, NY, 841
Finland, 160, 177, 321-22; geology, 321-22
Finlayson, James, cartographer, 934
Finley, Anthony (ca. 1790-1840, cartographer and publisher), publisher, 574
Fire Hole River, WY, 1433
Firmin Didot, publisher, 1686
First U.S. Bank, Phila., PA, 1001
Fish drying, Newfoundland, Canada, 89
Fisher, Donald William (b. 1922, geologist), cartographer, 839
Fisher, J. C., 9 (1)
Fisher, Joshua (1707-1783, Delaware/Pennsylvania merchant and surveyor, cartographer), cartographer, 32 (8), (92), 762
Fisher, Joshua Francis (1807-1873), donor, 1692
Fisher, Thomas (original member of APS), cartographer, 760-60a
Fisher, Thomas (1801-1856, publisher and author), publisher, 1013; donor, 1013
Fisher, William (fl. 1669-1691, British publisher), 858
Fisk, ——, engraver, 857
Fisk & Rusell, engravers, 857
Fitch, Charles Hall (b. 1854, geologist), cartographer, 1266
Fite, Emerson David (1874-1953, author, educator), *A book of old maps delineating American history,* with Archibald Freedman, 1734
Fitz-Carrald, ——, cartographer, 1570
Fitz Roy, Robert (1805-1865, English naval

officer, navigator), 1561, 1575
Fitzwilliams, Richard, cartographer, 1094
Five Forts battlefield, VA, 774
Five Nations, Indians, 752-53, 783
Flanders, Belgium, 188, 190, 214, 251
Flattery, Cape, and Neé-ah harbor, WA, 1466
Flemington, NJ, 886
Flensborg, Denmark, 300
Flexner, James Thomas (b. 1908, author), donor, 392, 394, 1720
Flexner, Simon (1863-1946, physician, pathologist, administrator), collection, 1720
Flexner, William Walsh, donor, 1720
Flinders, Matthew (1774-1814, navigator and hydrographer), 79
Fling, Richard F., cartographer, 465, 625
Floder, A., lithographer, 277
Flood, Margah Toogood, donor, 1022
Florence, Italy, 346
Florida (east and west, and the modern state), (29a), 32 (5), 449-50, 455, 550-52, 554, 569, 572, 575-77, 1081, 1083-87, 1093, 1184-99, 1217, 1504, 1536, 1668; Alachua-Savana, 4 (1); Atlantic coast (north), 1178; Canaveral, Cape, 1186; canals, 1192; Colinton, 1189; Espiritu Santo, 1190, Gulf of Florida, 1194, Hibernia, 1196; Hibernia to Jacksonville, 644; hydrography, 1081, 1093; Indians, war, 1193; Jacksonville, 1196; Jacksonville to Hibernia, 644; Key West, 1194; Key West harbor, 644, 1194; Middle River, 17 (1); Pensacola, 32 (5), Pensacola bar, 1185; Pensacola bay, 1190; Pensacola harbor and bar, 1191; St. John River, 4 (1), 1186, 1196; St. Mary's River to Savannah River, Atlantic coast, 1178; St. Augustine, 4 (1), 1081; Tampa Bay, 644, 1184, 1197; Yellow River, 17 (1)
Flora of Mt. Desert Island, E. L. Rand and J. H. Redfield, 649
Flower, John, (20)
Flower, William G., 46 (20)
Fohs, F. Julius, cartographer, 1249
Fond du Grange Bay, St. Domingue, 32 (57)
Fond du Lac, WI, 1277, 1358, 1372
Foot, Thomas (fl. 1790-1827, British engraver), engraver, 557
Forbes, John, & Co., 1188
Fores, Samuel W. (fl. 1789-1836, British publisher), publisher, 137
Forests, NJ, 891
Formosa, 380, 441, 1699
Fornaro, A., cartographer, 1230
Forshey, Caleb Goldsmith, cartographer, 1 (3)
Forster, J. H., cartographer, 1278-79
Forsyth, John (1780-1841, U.S. statesman), 1629
Fort Allen, PA, 33 (9)
Fort Augusta, PA, 45 (1)
Fort Burd, PA, 45 (3)
Fort Christina, DE, 1048
Fort Churchill, Canada, 92
Fort Dauphin, St. Domingue, 32 (45)
Fort de France, harbor, Martinique, 1542
Fort Erie, NY, 813
Fort la Constitution, La Croix des Bouquets, St. Domingue, 32 (50)
Fort Monroe, VA, 1125
Fort Niagara, NY, 813
Fort Osage, 8 (2); 28 (13)
Fort Oswego, NY, 813
Fort Pitt, PA, 12 (1)
Fort Schlosser, NY, 813
Fort Schuyler, NY, 813
Fort Severn, Ontario, Canada, 548
Fort St. Elisabeth (Orsova), 29 (54)
Fort Shirley, PA 33 (4)
Fort Vancouver, Columbia River, 32 (82)
Fortif, ——, engraver, 896
Fortified town, 29 (29)
Fossils (sea shells), 32 (3)
Foster, J. H., cartographer, 1287
Fougeroux de Bondaroy, Auguste Denis (1732-1789, botanist, traveller), collection, 14; cartographer, 14
Foulke, Capt., 32 (65)
Four Mile Creek, NY, 828, 831
Fourth Annual Report, Pennsylvania Railroad, 941
Fowler, John (b. 1833, cartographer), cartographer, 887
Fox, Charles Pemberton and Mary Fox, donors, 15
Fox, D., cartographer, 1600
Fox, Henry Wilson (b. 1863, Australian geologist), cartographer, 1593-94; lithographer, 1586, 1589, 1592
Fox Indians, 1378
Fox River, WI, 1281
Fra[], Joseph, cartographer, 821
Franca, Brazil, 1572
France, 52,112, 245-75, 325, 1654, 1670; Arras, 261; Artois, 248, 251; Béthune, 271; Biscay Bay, 1663, 1674; Brie, 245; Bureau des Longitudes, 116; Bureau des Travaux Publiques, 271; canals, 252; Cap de la Hève, 270; Capestang, 252; Champagne,

249; Corsica, 257, 269; Dépôt Général de la Guerre, 430; Etang du Thau, 252; geology, 271; geology of the area of Mont Blanc, 279; Gex, 230; hydrography, 268a; Isle St. Louis, Paris, 13 (3); Languedoc, 252; Lyon, 255; Meuse River, 13 (2); Paris, 247, 256, 263-65; Paris, Revolution (1790-94), 273-74; roads, 250, 253, 257; Renneville, 252; Rhône River, 267; Rhône River valley, geology, 272; St. Barthélemy Island, 1538; St. Jean de Luz, 268a; Savoie, geology, 279; Toulon, 262; Toulouse, 252; Trebes, 252; Verdun, 13 (2)
Franco-German frontier, 283-84
Frankford, PA, to N.Y.C., NY, 906
Franklin, Benjamin (1706-1790, signer of the Declaration, diplomat, printer, APS founder), 21-22, 32 (3), 33 (9), 51, 145, 253, 551, 579, 751, 1033, 1653, 1752; cartographer, 15, 1622-24; collection, 15; "A letter from Dr. Benjamin Franklin to Mr. Alphonsus Le Roy . . . ," APS *Transactions,* vol. 2, 1622-23; library, 29, 1651; publisher, 752; "Remarks upon the navigation . . . Gulf Stream," *American Museum,* 1624; residence, 32 (90)
Franklin, Benjamin, Parkway, Phila., PA; *See:* Benjamin Franklin Parkway
Franklin Court, Phila., PA, 15 (1-2)
Franklin County, PA, 955
Franklin, Sir John (1786-1847, English explorer and hydrographer), 23, 32 (79), 94
Franklin, TN, battlefield, 1261
Franklin Township, Luzerne County, PA, 16 (1), 947
Franklin, VT, 666
Franklin, William (1731-1813, governor of New Jersey, APS 1768), collection, 16
Franklin-Bache purchase, 253
Franklinia, state of, 564
Franks, J. H. (fl. 1821-1830, British engraver), engraver, 139
Franks and Johnson, engravers, 139
Franklinville, Mason County, KY, 1242
Franz Josef Land, U.S.S.R., 100
Frazer, Persifor (1844-1909, chemist, APS 1872), 115; cartographer, 955
Frederick County, VA, 774, 1120
Frederick, John L. (fl. 1814-1845, engraver), engraver, 923
Fredericksburg, VA, to Annapolis, MD, 1098-99
Freeman, Archibald (b. 1866, author), and Emerson D. Fite, *A book of old maps delineating American history,* 1734
Frehold, Edward, cartographer, 1461
Freiburg, Switzerland, 222
Frémont, John Charles (1813-1890, explorer, soldier), cartographer, 1426, 1758
French, G. H., cartographer, 1336
French & Indian War, 33 (4); 44 (1); 45, 501, 1202
French, British, and Spanish empires in North America, indexed under Canada, U.S., Mexico, and Florida
French empire, 266
French empire in America. *See:* Canada and Louisiana
French Neptune. See: Atlas Maritime
French revolution (1790-1794), Paris, France, 273-74
Frenchman's Bay, ME, 645
Frèreault, P., cartographer, 489, 548
Freyhold, Edmund von (b. 1845, cartographer), cartographer, 1416
Fridrich, Jacob Andreas (1714-1779, German engraver), engraver, 166
Friedenwald, Isaac, publisher, 598
Friedenwald company, publisher, 1222
Friedenwald Lithographic Company, lithographer, 1367
Friend, Norman (1815-ca. 1871, lithographer), engraver, 941, 1015, 1017
Friend & Aub, lithographers, 941, 1017
Friends of the Wissahickon, Inc., donor, 1034
Friesland, The Netherlands, 195
Friex (Fricx), Eugène Henri (d. 1733, publisher, bookseller), cartographer, 1642; *Grande Carte des Pays Bas Catholiques,* 1642
Fritsch, Karl von, cartographer, 231
Frobisher Bay, N.W.T., Canada, 487
"From the kings road near *Bethlehem* to the *Mahoning Creek* passing near the *Healing Waters* beyond the Blue Mountains," 899
Frost, H. (fl. 1816-1829, British engraver), engraver, 1628
Frühwirth, E. K., lithographer, 280
Fry, Joshua (ca. 1700-1754, professor of mathematics at William and Mary), cartographer, 32 (13), 469, 759, 1094, 1750; and Peter Jefferson, *A Map of the inhabited part of Virginia,* 1750
Fuerteventura, Canary Islands, Spain, 336
Fullearton, CO, 401
Fuller, M. L., cartographer, 395
Fullerton harbor, Canada, 489
Fulton County, PA, geology, 966

Funchal, Madeira islands, Portugal, 328
Furnace (ship), 1645
Furneaux, ——, 80, 82-83
Furrer, H., engraver, 237
Fyen, Denmark, 171-73

Gabb, William Moore (1839-1878, geologist), cartographer, 1551
Galapagos Islands, Ecuador, 1561
Galashiels, Scotland, 7 (4)
Galena, IL, 1322, 1324
Galicia, Spain, 326
Gallatin River, 1425
Galveston bay, TX, 644; harbor, 1230-31
Galway, Ireland, 145
Ganges River, India, 399
Gannett, Henry (1846-1914, cartographer, chief geographer of the U.S. Geological Survey), cartographer, 599-603, 610, 613-14, 616, 781, 837-38, 888, 1024, 1026, 1406-10, 1425, 1442
Garcia, Diego, cartographer, 32 (71)
Gardiner, James Terry (1842-192?, geologist, geographer), cartographer, 1408-10
Gardner, Alexander (1821-1882, photographer), 1461
Gardner, J., engraver, 1575
Gardner, James, cartographer, 591
Gardner, James Henry, 966
Gardner, Robert M., donor, 641
Garnett, John (APS 1802), donor, 79, 1628
Garrett County, MD, geology, 1076
Garrett Road, Phila., County, PA, 43 (2)
Garrigue, Mary, 46 (7)
Gas fields, PA, 978; Meade County, KY, 1250; WV, 1141, 1143
Gaspari, Adam Christian (1752-1830, geographer), 1677; *Allgemeiner Hand-Atlas der Ganzen Erde*, 1677
Gaspé Peninsula, Quebec, Canada, 508
Gates, ——, publisher, 1376
Gates and Seaton, publishers, 1376
Gatineau, Quebec, Canada, 503
Gatterers, Johann Christoph (1727-1799, European writer), 91
Gatun, Panamá, 1516a-1516b; dam site, 1516b
Gauld, George (1732-1782, surveyor, APS 1774), collection, 17; donor, 17
Gauthier-Villars, publisher, 116
Gaw, Robert M. (b. ca. 1805, engraver), engraver, 1194
Gay, E. F., cartographer, 942
Gay Head, Long Island, NY, to Cape Henlopen, DE, 773
Gedney, Joseph F. (fl. 1862-1869, lithographer, engraver, publisher), lithographer, 1133; publisher, 1129
Gedney, Thomas R., cartographer, 847, 851, 1094
Gelders, The Netherlands, 193, 195
The general atlas for Carey's edition of [William Guthrie's] geography improved, 32 (18a), 66-7, 73, 92, 107, 128, 160, 192, 195, 233, 254, 308, 350, 361, 367, 424, 554, 1542, 1554
General collection of voyages and travels, John Pinkerton, 1683
A general outline of the U.S., 1009
General Staff Lithographic Institute, 163
General topography of North America and the West Indies, T. Jefferys, 651
Genesee County, NY, 816, 819
Genesee Indians, NY, 764-65
Geneva, Switzerland, 340, 239
Gentleman's magazine, 993
A geographical description of Louisiana, William Darby, 1218
Geographical Institution, publisher, 1422
"Geographical intelligence . . . ," John Melish, 923
Geographical journal, 101
The geographical ledger and systematized atlas, Christopher Colles, 71, 863-68, 906-907
Geographical Publishing Co., publisher, 86, 1743
Geographical Review, 101
Geographical Society of Philadelphia, *Bulletin,* 99; publisher, 97
Geologic structure and orogenic history of Venezuela, W. H. Buchner, 1569
Geological Association of Canada, 481
Geological and Natural History Survey of Wisconsin, 1370
Geological and topographical atlas [U.S.], Clarence King, 1706
Geological Society of America, 466, 481, 623, 625, 1569, 1632-33, 1635; publisher, 465, 1569; *Special Papers,* 1632
Geological Society of London, donor, 142; publisher, 142
Geological Survey of Canada, 473, 489, 491, 548
Geological Survey of Japan, 382-386; *Outlines of the geology of Japan,* 382-86
Geological Survey of Kansas, 1401
Geological Survey of Kentucky, 1249-53
Geological Survey of Missouri, 1384

Geological Survey of New Hampshire, 659
Geological Survey of New Jersey, 889-92
Geological Survey of N.S.W., Australia, 1607-8, 1610
Geological Survey of Pennsylvania, 959, 978, 985; *Bulletin*, 985; Second, *Grand Atlas*, 1712. *See also*, Pennsylvania, Geology
Geological Survey of Queensland, Australia, 1586-1605
Geological Survey of the U.S. *See:* U.S. Geological Survey
Geological Survey of West Virginia, 1140-44
Geological Survey of West Australia, Australia, 1584-85
Geology, 26 (30); Alaska, 623; Alps, 281; *Australia*, N.S.W., 1607-10, Broken Hill district (N.S.W.), 1610, Sydney (N.S.W.), 1608, Queensland, 1586-1605, South Australia, 1582-83, Western Australia, 1585; *Austria*, Krems, 278, Manhardsberger, 278; *Brazil*, 1572, 1574, Natal, Rio Grande do Norte, 1574; *California*, 1493-94, San Francisco peninsula, 1489, Shasta County, 1486; *Canada*, 471, 488-89, 623, British Columbia, 537-38, Nova Scotia, 494, Ontario, 507; China, 395; *Colorado*, 1420-21, Red Mountain, Sneffels and Telluride districts, 1418-19; Connecticut, 749; Egypt, 433; England, 7 (1), 138-39, 142; Europe, 114-15; Finland, 321-22; *France*, 271, 279, Mont Blanc area, 279, Rhone River valley, 272; *Germany*, 303, 305, Rhineprovinz, 303, Saxony, 304, (Saxony) Hainichen, 304, Westphaliaprovinz, 303; Idaho, 1403; Indian territory, 32 (87); Japan, 384; Kansas, 1401; Luxembourg, 217; Maryland, Garrett County, 1076; *Michigan*, 1351, northern peninsula, 1352, southern peninsula, 1353; Middle Atlantic States, 772; Missouri, 1384; Montana, 1403, New Hampshire, 659, *New Jersey*, 886, 889-90, 892, 946, Bordentown sheet, 888; *New York*, 836, 839-42, Niagara sheet, 843; *Pennsylvania*, 928, 946, 960-61, 984-85, 1712, Bedford County, 954, 966, Berks County, 959, Fulton County, 966, Huntingdon County, 960, 966; Peru, Morococha, 1570; *Philippine Islands*, Mindanao Island, 420, Panay Island, 413; Portugal, 339; Scotland, 7, 138-39; South Dakota, Black Hills, 1394; *Switzerland*, 32 (94), 236, 281, Basel, 238, Brugge (Aargau), 237, St. Gotthard, 231, Sentis, 240-41, Valais, 242, Vaudois Alps, 242; Sweden, 161-63; Texas, 1234; *U.S.*, 466, 570, 623, 1706, Central U.S., 1377; U.S.S.R. (European), 319; Venezuela, 1569; Virginia, Wythe County, 32 (80); Wales, 138-39, 142; West Virginia, 1139, 1141, 1143; Wisconsin, 1359, 1370; world, 1730; Wyoming, 1403
Geology chemical, physical & stratigraphical, Sir Joseph Prestwick, 114
Geology of the South Mountain Belt of Berks County, E. V. D'Invillier, 959
George II (1683-1760, king of England), 29
George, Lake, NY, 784
The George Washington atlas, Laurence Martin, 1736
George Washington Bicentennial Commission, 1736; donor, 1736
Georges Bay, South Shetland Islands, Great Britain, 101
Georges Shoal and Banks, 1629
Georgian Bay, Ontario, Canada, 32 (98)
Georgetown, DE, 1052
Georgetown, D.C., 644; and Washington harbors, 1080
Georgia, 89, 455, 1082-84, 1093, 1134, 1159, 1170-83, 1504; Ackworth, 1181; Altamaha River, 32 (12), Atlanta, military campaign, 1180-81; Buttermilk Sound, 32 (12); Cassville, 1181; Chattahoochee River, 1181; Dallas, 1181, Glenn County, 32 (12); Hampton Creek, 32 (12), Jekyl Island, 1170; Jonesboro, 1181; Kennesaw Mountain, 1181; Kingston, 1181; Lost Mountain, 1181; Lovejoy's Station, 1181; McIntosh County, 32 (12); Marietta, 1181; Middle River, 32 (12); Mud River 32 (12), Oostanaula River, 1181; Peach Tree Creek, 1180; Pine Mountain, 1181; Resaca 1181; Sapelo Island to Savannah, 644, 1182; Savannah, 576; Savannah, siege, 1171; to Sapelo island, 644, 1182; Savannah River to St. Mary's River, Florida (Atlantic coast), 1178
Geraghty, T. R. (Australian surveyor), cartographer, 1587
Gerdes, Ferdinand H., cartographer, 742, 832-33, 851, 854, 1049, 1073, 1203-205, 1209
Gerhard, Peter, cartographer, 377
Gerig, William (b. 1866, soldier, army engineer), donor, 1516a-b
German atlas, 1641
Germania, (ship), 6
Germany, 251, 288-306, 1641, 1648, 1654, 1741; Bingen, 29 (9); Brunswick-Lune-

burg, 293; Carlsruh to Coblentz, Rhine River, 29 (1); Coblentz to Carlsruh, Rhine River, 29 (1); geology, 303, 305; Holstein, 175; Landau, 297; Mayence, 29 (8); meteorology, 302; Mülheim, 29 (15); Münster, 29 (9); Palatine, 290; Philippsburg, 296, 298; Rhine River, Carlsruh to Coblentz, 29 (1); Rhineprovinz geology, 303; roads, 291; Schleswig, 184, 300; Saxony, geology, 304; Saxony, Hainichen, geology, 304; Simmern, 29 (12); Westphaliaprovinz, 303; Westphalia, Soest, 299
Germany, Somerset County, PA, 920
Gerringong district, N.S.W., Australia, 1607
Gerstner, Francis Anthony, 276
Gettysburg, PA, 775; battlefield, 950, 956, 970; hospitals, 969; U.S. National Military Park, 971
Gex, France, 230
Geyer, A. R., cartographer, 984
Ghys, Martin, cartographer 216; publisher, 216
Giants Causeway, Ireland, 145
Gibbons, Charles (?) (1814-1885), donor, 938
Gibbs, Joseph, 43 (4)
Gibraltar, Straits of, 1650
Gibson, John (fl. 1750-1792, geographer, engraver, draughtsman), engraver, 1503, 1513, 1520-21, 1531, 1534, 1567
Gibson, William, donor, 1692
Gifford, John (John Richard Green, 1758-1818, author), 1798; *The history of France,* 260
Gilbert, G. K., cartographer, 1449
Gilbert, Joseph, cartographer, 1749
Gilbert, Samuel, cartographer, 644, 705, 853-54, 1167-68
Gilbert, William Wallace, cartographer, 837
Gillingham, Edwin, engraver, 703
Gillman, Henry, cartographer, 1282, 1286
Gilpin, John, "Observations on the annual passage of herrings, "APS *Transactions,* 1623
Gilpin, Thomas (original member of APS), 760
Gilpin, Thomas (1776-1853, papermaker), machine paper, 316
Gimbel, Richard, estate of, donor, 437, 637, 654, 662, 672, 713, 761, 813, 875, 915, 1043, 1066, 1084, 1117, 1146, 1157, 1172, 1239, 1254, 1271, 1621, 1680, 1727
Gingerbread age, James Mease, 704
Girard, Anthony, publisher, 1215
Girard, Stephen (1750-1831, merchant, banker, philanthropist), 1033
Girard College, Phila., PA, 1019, 1032; donor, 1032
Girgeh, Egypt, 427
Giroux, N. J., cartographer, 485
Glacier National Park, MT, 1432
Glaciers, North America, 465; Switzerland, 235, 243, Switzerland, Mont Blanc, 243; U.S., Atlantic coast to the Rocky Mountains, 625, Eastern U.S., 625
Gladstone, Queensland, Australia, 1599
Glarus, Switzerland, 225
Glasgow, Scotland, 148
Glegg, J. G., cartographer, 501
Glenn County, GA, 32 (12)
Glenwood Cemetery, Phila., PA, 1039
Globes and model, 1772-1774
Glot, C. B., cartographer, 230; engraver, 112, 265
Gloucester harbor, MA, 691
Glück, J. B., cartographer, 690, 705, 741-42, 832-33, 1075
Gnadenhütten, PA, 33 (9)
Goktscha Lake, U.S.S.R., 320
Gold Coast, Africa, 422-23, 435
Gold fields, Australia, Queensland, 1586-87, 1589, 1594-1603; Nevada, 1458-60
Gold mines, Australia, Western Australia, 1584
Goldsboro, NC, 1135
Goldsborough, J. R., cartographer, 717, 746, 849
Goldthwait, J. W., cartographer, 465
Goldthwaite, J. H., engraver, 1204-205
Goldthwaite, Richard P., cartographer, 625
Goll, Johann Jakob (1809-1860, Swiss cartographer and engraver), cartographer, 239
Gonaives Bay, St. Domingue, 32 (56)
Gonave, St. Domingue, 32 (27)
Gonne, George (1764-1839, British engraver), engraver, 1542
Good Hope, Cape of, 421, 439
Goodchild, J. G., cartographer, 114
Goodfellow, E., cartographer, 1021-21
Goodrich, A. T., & Company, publisher, 826
Goodrich, B. G., cartographer, 18 (1); collection, 18
Goos, Pedro (Pieter, ca. 1616-1675, Dutch cartographer, engraver, publisher), 1638
Gopsill, ——, 1023, 1025, 1027
Gordon, Thomas Francis (1787-1860, historian), cartographer, 879; publisher, 879
Gordon, William (1728-1807, author, clergy-

man), *History of the rise of the U.S.*, 761, 1084
Gore, ——, 80, 82
Gorman, Fred J., cartographer, 32 (90), 974, 976; donor, 976
Gorshkov, Sergei G., cartographer, 1755; [*World ocean atlas*], 1755
Gouadaquini, GA; *See:* Jekyl Island, GA
Goujon, J., cartographer, 266
Gould, L. M., cartographer, 465
Gourov, A., cartographer, 319
Graeme, John, 46 (20)
Graham, publisher, 1571
Graham, Andrew B., lithographer, 418, 1381, 1396, 1398, 1428, 1443, 1447, 1458, 1471, 1490, 1495, 1508; publisher, 390, 418, 1267, 1318, 1398, 1428, 1443, 1447, 1458, 1471, 1490, 1495, 1508
Graham, Campbell (d. 1866, soldier), cartographer, 1310
Graham, Curtis B., lithographer, 578, 1277, 1506; publisher, 578, 1506
Graham, James Duncan (1779-1865, cartographer), cartographer, 631-32, 679, 682-83, 1166, 1191, 1225, 1228, 1325-26, 1344-49, 1360, 1362, 1364; donor, 631-32, 679, 682-83, 1151-52, 1225, 1227, 1310, 1325-26, 1344-49, 1360, 1362, 1364
Graham, W. J., cartographer, 532
Grahams, ——, lithographer and publisher, 1631
Grand atlas [*of Pennsylvania*], Second Geological Survey of Pennsylvania, 1712
Grand atlas universel, 1682
Grand Haven, MI, 1345
Grand Island, Lake Superior, 1297
Grand Junction Canal, Northamptonshire, England, 136
Grand Manan Islands, New Brunswick, Canada, 493
Grand Rapids, Saskatchewan, Canada, 520
Grand Rapids and Indiana Railroad, 962-63
Grand River harbor, MI, 1345
Grande Baye du Mesle to Abacou Point, Vache Island, St. Domingue, 32 (53)
Grande Carte des Pays Bas Catholiques . . . , E. H. Friex [Fricx], 1642
Grant, ——, cartographer, 32 (8)
Grant, Ulysses Simpson (1822-1874, president of the U.S., general), 1259, 1262
Graphic Company, publisher, 1180, 1260, 1262
Graphic Company Photo-Lithographer, lithographer, 1707
Gravelot, Hubert François Bourguignon (d'Anville), (1699-1773, French engraver and designer of Paris), 65
Graves, W. H., cartographer, 1449
Gravity map. *See:* Bouguer gravity map
Gray, ——, publisher, 1265
Gray, Andrew B. (fl. 1838-1855, surveyor and civil engineer), cartographer, 1225-27, 1277
Gray, Asa, (1810-1888, American botanist, APS 1848), donor, 32 (67)
Gray, Carlisle, cartographer, 984
Gray, George (1725-1801[?], surveyor), cartographer 32 (1), 899
Gray, O. W., and H. F. Walling, *New topographical atlas of the state of Ohio,* 1312
Gray & Bunce, publishers, 1265
Gray's Ferry, Phila., PA, 46 (10)
Gray's harbor, WA, to Admiralty Inlet, Canada, 1462
Great Britain *(see also:* England, Scotland and Wales), 88, 1639-40, 1657, 1662; Admiralty, 1461; Board of Ordnance, donor, 139; Board of Trade, 757; Boundary commission, 1461; Falkland Islands, 1560-61; Hydrographic Department of the Admiralty, London, publisher, 93, 101; Lords Commissioners of the Admiralty, 1658-59; Lords of Trade, 1656; Northeast Boundary Dispute with the U.S., 760-60a; Orkney Islands, 1662; Scilly Islands, 1662; Shetland Islands, 1662; War Office, Geographical Section, General Staff, 362
Great Britain's coastal pilot, Greenville Collins, 1662
Great Chester Valley, MA, 1072; PA, 1072
Great Falls, Columbia River, 28 (3)
Great Falls, or Bellows, Connecticut River, 728
Great Falls, MT, 1430-31
Great Falls and portage, Missouri River, 28 (2)
Great Lakes, 831; area, 1274-1301; canal, 1276
Great Rapids, Columbia River, 28 (5)
Great Smoky Mountains National Park, NC-TN, 1156
Greece, 52
Greeg, Robert, cartographer, 32 (1), 899
Green, ——, publisher, 897
Green, Clerkenwell, engraver, 452
Green, John Richards, 260
Green, L. C., cartographer, 1588
Green & Thornton, publishers, 897

Green Bay, WI, 1281-1372
Green County, KY, 1249
Green Mountains, Canada, 1298; Canada to Vermont, 32 (84), Vermont, 1298
Greene, F. B., cartographer, 99
Greene, Robert, donor, 897
Greenfield, W. H., cartographer, 1596-1600, 1602
Greenland, 6 (2), 100, 107, 160, 177, 182, 546
Greenleaf, Moses (1777-1834, U.S. mapmaker and author), cartographer, 641; *Statistical view of the district of Maine,* 641
Greenough, George Bellas (1788-1855, F. R. S. and first president of the Geological Society of London), cartographer, 142
Greenwell, W. E., cartographer, 1203-205, 1209
Greenwich Observatory, London, donor, 63
Gregory, Edmund, publisher, 1589
Gregory, Herbert Ernest (b. 1869, geographer, geologist), cartographer, 749
Gridley, Enoch G. (fl. 1800-1818, engraver), engraver 1173
Griffith, Dennis, cartographer, 1065
Grimshaw, Percy H., cartographer, 1722
Gritzner, M. C., cartographer, 690
Groningen, The Netherlands, 195
Gros, C., cartographer, 1684
Gross, Alexander, cartographer, 380
Grosvenor, Gilbert Melville Beer (b. 1901, editor, National Geographic Society), 285
Grouse Creek, B. C., Canada, 537
Groux, ——, engraver, 359
Groves, D.(?), cartographer, 46 (17)
Growdon, Lawrence, 15 (4)
Gruner, Gottlieb Siegmund (1717-1778, Swiss naturalist), cartographer, 235
Guadalajara, Mexico, 32 (17)
Guam Island, U.S., 608
Guañape Islands, Peru, 35 (8)
Guatemala, 1508; agriculture, 1508
Guedes, P., cartographer, 340
Guerra, Giuseppe (fl. 1794, engraver), engraver, 353
Güssefeld (Gussefeldt), Franz Ludwig von (1744-1807, German cartographer), cartographer, 460
Guiamas, Sonora, Mexico, 32 (70)
Guide de la carte géologique des Grand-Duché de Luxembourg, N. Wies, 217
Guilford, CT, 730
Guillot, A. H., cartographer, 1223
Guinea (Africa), 182, 435
Guinea, Gulf of, West Africa, coastline, 441
Guiter (Guittair), Ch. A. (d. 1787, French engraver in Denmark), engraver, 169-73
Gulf of Florida. *See:* Florida, gulf
Gulf of Mexico. *See:* Mexico, gulf
Gulf of Tehuantepec. *See:* Tehuantepec, gulf
Gulf Stream, 78, 1622-26
Gullick, W. A., lithographer, 1608; publisher, 1607-608
Gunnison, John Williams (1812-1853, soldier, author), cartographer, 1278-79, 1281
Gutekunst, Frederick (1831-1917, photographer, APS 1885), photographer, 1550
Guthorn, Peter Jay (b. 1916, author), 1768; *U.S. Coastal Charts 1783-1861,* 1768
Guthrie, William (1708-1770, Scottish geographer and historian), 32 (18a), (19), (23-4), 306, 400; *see also: The general atlas for Carey's edition of [William Guthrie's] geography improved*
Guthrie, OK, 1267
Guyana, Berbice River, 1549; coast, 1549; Essequabo, 1549
Guyot, Arnaud, cartographer, 462, 586

Haack, Homan, cartographer, 1733, 1739
Haas, P. (fl. 1837-1845, lithographer), lithographer, 631, 1225
Hackensack, NJ, 890
Hadassah (ship), 1629
Haderslewhuus, Denmark, 184, 300
Hadley, CT, 732-33
Hainault, Belgium, 215, 251
Haines, D., engraver, 140
Hainichen, Saxony, Germany, geology, 304
Haiti (*see also:* Saint Domingue), 1548, 1551, Cap François, 1523; Môle St. Nicolas, 1526; Port au Prince, 1524; Port Paix roads, 1522
Halds, Denmark, 176
Hale, H. C., *Notes on Panama,* 1514-15
Hale, John S., cartographer, 1761
Hale, John S., donor, 1761; *A historical atlas of colonial Virginia,* 1761
Hales, John Groves (1785-1832, cartographer), cartographer and publisher, 703
Halifax, NC, 32 (91)
Hall, ——, engraver, 326
Hall, Harrison, publisher, 702
Hall, James (1811-1898, geologist, APS 1854), cartographer, 772
Hall, Norman Jonathan (d. 1867, cartographer), cartographer, 1131
Halley, Edmund (1656-1742, English astronomer), cartographer, 63, 77

Hallowell, Mrs. A. I., donor, 380, 480, 482, 489-90, 495, 502-503, 510-15, 520-29, 535, 636, 648, 660, 667, 694-95, 706, 718, 748, 835, 893, 1356, 1371-73, 1378-80
Hamilton, Fletcher McNutt (b. 1882, geologist), cartographer, 1494
Hamilton College, CT, Litchfield Observatory, 57
Hammond, ——, cartographer, 936; engraver, 936
Hammond, S. A., cartographer, 970
The Hammond-Harwood House atlas of historical maps of Maryland, Edward C. Papenfuse and Joseph M. Cole III, 1764
Hampton Creek, GA, 32 (12)
Hampton roads, York River, Chesapeake Bay, 1138
Hanckel & Son, lithographer and publisher, 945
Handatlas, Andrée, 100
Handatlas über alle Theile der Erd, Adolf Steiler, 1720
Handy reference atlas of the world, John Bartholomew, 1715
Hanna, Andrew (d. 1812, cartographer), 1070
Hanna and Warner, 1070
Hannibal (247-182 B.C., Carthaginian soldier), Hannibal's route across the Alps, 267
Hanover and Newcastle, VA, to Annapolis, MD, 1100-10
Hanus, G. C., cartographer, 1050
Harboe, P. (F. C. L.? 1758-1811, Danish hydrographer), cartographer, 178-81, 183
Harbor Island bar, NC, 1151
Hardcastle, Lt., cartographer, 1506
Harden, E. B., cartographer, 960
Harden, O. B., cartographer, 958
Hardie, James (1750?-1826?, Philadelphia author and publisher), 46 (13)
Harding, William Wilkins, cartographer, 882, 1138
Hardscrabble Creek, B.C., Canada, 537
Hardy, ——, publisher, 230
Harman, E. L., cartographer, 1336
Hare, C., cartographer, 19 (1)
Hare, Horace Binney, cartographer, 19 (2)
Hare, Robert (1781-1858, chemist, APS 1803), 19
Hare Street, Phila., PA 19 (1)
Hare-Willing family collection, 19
Harney, William Selby (1800-1889, soldier), cartographer, 32 (82)
Haro, canal de, and Strait of Rosario, WA, 1468
Harper, ——, publisher, 375
Harper's Ferry, WV, 774
Harris, Caleb, cartographer, 32 (24), 712
Harris, Harding, cartographer, 32 (24), 712
Harris, Jeremiah, 33 (3)
Harris, John (1726-1791, founder of Harrisburg, PA), 33 (3)
Harris, Joseph, 33 (3)
Harris, Thomas, donor, 1692
Harris, William J., cartographer, 1728
Harrisburg, PA, 936, 942
Harrison, A. M., cartographer, 717, 853-54, 885, 1182, 1479-80, 1482-85
Harrison, Carter H., 22 (1)
Harrison, Charles P. (1783-ca. 1850, engraver), engraver, 683, 923, 1150; publisher, 1150
Harrison, D. R., engraver, 770
Harrison, John, publisher, 70, 1669
Harrison, Samuel (ca. 1789-1818, engraver and publisher), engraver, 80, 568, 581, 1218, 1274; publisher, 80
Harrison, William (d. 1803, engraver), engraver, 554-55, 568, 770, 1004
Harrison, W. & S., engravers, 568
Hart, Lane S., publisher, 955, 959, 1712
Hart Island, NY, 849
Harvard University, Cambridge, MA, 696; Press, publisher, 1734, 1740; Yenching Institute, 1740
Harvey Creek, B.C., Canada, 537
Hassan, A. F., cartographer, 668, 1155, 1183, 1211, 1214, 1264, 1319, 1357, 1382
Hassler, Ferdinand Rudolph (1770-1843, geodist, mathematician, APS 1807), cartographer, 741-46, 773, 832-33, 847-48, 850-51, 881, 1020-21, 1049; donor, 218-229, 232, 234-35, 1774
Hassler, J. J. S., cartographer, 1138
Hatch & Co., publishers, 853
Hatteras, Cape, NC, to Boston harbor, MA, 750
Hauchecorne, G. (fl. 1862-1913, German cartographer), cartographer, 115, 1068
Hauducoeur, C. P., 1068
Haureballegards, Denmark, 174, 176
Haupt, Lewis Muhlenberg (1844-1937, engineer, APS 1878), 32 (14)
Havanna, Cuba, 32 (26), harbor, 1542, coast to Cape St. Antonio, 32 (72)
Havre de Grace, MD, 1068
Hawaii, 82, 608, 1495, Hawaii (island), 35

(17), Mauna Loa, 35 (17) Government surveys, 1495
Hawkins, George T., cartographer, 1315
Hawley, J. M., cartographer, 1197
Hayden, Ferdinand Vandiveer (1829-1887, geologist), cartographer, 1403, 1406-407, 1409, 1411, 1425, 1433-37, 1440-41, 1377; donor, 1434-35, 1437
Hayes, Isaac Israel (1832-1881, explorer, APS 1863), 96
Hayward, James (1786-1866, cartographer), 678
Hazard, Samuel (1784-1870, editor, antiquarian), 453
Hazzard, John L., cartographer, 1422; engraver, 691, 1463
Head Lynchs ordinary, VA, to Annapolis, MD, 1102
Head of Passes, Mississippi River, 1337
Heap, George (fl. 1715-1760, surveyor), cartographer, 993, 1013, 1016, 1038
Hearding, W. H., cartographer, 504, 1287
Hearn, ——, 92
Heart, Jonathan (1748-1791, soldier), cartographer, 1305
Heaslitt, C. F., donor, 1186
Hebert, L., engraver, 751
Hebrides Islands, Scotland, 157
Heckewelder, John Gottlieb Ernestus (1743-1823, Moravian missionary to the Indians, APS 1797), cartographer, 32 (22)
Heeren, Lillian, cartographer, 978
Heezen, Bruce Charles (b. 1924, geologist), cartographer, 1632-33, 1635
Heilprin, Angelo (1853-1907, naturalist, APS 1883), cartographer, 97
Heintz, ——, lithographer, 481, 625; publisher, 481, 1569, 1633, 1635
Helena, MT, 1430-31
Heliotype Printing Co., publisher, 609, 709
Hells Gate, N.Y.C., NY, 850
Helm, Charles H., cartographer, 1267, 1327, 1350, 1355, 1447-48, 1458, 1460
Hembel, William (1764-185, APS 1813), donor, 1692
Henderson, A. A., cartographer, 960
Henderson, John, cartographer, 156
Hendges, M., cartographer, 607, 611, 615, 1267, 1318, 1367, 1388-90, 1396-97, 1429, 1475-76, 1508
Henlopen, Cape, DE, lighthouse, 1003; to Gay Head, Long Island, NY, 773
Hennen, Ray V., cartographer, 1141-43
Henry, Alfred J., cartographer, 604
Henry, D. F., cartographer, 1282, 1288-90
Henry, Mathew Schropp (1790-1862, historian), cartographer, 20 (1-8); collection, 20; donor, 20
Henry, Cape, VA, to Cape Romain [Roman], SC, 1148
Henry County, KY, 1249
Herald, Frank A., cartographer, 395
Herbertson, A. J., cartographer, 1722
Herbst, F., cartographer, 504, 1281-82
Hercules (constellation), 50 (1)
Hergesheimer, E. (fl. 1862-1872, cartographer), cartographer, 1050, 1182, 1403, 1433-34, 1436, 1462
Hergesheimer, J., cartographer, 1020-21, 1197
Hérisson, Eustache (b. 1759, French hydrographical engineer, geographer), cartographer, 112, 266
Hérissont, Veuve, publisher, 1670
Hermann, ——, cartographer, 1707
Heroguel, Robert, cartographer, 13 (1)
"Hero-Transformer myth," Indian, 48 (13)
Herr, Laurenz, 280; and H. F. Müller, publishers, 280
Herrlein, E. (Julius, fl. 1855-1860, engraver), 1019
Herrmann, Albert, cartographer, 1740; *Historical and commercial atlas of China,* 1740
Herschel, Sir William (1738-1822, English astronomer, APS 1785), 1772
Hertugdummet, Denmark, 173
Heslop, ——, cartographer, 815
Hewitt, E. M., cartographer, 969
Hewitt, N. R. (fl. 1812-1828, British engraver), 1088
Heymann, Ignazio (1765-1815, Austrian geographer), cartographer and publisher, 355
Hibernia, FL, 1196; to Jacksonville, FL, 644
Higginson, J. H., cartographer, 1134
High Bridge, VA, 774; battlefield, 774
Hildreth, C. T., cartographer, 633
Hilgard, Julius Erasmus (1825-1891, hydrographer), cartographer, 747, 885, 957, 1020-21, 1046, 1050, 1182, 1195, 1209
Hill, ——, engraver, 458, 1544
Hill, Benjamin Felix (b. 1875, geologist), cartographer, 1234
Hill, Frank A., cartographer, 958
Hill, John, cartographer, 852
Hill, Richard R., donor, 966
Hill, Robert Thomas (b. 1858, geologist), cartographer, 1232

Hill, Samuel (fl. 1789-1804, engraver), engraver, 653, 712, 1085
Hilliard, ——, publisher, 641
Hills, John (1770-1850, cartographer, engraver), cartographer, 32 (8), 766, 998, 1003
Hinman, John, publisher, 577, 862
Hinton, John (d. 1781, British publisher), donor, 862
"Hiram," cartographer, 1 (1)
Hirschfeld, ——, 29 (13)
Hispanolia. *See:* Santo Domingo and Haiti
Histoire de Kentucke, John Filson, 1237
Histoire et description des voies de communication aux Etats-Unis . . . , Michel Chevalier, 1696
Historiae Canadensis seu Novae Franciae Libre dece ad Annum usque Christi, 32 (98)
Historic Publishing Co., NJ, 861
Historic Urban Plans, 996
An historical account of the expedition under the command of Henry Bouquet, William Smith, 901, 1302
Historical and commercial atlas of China, Albert Herrmann, 1740
A historical atlas of colonial Virginia, John S. Hale, 1761
An historical atlas of modern Europe, 1731
Historical Society of Pennsylvania, 32 (96), 449, 1053; *Memoirs,* 896, 898, 1048
An historical view of the United States, William Winterbotham, 1242-43
The history of England, Paul Rapin de Thoyas and N. Tindal, 450
The history of France, John Gifford (John Richards Green), 260
History of Kentucky, Z. F. Smith, 1236
History of New Hampshire, Jeremy Belknap, 653
History of Pennsylvania, Robert Proud, 763
History of Polperro, Jonathan Couch, 10 (1)
History of South Carolina, David Ramsay, 1163
The history of the county of Worcester, Peter Whitney, 671
The history of the district of Maine, James Sullivan, 638
History of the expedition under the command of Lewis and Clark, M. Lewis and W. Clark, 566
History of the rise of the United States, William Gordon, 761, 1084
A history of three of the judges . . . Ezra Stiles, 741, 729-737
Hitchcock, Charles Henry (1836-1919), geographer), cartographer, 659
Hobbs, Cecil, 32 (97)
Hockridge, T. C., cartographer, 1336
Hodge, ——, publisher, 761, 1084
Hodge, Allen, & Campbell, publishers, 761, 1084
Hodgeson, H. H., cartographer, 1421
Hoeing, J. B., cartographer, 1248-49, 1251
Hoen, August (1817-1886, lithographer, map-printer), lithographer, 1076, 1141-44, 1252-53, 1351, 1370; publisher, 102, 285, 1141-44, 1253, 1562
Holland, Samuel Jan (1728-1801, first Surveyor-General of British North America), cartographer, 112, 469, 483, 500-1, 627, 634, 652, 1655-56, 1741; *Atlas,* 634
Holland (*see also:* The Netherlands), 188, 193, 195
Holland Land Company, 816, 819
Hollis Island, NY, 35 (14)
Hollwood, Kent, England, 129
Holman, James D., cartographer, 592
Holme, Thomas (1624-1695, surveyor), cartographer, 897, 981, 987
Holmes, ——, publisher, 1756
Holmes & Meier, publishers, 1756
Holmes' Hole, MA, 689; harbor, 685
Holstein, Germany, 175
Holtedahl, Olaf (b. 1885), 165
Homann, Johann Baptist (1663-1724, Swiss engraver, mapseller, publisher), cartographer, 220-25, 227-29, 341, 460, 562, 1644, 1648; *Atlas Germaniae,* 1648
Home Insurance Company of NY, 1134
Homestead Act of 1909. *See:* U.S. Homestead Act of 1909
Honduras, 1509-10; San Fernando de Omoa, 1509-10
Hong Kong, China, 388-89
Hood, John Bell, 1261
Hood, Wash(ington? D. 1840, cartographer), cartographer, 679, 740, 1152, 1310
Hooe's Ferry, VA, to Williamsburg, VA, 1112-13
Hooker, William (fl. 1804-1843, engraver), 675, 683
Hooper, Robert Lettis, Jr., cartographer, 16 (1)
Hooper, Samuel (fl. 1770-1793, British bookseller, publisher, stationer), publisher, 1170
Hoopes, Alban Williamson, donor, 285, 621
Hopi and Zuñi ceremonialism, E. C. Parsons, 37
Hopi Indians, 37
Hopi Journal, Alexander M. Stephen, 37
Hornemann, ——, 438

Horsfield, Timothy, collection, 21; donor, 21
Horsley, ——, cartographer, 1654
Hosack, David (1769-1835, physician, botanist), 844
Hoskins, D. M., cartographer, 984
Hosmer, C., cartographer, 717, 747, 1182
Hospitals, Civil War, Gettysburg battlefield, 969
Hot Springs, AZ, 1263
Hough, Benjamin, cartographer, 1309; publisher, 1309
Houghe (Hooge, Hogius), Romain (Romanns, Romein) de (1645-1708, engraver), cartographer, 1639-40
Houghton, ——, cartographer, 1277
Howe, Charles Edward, Company, publisher, 1029
Howe, Sir William, 5th viscount Howe (1749-1814, British general), 32 (6), 702, 996, 1656
Howell, Charles Wagoner (d. 1882, cartographer), cartographer, 1231
Howell, George Pierce, cartographer, 1342
Howell, George W., cartographer, 884
Howell, Ralph Preston, Jr., cartographer, 1223
Howell, Reading (1795-1827, county surveyor of Philadelphia, cartographer), cartographer, 32 (23), 912, 914, 920-22; donor, 914
Huallago River, Peru, 1570
Hudson Bay, Canada, 6 (2), 48 (4), 88, 90, 452, 489, 548, 1645, 1661
Hudson County, NJ, 886
Hudson, Edward, donor, 699
Hudson, G. Donald, cartographer, 1744
Hudson River, 32 (3), 469, 826, 842, 1088
Hudson's Bay Company, "Covington Map," 32 (82)
Hughes, E. M., cartographer, 1231
Hulbert, ——, 46 (20)
Hulings, William E., cartographer, 2 (4); donor, 2 (4)
Hull, England, 123, 125
Humbertson, CO, 401
Humboldt, Alexander von (1769-1859, German naturalist, APS 1804), 32 (16), 317; cartographer, 1751; *Atlas géographique et physique du royaume de la Nouvelle Espagne,* 1751
Humphreys, Andrew Atchison (1810-1883, soldier), cartographer, 679, 775, 950, 1126b, 1341, 1706
Hunningen, The Netherlands, 196
Hunsingoo, The Netherlands, 200
Hunt, Samuel Valentine (1803-1892, engraver), engraver, 1479, 1481
Hunter, Thomas, publisher, 983
Hunter, William (1774-1849, U.S. Senator), 453
Huntingdon County, PA, geology, 960, 966; Juniata valley, 960
Huntington Bay, NY, 833
Hungary, 277, 361
Huron Indians, 32 (98)
Huron, Lake, 500, 758, 1283, 1291, 1296; False Presqu'ile, MI, 1292; Middle Island, MI, 1292; Presqu'ile harbor, MI, 1292; Saginaw Bay, MI, 1293; Thunder Bay, MI, 1294
Huron River, OH, Indian ruins, 1307
Hurricanes, U.S., 1092
Hutchins, Thomas (1730-1798, English surveyor, ass't. engineer in British Army in America, British geographer to the U.S., APS 1772), cartographer, 17 (1), 758-59, 901, 1269, 1275, 1302, 1304, 1321; publisher, 758; *A topographical description of Virginia,* 1269, 1304, 1321
Hyannis harbor, MA, 688
Hyde, E. Belcher, publisher, 1386
Hydrographic Office, Washington, D.C. *See:* U.S. Hydrographic Office
Hydrographical basin of the Upper Mississippi River, J. N. Nicollet, 1758
An hydrographical collection of . . . the West Indies, 1668
Hydrography, Canada, 472; Florida, 1081, 1093; France, 268a; Oceans, 1615; Poland, 310; Portugal, 328; U.S.S.R., 318
Hydrometry, Argentina, 1580
Hypsometry, Japan, 385; Portugal, 340
Hyslop, J. M., cartographer, 374

Iardella, C. T., cartographer, 957, 1046
Iceland, 107, 160, 182
Idaho, 1406-07, 1427, 1441, 1447-48; Boise, 1447-48; geology, 1403; Lewiston, 1447-48; Pocatello, 1447-48; Wallace, 1447-48
Idiaquez, Eduarto (fl. 1894-1901, Spanish geologist, geographer), 1571
Ile St. Louis, Paris, France, 13 (3)
Iljin [Il'in], Aleksyei Aleksyeevich (1837-1899, Russian cartographer), 320
Illinois, 578, 942, 1245, 1321-28; Alton, 1324; Cairo, 1337; Chicago, 1365, harbor and bar, 1326; Chicago River and harbor,

1325; Indians, 1269; Galena, 1322, 1324; railroads, 1771; roads, 1322, 1324
Illinois Central Rail Road Co., 1770
Illinois River, 758
Imlay, Gilbert (ca. 1754-1828?, author and political adventurer), *A topographical de scription of Kentucky,* 1238, 1270, 1308
Imperial Oil Company, publisher, 490
Independence Hall, Phila., PA, 449, 988, 990, 1001, 1013, 1016, 1019, 1033, 1038, 1652
Independence National Historic Park, Phila., PA, 1036
Independence Square, Phila., PA, 15 (3), 1012
"Independent Confirmation of the Magnetic Declination 1737-1738," by John Robert Connelly, 32 (1), 899
Independent State of the Congo, 445
India, 399-400, 1660; Agra, 399; Allahabad; 399; Bahar, 399; Bengal, 182, 399, 1660; Cape Comorin, 401; Delhi, 399; Ganges River, 399; Madras, 401; Mysore, 182; Nicobar Islands, 182; Oude, 399
Indian Ocean, 380, 1634-35; bathymetry, 1612; physiographic diagram, 1635; south, bathymetry, 102
Indiana, 942, 1245, 1318-20
Indians, 20, 33 (4), 502, 513, 525-26, 529, 752-53, 1302, 1378-80; Algonquian, 48; Aquanishuonîgy, 752-53; California, 1484; Catawba, 48 (2), (14); Chickasaw, 32 (5), Chippewa, 1277; Choctaw, 32 (5); Creek, 32 (5), war, 1177; Colombia, 35 (13), Couxsaxráge, 752-53; Eastern Woodlands, 48; Five Nations, 752-53, 783; Florida, war, 1188, 1193; Fox, 1378; Genesee, NY, 764-65; "Hero-Transformer" myth, 48 (13), Huron 32 (98); Illinois, 1269; Iroquois, 752-53, 783; languages, North America, 464, 468; Malecite, 48 (6); Mandan, 576; Mexico, 89; Miami, 901; Mingoes, 901; Mistassini, 48 (14); Mohicans, 901; Montagnais, 48 (14); Montagnais-Naskapi, 48 (1), (4), (5), (8), (11), (12); mound, 1 (2-3); Naskapi Montagnais, 48 (1), (4), (5), (8), (11), (12); New York, 49, 816, 819; Ohio, 752-53; Ottawas, 901; Panama, 35 (13); Penobscot, 48 (3), (9), Penn's treaty with, 579, 936; Pennsylvania, 49; Peru, 89; Rappahannock, 48 (16); ruins, Huron River, OH, 1307; ruins, Muskingum River, OH, 1305; Sac, 1378; Shawaneses, 901; Sioux, 48 (2); Skaniadarâde, 752-53; Southwest, 37; Tennessee, 32 (21), Tiiuxsoxrúntie, 752-53; town, Kittanning, PA, 33 (4); territory (see: Oklahoma), geology, 32 (87); U.S., 587, 598, 607, 611, 615, 621-22, 638, 1375-76; Virginia, 89; Wyandots, 901
Indo-China, 402-403; ethnolinguistic, 403
Indonesia (East Indies), 404-407
Ingersoll, C. J., donor, 451
Ingersoll, Joseph R., donor, 686, 743-44, 828, 830-31, 847-48, 1229, 1276, 1376, 1474
Inglis, John, 46 (20)
Inland navigation. *See:* canals.
Inman, ——, lithographer, 1047, 1072; publisher, 1047
Inrättin, ——, 163
Inskip, Abraham H., publisher, 566
Inskip, Abraham H., & J. Maxwell, publishers, 566
Institut de France, 247
Institut des Sciences, 186-87
Institut National de Géographie, Brussels, publisher, 443-44
Institut Royale Grand-Ducal, Luxembourg, 217
Institution d'Ethnographie, 118
International Bureau of the American Republics, 1571
Iowa, 1378-82
Ipiakdiuak, from Arsebémiling, Canada, 6 (3)
Iran, 373, 377
Iraq, 376
Ireland, 144-47, 152, 1657; Cork, 145; Dublin, 145; Galway, 145; Giants Causeway, 145; Isle of St. Patrick's purgatory, 145; Kinsale harbor, 145; Limerick, 145; roads, 145; to N.Y.C., track, 5 (1); Waterford, 145
Irois Bay, St. Domingue, 32 (35)
Iron ore, WV, 1141, 1143
Ironton, CO, 1420
Iroquois Indians, 752-53, 783
Irving, Alexander, cartographer, 1094
Isachsen, Yngvar W., cartographer, 842
Isle of St. Patrick's purgatory, Ireland, 145
Islington Lane, Phila., PA, 1039
Isogenic map, U.S., 595-96
Isothermal lines, world, 83
Istanbul, Turkey, 52, 362
Italy, 52, 251, 325, 341-58, 1641, 1654, 1691, 1741; Civitá Vecchia, 14 (1); Corsica, 347; Elba Island, 344, 357, Porto-Ferrajo, 357; Florence, 346; Milan, 356; Naples, 14 (2), 256, 342, 353, 1667; Pontine Marshes,

1686; Ponza Island, 354; roads, 347, 355; Rome, 256, 341, 343; Sardinia Island, 347, 350; Sardinia (kingdom), 345; Sicily, 347; Turin, 267, 349; Two Sicilies kingdom, coastline, 358; Venice, 348
Itasca, Lake, basin, MN, 1340
Itzenplitz, Graf von, 303
Ivanoff, ——, donor, 316
Iylland, Denmark. *See:* Jylland

Jack, Robert Logan (1845-1921, Australian geologist), 1586, 1588
Jäck, Carl (1763-1808, German traveller and publisher), publisher, 292
Jackson, ——, cartographer, 784, 1655
Jackson & Alger, donor, 494
Jackson, Charles Thomas (1805-1880, Canadian cartographer), 494
Jackson, Sir Frederick George (1860-1938, English explorer), 97
Jackson, MS, 1213
Jackson Purchase, KY, agriculture, 1248
Jacksonville, FL, 1196
Jacksonville, FL, to Hibernia, FL, 644
Jacmel, St. Domingue, 32 (47), (60)
Jacobsen, N. Kingo, cartographer, 1759
Jacotin, Pierre (1765-1827, French cartographer), cartographer, 432
Jacques, Herbert, cartographer, 649
Jagodina, 29 (23)
Jaillot, ——, cartographer, 250
Jalapa, Mexico, 32 (18), (71); and Orizaba to Vera Cruz, Mexico, 32 (71)
Jamaica, 1661, 1668; Bluefield harbor, 1525; Kingston harbor, 1542; Lucea harbor, 1527; Montego Bay, 1528; Port Antonio, 1529; Port Royal harbor, 89, 1530, 1533, 1542
James, ——, publisher, 1186
James, Commander, 80
James, J. E., cartographer, 1769
James, John A., cartographer, 650
James, Thomas P., 32 (67)
James and Johnson, publishers, 1186
Jan Mayen Island, Norway, 100
Janesville, WI, 1372
Janson, Charles William (fl. 1793-1807, British traveller and author), *The stranger in America,* 1002
Janson-La Palme, Robert J. H., 1767
Jansz, L. Schenk, engraver, 1496
Japan, 30, 378, 381-86, 1699; Ani mines, 30; bathymetry, 385; geological survey—*see:* Geological Survey of Japan; geology, 384; hypsometry, 385; Krafto Island, 1699; Magama, 30; Ryukyu Islands, 1699; Tokyo Bay, 389; Tootoomi oilfields, 30; volcanoes, 383; War Ministry, 381
Jasper, Alberta, Canada, 543-44
Jean, ——, (fl. 1800-1837, French cartographer, publisher), publisher, 264
Jeff Davis County, TX, 1234
Jefferson, Mrs. Martha Skelton, 22 (1)
Jefferson, Peter (1707/8-1757, Virginia surveyor, father of Thomas Jefferson), cartographer, 32 (13), 469, 759, 1094, 1750
Jefferson, Peter, and Joshua Fry, *A map of the inhabited part of Virginia,* 1750
Jefferson, Thomas (1743-1826, president of the U.S., author of the Declaration of Independence, APS 1780), 27, 28; cartographer, 22 (1), 32 (54); collection, 22; dedicated to president of the APS; 1 (1): donor, 28, 1330; *Notes on the State of Virginia,* 32 (19)
Jefferson, LA, 1333
Jefferson County, VA, 1120
Jefferson River, 560
Jefferys, Thomas (ca. 1710-1771, English engraver, geographer, publisher), *The American atlas,* 550; cartographer, 32 (7), 124, 293, 347, 404, 434, 451, 469, 550, 1663-64, 1674; *A compleat pilot of the West-Indies,* 1674; engraver, 988, 1749-50; *General topography of North America and the West Indies,* 651; *Neptune Occidental,* 1663; publisher, 124, 451, 651, 988, 1094; *The West-India atlas,* 1664
Jefferys & Faden, publishers, 124
Jekyl, GA, 1170
Jenkins, Olaf Pitt (b. 1889, cartographer), cartographer, 1574
Jenkins, T. A., cartographer, 847-48
Jennings, J. H., cartographer, 838, 1024, 1026
Jenny, W. L. B., cartographer, 1259
Jenny, W. P., cartographer, 1394
Jensen, Kr. Marius, cartographer, 1759
Jenson, Ruth Helkiaer, cartographer, 1759
Jeremie, St. Domingue, 32 (40)
Jersey City, NJ, 890
Jerusalem, Israel, 372
Jetersville battlefield, VA, 774
Jewett, J. P., & Co., publisher, 1399
Jillson, Willard Rouse, 1236
Jobbins, J. R., cartographer and engraver, 141

John Carter Brown Library, 975
Johns Hopkins Hospital, 1720; Press, publisher, 1764
Johnson, ——, engraver, 139; publisher, 921-22, 1186
Johnson, Guy (ca. 1740-1788, loyalist, super intendent of Indian affairs), cartographer, 1302
Johnson, James, publisher, 74
Johnson, Willard, J., cartographer, 1232
Johnson, William. *Sketches of the life and correspondence of Major General Greene*, 1089
Johnson & Warner, publisher, 921-22
Johnston, ——, 462
Johnston, Alexander Keith (1804-1871, British engraver and geographer), cartographer, 1713, 1729; engraver, 448; publisher, 114, 1713, 1729; *The royal atlas of modern geography*, 1713, 1729
Johnston, J., cartographer, 545
Johnston, John Eggleston (1807-1891, soldier), cartographer, 679, 1049, 1074
Johnston, T. B., cartographer, 1713
Johnston, William (fl. 1798, engraver), engraver, 448, 1148
Johnston, William (fl. 1850-1860, British engraver), cartographer, 328; publisher, 114, 1148, 1713, 1729
Johnston, W. & A. K., publishers, 114, 448, 1713, 1729
Johnston's royal atlas of modern geography, A. K. Johnston, 1729
Jomard, Edmé François (1770-1862, French cartographer, geographer), cartographer, 371, 427-29, 431-32, 438; donor, 432; *Mémoire sur la Communication du Niger et du Nile;* 429
Jones, ——, engraver, 557
Jones, Benjamin (fl. 1798-1815, engraver), engraver, 1120
Jones, Felix (fl. 1855-1857, British cartographer for the East India Company), cartographer, 374
Jones, John, cartographer, 51 (1)
Jones, N. F., cartographer, 952
Jones, Ralph Weston, cartographer, 1034
Jones, S. C. cartographer, 1249-1252
Jones, Samuel, English cartographer, 1772-73
Jones, Thomas H., cartographer, 978
Jones, William (1763-1831, English optician, APS 1801), donor, 1772-73
Jonesboro, GA, 1181
Journal, Andrew Ellicott, 1273, 1329
Judges cave, CT, 734
Juniata Valley, Huntingdon County, PA 960
Jukes, Francis (1747-1812, English engraver), engraver, 699
Jung, Jean, engraver, 1480
Junken, C., cartographer, 645, 747, 822, 1182
Jylland, Denmark, 171-72

K.K. Hof- u. Staats Druckerei, publisher, 1634
K. Viidenskabernes Societats Copenhagen, 167-74, 178-81, 183-85
Kalamazoo River, MI, 1346
Kaløe, Denmark, 176
Kamchatka Peninsula, U.S.S.R., 378, 456, 1647
Kanawha County, WV, 1140
Kane, Elisha Kent (1820-1857, Arctic explorer, APS 1851), cartographer, 23 (1-12), 32 (79), (94); collection, 23; donor, 94
Kane, John Kintzing (1795-1858, judge and lawyer, APS 1825), donor, 1692
Kansas, 1378, 1399-1401; geology, 1401; Lawrence, Eldridge House, 1399; Topeka, Constitution Hall, 1399
Kansas River, 1426
Karpinsky, A., cartographer, 319; donor, 319
Katanga, Zaire, 446
Kawartha Lakes, Ontario, Canada, 490
Kay, George Frederick (b. 1873, geologist), cartographer, 465
Kayser, Henry, cartographer, 1331-32, 1383
Kearny, ——, publisher, 461
Kearny, F., cartographer and engraver, 934
Kearney, James (d. 1862, soldier), cartographer, 1151, 1191, 1227, 1278-79, 1283-84
Kearny & Company, engraver, 1006
Keeling Islands, 1620
Keen, Gregory S., cartographer, 968
Kegonsa Lake, WI, 1358
Kehn, T. M., cartographer, 984
Keim, ——, cartographer, 986
Keithley Creek, B.C., Canada, 537
Keizer, Frank, cartographer, 650
Keller, Charles (b. 1868, soldier), cartographer, 1300
Kelley, Vincent C., cartographer, 1419
Kelly, Col., cartographer, 401
Kelly's Island, Lake Erie, 1278
Kelp, potash from, 1478
Kendall, Ezra Otis (1818-1899, astronomer),

cartographer, 1020-21
Kenhardt, South Africa, 448
Kennedy, James (?) (fl. 1797-1831, Philadelphia lithographer), lithographer, 932
Kennedy & Lucas, lithographers, 932
Kennemerland, The Netherlands, 201
Kennerville [Kenner], LA, 1333
Kennesaw Mountain, GA, 1181
Kenosha, WI, 1372
Kent, England, Hollwood, 129
Kentucky, 1236-53; Adair County, 1249; Big Sandy Valley, 1251; Boyle County, 1249; Breckenridge County, 1249; Christian County, 1249; Clark County, 1249; Clinton County, 1249; coal, 1247; Crittenden County, 1249; Crooked Creek, 2 (1-2); Cumberland River, 1253; geological survey, 1236; Green County, 1249; Henry County, 1249; Jackson Purchase, agriculture, 1248; Lexington, 1 (2), 1246; Lincoln County, 33 (8), 1249; Madison County, 1249; Mammoth Cave, 2 (5); Marion County, 1249; Mason County, Franklinville, 1242; Maysville, 1246; Meade County, 1249, gas, 1250; Mercer, 1249; Montgomery County, 1249; Nelson County, Lystra, 1243; nitre cave map, 2 (1); Ohio Cincinnati quadrangle, 1315; Oldham County, 1249; roads, 1246; Shelby County, 1249; Taylor County, 1249; Trigg County, 1249; Webster County, 1252; Warren County, 1249
Kereven, Patrick, publisher, 1012
Kervand, J. L., lithographer and publisher, 585
Key, J. R., cartographer, 1468
Key West, FL, 1194; harbor, 644, 1194
Kildinghuus, Denmark, 174
Kimber, ——, publisher, 921-22
Kimber, Emmor., donor, 568; publisher, 568
Kimber & Conrad, publishers, 921-22
King, Charles Francis (b. 1843, geographer), cartographer, 489
King, Clarence (1842-1901, scientist), cartographer, 1706, 1769; *Geological and topographical atlas of the U.S.*, 1706
King, Harry, cartographer, 1207, 1222, 1367, 1396, 1495
King, Henry, cartographer, 599
King, Phillip Parker (1793-1856, F.R.S., naval officer), cartographer, 1575
King, Thomas, publisher, 952
King William Land, N.W.T., Canada, 487
Kingawa, from American harbor, Canada, 6 (4)
Kingston, GA, 1181
Kingston, NJ, to N.Y.C., NY; 866
Kingston harbor, Jamaica, 1542
Kinsale harbor, Ireland, 145
Kirsna, 29 (19)
Kitchin, Thomas (1718-1784, geographer), cartographer, 108, 126, 146, 150, 669, 786, 995; engraver, 452, 1649
Kittanning, Armstrong County, PA, 919; Indian town, 33 (4)
Kittatinny Valley and Mountain, NJ, 886
Kleckhoff, K., engraver, 720
Klemp, Egon, *America in maps,* 1756
Kluane River, Yukon Territory, Canada, 547
Kneass, William (1781-1840, engraver), cartographer, 32 (68); engraver, 1003, 1007; publisher, 1007
Knecht, Jakob (1824-1894, French lithographer, engraver), lithographer, 113
Knecht & Roissy, lithographers, 113
Knight, A., engraver, 1049
Knight, H. M., engraver, 645, 693, 854, 1182, 1204
Knight, John (b. 1802, engraver), engraver, 269, 687, 693, 743, 848, 850-51, 881, 1138, 1167, 1204-205, 1209, 1463, 1485
Knight, John George David (d. 1895), cartographer, 1341
Knight, J. J., engraver, 1464
Knight, R. T., engraver, 849, 1203-205, 1469
Knobs Oaks Rush, Queensland, Australia, 1597
Knorr, E. K., cartographer, 1195
Knott, W. T., cartographer, 1249
Knowles, ——, 46 (2)
Knoxville, TN, 32 (21), 1258, 1260
Königsfeld, 29 (28)
Köpp, Henry, cartographer, 29 (1-60)
Koldinghuus, Denmark, 183-84
Kolecki, Theodor, cartographer, 1461
Kollar, 29 (18)
Komarzewsky, ——, donor, publisher, 310
Konchin, A., cartographer, 319
Kondrup, J. C., engraver, 1138
Konigseck, Count, 29 (45)
Korea, 379-81, 1699; Quelpart Island, 381
Kort Beskrifning om Provincien nya Swerige uti America, Campanius Holm, 896, 898, 1048
Kosak, William, cartographer, 1091
Koversky, Edward (1837-1916, Polish military cartographer), cartographer, 311
Krafto Island, Japan, 1699
Krebbs, ——, publisher, 1140

Krebs, Charles G. (fl. 1862-1867, lithographer); engraver, 1130; lithographer, 1497, 1259; publisher, 1497
Krems, Austria, geology, 278
Krotska (Krozka), 29 (17), (60)
Krusenstern, Adam Johann von, baron (Ivan Fedorovic von) (1770-1846, Russian admiral, explorer), *Atlas de l'Océan Pacifique,* 1688; cartographer and donor, 1688
Kübul, S. J., 1449
Kümmel, Henry B., cartographer, 884, 892
Kündig, Andreas (fl. 1830-1862, Swiss cartographer), cartographer, 238
Kuiper, G. P., 1747; *Photographic lunar atlas,* 1747
Kuril Island, U.S.S.R., 1699
Kwantung peninsula, China, 390

Laaland, Denmark, 170-71
Labrador, Canada, 452; Attaslopi (?), 48 (8); Eskimos, 48 (11); Lake St. John, 48 (10), (14)
Lac de Flambeau, WI, quadrangle, 1371
Lackawanna & Pittsburgh Railroad, 943
Lacoste, Louis, *Précis historique du canal du Languedoc ou des Deux Mers,* 252
La Croix des Bouquets, St. Domingue, 32 (50); Fort la Constitution, 32 (50)
La Crosse, WI, 1372
Ladd, Samuel B., cartographer, 1410
Lafon, Benjamin, cartographer, 1217, 1219; donor, 1217
La France, Joseph, cartographer, 1645
Laguimanor, Luzon Island, Philippines, 411
La Hève, Cap de, France, 270
Laird, David, (b. 1883), 545
Lake Erie. *See:* Erie, Lake
Lake Huron. *See:* Huron, Lake
Lake Michigan. *See:* Michigan, Lake
Lake of the Woods, MN, 568
Lake Ontario, *See:* Ontario, Lake
Lake Pontchartrain, LA. *See:* Pontchartrain, Lake, LA
Lake Superior. *See:* Superior, Lake
Lamaire, ——; 1619
Lamb, Francis (fl. 1665-1700, engraver, publisher), engraver, 897
Lamb, George, cartographer, 709
Lambert, John, cartographer, 642, 849, 1278-81, 1466, 1480, 1485
Lambert Herbarium, 32 (67)
Lamont Geological Observatory, Columbia, University, N.Y.C., NY, 1632-33, 1635
Lamson, A. G., cartographer, 1286
Lamson, G. W., cartographer, 1286
Lancaster County, PA, 929; Nabron(?) twp., 33 (3)
Land of Lakes Publisher, 1373
Landau, Germany, 297
Lansdowne, PA, quadrangle, 980
Lane, Michael, cartographer, 1749
Lang, cartographer, 1707
Langeland, Denmark, 171, 173
Languedoc, France, 252
Languinais, Jean Denis, comte de (1753-1827, French geographer), donor, 268
Lanzarote Island, Canary Islands, Spain, 337
La Pérouse, Jean Françoise de Galaup, comte de (1741-1788(?), French explorer), 80, 82, 370, 1673; *Chart atlas. A voyage around the world,* 1673
Lapham, Increase Allen (ca. 1810-1875, geologist), cartographer, 1359, 1361; donor, 1359
Lapie, Pierre (1779-1851, geographer, cartographer), cartographer, 182
La Plata, CO, 1411
La Plata River, track survey of tributaries, 1563-64
Laporte, ——, cartographer, 273-74
Lapova, 29 (21)
Lapsworth, Charles (b. 1842), cartographer, 7 (2-3)
Laramie, WY, 1443-44
La Rouchefoucauld Liancourt, François Alexandre Frédéric, duc de (1747-1827, French traveller, writer), 557; *Travels through the United States,* 557
Las Casas, E. A. D., cartographer, 1676; *Atlas historique, géologique, chronologique, et géographique,* 1676
Latrobe, Benjamin Henry (1764-1820, architect, engineer, traveler, naturalist), cartographer, 33 (7), 760
Lattré, Jean (fl. 1743-1793, French engraver, publisher), cartographer, 551
Latvia, 313
Laurel Hill Cemetery, Phila., PA, 1019
Laurie, Robert (1755-1836, English publisher, cartographer); publisher, 90, 193, 332, 370, 762, 1672, 1674
Laurie, Robert, & James Whittle, publishers, 32 (92), 90, 193, 332, 370, 762, 1672-74
Lavoisne, C. V., 316; cartographer, 1684; *A complete genealogical, historical, chronological, and geographical atlas,* 1684
Law, S. A., cartographer, 9 (1)

Law, W. W., 58
Lawrence, KS, Eldridge House, 1399
Lawson, Alexander (1773-1846, Scottish-American engraver), engraver, 261, 298, 1273, 1329
Lawson, J. S., cartographer, 1469-70
Lawson, O. A., engraver, 687, 743, 746, 848, 851, 881, 1049, 1073-74
Lawton, OK, 1267
Lay, Amos (b. 1765, geographer, map publisher), cartographer, 483, 575, 823; donor, 823, 483; publisher, 483, 575, 823
Lay, Amos, & J. Webster, publishers, 483
Lea, Isaac (1792-1886, naturalist), 1685; donor, 1685
Leach, Smith Stallard, cartographer, 1341
Leadville, CO, 1412-14
League Island, Delaware River, Phila., PA, 46 (6)
Leaõ, Spain, 329
Leavenworth, A., cartographer, 1701
Lebas, Philippe, cartographer, 1613
Lecesne, ——, cartographer, 432
Le Conte, John Eatton (1784-1860, engineer, naturalist, APS 1851), cartographer, 852
Le Conte, John Lawrence (1825-1883, entomologist, physician, APS 1853), collection, 24
Lee, Richard Henry (1732-1794, statesman), collection, 25
Lee, Richard Henry II (1794-1865), donor, 25
Lee, Robert Edward (1807-1870, soldier), cartographer, 1331-32, 1383
Lee, Thomas Jefferson (1808-1891, soldier), cartographer, 632, 882, 1151-52, 1225, 1228, 1280
Leeward Islands, 1518
Lehigh and Schuylkill, PA, coal regions, 936
Lehigh River, PA, 32 (75), (76), 33 (9)
Lehman, Ambrose Edwin (b. 1851, geologist), cartographer, 955
Lehman, George M., cartographer, 955
Lehman Street, Phila., PA, 19 (1)
Leicestershire, England, Union Canal, 136
Leidenfrost, E., engraver, 56
Leighton, Morris Morgan (b. 1887, geologist), cartographer, 465
Lemercier, —— (fl. 1838-1875, French publisher), cartographer, 270; lithographer, 270
Lemercier & Cie, publishers, 217, 442
Lempriere, Clement (fl. 1716-1786, British engineer), cartographer, 449
Lenni Lenape-English and English-Lenni Lenape dictionary, 20
Lenni Lenape names, Northampton County, PA, 20 (8)
Leon, Manuel Ponce de, cartographer, 1514
Lepontina, Switzerland, 226
Le Rouge, Georges Louis (fl. 1740-1780, French cartographer, publisher), publisher, 249, 904, 994, 1622; *Atlas Américain Septentrionale,* 904
Le Roy, Alphonse Vincent Louis Antoine (1741?-1816, French philosophe), 1622
Le Sage, A., 1676
Lesley, J. Peter (1819-1903, geologist), cartographer, 32 (80), 954, 958-60, 1247, 1618, 1712; collection, 26, 30
Lesley, Joseph Jr., (geologist), 1247
Lesoüef, A., donor and publisher, 118
Lesser Slave Lake, Alberta, Canada, 531
Lesueur, Charles Alexandre (1778-1846, French naturalist, traveller, APS 1817), donor, 270
Leth, Henry de (Hendrick de, 1703-1766, Dutch engraver, mapseller, publisher), cartographer, 213; engraver, 213
"A letter from Dr. Benjamin Franklin to Mr. Alphonse Le Roy" . . . APS Trans., 1622-23
Letters from an American farmer, M. G. S. Crêvecoeur, 670
Levant, 52
Levees, Mississippi River, 1339, 1341
Lévêque, Pierre (1746-1814, French mathematician, hydrographer), cartographer, 1613
Leverett, Frank (b. 1859), cartographer, 465
Levrault, F. G., publisher, 113
Lewis, ——, cartographer, 1678, 1680
Lewis, Abraham, 32 (54)
Lewis, F., cartographer, 933
Lewis, I. W. P., cartographer, 682
Lewis, Meriwether (1774-1809, explorer, APS 1803), 32 (54), 566; cartographer, 27, 28 (2), (8), (9), 1330, 1765; collection, 27
Lewis, Meriwether, and William Clark Expedition, 1461, 1765; map, 560, 566; *History of the Expedition under the Command of Lewis and Clark,* 566; route, 560
Lewis, Meriwether, and William Clark, *Journals,* 28
Lewis, Nicholas, 22 (1)
Lewis, Samuel (fl. 1774-1807, cartographer, geographer, draughtsman to the Great Britain Plantation Office), cartographer

32 (18a), (19), (23), 554, 566, 568, 655, 768, 1257, 1274
Lewis, S. M., engraver, 933
Lewis River, 560
Lewiston, ID, 1447-48
Lexington, KY, 1 (2), 1276
Leyden, The Netherlands, 194
Leyte Island, Philippines, 417
Li, C., cartographer, 395
Liberia, 82, 440; Monrovia, 440
Library Company of Philadelphia, Phila., PA, 32 (3), 449, 1001, 1037
Libya (Tripoli), 436
Lick Observatory, 58, 1747
Liège, Belgium, 189
Lighthouses, U.S., Atlantic coast, 624; Cape Henlopen, DE, 1003; Gulf of Mexico coast, 624; lake coasts, 1295; Pacific coast, 1465
Lightning Creek, B. C., Canada, 537
Ligowsky, Capt., cartographer, 1259, 1262
Lima, Peru, coastal valley, 35 (19)
Limerick, Ireland, 145
Linant de Bellefonds, Louis Maurice Adolphe (1800-1883, French cartographer), cartographer, 442; *L'Etbaye: Pays inhabité par les Arabes Bicharieh,* 442
Lincoln County, KY, 1249
Lincoln, Sea, Canada, 99
Lindenkohl, Adolph (1833-1904, cartographer, oceanographer), cartographer, 645; engraver, 854
Lindenkohl, H., cartographer, 645, 1080; engraver, 854
Lindenkohl, W., cartographer, 882, 885
Lindeström, Peter M., (1632-1691, author), cartographer, 895-96
Lindsay, David (1856-1922, Australian explorer), cartographer, 1581
Lindsay, John, earl of Crawford and Lindsay (1702-1749, soldier), 29; collection, 29
Linney, W. M., cartographer, 1249
Lipoda (Lipova), Rumania, 29 (20)
L'Isle, Guillaume de (Insulanus) (1675-1726. Premier géographe du roi, 1718), cartographer 214, 247, 258, 783, 1374, 1647; *Carte des nouvelles découvertes au nord . . . Siberie,* 1647
Litchfield Observatory, Hamilton College, CT, 57
Lithographers: See also: Engravers. *(See)* Acheson, H.; American Photo-Lithographic Co.; Andrews, A. B., Co.; Aub, ——; Bach, J. G.; Bauman, William, Jr.; Berliner Lithe. Instituts; Bien, Julius; Bien, Julius & Co.; Bien & Sterner; Bourquin, F.; Bouvé, E. W.; Bower, J. T.; Bradford, L. H.; Brett Lithograph Co.; Bulens, L.; Burland Lithograph Co.; Century Lithograph Co.; Charles, F.; Childs, Cephus G.; Childs & Inman; Colton, C. B.; Colton, G. W. and C. B., & Co.; Diessler, Oscar; Doms, P.; DuVal, P. S.; Eckert Lithograph Co.; Endicott, ——; Floder, A.; Fox, H.W.; Friedenwald, Lithograph; Friend, ——; Friend & Aub.; Frühwirth, E. K.; Gedney, J. F.; Graham, Andrew B.; Graham, Andrew B., Co.; Graham, Curtis B., Co.; Grahams, ——; Graphic Co. Photo-Lithograph; Gullick, W. A.; Haas, P.; Hanckel & Son; Heintz, ——; Hoen, A.; Hoen, A. & Co.; Hoen, A., Lithograph; Inman, ——; Kennedy, ——; Kennedy & Lucas; Kervand, J. L.; Knetch, ——; Knecht & Roissy; Krebs, Charles G.; Lemercier, Lithographic Ass'n. of New York; Lucas, ——; McGuigan, ——; McLaren, A.; Mayer, Ferd., & Co.; Mendel, Edward; Mesier, P. A.; Mortimer & Co.; Pendleton, ——; Peters, Norris, Lithograph Co.; Pether, H. J.; Probst, J.; Rehn, J.; Roissy, ——; Rosenthal, L. N.; Selves fils; Sherman, ——; Sherman & Smith; Sinclair, T.; Smith, ——; Standard Map Co.; Sterner, ——; Stone, Rolph Clark, Ltd.; Union Lithograph Co.; U. S. Geological Survey; Wagner, ——; Wagner & McGuigan; Watson, ——; Weber, E. & Co.; Weller, E.; Weller, E. & Graham Lithograph; Western Printing and Lithographing Co.; Williams, ——; Williams & Heinz Lithograph Co.; Wright, A. A.
Lithographic Association of NY, lithographers, 1259
Lithuania, 309
Little Egg Harbor, NJ, 881
Little Forest, N.S.W., Australia, 1607
Little Rock, AR, 1263
Little Snowshoe Creek, B.C., Canada, 537
Littrow, J. J., cartographer, 54
Liverpool, England, 125
Livingston, H., cartographer, 1791
Livonia, 159
Lizzard Creek, PA, 32 (75-76)
Lloyd, H. H., & Co., publisher, 545
Lloyd, J. T., publisher, 584
Lobos de Afuera Island, Peru, 35 (9), (11)
Locke, John (1792-1856, soldier), cartogra-

pher, 1020-21
Lockport route canal, Great Lakes, 1276
Locust Street, Phila., PA, 997
Loess, Alaska, 623; Canada, 623; U.S., 465, 623
Logan's Ferry, LA, 1225-27
London, England, 123, 125-26, 129, 143; London Docks, 132; Cornhill Street, 119; Royal Exchange, 131; Society of Arts & Commerce, 1535; Tower, 143
London Magazine, 669, 786, 995
Long, ——; publisher, 1695
Long, Colonel, donor, 1166
Long Beach, NJ, Barnegat City, 894
Long Branch, NJ, 890
Long Island, NY, 20 (6), 35 (14), 635, 835, 851; Gay Head to Cape Henlopen, DE, 773; Riverhead tract, 35 (20); roads, 780
Longfellow, A. W., cartographer, 643, 647, 692, 1182, 1502
Longitude finding, 53
Longshow, Robert, cartographer, 21 (1)
Longworth, D., publisher, 845, 1163
Lopez, Juan (fl. 1784-1812, geographer to the king of Portugal), cartographer, 32 (10), (11), (26), (42)
Lopez de Vargas Machuca, Don Tomas (1730-1802, Spanish geographer), cartographer, 32 (28), (63)
Lorenz, W., cartographer, 945
Loreto Department, Peru, 1570
Los, Isle of, 439
Los Angeles, CA, 1490
Los Coronados, CA, 1482
Los Llanos, Mexico, 32 (18)
Losanitz (?), 29 (19)
Lost Mountain, GA, 1181
Lotter, Matthieu Albrecht (1741-1810, German engraver, publisher), cartographer, 455; engraver, 455
Lotter, Tobias Conrad (1717-1777, German engraver, publisher), engraver, 218-19
Loudon, earl of. *See:* Campbell, John, 4th earl of Loudon
Loughridge, Robert Hills (b. 1843, geologist), cartographer, 1248-50
Louise, Lake, Alberta, Canada, Rocky Mountains, 533-34
Louisiana, 1215-24, 1374, 1504; canals, 1221; Jefferson, 1333; Kenner, 1333; Longan's Ferry, 1227; *New Orleans,* 85, 576-77, 644, 1216, 1219-21, 1224, battle of, 1220, Esplanade or Place of Arms, 1216, to Soniat Plantation, 1333, to Washington, D.C., 1090; Lake Pontchartrain, 1220; roads, 1221; Spanish 550, 552
Louisiana Purchase, 561, 1216
Louisiana Purchase Exposition, St. Louis, MO, 1385
Lovejoy's Station, GA, 1181
Lovell, William H., cartographer, 1024, 1026
Low, Albert Peter, cartographer, 485, 489
Low Countries *(see also:* The Netherlands, Belgium, and Luxembourg), 186-93, 251-94
Lower Geyser Basin, Yellowstone National Park, NY, 1433, 1438
Lownes, Caleb, engraver, 698
Lowthian, Joseph, cartographer, 7 (1)
Lucas, ——, lithographer, 932
Lucas, Claude, cartographer, 1643
Lucas, Fielding, Jr. (1781-1854, artist), publisher, 166, 1274
Luce, W., cartographer, 745, 668, 1074
Lucea harbor, Jamaica, 1527
Lucerne, Switzerland, 220
Ludlow, William (1843-1901, soldier), cartographer, 1391-93
Lugumcloster, Denmark, 300
Lummis, ——, cartographer, 887
Lunar orbital materials, 61
Lundenaes, Denmark, 181, 183
Luneau de Boisjermain, Pierre Joseph François (1732-1804, French writer), 65
Lunny, Robert M., *Early maps of North America,* 1748
Lupit, 29 (36)
Lurlanetto, Lodovico, ——, publisher, 348
Luxembourg, geology, 217, 251, 283
Luzerne County, PA, 958; Franklin Twp., 947
Luzon Island, Philippines, Laguimanor, 411; Manila, 418-19
Lydecker, Garret J. (soldier), cartographer, 1300
Lyell, Sir Charles, bart. (1797-1875, geologist), 11 (1)
Lyman, Benjamin Smith (1835-1920, geologist, APS 1869), collection, 30; cartographer, 961; donor, 961
Lynch, Francis Christopher Chisohlm (b. 1844, Canadian railroad man), cartographer, 508, 510
Lynhaven Bay, MA, 676
Lyon, James B., publisher, 836, 1312
Lyon, France, 255
Lyon Inlet, N.W.P., Canada, 487
Lystra, Nelson County, KY, 1243
Lytle, W. S., cartographer, 984

Macabi Islands, Peru, 35 (10)
Macao, China, 387
McAdoo, J. F., cartographer, 1249
McAlester, OK, 1267
McArthur, James Joseph (1856-1925, Canadian-U.S. surveyor), cartographer, 547
McArthur, W. P., cartographer, 1074, 1153
McBlair, C. H., cartographer, 690, 693
McCall, Lawrence, 46 (20)
McCalls Ferry-Penna. quadrangle, 973
McCalpin, William, cartographer, 822
McCaskill, Murray, cartographer, 35 (16)
McClellan, D. (fl. 1850-1880, engraver, publisher), engraver, 777
McClellan, George Brinton (1826-1885, soldier), 1126-28, 1131, 1506
McClellan, James E., 25
McClery, M. I., cartographer, 685-86, 743-44, 849, 881
MacClintock, Paul, cartographer, 465
McComb, John Navarre (d. 1889, soldier), cartographer, 1133
McConnell, R. G., cartographer, 531
McConomy, John A., cartographer, 33 (9)
McCorkle, Spencer C., 644, 705, 717, 854, 882, 885, 957, 1050, 1075, 1080, 1137-38, 1168, 1182, 1195-97, 1204, 1231, 1333, 1470, 1485; cartographer, 717, 747, 957, 1020-21, 1046
McCoy, G., engraver, 850, 1167, 1462-64, 1482
McCoy, Isaac (1784-1846, Indian scholar), cartographer, 1265; *Remarks on the practicability of Indian reform,* 1265
McCoy, James Culbertson (d. 1875, soldier, surveyor), cartographer, 1265
McCoy, Lee H., cartographer, 59; donor, 59
McCrackin, A., cartographer, 1333
McDonald Observatory, 1747
Macdonough, Thomas (1783-1825, naval officer), 824
McDowell, Irvin (d. 1885, soldier), cartographer, 1259, 1262
McElfatrick, S., cartographer, 1262
McElroy, A., publisher, 1015
McElroy, A., and P. S. Duval, publishers, 1015
McElroy, Robert, cartographer, 1741
Macevoy, James, cartographer, 537
M'Gillycuddy, V. T., cartographer, 1394
McGoffin, John, 46 (20)
McGuigan, ——, lithographer, 852
McIlhenny, Francis S., 967
McInnes, William (1858-1925, Canadian geologist), cartographer, 485, 496
McIntosh County, GA, 32 (12)
McIntyre, Henry B., cartographer, 1035
MacKay, Alexander D. (d. 1836), cartographer, 832
MacKay, John (d. 1848, soldier), cartographer, 1193
McKendree, William G., cartographer, 842
McKinnon, I., cartographer, 1189
Mackenzie, Murdoch (d. 1797, hydro-surveyor to the Admiralty), cartographer, 1657; *A maritime survey of Ireland and the west of Great Britain,* 1657
Mackenzie River, Australia, 1599
Mackinac, Strait, MI, 1283
McLain, William (1806-1873, Presbyterian minister), 440
McLaren, A., lithographer, 1589, 1595
McLaughlin, D. B,. cartographer, 984
McLaughlin, Fred, cartographer, 1235
Maclure, William (1763-1840, explorer, geologist), cartographer, 570
McMahon, B., donor, 711, 729-37
Macmillan, publishers, 156, 816, 819
McMurtrie, W. B., cartographer, 1463-64, 1479, 1481
Macomb, John Navarre (d. 1889, soldier), cartographer, 504-506, 679, 1278-83, 1285-86, 1288-90, 1293, 1422
Madeira Islands, Portugal, 328; Funchal, 328
Madison, WI, 1358, 1365-66, 1372
Madison County, KY, 1249
Madison River, 1404-405, 1425
Madras, India, 401
Madrid, Dirección de Hidrografia. *See:* Dirección de Hidrografia, Madrid
Maedel, E. A., engraver, 643, 645, 854, 882, 885, 1182, 1196
Mädler, Johann Heinrich (1794-1874, astronomer and physician), cartographer, 56
Maffitt, John Newland (1819-1886, naval officer), cartographer, 688, 1167
Magama, Japan, 30
Magdalen Island, Quebec, Canada, 633
Magellan, Strait of, Argentina, 1575; Chile, 1575
Magellanic Premium of the APS, 1 (1)
Magnetic compass needle, useless in Mammoth Cave, KY, 2 (5)
Magnetic variations, *Canada,* Norway House, Manitoba, 522, Oxford House, Manitoba, 523, Selkirk, Manitoba, 524; U.S., 581-82, 593, 603; world, 63, 72, 74, 77, 79

Magnetism, Europe, 1770; terrestrial, 87; U.S., 597
Mahlo, Emil, cartographer, 1394
Mahon, Charles, cartographer, 692, 850, 1074, 1204, 1209
Mahoning Creek, PA, 32 (1)
Mahony Shamokin coal basin, PA, 958
Maine, 48 (9), 577, 637-50, 660; Bar Harbour, 636; Blue Hill Bay, 645; Eastport harbor, 644; Frenchman's Bay, 645; Mount Desert Island, 644-45, 649; Penobscot Bay, 644, 646; Portland, 650; Portland harbor, 644, 647; railroads, 650; Richmond's Island harbor, 642; Waldo Patent, 32 (20); Winthrop, 51 (1); York River harbor, 643
Maine Historical Society, 1502
Maire, ——, de la, 1374
Maitland, Andrew Gibb (b. 1864, Australian geologist), cartographer, 1586, 1588-89
Maltby, F. B., cartographer, 1516b
Malden, MA, 699
Malecite Indians, 48 (6)
Malesherbes, Chrétien Guillaume de Lamacienon de (1721-1794, statesman, politician), collection, 31
Malpelo Island, 35 (15)
Malta, 355
Malvaux, J., publisher, 446
Mammoth Cave, KY, 2 (5)
Man, Thomas, publisher, 990
Man of War Shoal, St. Martin, 32 (65)
Manassas battlefield, VA, 1123
Manchester, MA, 675
Manchuria, China, 379
Mandan Indians, 576
Mandao, Masbate Island, Philippines, 408
Manhardsberge, Austria, geology, 278
Manila, Luzon Island, Philippines, 418-19
Manitoba, Canada, 515-28, 545; elevators (grain), 519; Grand Rapids, 520; Lake Winnipeg, 525; Norway House, 522, magnetic variations, 522; Oxford House, 523, magnetic variations, 523; Selkirk, 524, magnetic variations, 524; York factory, 548
Manitonoc, WI, 1372; harbor, 1360
Mannert, C. (Karl Konrad, 1756-1834, German geographer), cartographer, 91
Mansfield, H. B. cartographer, 957, 1046, 1197, 1231
Manú River, Peru, 1570
Manuscript maps, 1-52
Manuscripts communicated to the APS, 2
Map collection (manuscript), 32
Map of the city and liberties of Philadelphia, John Reed, 1652
A map of the inhabited part of Virginia, Joshua Fry and Peter Jefferson, 1750
Map of the Lower Mississippi River, Mississippi River Commission, 1711
Map of the Missouri River, Missouri River Commission, 1718
A map of Wayne and Pike Counties, Pennsylvania, Jason Torrey, 1681
Maps of Holland, Germany, Bohemia, Austria and Italy, 1641
Maracaybo, Venezuela, 32 (69)
Marblehead, MA, 675
The March of Man, 1741
Marchant, J., engraver, 579
Marcus Hook quadrangle, PA-DE-NJ, 782
Margery, Harry, publisher, 469
Maria (ship), 1629
Mariager, Denmark, 176
Marietta, GA, 1181
Marietta, OH, Campus Martius, 1306
Marinelli, O., cartographer, 1737
Marinden, H. L., cartographer, 957, 1020-21, 1046
Marino County, KY, 1249
Marion, Israel, cartographer, 538
Maritime atlas, 1639
A maritime survey of Ireland and the west of Great Britain, Murdoch Mackenzie, 1657
Markham, Clements R., 1716
Marlette, ——, cartographer, 1769
Marriott, ——, 46 (20)
Mars (planet), 50 (1-2)
Marshall, Richard (fl. 1764-1786, British publisher, bookseller), 125
Marshall, R. B., cartographer, 668, 1151, 1155, 1183, 1211, 1214, 1264, 1319, 1357, 1382
Marshall Road, Delaware County, PA, 43 (2); Philadelphia County, 43 (2)
Martha's Vineyard, MA, 695
Martin, Claude Emmanuel (fl. 1768-1772, French engraver at Copenhagen), 167-68
Martin, David (fl. 1796-1797, engraver), engraver, 917
Martin, G. C., cartographer, 1076
Martin, Helen M., cartographer, 1352-53
Martin, Lawrence (1880-1955, geographer), 1736; donor, 1077; *The George Washington atlas,* 1736
Martin, Richard A., cartographer, 376
Martin, William Thomas, Cartographer, 691
Martinez, J. de la C., cartographer, 1551
Martinez, Juan J., cartographer, 32 (72);

donor, 32 (72)
Martinique, Fort de France harbor, 1542
Marvin, Mrs. Mary Vaughan, donor, 51
Maryland, 32 (13), 455, 752, 755-56, 758-60a, 763, 766-67, 777, 779, 973, 1053-78, 1084, 1094, 1133, 1661, 1767; Alexandria, to Annapolis, MD, 1054-55; Annapolis harbor, 1073; Annapolis to Alexandria, MD, 1054-55; Annapolis to Bladensburg, MD, 1056-57; Annapolis to Bowling-green ordinary, VA, 1096; Annapolis to Dumfries, VA, 1097; Annapolis to Fredericksburg, VA, 1098-99; Annapolis to Hanover and Newcastle, VA, 1100; Annapolis to Hanover Court House, VA, 1101; Annapolis to Head Lynchs ordinary, VA, 1102; Annapolis to New Kent Court House, 1103-104; Annapolis to Philadelphia, PA, 908-11, 1058-63; Annapolis to Todd's Ordinary, VA, 1105; Annapolis to Williamsburg, VA, 1106-107; Annapolis to York, VA, 1108-109; Antietam, 774; Baltimore, 576-77, 766, 769, 1070, 1072, harbor, 644, 1075; Bladensburg to Annapolis, 1056-57; boundary, 902; Canals, 767, 1065; Chesapeake and Ohio canal, 576; Chesapeake Bay & Susquehanna River, 1068; Chesapeake Bay, to Delaware Bay, DE, canal, 760; Chester River, 760, 1074; Cumberland, Potomac River, 1069; Elk River, 766; Garrett County, geology, 1076; Geological Survey, 1076; Great Chester Valley, 1072; Havre de Grace, 1068; Patapsco River, 664, 1075; railroads, 1072; roads, 767, 769, 777, 1065; to Philadelphia, PA, 1064; west of Port Cumberland, 1067
"Maryland-Pennsylvania boundary survey," T. D. Cope, 1053.
Marzolla, Benedetto (fl. 1836-54, geographer of Naples), 1691; *Atlante corografico storico . . . Due Sicilie,* 1691
Masbate Island, Philippines, Mandao, 408; Nin, 408
Maskelyne, Nevil (1732-1811, English astronomer royal; APS 1771), 1772
Mason, Charles (1730-1787, English cartographer, surveyor), 754, 902
Mason and Dixon Line, 754, 920, 973, 1122
Mason County, KY, Franklinville, 1242
Massachusetts, 32 (18a), 455, 583, 667, 669-97, 761; Beverly, 675; Barnstable Bay, 677; *Boston,* 89, 576-77, 633, 636, 697-710, 718, American Revolution, 700; fortifications on Boston neck, 25 (1), harbor, 644, 698, 705, to Cape Hatteras, NC, 750; Mount Auburn Cemetery, 704; Brookline, 699; Buzzards Bay, 677; Cape Cod, 636, 679, 695, harbor, 679, 782-83, Isthmus, 677; roads, 695; Cambridge, 699; Cambridge, Harvard University, 696; Radcliffe College, 696; Charlestown, 699, 702; Chelsea, 699; Dorchester, 699; Edgartown harbor, 686; Gloucester Harbor, 691; Holmes' Hole harbor, 685, 689; Hyannis harbor, 688; Lynhaven Bay, 676; Malden, 699; Marblehead, 675; Martha's Vineyard, 695; Medford, 699; Milton, 699; Nantucket Island, 670, 681, 695; Nantucket harbor, 687; New Bedford, 684; Newburyport, 692; Provincetown, 679; railroads, 678; roads, 780; Roxbury, 699; Salem, 675; Salem harbor, 693, Sandwich, 677 Tarpaulin Cove, 685; Taunton, 680; Truro, 679 682; Tuckernuck Island, 681; Worcester, 671; Wellfleet harbor, 690
Matanzas, Cuba, 1531
Mathewson, A. J., cartographer, 1324
Matsudaira, N., engraver, 382, 384
Matthew, George Frederick, 485
Matthews-Northrop Works, publishers, 1234
Maull, James L., cartographer, 955
Maulo, ——, cartographer, 1707
Maumee Bay, OH, 1287
Mauna Loa, Hawaii Islands, 35 (17)
Maury, Matthew Fontaine (1806-1873, American hydrographer, APS 1852), 1612, 1618, 1697; donor, 1697; *Wind current and thermal charts,* 1697
Maverick, Peter Rushton (1780-1831, publisher, engraver, lithographer), cartographer, 1121; engraver, 765, 821, 845
Maxwell, John (fl. 1708-1712, British cartographer, publisher), cartographer, 186-87, 215, 246, 288; publisher, 214, 246-48, 288, 566
May, W., cartographer, 1629
Mayence, Germany, 29 (8)
Mayer, Christian (1719-1783, German Moravian astronomer, cartographer), 290; donor, 290
Mayer, Ferd., & Co., lithographers, 951, 1701
Mayer, John R. (fl. 1856-1865, surveyor), cartographer, 1844-49, 1364
Maynadier, Henry Eveleth (d. 1868, soldier), cartographer, 1402
Mayo, George U., cartographer, 1318
Mayo, William (b. 1684, British surveyor),

cartographer, 1094, 1519
Maysville, KY, 1246
Mazatlan, Mexico, to Cape Corrientes, Mexico, 23 (15)
Meade, George Gordon (1815-1872, soldier), cartographer, 505-506, 834, 1225, 1284, 1287-97; donor, 504-506, 1278, 1280-90
Meade County, KY, 1249; gas, 1250
Mease, James (1771-1846, physician, APS 1802), 2 (3), 751, 1692; donor, 902
Mease, James, *Gingerbread age,* 704
Mechan, J., cartographer, 854, 1138
Mechlin, J. R. P., cartographer, 850
Medford, MA, 699
Media quadrangle, PA, 979
Mediterranean Sea, 112, 118, 258; basin, 323-25; pilot charts, 1631; world, 323-25, 1665-66
Meekatharra, Western Australia, Australia, 1585
Mehemet 'Ali (ca. 1769-1849, Viceroy of Egypt), 371
Meier, ——, publisher, 1756
Meigs, Montgomery Cunningham (1816-1892, soldier), cartographer, 1331-32, 1383
Meil, Johann Wilhelm (1732-1805, German painter, engraver), cartographer, 292
Melish, John (1771-1822, geographer, traveller, merchant), cartographer, 80, 501, 565, 569, 571-72, 676, 923, 930, 934, 1006, 1071, 1086-87, 1177, 1201; donor, 80, 569, 571, 1309, 1681; publisher, 80, 501, 565, 569-72, 676, 703, 928, 930, 1006, 1085, 1177, 1201, 1218; *Documents relative to the negotiations for peace between the U.S. and Great Britain,* 565; "Geographical Intelligence . . .", 923; *A military and topographical atlas of the U.S.,* 501, 676, 1071, 1086
Melish, John, & Samuel Harrison, publisher, 80
Melish, John G., cartographer, 934
Melville Bay, Canada, 23 (1)
Melville Sound, Canada, 95
"Memoir on the Use of the Thermometer in Discovering Banks, Soundings, etc." Jonathan Williams, Jr., 1625-26
Mémoire sur la communication du Niger et du Nil, E.F. Jomard, 429
Mémoires sur la formation des os, Duhamel du Monceau, 14
Memoirs, Historical Society of Pennsylvania, 896, 1048
Memoma Lake, WI, 1358
Mendarina, 1619
Mendel, Ed (fl. 1856-1864, lithographer, publisher), lithographer, 1326, 1344-49, 1362, 1364; publisher, 1326, 1344-49, 1362, 1364
Mendenhall, Thomas Corwin (1841-1924, physicist, APS 1899), cartographer, 593, 595, 596-97
Mendonna, 1619
Mendota, Lake, WI, 1358
Menzel, Donald Howard (1901-1976, astronomer, APS 1943), donor, 61
Mercator, Gerhardus (Gerhard Kremer, 1512-1594, Flemish mathematician, geographer, cartographer), 32 (93), 66, 73, 80, 83
Mercer County, KY, 1249
Merchants Exchange, Phila., PA, 1011, 1019
Meredith and Clymer lands, 9 (2)
Meriden Gravure Co., publisher, 32 (8), 467, 560, 753, 858, 975
Meriwether, Nicholas, 32 (54)
Merrick, George, 1009; donor, 925
Merrill, Frederick James Hamilton (1861-1916, geologist), cartographer, 836
Merrill, William Emery (d. 1891, soldier), cartographer, 1179, 1261
Merrymount Press Publishers, publisher, 696
Mesa Verde National Park, CO, 1415
Mesier, P. A., lithographer, 1275, publisher, 1275
Messinger, John (fl. 1835-1853, cartographer), cartographer, 1324
Meteorology, Alaska, 1498; Argentina, 1580; Germany, 302; Ohio, 1312; U.S., 585-86, 604-605; world, 1722
Metzeroth, G. B., engraver, 647, 691, 1463, 1485
Meuse River, France, 13 (2)
Mexican War, 32 (71)
Mexico, 32 (11), (18), 450, 453-54, 569, 572, 577, 598, 1215, 1501-507, 1536, 1562, 1751; Antigua, 32 (18); Baja California, 1507; Chalco Lake, 32 (11); Chiapas intendency, 32 (16); Chihuahua, 1562; Cordoba, 32 (18), Cosamaluapan, 32 (18), Guadalajara, 32 (17), Guanajuato to Manzanilla to Mexico City, 41 (1); Guimas, Sonora, 32 (70); Indians, 89; Los Llanos, 32 (18), Manzanilla to Guanajuato to Mexico City, 41 (1); Mazatlan to Cape Corrientes, 32 (15); Mexico City, 32 (18), 1507; Mexico City to Guanajuato to Manzanilla, 41

(1); Mizantla, 32 (18); Monterrey, 1507; Nueva Galicia, 32 (15); Oaxaca intendency, 32 (16), (18); Orizaba, 32 (18), (71); Pachuca, 1507; Puebla, 32 (18), 1507; railroads, 594, 1507; Salamanca, 32 (17); Santa Cristobal Zumpango lake, 32 (11); Sonora, 1454; Tehuacan, 32 (18); Tehuantepec, gulf of, 32, (16); Tepeaca, 32 (18); Tepixi de la Seda, 32 (18); Tetela y Xonotola, 32 (18); Texcoco lake, 32 (11); Tezuitlan, 32 (18); Toluca to Vera Cruz, 32 (18), Torreon, 1507; Vera Cruz, 1503; Vera Cruz intendency, 32 (16); Vera Cruz to Orizaba and Xalapa, 32 (18), (71); Vera Cruz to Toluca, 32 (18); Xalapa, 32 (71); Xochimilco Lake, 32 (11); Yucatan, 32 (16), 1507; Zacatlan, 32 (18); Zalapa. *See:* Xalapa
Mexico, Gulf of, 32 (3), (16), (25-29), (32-53), (55-64), 1187, 1192, 1225, 1228, 1341, 1545, 1553; coast, 40; coast, U.S., 32 (86); coast, U.S., lighthouses, 624; coast, U.S., south of New Orleans, 554
Miami Indians, 901
Michels, Ivan C., 1701
Michigan, 1344-53; Bear Island, 1285; Black Lake harbor, 1344; Department of Conservation–Geological Survey Division, 1352-53; False Presqu'ile harbor, Lake Huron, Geological Survey, 1292; geology, 1351; geology, northern peninsula, 1352; geology, southern peninsula, 1353; Grand Haven, 1345; Grand River harbor, 1345; Kalamazoo River, 1346; Middle Island, Lake Huron, 1292; New Buffalo, 1349; Ontonagon, 1299; Ontanogon harbor, 1290; Oronto Bay, 1299; Pottowottomee, Lake, 1349; Presqu'ile harbor, Lake Huron, 1292; Ronald, 47 (1); Saginaw Bay, Lake Huron, 1293; St. Clair flats, 1284; St. Clair River, 1296; St. Joseph harbor, 1347; St. Mary's River 1283; Sault Ste. Marie, 19 (20); South Black River, 1348; Tawas Harbor, 1286; Thunder Bay, Lake Huron, 1294
Michigan, Lake, 758, 775, 1275, 1283, 1285
Middle Atlantic States, 32 (3), (78), 628, 750-82
Middle Island, MI, Lake Huron, 1292
Middle River, GA, 32 (12)
Middle River, West Florida, 17 (1)
Mifflin, ——, 46 (20)
Mifflin, Thomas (1744-1800, governor of Pennsylvania, APS 1768), 914
Mikhalsky, A., cartographer, 319
Milan, Italy, 356
Milet-Mureau, L. A., cartographer, 1673
Milford, CT, 737
A military and topographical atlas of the U.S., John Melish, 501, 676, 1071, 1086
Miller, ——, 33 (5), 984, 1089, 1249
Miller, Clarence William (b. 1914, scholar, professor of English), donor, 753
Millett's Wellington, Queensland, Australia, 1586
Milliadge, T., cartographer, 32 (8)
Mills, Edwin, cartographer, 39 (1)
Mills, Robert (1781-1855, architect, surveyor), cartographer, 1178, 1689; donor, 1689; *Atlas of the state of South Carolina*, 1689
Milne, Thomas, cartographer, 129
Milner, Thomas, cartographer, 1698; *The atlas of physical geography*, 1698
Milton, MA, 699
Milwaukee, WI, 1361, 1372
Milwaukee and Mississippi Rail Road, 1365
Milwaukee, Watertown and Madison Rail Road, WI, 1363
Mindanoa Island, Philippines, geology, 420
Mingo Indians, 901
Mining, 30
Mining centers, Queensland, Australia, 1596
Ministerie van Oorlog. Topographish Bureau, The Netherlands, 197-212
Minneapolis, MN, 1355-56
Minnesota, 1354-57, 1379-80; Cook County, 1356; Lake Itasca basin, 1340; Minneapolis, 1355-56; St. Paul, 1355-56; St. Peter's River, 1354
Mintzer, William A., 6 (1), 32 (85)
Miscellaneous Manuscript Collection, 33
Missionary Education Movement of the U.S. and Canada, publisher, 394
Mississippi, 1209-14, 1217-18, 1504; Cat Island harbor, 1209; Carrollton, 1333; Jackson, 1213; Ship Island harbor, 1209; Washington, Adams County, 1 (3)
Mississippi River, 560, 566, 758, 1220, 1269, 1277, 1321, 1329-43, 1374-75, 1705, 1710-11; alluvial valley, 1339-40; confluence with St. Peter's River, 28 (1); delta, 40, 1192, 1335-42; Des Moines rapids, 1331; Falls of St. Anthony, 1339; levees, 1339, 1341; Rock Island rapids, 1332; St. Francis basin, 1341-42; St. Louis harbor, MO, 1383; source, 39; up from New Orleans to Natchez, MS, 2 (4)

Mississippi River Commission. *Detail map of the Lower Mississippi River,* 1710; *Detail map of the Upper Mississippi River,* 1705; *Map of the Lower Mississippi River,* 1711
Missoula, MT, 1430-31
Missouri, 1376, 1378-80, 1383-88, 1504; Geological Survey, *Reports,* 1384; geology, 1384, St. Louis, 942, 1324, 1386; St. Louis harbor, Mississippi River, 1383; St. Louis, Louisiana Purchase exposition, 1385, World's Fair, 964, 1385, 1387; to Oregon (Trail), 1426
Missouri River, 560, 1402, 1426, 1718; Great Falls and portage, 28 (2)
Missouri River Commission, *Map of the Missouri River,* 1718
Mistassini Indians, 48 (14)
Mitchell, A. M., cartographer, 1151-52
Mitchell, H., cartographer, 717, 854
Mitchell, John (1711-1768, botanist), 751
Mitchell, John (1711-1768, British botanist, cartographer, naturalist), cartographer, 469
Mitchell, John, cartographer, 452, 469
Mitchell, Samuel Augustus (1792-1868, publisher), cartographer, 1505; donor; 577; publisher, 577, 1322, 1505, 1702; *Mitchell's new general atlas,* 1702
Mitchell, William (fl. 1838-1856, surveyor), cartographer, 681
Mitchell and Hinman, publishers, 577
Mitchell's new general atlas, S. Augustus Mitchell, 1702
Mizantla, Mexico, 32 (18)
Mobile, AL, 1208; bar, 1200; Bay, 23 (10), (11), 644, 1190, 1203-5
Mococa, Brazil, 1572
Model, 1774
A modern atlas . . ., John Pinkerton, 1683
Modern Club, Phila, PA, 1035
Moehlman, R. S., cartographer, 1419
Moellendorff, G. von, 302
Möen, Denmark, 171
Moesch, Casimir (1827-1898, German geologist), cartographer, 237
Moffat, John (fl. 1795-1820, Scottish engraver), engraver, 1627
Mohammed Ali. *See:* Mehmet 'Ali
Mohawk River, NY, 842
Mohican Indians, 901
Moithey, P., l'Ainé, publisher, 255
Moldau River, 276
Môle St. Nicolas, Haiti; 1526
Molitor, Edward, cartographer, 1301-1341
Molkow, E., engraver, 854
Moll, Herman (d. 1732, geographer), cartographer, 122, 145, 148, 151
Monaco, Bureau Hydrographique Internationale, 1614
Monmouth shore, NJ, 886
Monongahela House, Pittsburgh, PA, 12 (1)
Monongahela River, PA, 45 (3)
Monrocq, Léon (fl. 1864-1876, French publisher), publisher, 1514
Monrovia, Liberia, 440
Monsfeld, 29 (14)
Montana, 1425, 1430-32; Butte, 1430-31; geology, 1403; Great Falls, 1430-31; Glacier National Park, 1432; Helena, 1430-31; Missoula, 1430-31
Mont-Blanc, France, geology in the area, 279; model, 1774; Switzerland, glaciers, 243
Montagnais-Naskapi Indiahns, 48 (1), (4), (5), (8), (11), (12), (14)
Montego Bay, Jamaica, 1528
Monterey, CA, 644; harbor, 1479
Montero y Gay, Claudio, cartographer, 416
Monticello, VA, 32 (54)
Montgomery County, KY, 1249
Montgomery County, PA, 33 (10), 983; commissioners, 974; Planning Commission, donor, 983
Montreal, Quebec, Canada, 32 (84), 490, 498, 502, 780, 1298
Montrose, Scotland, 148
Moon, 53-54, 56, 58, 1747
Moore, ——, 434; publisher, 1019
Moore, E., 1747
Moore, Francis, 46 (20)
Moore, John Hamilton (d. 1801, chart-seller, publisher), cartographer, 1182
Moore, Raymond Cecil (1892-1974, geologist), cartographer, 1401
Moore, Samuel, donor, 1665
Moore, Willis Luther (1856-1927, meteorologist), cartographer, 1092
Moorsfoot Hills, Scotland, 7 (6)
Moraines, NY (west), 843
Morara River (creek?), 29 (26)
Moravian road, PA, 32 (1), 899
Morea coal lands, Schuylkill County, PA, 961
Moreau de St. Méry, Médéric Louis Elie (1750-1819, historian, publisher), 1543, donor, 1543; *Description topographique et politique de la partie espagnole de l'Isle St. Dominque,* 1543
Morehead, J., S., cartographer, 1331, 1383

Moreledo, Don Severiano, donor, 408-12, 441, 639
Morelli, ——, donor, 358, 1667
Morgan, B., cartographer, 32 (8)
Morgan, John, 903
Morgan, John H., cartographer, 1206
Morghen, Filippo (1730-post 1807, Italian engraver), engraver, 342
Morhardt, Major, cartographer, 1259, 1262
Morehead, J. S., cartographer, 1332
Morocco, 436; Fez, 436
Morococha, Peru, geology, 1570
Morono River, Peru, 1570
Morrell, ——, cartographer, 1613
Morrelli, ——, 1691
Morris, Edmund (1804-1874, cartographer), cartographer, 1047
Morris, Israel, 46 (9)
Morris, Robert (1734-1806, financier of the American Revolution), 816, 819
Morris, Samuel, 46 (9)
Morris Canal, NJ, 576, 878
Morrison, Russell, 1767; donor, 1767; *Washington College Presents: On the Map,* 1767
Morristown, NJ, 890
Morrow, A. T., cartographer, 1336
Morse, Jedidiah (1761-1826, Congregational clergyman, "Father of American geography"), 1680; *The American gazetteer,* 406, 458, 629, 1085, 1540, 1557; *American universal geography,* 76, 368, 628, 1083, 1544, 1558
Mortier, Pierre (?), publisher, 213, 278, 287, 720, 1639-40
Mortimer, ——, publisher, 1499
Mortimer & Company, lithographers, 537
Morton, Cecil C., cartographer, 1603
Morton, Robert H., cartographer, 1207, 1210
Mosbate Island, Philippines, 417
Moser, J. F., cartographer, 882, 885
Moskaira, A., engraver, 420
Mosman, A. T., cartographer, 885, 1046, 1050
Mosquito Shore. *See:* Belize
Monterrey, Mexico, 1507; battlefield, 1505
Moulton, Gary E., *Atlas of the Lewis and Clark expedition,* 1765
Mount, J., publisher, 1650, 1661-62
Mount, J., & T. Page, publishers, 1650, 1661-62
Mount, William, publisher, 1661-62
Mount, William, & T. Page, publishers, 1661-62
Mount Auburn cemetery, Boston, MA, 704
Mount Desert Island, ME, 644-45, 649, 1747
Mount Holly, NJ, 886, 890; to N.Y.C., NY, 872
Mount Rainier National Park, WA, 1473
Mountain States, U.S., 1402-407
Mountaine, William (fl. 1756, English cartographer, hydrographer), cartographer, 77
Mouschketov, J., cartographer, 319
Moustique Bay, St. Domingue, 32 (58)
Mouzon, Henry, Jr. (1741-1807, British geographer), cartographer, 1081, 1093
Mowbray Company, publisher, 712
Moyamensing Road, Phila., PA, 46 (12)
Moyer, J. M., 33 (5)
Mud River, GA, 32 (12)
Muddy River, PA, 45 (3)
Mülheim, Germany, 29 (15)
Müller, Albricht (1819-1890, Swiss geographer), cartographer, 238, 287
Müller, H.F., publisher, 280
Mueller, J. U., cartographer, 505-506, 1282, 1285, 1288-91, 1293
Münster, Germany, 29 (9)
Muhlenberg, Gotthilf Heinrich Ernst (1753-1815, Lutheran clergyman, botanist), collection, 34; cartographer, 34 (1)
Muir, Ramsay, cartographer, 1741
Muldrow, Robert, chartographer, 1024, 1026
Mulgrave, Capt. *See:* Phipps, Constantine John, 2nd. baron Mulgrave
Muller, Ernest H., cartographer, 843
Muller, John B., cartographer, 1091
Multhauf, Robert P., *A catalogue of instruments and models in the possession of the American Philosophical Society,* 1772
Multonomah River, 28 (12); and confluence with Columbia River, 28 (9)
Munroe, William H., cartographer, 649
Munsell, Luke (1790-1854, surveyor, publisher), cartographer, 1245; donor, 1245; publisher, 1245
Munsterland, The Netherlands, 202
Murphy, Robert Cushman (1887-1973, zoologist, ornithologist, APS 1946), cartographer, 35 (1-15, 17-20), 1695; donor, 35, 1695; collection, 35
Murray, G., engraver, 80
Murray, John, 4th earl of Dunmore (1732-1809, British governor of New York and Virginia), 1095
Muscle Shoals, Tennessee River, AL, 32 (66); canal, 1206
Muskingham River, OH, 1302; Indian ruins,

1305
Muskoga, Ontario, Canada, 490
Muskogee, OK, 1267
Mutlow, I (John) (fl. 1794-1840, English engraver), 128, 160, 192, 308, 350, 361, 367, 400; publisher, 73, 308, 350, 361, 367, 400, 424
Myers, Albert Cook (1874-1960, historian), cartographer, 32 (89); donor 32 (89)
Mynde, James (fl. 1753, British engraver), engraver, 1517-18
Mysore, India, 182

Na, J. F., cartographer, 395
Nabron (?) Twp., Lancaster County, PA, 33 (3)
Nagra ord till Upplysning om Bladet. Sveriges Geologiska Undersökning, 161
Nagy, Károly (1797-1868, Hungarian patriot, scientist), donor, 54, 277
Namur, Belgium, 215
Nansen, Fridtjof (1861-1930, Arctic explorer), 97; cartographer, 32 (88); *Farthest North,* 32 (88)
Nantucket Island, MA, 670, 681, 695; harbor, 687
Naples, Italy, 14 (2), 256, 342, 353, 1667
Napo River, Peru, 1570
Napoleon I (1769-1821, emperor of the French), 301
Narragansett Bay, RI, 644, 676, 711, 716-17
Narrative of the surveying voyages of his majesty's ships Adventure and Beagle, 1561, 1575-78, 1620
Narrows, Columbia River, 28 (6)
Narstin, John L., publisher, 1504
Narvaez, Jose Maria, cartographer, 32 (15)
Nashville, TN, 32 (21), 1258
Naskapi-Montagnais Indians, 48 (1), (4-5), (8), (11-12)
Natal, Rio Grande do Norte, Brazil, geology, 1574
Natchez, MS, 2 (4)
The national atlas of the U.S., U.S. Dept. of the Interior, 1753
National Geographic Society, 285
National Map Co., publisher, 972
National Railway Publications Co., 594
National Research Council, U.S., 623
The natural and civil history of Vermont, S. Williams, 661
Naumann, Carl Friedrich (1832-1910(?), German cartographer), cartographer, 304
Nav, Anthony (fl. 1806-1810, explorer), cartographer, 1330
Navesink, NJ, 890
Naylor, T. C., cartographer, 1605
Near East, 371-77
Nebraska, 1398; Omaha, 1398
Neé-ah harbor to Cape Flattery, WA, 1466
Neele, Samuel John (1758-1824, English engraver), engraver, 70, 137, 155-56, 759
Negrete, Pedro Celestino, cartographer, 32 (70)
Negros Island, Philippines, 417
Neilson, J. C., cartographer, 1074, 1153
Nell, Louis, cartographer, 1707
Nelson County, KY, Lystra, 1243
Nelson Creek, B.C., Canada, 537
Nelson River, Canada, 1645
Neptune (ship), 489
Neptune Occidental, Thomas Jeffrey, 1663
Nes, F. F., cartographer, 854
The Netherlands, 186-213, 254, 262, 1641, 1742; Amstelland, 213; Amsterdam, 213; Bargeveen, 197; Betuwe, 198; Biesbosch, 199; Drent, 191; Friesland, 195; Gelders, 193, 195; Groningen, 195; Holland, 188, 193, 195; Hunningen, 196; Hunsingoo, 200; Kennemerland, 201; king, 453, 632; Leyden, 194; Ministerie van Oorlog. Topographisch Bureau, donor, 197-212; Munsterland, 202; Ostergoo, 203; Overyssel, 195; Peel, 204; Schouwen, 205; Texel, 206; Twenthe, 207; Utrecht, 193, 195; Veluwe, 208; Wadden, 209; Walcheren, 210; Westerwolde, 211; Zealand, 188, 190, 193, 195; Zuiderzee, 212
Neuman, Louis E., engraver, 956
Nevada, 1427, 1457-60; Carson City, 1460; Columbia, 1460; Elko, 1460; Ely, 1460; Goldfield, 1458-60; North Goldfield, 1458-59; Reno, 1458-60; Tonopah, 1460; Washoe mining district, 1457; Comstock Lode, 1457; Winnemucca, 1460
Neversink River, NY, 766
Nevil, James, publisher, 903
Neville, Rev. E., donor, 680
A new American atlas, Henry S. Tanner, 1687
A new and complete system of universal geography, John Payne, 78a-b, 262, 323, 369, 407, 426, 459, 558, 640, 656, 665, 674, 715, 739, 817, 877, 918, 1069, 1119, 1149, 1160, 1176, 1244, 1256, 1546, 1559, 1616
A new and elegant general atlas, Arrowsmith and Lewis, 1678, 1680
New atlas of the mundane system, Samuel

Dunn, 1653, 1672
New Bedford, MA, 684
New Boston coal lands, Schuylkill County, PA, 961
New Brunswick, Canada, 485, 488, 492-96; Campobello Island, 493; Deer Island, 493; Grand Manan Islands, 493; Passamaquoddy Bay, 493; St. Croix River, 493
New Brunswick, NJ, 892
New Buffalo, MI, 1349
New England, 455, 469, 499, 627, 629-30, 634-36, 651, 695, 756-57, 1656, 1661; railroads, 590, 636; roads, 636; seismotechtonic study, 633; southern shore, 635
New France, 32 (98)
A new, general, and universal atlas . . ., 1649
New general atlas, Thompson, 1088
New Guinea, 406
New Hampshire, 455, 583, 648, 651-60, 761: donor, 659; Geological Survey, 659; geology, 659; Shelburne, 657; White Mountains, 657
New Haven, CT, 729-30, 735-36; harbor, 741
New Holland, 406
New Jersey, 32 (8), (14), 455, 469, 627, 751-53, 756, 761, 763, 777, 779-82, 858-94, 973, 1024, 1026, 1094, 1661; Absecon Inlet, 882; Amboy, 890; Atlantic City, 890; Barnegat Bay, 886; Barnegat City, Long Beach, 894; Barnegat Inlet, 882, 885; Barnegat Lighthouse, 894; Barnegat Lighthouse to Sandy Hook, 644; Bergen County, 886; Black Horse Pike, to N.Y.C., NY, 871; Boonton, 892; Bordentown sheet-geology, 888; Bridgeton, 886; Brunswick to N.Y.C., NY, 864-65; Camden, 886, 890, 1014, 1029; canals, 878, 935; Cranberry to N.Y.C., NY, 869; Cape May City, 644, 883, peninsula, 886, roads, 880; Chester, 892; Dover-Stanhope, 892; Egg Harbor, 886; Elizabeth, 890; Elizabethtown to N.Y.C., NY, 863; Essex County, 886; Flemington, 886; forest, 891; geology, 886, 889-92, 946; Hackensack, 890; Hudson County, 886; Jersey City, 890; Kingston to N.Y.C., NY, 866; Kittatiny valley and mountain, 886; Little Egg Harbor, 881; Long Branch, 890; Monmouth shore, 886; Morris canal, 576, 878; Mount Holly, 886, 890, to N.Y.C., NY, 872; Navesink, 890; Newark, 890; New Brunswick, 892; NJ-N.Y.C. sheet, 781; Newbold, 887; Norristown, 890; north, 860-61, 884; North Mountain, 859; Passaic valley, 886; Paterson, 890; NJ-PA Bridgeport quadrangle, 782; NJ-PA-DE Marcus Hook quadrangle, 782; NJ-PA-DE quadrangle, 973; NJ-PA-Phila. quadrangle, 973; Plainfield, 890; Pluckemin, 892; railroads, 779, 935; Raritan River, 859; roads, 777, 780; Salem, 886; Sandy Hook, 826, 885; Sandy Hook to Barnegat lighthouse, 644; Shark River, 890; Somerville, 892; Taunton, 890; Trenton, 576, 886, 890; Trenton to N.Y.C., NY, 867; Woodbury, 890
New Jersey Historial Society, Ely Collection, 861; publisher, 1748
New Kent Court-House, VA, to Annapolis, MD, 1103-104
New London, Conn., 747; harbor, 644, 742
A new map of part of North America . . ., 1645
New Mexico, 39, 1422-24, 1427
New Orleans, LA, 85, 576-77, 644, 1090, 1216, 1219-21, 1224; battle of. *See:* Battle of New Orleans; Esplanade or Place of Arms, 1216; meridian, 2 (4)
New Philadelphia, Schuylkill County, PA, 940
New revised atlas of the world, World Syndicate Publication Co., 1742
New South Britain, 101
New South Wales, Australia, 1607-10; Broken Hill district, geology, 1610; Conjola, 1607; Department of Mines, 1609; Geological Survey. *Records,* 1607; geology, 1607-1610; Gerringong district, 1607; Little Forest, 1607; Sydney, geology, 1608
New Sweden. *See:* Pennsylvania
New topographical atlas of the state of Ohio. H. F. Walling and O. W. Gray, 1312
New York (state), 455, 469, 499, 627, 751-53, 756, 761, 764-66, 776-78, 780-81, 783-843, 861, 1071, 1661; Adirondack Mountains, 32 (84), 842, 1298; Albany, 576-77, 826, 838, to Newborough, NY, 787-94, to Poughkeepsie, NY, 804-9; Buffalo harbor, 834; Bouguer gravity anomaly map, 841; canals, 829-31, 1300; Cambrian rocks, 839; Chautauqua lake, 816, 819; City Island, 849; Crown Point, 784; Erie Canal, 573, 576, 827; Finger Lakes, 841; Four Mile Creek, 828, 831; Fort Erie, 813; Fort Niagara, 813; Fort Oswego, 813; Fort Schlosser, 813, Fort Schuyler, 813; Genessee County, 816, 891; Genessee Indians, 764-65; geology, 836, 839-42; geology of Niagara sheet, 843; George Lake 784; Hart Island, 849; Hollis Island, 35 (14); Hudson River, 469,

826, 842, 1088; Huntington Bay, 833; Indians, 49, 816-19; Long Island, 20 (6), 35 (14), 835, 851, Gay Head to Cape Henlopen, DE, 773, Riverhead tract, 35 (20), roads, 780; Mohawk River, 842; New York City. *See:* New York City, New York, following New York (state); Niagara Falls, 841, canal, 828; Neversink River, 766; NY-NJ New York sheet, 781; Newborough, to Albany, NY, 787-794; Onondaga Lake, 32 (30); Ontario County, 765, 815; Oyster Bay, 832; Plattsburg siege, 824; Poughkeepsie, 725-29, to Albany, NY, 804-9, to N.Y.C., NY, 798-803; railroads, 590; roads, 777, 780, 822, 826; Rochester, 577; Sachem's Head harbor, 849; St. John, 784; St. Regis, 632; sheet, NY-NJ, 781; south, 786; Steuben County, 765; 815; Stratford to Poughkeepsie, NY, 810-12; Tonnawanta Creek, 830; Tryon County, 16 (2); (west) moraines, 843

New York City, New York, 85, 576-77, 721-24, 766, 826, 844-58, 945; bay, 644, 890; bay and harbor, 848, 852-54; Central Park, 855; City Hall ground plan, 32 (2); Department of Parks, 855; from Ireland, track of ship, 5 (1); harbor, 449; Hell Gate, 850; Revolutionary War, 38 (1); to Allentown, PA, 870; to Barnstable, MA (shore line), 635; to Black Horse Pike, NJ, 871; to Bristol, PA, 868; to Brunswick, NJ, 864-5; to Cranberry, NJ, 869; to Elizabethtown, NJ, 863; to Frankford, PA, 906; to Kingston, NJ, 866; to Mt. Holly, NJ, 872; to Phila., PA, 873-74, 907, 911; to Poughkeepsie, 798-803; to Stratford, CT, 795-797; to Trenton, NJ, 867

New-York Historical Society, 1236; *Quarterly,* 372

New York, Philadelphia, and Norfolk Rail Road, 962-63

New York State Museum and Education Department, 633, 836; donor, 840

New Zealand, 35 (16)

Newark, NJ, 890

Newberry Library, Chicago, IL, 551

Newbold, NJ, 887

Newborough, NY, to Albany, NY, 787-94

Newburyport harbor, MA, 692

Newcastle, England, 123, 125

Newcastle and Hanover, VA, to Annapolis, MD, 1100

Newfoundland, Canada, 550, 627, 1661, 1749; fish drying, 89

Newnan, John, cartographer, 36 (1); collection, 36

Newton, Henry (1845-1877, geologist), cartographer, 1394; *Report,* 1394

Niagara Falls, NY, 577, 630, 841; canals, 828

Niagara River, 768, 816, 819, 828, 830-31, 834, 837, 1276, 1279

Niagara sheet, geology, 843

Niantilic harbor, Baffin Bay, Cumberland Sound, Canada, 6 (1), 32 (85)

Nicholls, Sutton (fl. 1680-1740, draughtsman, engraver), engraver, 149

Nichols, Beach, cartographer, 1701

Nichols, D. A., cartographer, 465

Nichols, James (1785-1861, British printer, theological writer), 139

Nicholson, John P., cartographer, 969, 971

Nicholson, W. L. (fl. 1863-1880, topographer), cartographer, 777, 1130

Nicobar Islands, India, 182

Nicollet, Joseph Nicolas (1786-1843, traveller, cartographer, APS 1842), cartographer, 1277, 1758; *Hydrographical basin of the Upper Mississippi River,* 1758

Nikitin, S., cartographer, 319; donor, 319

Nile River, 3 (1-3), 428-29

Nimrud, Assyria, 374

Ninevah, Assyria, 374

Nin, Masbate Island, Philippines, 408

Nissa(?), 29 (30)

Niter cave, KY, 2 (1-2), (5)

Noble, A., 46 (20)

Noetzel, G., cartographer, 84

Noguera, J., cartographer, 338, 408, 416; engraver, 268a, 335-37, 409-11, 416, 441

Nolan, James Bennett, donor, 119

Nolin, Jean Baptiste (1686-1762, French engraver, author), cartographer, 256

Nordenskiöld, Nils Adolf Erik (1832-1901), cartographer, 1716; *Facsimile-atlas to the early history of cartography,* 1716

Norfolk Canal, 33 (7)

Norfolk harbor, VA, 644, 1137

Norie, John William (1722-1843, hydrographer, publisher, teacher of navigation), cartographer, 439; publisher, 439

Norman, John (fl. 1787-1795, publisher), engraver, 1269; publisher, 553

Norman, G. W. H., 465

Normand, ——, cartographer, 1613

Norris, Joseph Parker (1763-1841, Philadelphia banker), donor, 751

Norris Peters Lith. Co. *See:* Peters, Norris, Lith. Co.

Norske Videnskaps-Akademii, Oslo, 165
North America, 32 (7), 74a, 88-91, 449-70, 559-60, 574, 1621, 1627, 1645, 1685, 1756; geology, 466; Indian languages, 464, 468; railroads, 1703
North American atlas, William Faden, 499, 1013
North American pilot, 762
North Anna, VA, battlefield, 774
North Carolina, 455, 758, 1081-84, 1093-94, 1135, 1145-56, 1661; Albemarle Sound, 32 (91); Belle Point shoal, 1151; Cape Fear Harbor, 1145; Cape Hatteras, to Boston harbor, MA, 750; Core sound, 1151; Drum shoal, 1151; Edenton, 32 (91); Geological and Economic Survey, 1154; Goldsboro, 1135; Great Smoky Mountains National Park, 1156; Halifax, 32 (91); Harbor Island bar, 1151; Pasquotank River, 1153; Piney Point shoal, 1151; Roanoke Island, 32 (91); Salisbury, 1135; VA, boundary, 32 (91)
North Dakota, 1389-90
North Goldfield, NV, 1458-59
North Keeling Islands, 1620
North Mountain, NJ, 859
North Pole, 69, 89, 92
North West Territory, Canada, 545; Akudnirn, 487; Boothia Isthmus, 487; Eclipse Sound, 487; Frobisher Bay, 487; King William Land, 487; Lyon Inlet, 487; Oqo, 487; Repulse Bay, 487
Northampton, County, PA, 20 (8), 32 (2), (75-76); Historical and Genealogical Society, "The Penn Patents in the Forks of the Delaware," 899; Lenni-Lenape names, 20 (8)
Northamptonshire, England, 121; Grand Junction Canal, 136; Union Canal, 136
Northeast U.S. and Canada, 626-633; boundary dispute with Great Britain, 760-60a
Northeast boundary, U.S., 631
Northern Canada, 545-49
Northern Hemisphere, 608
Northrup, William P., publisher, 1199
Northumberland County, PA, 32 (31), 958
Northwest U.S., 1461
Norway, 159-60, 164-65, 177; bathymetry, 165; Bear Island, 100; Jan Mayen Island, 100; Sptitzbergen, 100, 107
Norway House, Manitoba, Canada, magnetic variations, 522
Norwich, CT 747
Norwood, Charles J., cartographer, 1249, 1251
Norwood, John Wall (b. 1876, cartographer, geologist), cartographer, 1249, 1252
"A Note on the William Penn Map of Pennsylvania," 975
"Note to accompany Maps . . . ," George M. Dawson, Geological and Natural History Survey of Canada, *Annual Report,* 537
Notes on Panama, H. C. Hale, 1514-15
Notes on the state of Virginia, Thomas Jefferson, 32 (19), 759
Nova Scotia, Canada, 30, 455, 488, 492-96, 550, 627, 761, 1658; geology, 494; Whiteburn Gold District, Queen County, 496
Novara (ship), 1634
Novaya Zemlya, U.S.S.R., 100
Nueva Galicia, Mexico, 32 (15)
Nyström, E. T., cartographer, 395

Oak Street, Phila., PA, 56 (13)
Oakford, Joshua L., donor, 388, 396-97
Oaxaca, Mexico, 32 (16), (18); intendency, 32 (16)
Oberholtzer, Ellis Paxton (1868-1936, author), *Philadelphia, a history of the city and its people,* 968; *Robert Morris, patriot and financier,* 816, 819
"Observations on the Annual Passage of Herrings," John Gilpin, APS, *Transactions,* 1623
Oceana, 117
Oceans, 1760, 1766; hydrography, 1615; world, 1611-15, bathymetry, 1613-14, currents and winds, 1611-15
Ockerman, J. A., cartographer, 1339
Ockerson, C. L., cartographer, 1336
Ocoa Bay, St. Domingue, 32 (36), 1521
Oczakov, 29 (56)
O'Donoghue, John. *See also:* Donoghue, John, cartographer, 1360
Ogden, H. G., cartographer, 717, 747, 1196-97
O'Hare, Daniel, cartographer, 1207-08, 1222, 1430-31, 1454-56, 1458, 1471, 1500
Ohio, 32 (22), 942, 1302-17; canal, 576; Cincinnati, 577, 942, 1311; Cleveland, 942; coal fields (bituminous), 32 (83); Huron River, Indian ruins, 1307; Indians, 32 (22), 752-53; Marietta, Campus Martius, 1306; Maumee Bay, 1287; meteorology, 1312; Muskingum River, Indian ruins, 1305; oil district, 32 (83); railroads, 1317; roads,

777; Sandusky Bay, 1310; Turnbull County, coal, 1314
Ohio Canal, 576
Ohio River, 32 (19), 33 (8), (45), 451, 752-53, 757-58, 828, 1273, 1302, 1335, 1711; rapids, 1304, 1308; falls, 1246
Ohio-KY, Cincinnati quadrangle, 1315
Oil, PA, 1701; WV, 32 (83), 1139, 1141, 1143
Oil district, OH, 32 (83); PA, 32 (83)
Oil fields, PA, 978
Oklahoma, 1265-68; Guthrie, 1267; Lawton, 1267; McAlester, 1267; to Muskogee, 1267; railroads, 1268
Old Dominions Publications, publisher, 1761
Old Northwest, 1269-73
Old Providence Island, Colombia, 1565
Oldham County, KY, 1249
Ollaki mines, 442
Omaha, NB, 1398
Omenak Fiord, Baffin Bay, Canada, 93
Onondaga Historical Association, 32 (67)
Onondaga Lake, NY, 32 (30)
Ontario, Canada, 488, 504-14; Algoma, 509; bands (chartered), 508; Fort Severn, 548; geology, 507; Georgian Bay, 32 (98); Kawartha Lakes, 490; Lake of Bays, 490; Mattagami, 509; Rideau Lakes, 490; Muskoga, 490; Sudbury, 509; Timiskaming, 509;
Ontario County, NY, 765, 815
Ontario, Lake, 32 (98), 752-53, 816, 819, 829-30, 1276, 1301
Ontonagon, MI, 1299; harbor, 1290
Oostanaula River, GA, 1181
Oostergoo, The Netherlands, 203
Oqo, N. W. T., Canada, 487
Orange, PA, 947
Ord, Edward Otto Cresap, cartographer, 1182
Ord, George (1781-1866, naturalist, philologist, APS 1817), 453, 499, 631
Oregon, 576, 1427, 1474-78; Crater Lake National Monument, 1477; Portland, 1475-76; to Missouri (Oregon Trail), 1426; Umpquah River to Canada, 1464; Umpquah River, to San Francisco, 1463
Organisada pela Commissão Geographica e Geologica, Brazil, 1573
Origin of species, Charles R. Darwin, 11
Oriskany formation, PA, 985
Orizaba, Mexico, 32 (18), (71); and Xalapa to Vera Cruz, 32 (71)
Orkney Islands, Scotland, 148, 1662
Orogeny, Venezuela, 1569
Orography, Canada, 473; U.S., 473
Oronto Bay, MI, 1299
Orr, William S., & Co., publisher, 1698
Orsova (Fort St. Elisabeth), Rumania, 29 (53-54)
Ortelius (Ortel, Ortell, Oertel, Wortels), Abraham (1527-1598, cartographer, publisher), 1636; *Theatrum Orbis Terrarum,* 1636
Orten, Switzerland, 228
Ørum, Denmark, 180
Oshkosh, WI, 1372
O'Sullivan, Owen, cartographer, 548
Ottawa Indians, 901
Oude, India, 55, 399
Ourand, Charles H., cartographer, 381, 1516; engraver, 417
Ouray County, CO, 1419
Outaovacs (Society of Jesus), 32 (74)
Outer Island, WI, 1299
Outlines of the geology of Japan, Geological Survey of Japan, 382-386
Overyssel, The Netherlands, 195
Owen, F. W., cartographer, 439
Oxford House, Manitoba, Canada, 523; magnetic variations, 523
Oxford University Press, publisher, 1731
Oyster Bay, NY, 832
Ozaizack, 29 (35)

Pacasmayo province, Peru, 1570
Pachitea River, Peru, 1570
Pachuca, Mexico, 1507
Pacific Coast, U.S., 1462-65; lighthouses, 1465
Pacific Ocean, 380, 456, 560, 566, 1616-18, 1688, 1755; bathymetry, 1612; north, 1497; south, 102; south, bathymetry, 102
Packard, Francis J., 1030
Page, T., publisher, 1650, 1661-62
Page, Thomas Jefferson (1808-1899, naval officer), cartographer, 1563-64; donor, 1563-64
Paguenard, Edward, cartographer, 934
Paige, James A., cartographer, 1336
Paita to Eten on the Alto Marañon River, Peru, 1570
Palancka von Czazack, ——, 29 (37)
Palanka Krujowatz, 29 (33)
Palatine, Germany, 290
Palaullo [Pataullo], Nicholas, donor, 1553
Palestine, 82, 324

Palfrey, Carl F., cartographer, 1336
Paliser, John, cartographer, 1461
Pallavona, GA, *See:* Jekyl, GA
Palmer, ——, publisher, 565, 676, 1071, 1086
Palmer, John T., Co., publisher, 970
Palmer, Thomas H. (1782-1861, educator, publisher), publisher, 316
Palmer, William Ricketts (1809-1862, topographical engineer, APS 1859), cartographer, 631, 679, 1127; donor, 853
Panamá, 1502, 1513-16b, 1562; Darien del Norte, golfo de, 1514; Gatun dam site, 1516b; Indians, 35 (13); Panama Canal, 608, 1510, 1516, 1516a-b; Portobelo, 1512; Uraba, golfo de, 1514
Panay Island, Philippines, 417; Batan, 409; geology, 413
Pancevo, Yugoslavia, 29 (47)
Pando, José Manuel, 1571
Panzowa (Pancevo), Yugoslavakia, 29 (47)
Papenfuse, Edward C., 1764, 1767; and Joseph M. Cole III, *The Hammond-Harwood House atlas of historical maps of Maryland,* 1764
Papers relating to an act of New York for encouragement of the Indian trade, Cadwallader Colden, 783
Para, Brazil, 1562
Paraguay, 23 (5)
Paramapanema, Brazil, 1562
Parham, ——, 46 (2)
Paris, France, 247, 256, 263-65, 1643, 1743; Académie Royale des Sciences, 1374; Isle St. Louis, 13 (3); Revolution (1789-1794), 273-74
Park, C. F., cartographer, 1419
Parke, John G., cartographer, 1461
Parker, Daniel, Jr., donor, 677, 770, 845, 987, 1090
Parker, Ely Samuel (1828-1895, soldier), 987
Parker, Richard Elliot (1783-1840, soldier, statesman, jurist), donor, 1090
Parkinson, T. D., cartographer, 1769
Parraud, M., translator, 1237
Parrish, H. E., cartographer, 958
Parsons, Elsie Clews (1875-1941, anthropologist, folklorist), cartographer, 37; collection, 37
Parsons family, donor, 37
Pasquotank River, NC, 1153
Passaic valley, NJ, 889
Passamaquoddy Bay, New Brunswick, Canada, 493
Passyunk Road, Phila., PA, 46 (13)
Pastaza River, Peru, 1570
Patapsa River, MD, 644, 1075
Pataschin, 29 (22)
Paterson, NJ, 890
Patos Island, 41 (3)
Patrick, Mason Matthews (b. 1863, soldier), cartographer, 1299, 1338-39, 1342
Patten, Richard, cartographer, 1187
Patterson, Carlisle Pollock (1816-1881, hydrographer), cartographer, 1203, 1209, 1333
Patterson, Col., cartographer, 45 (5)
Patterson, Rev. Robert, donor, 822, 925
Patterson, Robert Maskell (1787-1854, scientist, director of U.S. Mint), donor, 1692
Paucartambo valleys, Peru, 1570
Paullin, Charles Oscar, cartographer, 1735; *Atlas of the historical geography of the U.S.,* 1735
Paxton, John Adams, 1005; cartographer, 1004; *Paxton's strangers guide,* 1004
Payne, John (fl. 1792-1800, author), *A new and complete system of universal geography,* 78a-b, 262, 323, 369, 407, 426, 459, 558, 640, 656, 665, 674, 715, 739, 817, 877, 918, 1069, 1119, 1149, 1160, 1176, 1244, 1256, 1546, 1559, 1616
Payne, M., publisher, 119
Paz, Manuel Maria (fl. 1864-1889, Colombian geographer, cartographer), cartographer, 1514
Peace River, Alberta, Canada, 531, 534, 536
Peach Tree Creek, GA, 1180
Peak Mountain, VA, 32 (80)
Peale, Albert Charles, cartographer, 1425
Peale, Charles Willson (1741-1827, artist, naturalist, APS 1786), cartographer, 38 (1-2), 42, 43; collection, 38
Peary, Robert Edwin (1856-1920, admiral and explorer), 97, 99
Peck, C. S., cartographer, 1701
Peck, James Mason (fl. 1835-1853, author), cartographer, 1324
Pedersen, Peder (Danish consul at Philadelphia, APS 1822), donor, 166-174, 176, 178-81, 183-85, 300
Peebleshire, Scotland, 7 (2)
Peel, The Netherlands, 204
Peirce, Benjamin (1809-1880, geologist, APS 1842), cartographer, 717, 1497, 1182
Peirce, John, cartographer, 671
Peking, China. *See:* Beijing, China
Pelegrad, 29 (16)
Pelham, Henry (fl. 1777-1787, British car-

tographer), cartographer, 699
Pendleton, John B. (1798-1866, engraver), engraver, 680; lithographer, 704; publisher, 704
Penn, John (1729-1795, Lt.-Gov. of Pennsylvania), 903
Penn, Richard (1735-1811, Lt.-Gov. of Pennsylvania), 900, 903
Penn, Thomas (1702-1775, Proprietor of Pennsylvania), 900, 903
Penn, William (1644-1718, English Quaker, founder of Pennsylvania), 32 (89), 981, 1033; cartographer, 975; manors, PA, 974; treaty with the Indians, 579, 936
"The Penn Patents in the Forks of the Delaware," A. D. Chidsey, Northampton County Historical and Genealogical Society, 899
Pennell, A., cartographer, 1551
Pennell, L., cartographer, 1551
Penner, G. F., engraver, 1299
Pennine fault, England, 7 (1)
Pennsbury Manor, PA, 981
Pennsylvania, 32 (1), (4), (19), (22-3), (31), (67), (77), (95), 455, 469, 627, 751-56, 758-61, 763, 767, 769-70, 774-77, 779-80, 782, 895-986, 1053, 1094, 1303; Adams County, 927, 955; Allegany County, 32 (31); Allentown, to N.Y.C., NY, 870; Armstrong County, Kittanning, 919; Bedford County, geology, 954, 766; Berks County, geology, 959; Birdsboro quadrangle, 977; Bethlehem, 32 (1); boundary, 902; Boyertown quadrangle, 982; Bristol, to N.Y.C., NY, 868; Broad Top coal field, 966; Bucks County, 15 (4), 897; Bushy Run, battle, 901; Cambria County, 937; canals, 767, 913, 935-36, 945; Catawissa, 32 (95), 986; Central valley of the Susquehanna River, 12 (2); Cheat River, 45 (3); Chester County 897, Darby Creek, 43 (1); Clearfield County, 937; Centre County, 937; coal, 937, 948, 952, 961, 965; coal, (bituminous), 32 (83); Cobb Creek, 43 (5); Columbia County, 958; Columbia Railroad, 576; Conestoga navigation, 939; county maps, 923; Danville, 32 (95), 986; Danville and Pottsville Rail Road, 932; Darby, 43 (4); Darby Creek to Young's Ford on the Schuylkill, 32 (68); DE-Coatesville quadrangle, 973; Delaware County, 1709, Garrett Road, 43 (2), Marshall Road, 43 (2), West Chester Road, 43 (2), (5); DE-West Chester quadrangle, 973; DE-NJ Marcus Hook quadrangle, 782; DE-NJ quadrangle, 973; Department of the Highways, 974; Department of Internal Affairs, publisher, 967; Fayette County, 933; Fell, 947; Fort Allen, 33 (9); Fort Augusta, 45 (1); Fort Burd, 45 (3); Fort Pitt, 12 (1); Fort Shirley, 33 (4); Frankford to N.Y.C., NY, 906; Franklin court, Philadelphia, 15 (1), (2); Franklin County, 955; Franklin Twp., 16 (1); Fulton County, geology, 966; gas fields, 978; Geological Survey, 959; 978, 985; Geological Survey, Second, 30, 955, 958, 960, *Report of Progress,* 958-60; geology, 928, 946, 960-61, 984-85, 1712; Gettysburg, 775, battlefield, 950, 956, 970, hospitals, 969; Gettysburg National Military Park, 971; Gnadenhütten, 33 (9); Great Chester Valley, 1072; Harrisburg, 936, 942; Huntingdon County, geology, 960, 966, Juniata valley, 960; Indians, 32 (22), (49), Inland navigation, 912-13; Kittanning (Indian town), 33 (4); Lancaster County, 929, Nabron Twp., 33 (3); Lansdowne quadrangle, 980; Lehigh and Schuylkill coal regions, 936; Lehigh River, 32 (75-76); Lizzard Creek, 32 (75-76); Luzerne County, 958; Franklin Twp., 947; McCalls ferry quadrangle, 973; Mahoning Creek, 32 (1); Mahony Shamokin coal basin, 958; Mason and Dixon Line, 973; Media quadrangle, 979; Monongahela House, Pittsburgh, 12 (1), 45 (3); Moravian road, 899; Montgomery County, 33 (10), 983, Commissioners, 974; Muddy River, 45 (3); NJ-Bridgeport quadrangle, 782; NJ-Phila. quadrangle, 973; New Philadelphia, Schuylkill County, 940; Northampton County, 20 (8), 32 (2), (31), (75-76), 958; Lenni Lenape names, 20 (8); oil 32 (83), 978, 1701; Orange, 947; Oriskany formation, 985; Perkiomen, 33 (10); Pennsbury Manor, 981; Philadelphia. *See:* Philadelphia, PA; Philadelphia County. *See:* Philadelphia County, PA; Philipsburg estate, 937; Pike County, 1681, roads, 1681; Pittsburgh, 770, 924, 936, 942, 965; English fort, 45 (4); Port Carbon, 948; Port Clinton, 931; 986; Pottstown, 52; Pottsville, 32 (75-76), 948; railroads, 32 (75-76), 576, 779, 932, 935, 943, 945, 949, 952, 962-65, 967, 972, 1072; roads, 767, 769, 777, 780, 899, 912-13, 939; Quarryville quadrangle, 973; Schuylkill canal, 32, (75-76); Schuylkill County, 32 (75-76),

958, Morea coal lands, 961; Schuylkill River, 32 (77), 43 (4), 576, 931, 987, 1035, Young's Ford to Darby Creek, 32 (68); Shippensburg, 33 (4); Silver Lake, 926; Somerset County, Germany, 920; Springfield, 43 (4); statistics, 938, 965; Susquehanna County, 926; Susquehanna River, 32 (77), 45 (2), West Branch, 45 (1), (5); Topographical and Geological Survey, 966; Trinity Fort, 898; Tuscarora Valley, 960; Union Canal, 576; Venango County, 951; Wayne County, 18, 1681, roads, 1681; Weisport borough, 33 (9); Wyoming coal basin, 958; York County, 927; York Haven, 923; Young's Ford on the Schuylkill to Darby Creek, 32 (68)
Pennsylvania Hospital, Phila., PA, 990
Pennsylvania Magazine of History and Biography, 497-98, 626, 697-98, 785, 1082, 1095
The Pennsylvania Rail Road Corp., 32 (78), 941-42, 962-64; *Fourth Annual Report*, 941
Penobscot Bay, ME, 644, 646
Penobscot Indians, 48 (3), (9)
Penrose, ——, 46 (20)
Pensacola, FL 32 (5); bar, 1185, 1190-91
Pergamon Press, 1755
Perkin, Henri, cartographer, 1692; *Atlas de l'Europe*, 1692
Perkins, F. W., cartographer, 882, 885, 1020-21
Perkiomen, PA, 33 (10)
Perrault, Paul Hyacinthe (d. 1834, soldier), cartographer, 677
Persia. *See:* Iran
Persian Gulf, 112
Perthes, Johann George Justus (1749-1816, German bibliophile), 310, 1720, 1733
Peru, 1570; Alto Puros, 1570; Alto Yorua, 1570; Bajo Marañon River, 1570; coastline, 35 (5); Chincha Islands, 35 (1-2), (7); Choco Island, 35 (12); Cuzco, 1570; Guañape Islands, 35 (8); Huallago River, 1570; Indians, 89; Lima, coastal valley, 35 (19); Lobos de Afuera Island, 35 (9), (11); Loreto department, 1570; Macabi Islands, 35 (10); Manú River, 1570; Morococha, geology, 1570; Morono River, 1570; Pacasmayo province, 1570; Pachitea River, 1570; Paita to Eten on the Alto Marañon River, 1570; Pastaza River, 1570; Paucartambo valleys, 1570; Pichis River, 1570; Pisco Bay, 35 (3); Putumayo River, 1570; San Gallan Island, 35 (6); Santa Rosita Island, 35 (4); Serjali River, 1570; Tigre River, 1570; Ucayali River, 1570; Viegas Island, 35 (4)
Pesniza, 29 (32)
Peter, J. F., cartographer, 1278-79
Peterhead, Scotland, 32 (81)
Peters, Chr. H. Fr. (1813-1880, astronomer), 57
Peters, Norris, Litho. Co., publisher, 84, 282, 379, 381, 447, 586, 1154, 1206, 1454, 1514-16
Peters, W. J., cartographer, 1315
Petersburg, VA, 1125; siege, 774
Petersen, A., engraver, 96, 647, 1469
Peterwardein, 29 (44)
Pether, H. J., lithographer, 1585; publisher, 1585
Le petit atlas maritime, 1651
Petit Goave, St. Domingue, 32 (46)
Pettit, J. S., engraver, 1468
Pettit, S. T., engraver, 743-45, 848, 1153, 1203
Pfyffer von Altshofen, Franz Xaver (Christoph, 1680-1750, German general), cartographer, 234
Phelps, T. S., cartographer, 882
Philadelphia, PA, 32 (68), (89), 46 (14), 576-77, 769, 897, 936, 987-1040, 1072, 1652; American Revolution, 995-96; Board of Marine Underwriters, 586; Catharine Street, 46 (15); Cedar Street, 46 (16-17); Center Square Water Works, 1007; Chestnut Street between Sixth and Seventh Streets, 33 (2); City Hall, 1001; Congress Hall, 1001; Darby Road, 19 (1); Delaware Bay and River to Phila.; 1051; Delaware River, 46 (1), (6), (19), 1020; Delaware River chart, 644; Fairmount Park, 1022; Fairmount Park Association, 1028; Fairmount Water Works, 1005; Fifth Street, 46 (14); First U.S. Bank, 1001; Fourth Street, 46 (14); Franklin Court, 15, (1-2); Franklin, Benjamin, residence, 32 (90); Franklin, Benjamin, Parkway, 1028; Garrett Road, 43 (2); Girard College, 1019, 1032; Glenwood Cemetery, 1039; Gray's Ferry, 46 (10); Hare Street, 19 (1); Independence Hall, 449, 988, 990, 1001, 1013, 1016, 1019, 1033, 1038, 1652; Independence Square, 15 (3), 1012; Islington Lane, 1039; Laurel Hill Cemetery, 1019; League Island, Delaware River, 46 (6); Lehman Street, 19 (1); Library Company of Philadelphia, 1001; Locust Street, 997; Marshall Road, 43 (2); May Road, 46 (12);

Merchants Exchange, 1011, 1019; Modern Club, 1035; Moyamensing Road, 46 (12), NJ quadrangle, 973; Oak Street, 46 (13); Old City Area, 1036; Passyunk Road, 46 (13); Pennsylvania Hospital, 990; Philadelphia Water Works, 1003; Queen Street, 46 (7); Ridge Road, 1039; Sassafras and Fourth Streets, 33 (1); Schuylkill River, 1021, bridge, 921; Second U.S. Bank building, 1011; 17-19th centuries, 32 (96); Shippen Lane, 46 (5), (15); Shippen Street, 46 (14); Solitude, 46 (9); South Street, Shippen estate, 46 (4); Southwark, 46; Spruce Street, 997; street plan, 5 (1); to Annapolis, MD, 908-11, 1058-63; to Maryland, 1064; to N.Y.C., NY, 873-74, 907-11; to Pittsburgh, PA, Pennsylvania Rail Road, 32 (78); to Pottsville, 32 (75-76); Turner's Lane, 46 (12); voyage to Cork, Ireland, track, 1 (1); Washington Square, 1012, east, 1040; water front, 1020-21; water works, 1003; West Chester Road, 43 (2); Wissahickon valley, Fairmount Park, roads, 1034

Philadelphia, a history of the city and its people, E. P. Oberholtzer, 968

Philadelphia coal oil circular and petroleum price current, 1701

The Philadelphia directory for 1797, C. W. Stafford, 1000

Philadelphia Water Works, Phila., PA, 1003

Philip, George (b. 1799, geographer, publisher, globe maker), cartographer, 1741

Philippeaux, ——. *See:* Bois St. Lys

Philippine Islands, 408-20, 608, 965; Bohol Island, 417; Cebu Island, 412, 417; Division of Geology and Mines. *Bulletin,* 414; jurisdiction, 32 (26); Leyte Island, 417; Luzon Island, Laguimanor, 411, Manila, 418-19; Masbate Island, Nin, 408, 417, Mandao, Masbate Island, 408; Mindanao Island, geology, 420; Negros Island, 417; Panay Island, 417, Batan, 409, geology, 413; Samar Island, 417; Viscayan group, 417

Philippsburg, Germany, 296, 298

Philipsburg Estate, PA, 937

Phillips, Henry Meyers (1811-1884, librarian, APS 1877), donor, 644, 1505

Phillips, P. L., *A descriptive list of maps and views of Philadelphia,* 998

Phillips, Sir Richard (1767-1840, British author, bookseller), publisher, 557

Phillips, William B., cartographer, 1235

Philosophical Transactions, Royal Society of London, 754

Phipps, Constantine John, second baron Mulgrave (1744-1792, naval commander), 69, 74c, 80

Phoenix, AZ, 1455-56

Photographer, Alexander Garden, 1461

Photographer, F. Gutekunst, 1550

Photographic lunar atlas . . . , ed. by G. P. Kuiper, 1747

Physiographic diagram, Atlantic Ocean, 1632-33; Caribbean Sea, 1633; Celebes Sea, 1635; Indian Ocean, 1635; Red Sea, 1635; Scotia Sea, 1633; South China Sea, 1635; Sulu Sea, 1635

Physical atlas, J. G. Bartholomew, 1722

Pic du Midi observatory, 1747

Pichardo, ——, cartographer, 1721

Pichis River, Peru, 1570

Pichon, ——, cartographer, 265

Picquet, Charles (1755-1824, French agronomist, diplomat), cartographer, 263, 357; engraver, 263; publisher, 263, 1215

Picuris Pueblo, AZ, 37

Pike, Zebulon Montgomery (1779-1813, soldier, explorer), *Account of a voyage up the Mississippi River,* 1330; *Account of expeditions to the source of the Mississippi,* 1330; cartographer, 1330; collection, 39

Pike County, PA, 1681; roads, 1681

Pike's Peak, CO, 29, 1411

Pilarski, Casimir von V., cartographer, 280

Pillans, P. J., cartographer, 1225, 1228

Pinchot, Gifford (1865-1946), forester, conservationist, governor of PA), 608

Pindamonhangaba, Brazil, 1572

Pine, T., engraver, 10 (1)

Pine Mountain, GA, 1181

Piney Point shoal, NC, 1151

Pingeling, T. A. (fl. 1776-1798, German engraver), engraver, 1545

Pinkerton, John, cartographer, 1683; *General collection of voyages and travels,* 1683; *A modern atlas,* 1683

Pintard, John (1759-1844. Merchant, philanthropist), 1236

Piraçununga, Brazil, 1572

Piri Reis, cartographer, 1738; *Haritası* [Piri Reis's Map], 1730

Pisco Bay, Peru, 35 (3)

Pitrat ainé, publisher, 272

Pitt, William, first earl of Chatham (1708-1778, statesman), 129

Pittman, Edward Fisher (b. 1849, Australian

geologist), cartographer, 1608-9
Pittsburgh, PA, 770, 924, 936, 942, 965; English fort, 45 (4); Monongahela House, 12 (1); to Phila., PA, Pennsylvania Rail Road, 32 (78)
Place of Arms or Esplanade, New Orleans, LA, 1216
Plainfield, NJ, 890
Plains of Abraham, Quebec, Canada, 44 (1)
[Plan de Paris], Louis Bretez, 1643
Plan of the environs of Philadelphia [copper plate], 1659
Plan of the London dock, William Vaughan, 132
Planet, Mars, 50 (1-2); Saturn, 50 (3)
Platt, R., cartographer, 1197
Plattsburg, NY, siege, 824
The plea of the Colonies brought against them by Lord Mansfield, 1621
Plowman, Thomas L., donor, 1504; publisher, 563
Pluckemin, NJ, 892
Plymouth, England, 123, 125
Pocatello, ID, 1447-48
Poe, Orlando Metcalf (1832-1895, soldier), cartographer, 1260, 1282
Poggiali, Giuseppe, cartographer, 346; engraver, 346
Point Adams, 28 (8)
Point Pinos to Bodega Head, CA, 1487
Poizat, Charles A., publisher, 1550
Poland, 307-11; hydrography, 310
Polaris (ship), 32 (85)
Pollock, William Wilson, cartographer, 32 (89)
Polynesia, 1619-20
Ponce de Leon, Manuel (Colombian geographer), 1514
Pontchartrain, Lake, LA, 1220
Pontine marshes, Italy, 1686
Ponza Island, Italy, 354
Poole, Charles Henry(?), cartographer, 777
Popple, Henry (d. 1743, British geographer, Clerk to the Board of Trade, Cashier to Queen Anne), cartographer, 449
Porpoise (ship), 1629
Port Alécu, St. Domingue, 32 (59)
Port Antonio, Jamaica, 1529
Port au Prince, St. Domingue, 32 (26), (43), 1524
Port Carbon, PA, 948
Port Clinton, PA, 931, 986
Port Cumberland, MD, western, 1067
Port Deposit, MD, 1072
Port Français, St. Domingue, 32 (51)
Port Paix roads, Haiti, 1522
Port Royal, Jamaica, 89, 1530, 1533, 1542
Port Royal, VA, to Williamsburg, VA, 1114-15
Port Townsend, WA, 85
Porter, D. D., cartographer, 850
Portland, ME, 650; harbor, 644, 647
Portland, OR, 1475-76
Porto-Ferrajo, Elba, Italy, 357
Porto Rico, U.S., 32 (28), 622, 965, 1520, 1534, 1661, 1668; San Juan, 32 (63)
Porto Santo Island, Portugal, 328
Portobelo, Panamá, 1512
Portsmouth, England, 123, 125
Portugal, 326- 341; Azores Islands, 422; Dezertas Islands, 328; geology, 339; hydrography, 328; hypsometry, 340; Madeira Islands, 328; Funchal, 328; Porto Santo Island, 328
Post roads. *See:* Roads
Potash from kelp, 1478
Potomac River, 1041; from Cumberland, MD, 1069
Pottowottomee, Lake, MI, 1349
Pottstown, PA, 52
Pottsville, PA, 32 (75-76), 948; to Phila., PA, 32 (75-76)
Poughkeepsie, NY, 725-27; to Albany, NY, 804-9; to N.Y.C., NY, 798-803; to Stratford, NY, 810-12
Poupard, James (fl. 1772-1814), engraver, 1623
Poussin, William Tell Lavallée (1794-1876, soldier, cartographer), cartographer, 769; donor, 1192
Powell, John Wesley (1834-1902, geologist, philosopher, administrator, APS 1889), cartographer, 600-602, 781, 888, 1394, 1449-50, 1484, 1708
Powers, Stephen, cartographer, 1484
Pownall, Thomas (1722-1805, colonial governor), cartographer, 88, 90, 453, 627, 756-57
Pratt, Joseph Hyde, cartographer, 1154-55
Précis historique du canal du Languedoc ou des Deux Mers, Louis Lacoste, 252
Presidio County, TX, 1234
Presqu'ile, MI, Lake Huron, 1292
Preston, Capt., 1259
Prestwich, Sir Joseph (1812-1896, British geologist), 114; *Geology, chemical, physical & stratigraphical,* 114
Preuss, Charles (1803-1854, explorer, cartographer), cartographer, 1426

Price, ——, cartographer, 1148
Price, J. Sergeant, 115
Price, James Warwick, 1035
Price, Jonathan, cartographer, 1150
Price, Paul H., cartographer, 1144
Priestley, Joseph, 139
Priestman, Charles, 1070
Priestman, William, 68-69, 88, 108, 124, 146, 150, 159, 175, 188-89, 190-91, 244, 249-50, 258, 286-87, 291, 293, 307, 312, 325, 327, 345, 347, 359-60, 363, 399, 401, 404, 423, 434-35, 552, 1535
Prime, F., cartographer, 959
Prinald, ——, engraver, 1509, 1511, 1522-30, 1533, 1565
Prince Edward Island, Canada, 485, 488
Princeton University Press, 1757
Prior, ——, publisher, 500
Prior & Dunning, publishers, 500
Pritchett, Henry S., cartographer, 1723
Probst, Johann Michael (d. 1809, German publisher, engraver), publisher, 232
Probst, J., engraver, 939; lithographer, 939
Proceedings, APS
Proceedings, Colorado Scientific Society, 1418-19
Proctor, John Robert (1844-1903, publisher), cartographer, 1248-49
Proselite (ship), 32 (65)
Proud, Robert (1728-1813, historian, educator, APS original member), donor, 763; *History of Pennsylvania*, 763
Providence, RI, 718
Provincetown, MA, 679
Prussia, 289, 306, 308-309
Pryor, N., 46 (8)
Ptolomaeus (Ptolomaei, Claudius, ca. 90-168 A.D., Egyptian astronomer, geographer), 62
Publication, Carnegie Institution of Washington, 1725, 1735
Publication, Queensland Geological Survey, 1594
Publishers: See: Abbeville Press Publishers; Acheson, H.; Aitken, Robert; Alexander, James; Allen, ——; Almon, J.; American Bank Note Co.; American Geographical Society of New York; APS; American Photo-lith Co.; Andrews, ——; Andrews, A. B. Co.; Arrowsmith, John; Ashford, ——; Ashford, Kent; Harry Margery, Lympne Castle; Ashmead, Henry B.; Augustin, J. J. Inc.; Bachelder, John D.; Baldwin, R.; Barnes, R. L.; Barre, ——; Bartholomew, John & Co.; Beers, F. W.; Benner H.; Bennett, J.; Bien, Julius, & Co.; Bien, Julius, Litho Co.; Black, A. C.; Black, Barbury & Allen; Blanchford, R.; Blockwood, William & Sons; Bourne, A.; Bourquin, F.; Bouvé, E. W.; Bowen, J. M.; Bowles, Carrington; Bowles, John; Bowles, T.; Bowles, T., Robert Sayer and John Bowles; Bowles, Carrington & Robert Wilkinson; Bradford, ——; Bradford & Inskip; Brett, Lithograph Co.; Bradley, William M. & Brother; Bridgman, E. C.; Britton, ——; Britton & Rey; Brookes, Vincent; Brown, ——; Buckingham, ——; Buisson, ——; The Bullard Co.; Bunce, ——; Burland, ——; Campbell, ——; Canada Bank Note Co.; Carey, Mathew; Carey, M. & Son; Carey, H. C.; Carey, H. C., and I. Lea; Carletti, Nicolaus; Carmelo y Bauermann; Carnegie Institute of Washington, D. C.; Carrington, Bowles; Cary, J.; The Century Co.; Chambers, ——; Childs, C. G.; Claredon Press; Clarke, S. J. Publishing Co.; Coggins, E. H.; Colburn, Henry; Collier, P. F. & Son Corp.; Colton, ——; Colton, C. B.; Colton, G. W.; Colton, J. H. & Co.; Columbia Planograph Co.; Columbus Lithograph Co.; Conrad, ——; Conrad, John; Cook, James; Corporation Commission; Covens, ——; Covens, J. Jr.; Covens et Mortier; Covens & Mortiere; Cowperthwait, Thomas; Cram, George F.; Cramoisy, ——; Cumming, Anthony J.; Cummings, ——; Cummings & Hilliard; Cundee, James; Darby, William; Darby, William, & B. Tanner; Davies, ——; Davis, ——; Davis, John; Davis & Dickson; Davis, M. L. & W. A.; Day & Son; De Silver, Charles; Desray, ——; Desray & J. Goujon; Detloff, C.; Devlet Basımevi; Dewing, Francis; Dezauche, ——; Dickson, ——; Didot l'Ainé; Dilly, C.; Dilly, C. & G. Robinson; Dirección de Hidrografia, Madrid; Dobson, Thomas; Dobson, Thomas & Son; Dodsley, R.; Doolittle, Amos; Dunbar, Samuel O.; Dury, Andrew; Duval, P. S.; Dybwad, Jacob; Eckert Lithograph Co.; Ehrgott, ——; Ehrgott & Krebs; Ellis, A. D.; Encyclopaedia Britannica; Endicott & Co.; Erhard Fréres; Engelmann, ——; Evans, Lewis; Faden, William; Fenner, R.; Finley, Anthony; Firmin Didot; Fisher, Thomas; Fisher, William; Fores, ——; Franklin, Benjamin; Friedenwald, Isaac;

Friedenwald Co.; Gates, ——; Gates and Seaton; Gauthier-Villars, ——; Gedney, J. F.; Geographical Institute Geographical Publishing Co.; Geographical Society of Philadelphia; Geological Society of America; Geological Society of London; Ghys, Martin; Girard, Anthony; Goodrich, A. T., & Co.; Goos, Pedro; Gordon, Thomas; Graham, ——; Graham, Andrew B.; Graham, Andrew B., & Co.; Graham, Curtis B. Graphic Co.; Gray, ——; Great Britain; Hydrographic Dept. of the Admiralty, London; Green, ——; Green & Thornton; Gregory, Edmund; Gullick, W. A.; Hales, John G.; Hall, Harrison; Hanckel & Son; Hardy, ——; Harper, ——; Harrison, C. P.; Harrison, John; Harrison, Samuel; Hart, Lane S.; Harvard University Press; Hatch & Co.; Heintz, ——; Heliotype Printing Co.; Hérissont, Veuve; Heymann, Ignazio; Hilliard, ——; Hinman, ——; Hinton, John; Hodge, ——; Hodge, Allen and Campbell; Hoen, A., & Co.; Holmes, ——; Holmes & Meier; Homann, John Baptist; Hooper, ——; Hough, B.; Howe, C. E., Co.; Hunter, Thomas; Hutchins, Thomas; Hyde, E. Belcher; Imperial Oil Co.; Inman, ——; Inskip, Abraham H.; Inskip, Abraham H., & J. Maxwell; Institut National De Géographie, Brussels; Jäck, Carl; James, ——; James & Johnson; Jean, ——; Jefferys, Thomas; Jefferys & Faden; Jewett, J. P., & Co.; The John Hopkins Press; Johnson, ——; Johnson, James; Johnson and Warner; Johnston, Alexander Keith; Johnston, W.; Johnston, W., & A. K.; K. K. Hof- u. Staats-Druckerei; Kearny, ——; Kereven, Patrick; Kervand, J. L.; Kimber, Emmor; Kimber and Conrad; King, Thomas; Kneass, W.; Komarzewsky, ——; Krebbs, ——; Krebs, Charles G.; Land of Lakes; Laurie, Robert; Laurie, Robert, & James Whittle; Lay, Amos; Lay, A., & J. Webster; Lea, I.; Lemercier et Cie; Le Rouge, ——; Lesoüef, A.; Levrault, ——; Lloyd, H. H. & Co.; Lloyd, J. T.; Long, ——; Longworth, D.; Lucas, Fielding, Jr.; Lurlanetto, Lodovico; Lyon, James B.; McElroy, ——; McElroy and P. S. Duval; McMillan, ——; Malvaux, J.; Man, T.; Margery, Harry; Marshall, R.; Marshall, T.; Matthews-Northrup Works; Maxwell, John; Melish, John; Melish, John, & Samuel Harrison; Mendel, Edward; Meriden Gravure Co.; Merrymount Press; Mesier, P. A.; Meier, ——; Miller, ——; Missionary Education Movement of the U.S. and Canada; Mitchell, Samuel Augustus; Mitchell & Hinman; Moithey, ——; Monrocq, ——; Moore, ——; Mortier, Pierre; Mortimer, ——; Mount, J.; Mount, W.; Mount, J., & T. Page; Mount, W., & T. Page; Mowbray, G.; Mutlow, ——; Munsell, Luke; Narstin, John L.; National Map Co.; Nevil, James; New Jersey Historical Society; Nichols, ——; Norie, J. W.; Northrup, William P.; Old Dominion Publication; Orr, William S., & Co.; Oxford University Press; Page, T.; Palmer, ——; Palmer, John T., Co.; Palmer, T. H.; Payne, M.; Pendleton, ——; Pennsylvania Dept. of Public Affairs; Pergamon Press; Perthes, Justus; Peters, Norris Co; Pether, H. J.; Phillips, R.; Picquet, Charles; Pitrat ainé; Plowman, T. L.; Poizzat, Charles A.; Princeton University Press; Probst, Michael; Queensland, Australia. Government Engraving and Lithographic office; Rakestraw, Joseph; Rand, McNally & Co.; Randegger, ——; Ranney, C.; Real Litografia Milinte, Naples; Rehn, J.; Reid, John; Reimer, Dietrich; Reitzels, C. A. Forlag; Rey, Marc Michel; Rinehart, ——; Robertson, W. A.; Robinson, G.; Robinson, G. G.; Robinson, G. G., & J.; Robinson, J.; Rochette, L. de la; Rook, T.; Rose, Robert R.; Routledge, George; Sage, J. & Sons; Savage, E.; Sayer, John; Sayer, Robert; Sayer, Robert & Co.; Sayer, R., & J. Bennett; Sayer, Robert, and Thomas Jefferys; Schedler, J.; Schiffer Publishing Co.; Schmitz, M.; Schmitz, M., & P. Kereven; Schreiber, Adam Gottlieb; Schreiber, Adam Gottlieb, u. Weigel; Schropp, Simon & Co.; Scott, ——; Scott & Moore; Seaton, ——; Seller, John; Seller & Fisher; Selves Fils; Senex, John; Senex & Maxwell; Service Géographique de l'Indochine; Sherman, ——; Sherman & Smith; Smith, ——; Smith, J. L.; Smith and Stroup; Smith & Wistar; Smith, Reid & Wayland; Society for the Diffusion of Useful Knowledge; Soule, G. G.; Sowle, Andrew; Speer, Joseph Smith; Spencer & Co.; Small, D.; Stanford, Edward; Stedman, ——; Stedman, Brown & Lyons; Stewart, J., & Co.; Stroup, ——; Sullivan, John; Tallis, F.; Tallis, J.; Tallis, J. & F.; Tanner, Benjamin;

Tanner, Henry Schenk; Tanner, Vallance, Kearny & Co.; Tardieu, P. A. F.; Thomas, Isaiah; Thomas & Andrews; Thomas, Cowperthwait, & Co.; Thornton, John; Thornton, John, and John Seller; Tichnor, William D.; Tiernan, ——; Tilgmann, ——; Tiller, S.; Times Publishing Co.; Toronto Lithograph Co., Ltd.; Touring Club Italiano; Tracy, Gibbs & Co.; Treuttel, ——; Treuttel et Würtz; Union Lithograph Co.; U. S. Army Map Service; U.S. Geological Survey; U.S. Government Printing Office; U.S. Hydrographic Office; U.S. Weather Bureau; The University of Chicago Press; University of Nebraska Press; University of Virginia Press; Vallance, J.; Vaughan, A. Lithograph; Vaughan, George Arthur; Wagner, W.; Walker, C.; Walker, George H.; Walker, John; Walker, John, & Co.; Wallis, James; Wankijfs, S.; Warner, ——; Washington College; Wayland, ——; Weber, E. & Co.; Webster, J.; Weigel, ——; Weigel-Schneider, Chr.; Weller, E.; Weller, E., & Grahams Lith.; Whittle, James; Wiley, ——; Wiley & Long; Wilhelms, ——; Wilkinson, Robert; Wilkinson, Robert, and Robert Sayer; Williams, ——; Williams, W.; Williams & Heintz Map Corp. Wilson, ——; Wistar, ——; Wit, Frederick de; Woodman, ——; Woodman & Mutlow; Woodward, ——; Woodward & Tiernan; World Syndicate Publishing Co.; Würtz, ——; Wuhrer, L.; Wurster, ——; Wurster, Randegger; Wyld, James, & Son; Wyman & Sons; Wynkoop, ——; Yoshii, G.; Young, J. H.; Zingg, A.
Puebla, Mexico, 32 (18), 1507
Pueblo Indian Journal, E. C. Parsons, 37
Pueblo Indian Religion, E. C. Parsons, 37
Pueblo, CO, 39
Puerto Rico. *See:* Porto Rico
Puget Sound, WA, 644, 1469-70
Puig, Victor, cartographer, 1571
Puissant, Louis (1769-1843, French geodosist), 357
Purcell, Henry D. (fl. 1774-1784, engraver), engraver, 1236
Purcell, Joseph, cartographer, 1083
Pursh, Frederick (1774-1820, botanist, horticulturalist, explorer), cartographer, 32 (67)
Putnam, Robert, *Early sea charts,* 1766
Putumayo River, Peru, 1570
Quarryville, PA, quadrangle, 973
Quarterly, New-York Historical Society, 372
Quebec, Canada, 44 (1), 485, 488, 497-503, 627, 756; banks (chartered), 508; battle of Montmorency, 44 (1); Gaspé Peninsula, 508; Magdalene Island, 633; Montreal, 32 (84), 490, 498, 502, 780, 1298; Plains of Abraham, 44 (1); Quebec (city), 490, 497, 501-2, 831; Ungava Bay, 48 (15)
Queen Street, Phila., PA, 46 (7)
Queens County, Nova Scotia, Whiteburn gold district, 496
Queensland, Australia, 1586-1605; Bowden, 1600; Cape York, 1588, 1600; Charters Towers gold field, 1586; coal, 1599, 1604-5; copper mining, 1595; Geological Survey, 1586, *Publication,* 1594; geology, 1586-1605; Gladstone, 1599; gold fields, 1586-7, 1589, 1594-1603; Government Engraving & Lithographic Office, 1586, 1594, 1597, 1599, 1601; Knobs Oaks Rush, 1597; Millett's Wellington, 1586; mining centers, 1596; railroads, 1596; Rockhampton, 1599
Queiroz, Gregorio Francisco de (1768-1845, Portuguese engraver), engraver, 329
Quelpart Island, Korea, 381

Rabout, P. C., cartographer, 1287-89
Racine, WI, 1372
Radcliffe College, Cambridge, MA, 696
Railroad economy; the revised and extended report on European and American railroads, Zerah Colburn, 580
Railroads, Argentina, 1579; Arizona, 1455-56; Beloit and Madison Rail Road, 1365; British Columbia, Canada, 540, 541; Buffalo and Western New York Rail Road, 952; Buffalo, New York and Philadelphia Rail Road, 776; Buffalo, Pittsburgh and Western Rail Road, 778; Canada, 476-77, 479, 580, 583-84, 594; Central Europe, 276; Chicago, Milwaukee and St. Paul Rail Road, 1395; China, 389, 393; Columbia Rail Road, 1072; Czechoslovakia, 280; England, 139, 141; Grand Rapids and Indiana Rail Road, 962-63; Illinois, 1770; Lackawanna and Pittsburgh Rail Road, 943; Maine, 650; Maryland, 1072; Massachusetts, 678; Mexico, 594, 1507; Milwaukee and Mississippi Rail Road, 1365; New England, 590, 636; New Jersey, 779, 935; New York, 590; New York, Philadelphia

and Norfolk Rail Road, 962-63; North America, 1703; Ohio, 1317; Oklahoma, 1268; Pennsylvania, 32 (75-6), 576, 799, 932, 935, 942-43, 945, 949, 952, 962-65, 967, 972, 1072; Pennsylvania Rail Road, 942, 964; Queensland, Australia, 1596; Scotland, 139; Texas, 1233; U.S., 578, 580, 583-84, 587-91, 594, 598, 607, 611, 615, 617, 1700; Wales, 139, 141; Watertown, Fond du Lac and Lake Superior Rail Road, 1365; West Virginia, 1142; Wisconsin, 1358, 1363, 1365; Wisconsin, Milwaukee, Watertown and Madison Rail Road, 1363

Rakestraw, Joseph, publisher, 1681

Ralegh, Sir Walter, in America, 20 (5)

Rameletti, Giuseppe, cartographer, 349

Ramsay, David (1749-1815, historian, APS 1803), cartographer, 567; *History of South Carolina,* 1163

Ramsay, V., cartographer, 319

Rand, Edward Lothrop (1859-1924, botanist), cartographer, 649; and J. H. Redfield, *Flora of Mt. Desert Island,* 649

Rand, McNally & Co., 468, 495, 515, 588, 617, 648, 660, 667, 694, 706, 718, 748, 835, 965, 1052, 1212, 1356, 1395, 1717, 1727, 1760, 1762; *Atlas of the oceans,* 1760; *Rand, McNally & Company's Neuer Familien-Atlas der Welt,* 1717; *The Rand McNally new concise atlas of the universe,* 1762; *Rand, McNally & Co.'s railway guide,* Robert A. Bower, 1703; *Rand, McNally & Co.'s universal atlas of the world,* 1727

Rande, Spain, 326

Randegger, Johannes (1830-1900, Swiss engraver, map lithographer), publisher, 231, 236, 239-43, 281

Rands, William Henry (Australian geologist), cartographer, 1586a, 1588

Ranney, A., publisher, 580

Rapin-Thoyas, Paul de (1661-1725, English historian), *The history of England,* 450

Rapkin, John, cartographer, 579

Rappahannock, Indians, 48 (16)

"Rapport sur les Archives de France, Relative à l'Histoire du Canada," J. Edmond Roy, 557

Raritan River, NJ, 859

Ratscha, 29 (52)

Ratzer [Ratzen], Bernard (fl. 1756-1778, cartographer, surveyor), cartographer, 861, 1302

Rausch, I (Joh. Lor. Jac., German engraver), engraver, 460

Raynolds, William Franklin (b. 1820, soldier), cartographer, 91, 1281-82, 1288-89, 1290, 1393, 1402

Reade, Charles, 16 (2)

Reade, J. A., cartographer, 1551

Real Litografai Milinte, Naples, publisher, 1691

Recherches sur les ruines d'Herculanéum, Duhamel de Monceau, 14

Records, N.S.W. Australia, geological survey, 1607

Red Mountain, Sneffels and Telluride districts, CO, 1418-19

Red River, 39

Red Sea, physiographic diagram, 1635

Redfield, John Howard (1815-1895, botanist), and E. L. Rand, *Flora of Mt. Desert Island,* 649

Reed, James G. (d. 1856, soldier), cartographer, 828-31, 1276

Reed, John (fl. 1774-1785, cartographer), cartographer, 990, 992, 1033, 1652; *An explanation of the map of the city and liberties of Philadelphia,* 989-92; *Map of the city and liberties of Philadelphia,* 1652

Reed, Joseph, donor, 989, 992, 1652

Reger, D. B., cartographer, 1141, 1143

Rehn, J., lithographer, 1247; publisher, 1247

Reichard, C. G. von, cartographer, 562

Reid, A. C., cartographer, 1605

Reid, John, *The American atlas,* 917, 1159, 1175, 1240; publisher, 917, 1159, 1175, 1240

Reid, John H., cartographer, 1603

Reimer, Dietrich (1818-1899, German publisher), publisher, 115

Reinagle, Hugh, 824

Reitzels, C. P. Forlag, publisher, 1759

Relief map, U.S., 602

Relations des Jesuits en la Nouvelle France, 32 (74)

Remarks, Instructions . . . , Thomas Truxtun, 78

Remarks on the practicability of Indian Reform, Isaac McCoy, 1265

"Remarks upon the Navigation . . . Gulf Stream," B. Franklin, *American Museum,* 1624

Remington, T. J. L., cartographer, 1261

Renard, C., cartographer, 847-48, 851, 881-82, 885

Renevier, Eugène (1831-1906, Swiss geologist), cartographer, 236, 242

Rennell, James, cartographer, 399, 1660; *A Bengal atlas,* 1660
Renneville, France, 252
Reno, NV, 1458-60
Renshawe, John Henry (b. 1853, geologist, cartographer), cartographer, 1315, 1415-16, 1432, 1445, 1473, 1477, 1492-93
Renwick, J., cartographer, 885
Report, Henry Newton, 1394
Report, Clarence Edward Dutton. U.S. Geological and Geographical Survey, 1450
Report, Second Geological Survey of PA, 958-60
"Report of a Geological Reconnaissance" (copy), 1354
Report of explorations and surveys . . . for a railroad from the Mississippi River to the Pacific, U.S. Army, 1700
Report of exploring expedition from Santa Fe to the juncture of the Grand and Green Rivers, 1422
"Report of a Geological Reconnaissance . . . ," G. W. Featherstonhaugh, 1375
Report of the Commission, Fairmount Park Association, Phila., PA, 1028
Report of the Superintendent of the Danville and Pottsville Railroad, 932
Reports, Missouri Geological Survey, 1384
Reps, John W., donor, 996
Repulse Bay, N.W.T., Canada, 487
Resaca, GA, 1181
Research in China, Carnegie Institution of Washington, 1725
Resolution (ship), 73, 80
Revetta, Frank A., cartographer, 841
Rey, Marc Michel, publisher, 1486, 1496
Rheinthal, Switzerland, 228
Rhine River, 275
Rhineprovinz, geology, Germany, 303
Rhode Island, 32 (24), 455, 583, 706, 711-18, 752-53, 761, 780; Naragansett Bay, 644, 676, 711, 716-17; Providence, 718
Rhode Island Historical Society, 712; donor, 712
Rhone River, France, 267; valley, geology, 272
Riberhuus, Denmark, 183-84
Ribnich, 29 (5)
Ricarte, Hipolito (fl. 1770-1775), engraver, 1553
Richard, Albert, cartographer, 54
Richardson, H. B., cartographer, 1341
Richardson, W. H., cartographer, 644
Richmond, VA, 1122, 1125, 1134-35; siege, 774
Richmond Canal, 33 (7)
Richmond's Island harbor, ME, 642
Richomme, ——, engraver, 825, 1215
Ricketts, J. J., cartographer, 690, 851, 1167, 1203, 1469, 1485
Rideau Lakes, Ontario, Canada, 490
Ridge Road, Phila., PA, 1039
Ridgely, Frederick, cartographer, 2 (5)
Ridley, J., engraver, 1187
Rigaud, André, 32 (47)
Rinehart, ——, publisher, 704
Rio Grande do Norte, Natal, Brazil, geology, 1574
River basin, 8 (1), 28 (14)
Riverhead tract, Long Island, NY, 35 (20)
Rivinus, Florens, donor, 449
Rizzi Zannoni, Giovanni Antonio (ca. 1736-1814), cartographer, 221, 353, 1667; *Atlante geografico del Regno di Napoli,* 1667
Road from Washington to New Orleans, U.S. Congress, 1090
Road map, 29 (31)
Roads, W., engraver, 364
Roads, Antigua, 1517; Austria, 280; Canada, 452, 780; Canada, east, 490; Cape Cod, MA, 695; Connecticut, 780; Czechoslovakia, 280; Delaware, 767, 769, 777; D.C., 777; England, 122, 124-25, 127-28, 133, 137; France, 250, 253, 257; Germany, 291; Illinois, 1322; Ireland, 145; Italy, 347, 355; Kentucky, 1246; Long Island, NY, 780; Louisiana, 1221; Maryland, 767, 769, 777, 1065; Massachusetts, 780; Middle Atlantic States, 766; Missouri to Oregon, 1426; New England, 636; New Jersey, 777, 780; New York, 777, 780, 822, 826; Ohio, 777; Pennsylvania, 767, 769, 777, 780, 899, 912-13, 939, Pike County, 1681, Wayne Co., 1681; St. Christopher, 1518; Scotland, 152-53; Spain, 330; Tennessee, 1258; U.S., 452, 490, 555, 559, 568, 570, 1090; Vermont, 663, 780; Virginia, 777, 1096-16; Wales, 122, 124-25, 127-28, 133, 137; West Virginia, 777; Wisconsin, 1372; Wissahickon Valley, Fairmount Park, Phila., PA, 1034
Roanoke Island, NC, 32 (91)
Robbins, Mrs. Caroline, donor, 844
Robbins, William Jacob (1890-1978, botanist, APS 1941), donor, 981
Robert, ——, cartographer, 359
Robert Morris, patriot and financier, Ellis Paxton Oberholzer, 816, 819
Roberts, A. C., cartographer, 84
Robertson, C. Grant, 1731

Robertson, John, cartographer, 686-87, 849
Robertson, W. A., publisher, 141
Robeson, M., donor, 1692
Robeson Channel, Canada, 100
Robinson, George (1737-1801, English bookseller), publisher, 66-67, 107, 254, 1554
Robinson, G.G., publisher, 1673
Robinson, G.G. and J., publishers, 1673
Robinson, Henry Hollister (1873-1925, geologist), cartographer, 749
Robinson, J., publisher, 1673
Robinson, John Henry (1796-1871, cartographer), cartographer, 1504
Robinson, Percy J., 32 (98)
Rochester, NY, 577
Rochette, L. de la, publisher, 289
Rock Island Lines, 617
Rock Island Rapids, Mississippi River, 1332
Rockefeller University, donor, 392, 394, 1720
Rockhampton, Queensland, Australia, 1599
Rockhill, William Woodville (1854-1914, U.S. orientalist and diplomat), 1508
Rockwell, C., cartographer, 854, 1182, 1483
Rocky Mountains, Canada, 549; Alberta, Banff, 532, Lake Louise, 533
Rocky Mountain National Park, CO, 1416
Rocky Mountains to the Atlantic Coast. U.S., glaciers, 625
Rodgers, A. F., cartographer, 1483, 1485
Rodgers, John (1812-1882, Supt. of the Naval Observatory), cartographer, 1195
Roeser, C., cartographer, 587, 1313
Rogers, Fred Baker (1926- , physician), donor, 630
Rogers, J., engraver, 579
Roggeveen, Arnold, cartographer, 1638
Roggewin, ——, 1619
Roisey, ——, lithographer, 113
Rollé, A., engraver, 692, 743, 745, 848, 850
Rollinson, William (1762-1842, English/American engraver), engraver, 78a-b, 369, 407, 426, 826, 1189
Rolt, Richard, *Memoirs of the life of the Right Hon. John Lindsay, earl of Crawford and Lindsay,* 29
Romain, Cape, SC, to Cape Henry, VA, 1148
Romans, Bernard (ca. 1720-1784, civil engineer, naturalist, APS 1774), cartographer, 32 (5), 720, 1081, 1093, 1184-85, 1200, 1655; *A concise natural history of East and West Florida,* 1184-85, 1200
Rome, Italy, 256, 341, 343
Ronald, MI, 47 (1)
Rook, Ternon (fl. 1784-1785, publisher), publisher, 1236
Roosevelt, Franklin Delano (1882-1945, president of the U.S.), 86
Roper, W. H., cartographer, 1336
Ropes, William, III, cartographer, 675
Rosario, Strait of, to Canal de Haro, WA, 1468
Rose, Gustav (1798-1873, German geologist), 317
Rose, Robert H., cartographer, 926; publisher, 926
Rosenthal, L. N., lithographer, 1016
Ross, J. W., cartographer, 97
Ross, Scotland, Channery town, 148
Rossland, B.C., Canada, 542
Rostan, ——, père, 310
Rouse's Point, Lake Champlain, NY, 632
Routledge, George, publisher, 1715
Roxbury, MA, 699
Roy, J. Edmond, "Rapport sur les archives de France, relative à l'histoire du Canada," 557
Royal Asiatic Society of Calcutta, donor, 374
Royal Astronomical Society of London, 1646
The royal atlas of modern geography, Alexander Keith Johnston, 1713
Royal Exchange, London, England, 131
Royal Geographical Society of Australia, 1581
Royal Society of Canada, 472
Royal Society of London, 186-87, 1773; *Philosophical Transactions,* 754
Rue, ——, cartographer, 32 (8)
Ruggeveen, Arnold, *Le Primera Parte del Monde de la Turba . . . India Occidental,* 1638
The Ruins . . . , Constantin F. C. Volney, 87a
Rumania, Fort St. Elisabeth, 29 (53); Lipoda (Lipova), 29 (20); Orsova, 29 (54)
Runnebaum, C., cartographer, 1551
Rusell, ——, engraver, 857
Russell, ——, 733
Russell, Richard Joel, cartographer, 40; collection, 40
Russia. *See* also: U.S.S.R., 159, 308; Crimea, 29 (40); Dneister River, 29 (42)
Russian Imperial army, 29 (40-42)
Russo-Turkish wars, 29
Ryukyu Islands, Japan, 1699

Sabatz, 29 (38)
Sabine, Edward (1778-1883, physician and

Arctic explorer), donor, 87
Sabine River, TX, 1225-27
Sac Indians, 1378
Sachem's Head Harbor, NY, 849
Sackett, ——, publisher, 607, 611, 615
Sackett & Wilhelms Co., publisher, 607, 611, 615
Sage, J., & Sons, publishers, 583
Saginaw Bay, MI, Lake Huron, 1293
Sagittarius (constellation), 50 (2)
Sahara Desert, 74b
Sailor's Creek battlefield, VA, 774
St. Andrews, Scotland, 148
St. Antonio Cape, Cuba, coast to Havana, 32 (72)
St. Augustine, FL, 4 (1), 1081
St. Barbot-de-Marny, N., cartographer, 319
St. Barthélemy Island, France, 32 (65), 1538
St. Charles River, Canada, 44 (1)
St. Christopher, 1518; roads, 1518
St. Clair flats, MI, 1284
St. Clair River, MI, 1296
St. Croix River, New Brunswick, Canada, 493, 631-32
St. Domingue. *See also:* Haiti and Santo Domingo, 32 (10), (25-29), (32-53), (55-64); Abacou Point to Grande Baye du Mesle, Vache Island, 32 (53); Acul Bay, 32 (33); Acul Bay to Petit Anse, 32 (52); Artibonite Bay to Baradaires Bay, 32 (27); Baradaires Bay to Artibonite Bay, 32 (27); Basse Terre anchorage, 32 (64); Cap François 32, (44), plain, 32 (52); Chouchou Cove, 32 (62); Cayes, 32 (39); Dame Marie Bay, 32 (32); Fort Dauphin, 32 (45); Fond du Grange bay, 32 (57); Gonaives Bay, 32 (56); Gonave, 32 (27); Grande Baye du Mesle to Abacou Point, Vache Island, 32 (53); Irois Bay, 32 (35); Jacmel, 32 (47), (60); Jeremie, 32 (40); La Croix des Bouquets, 32 (50), Fort la Constitution, 32 (50); Moustique Bay, 32 (58); Ocoa Bay, 32 (36); Port Alécu, 32 (59); Port Français, 32 (51); Petit Anse to Acul Bay, 32 (52); Petit Goave, 32 (46); Port au Prince, 32 (26); St. Marc, 32 (41), (61); Samana, 32 (48), Samana bay, 32 (34); Tiburon, 32 (37); Vache Island, Grand Baye du Mesle to Abocou Point, 32 (53)
St. Florentin, ——, comte de, 65
St. Francis basin, Mississippi River, 1341-42
St. Gallen, Switzerland, 228
St. Gotthard, Switzerland, geology, 231
St. Jean church, 13 (1)
St. Jean de Luz, France, 268a
St. Jean de Porto Rico. *See:* San Juan
St. John, NY, 784
St. Johns River, FL, 4 (1), 631, 1186, 1196
St. Joseph harbor, MI, 1347
St. Lawrence River, 32 (3), 48 (1), (14), 632, 831; gulf, Canada, 32 (84), 1298; gulf, canal, 32 (84); valley, 1071
St. Louis, MO, 942, 1324, 1386; harbor, Mississippi River, 1383; Louisiana Purchase Exposition, 1385; World Fair, 964, 1385, 1387
St. Marc, St. Domingue, 32 (41), (61)
St. Martin, Man of War Shoal, 32 (65)
St. Mary's Loch, Scotland, 7 (3)
St. Mary's River, FL, to Savannah River, GA, Atlantic coast, 1178
St. Mary's River, MI, 504-6, 1283
St. Nicholas Channel, 1187
St. Paul, MN, 1355-56
St. Peters River, MN 1354; confluence with Mississippi River, 28 (1)
St. Petersburg, Russia. Académie Impérial des Sciences, 312, 1496
St. Regis, NY, 632
St. Vincent Island, 1541
Salamanca, Mexico, 32 (17)
Salem, MA, 675; harbor, 693
Salem, NJ, 886
Salisbury, NC, 1135
Salmerohr, 29 (5)
Salter, Capt. ——, 32 (81)
Saltpeter. *See:* Niter
Samana, St. Domingue, 32 (48); bay, 32 (34)
Samar Island, Philippines, 417
Samoa, Tutuila archipelago, 608
Samsöe, Denmark, 171
San Carlos, Argentina, 1561
San Diego, CA, 1490; entrance, 1482; to San Francisco, CA, 1481
San Fernando de Omoa, Honduras, 1509-10
San Francisco, CA, 85, 1490; bay, 1491; bay entrance, 644, 1485, 1488; Peninsula, 1483, geology, 1489; to San Diego, CA, 1481; to Umpquah River, OR, 1463
San Gallan Island, Peru, 35 (6)
San Juan, Puerto Rico, 32 (63), 1534, 1562
San Miguel County, CO, 1419
San Pablo Cape, 441
Sanders, R. H., cartographer, 959, 960
Sanderson, Ivan Terence (1911-1973, naturalist), cartographer, 41; collection, 41
Sands, Benjamin Franklin (1812-1883, naval officer), cartographer, 848, 881, 1049,

1204-205
Sandusky Bay, OH, 1310
Sandwich, MA, 677
Sandwich Islands. *See:* Hawaii
Sandy Hook, NJ, 826, 885; to Barnegat Lighthouse, NJ, 644
Sanson, Nicolas (1600-1677, geographer), cartographer, 149, 251
Santa Barbara, CA, 1480
Santa Cristobal Lake, Zumpango, Mexico, 32 (11)
Santa Fe-Cordoba, Argentina, 1562
Santa Lucia, Windward Islands, 1537
Santa Rosita Island, Peru, 35 (4)
Santaren Channel, 1187
Santee Canal, SC, 1162
Santiago de Cuba, Cuba, 32 (26), (38), (49)
Santo Domingo. *See also:* Saint Domingue, 1543, 1550, 1661, 1668; Axua province, 1551; Dominican Republic, 32 (42); geology, 1551; Ocoa Bay, 1521; San Juan, 1562
São Bento, Brazil, 1572
São Paulo (city), Brazil, 1573
São Paulo (state), Brazil, 1572-73
São Sebastiño do Paraizo, Brazil, 1572
Sapelo Island, GA, to Savannah, GA, 644, 1182
Sardinia (island), Italy, 347, 350
Sardinia (kingdom), Italy, 345
Sargeant, Thomas, donor, 1692
Sartain, Samuel, cartographer, 1422
Saskatchewan, Canada, 516-19, 527-30; Assiniboia, 516; elevators (grain), 519
Satow, Sir Ernest Mason (1843-1929), 381
Saturn (planet), 50 (3)
Sauerwein, Charles, cartographer, 1613
Sault Ste. Marie, MI, 19 (2)
Saunders, Sir Charles (1713?-1775. Admiral of the Fleet), 44 (1)
Sauthier, Claude Joseph (fl. 1768-1779. British surveyor, cartographer), cartographer, 469, 499
Savage, E., publisher, 766
Savannah, GA, 576; siege, 1171; to Sapelo Island, GA, 644, 1182; to St. Mary's River, FL, Atlantic Coast, 1178
Savoie, France, geology, 279
Sawatch Range, CO, 1411
Saxony, Germany, geology, 304; Hainichen, geology, 304
Saybrook harbor, CT, 740
Sayer, Robert (1725-1794, British publisher), cartographer, 149, 1094; publisher, 32 (7), 88, 108, 120-23, 127, 146, 149, 152, 159, 250-51, 307, 312, 327, 347, 363-65, 421-23, 434, 453-54, 456, 550, 627, 651, 756, 784, 1081, 1093, 1536, 1653-55, 1663-64, 1668, 1750
Sayer, Robert, & Co., publisher, 89
Sayer, Robert, & John Bennett, publishers, 152, 453, 550, 627, 756, 784, 1081, 1093, 1536, 1654-55, 1663-64
Sayer, Robert, & Thomas Jefferys, publishers, 651, 1094
Scammon, Eliakim Parker (d. 1894, soldier), cartographer, 504, 1278-79
Schade, Theodor, cartographer, 302
Schapeljack, 29 (25)
Schardt, Hans (1858-1931, Swiss geologist), cartographer, 236
Schedler, Joseph (fl. 1852-1862, engraver, lithographer), engraver, 1124-25; publisher, 1135
Scheldt River, Belgium, 188, 190
Schenk Jansz, Leonard (1732-1800, Dutch engraver), 1496
Schieble, Erhard, engraver, 1514
Schielde, Georges Erhard, engraver, 217
Schiffer Publishing Co., publisher, 1768
Schleifer, Rudolf, cartographer, 1739
Schleswig, Denmark, 172-73, 184, 300; Germany, 172-73, 184, 300
Schmettau, Fried. Wilhelm Carl, baron von, cartographer, 289
Schmitz, M., cartographer, 1012; publisher, 1012
Schmitz, M. & P. Kereven, publishers, 1012
Schönborn, A., cartographer, 1433-34, 1436
Schofield, John McAllister, 1263
Schomburck, Sir Robert Hermann (1804-1865, British explorer), 1550-51
Schott, Charles A., cartographer, 96, 593, 595-97, 882, 885, 1020-21
Schouwen, The Netherlands, 205
Schreiber, Adam Gottlieb, publisher, 91; u. Weigel, publishers, 91
Schropp, Simon, publishers, 303; & Co., publisher, 1699
Schruers, Edwin J., cartographer, 696
Schubert, ——, 317
Schulz, Edward Hugh, cartographer, 1223
Schuylkill Canal, PA, 32 (75-6)
Schuylkill and Lehigh coal regions, PA, 936
Schuylkill County, PA, 32 (75-6), 958; Morea coal lands, 961; New Boston coal lands, 961; New Philadelphia, 940
Schuylkill River, PA, 32 (77), 43 (4), 576, 931, 987, 1021, 1035; bridge, Phila., PA,

921; Young's Ford to Darby Creek, 32 (68)
Schweiz, Switzerland, 223
Schweizerischen Naturforschenden Gesellschaft, 241; *Beiträge zur Geologischen Karte der Schweiz,* 238
Schwemo district, Burma, 32 (97)
Scilly Islands, Great Britain, 1662
Scioto River, 1303
Scobey, Frank C., cartographer, 1493
Scoles, John, (fl. 1793-1844, engraver), engraver, 1069, 1160, 1176, 1224, 1559
Scot, Robert (1744-1841, engraver), engraver, 74a, 130, 154, 315, 425, 457, 1556
Scot & Allardice, engravers, 154, 315
Scotia Sea, physiographic diagram, 1633
Scotland, 139, 148-158, 1662; Aberdeen, 148; The Bass, 148; canals, 139; Channery town, 148; Dunotyr castle, 148; Edinburgh, 148, 152, castle, 148-49; Esk River, reservoir, 7 (5); Galashiels, 7 (4); geology, 7, 138-39; Glasgow, 148; Hebrides Islands, 157; Montrose, 148; Moorsfoot Hills, 7 (6); Orkney Islands, 148; Peebleshire, 7 (2); Peterhead, 32 (81); railroads, 139; roads, 152-53; St. Andrews, 148; St. Mary's Loch, 7 (3), Shetland Islands, 148, 152-54; Sterling, 148, castle, 148-49; Sutherland, 155
Scott, ——, engraver, 1001; publisher, 1019
Scott, J., cartographer, 939
Scott, James D.? (fl. 1850-1877, cartographer), cartographer, 983
Scott, Joseph. *The United States gazeteer,* 637, 654, 662, 672, 713, 813, 875, 915, 1043, 1066, 1117, 1146, 1157, 1172, 1239, 1254, 1271
Scott, Joseph T. (fl. 1795-1806, engraver), engraver, 436, 1548
Scott, Joshua, cartographer, 929
Scott, Robert (1744-1841, engraver), engraver, 1001
Scott, Winfield (1786-1866, soldier, pacificator, general), 1506
Scott & Moore, publishers, 1019
Scrivener, C. P., cartographer, 1234
Scudder, Samuel H., cartographer, 462
Scull, John, cartographer, 33 (5)
Scull, Nicholas (1687-1761, Pennsylvania surveyor-general), cartographer, 15 (4), 33 (1), 759, 900, 988, 993, 1013, 1016, 1038
Scull, William (fl. 1765, cartographer), cartographer, 46 (5), 903-904
The Sea Coast of Nova Scotia, J.F.W. Des Barres, 1658
Sea shells (fossils), 32 (3)
Seale, Richard William (1732-1785, British engraver, draughtsman), cartographer, 450; engraver, 450
Searl, A. D., cartographer, 1399
Searle, Frederick, cartographer, 677
Sears, Clinton Brooks, cartographer, 1341
Seaton, ——, publisher, 1376
Seattle, WA, 1471
Second American Grinnell Expedition, 94
Second Geological Survey of Pennsylvania, 955, 960
Second U.S. Bank building, Phila., PA, 1011
Sectional maps showing 2,500,000 acres farm and wood lands, of the Illinois Central Rail Road Company, 1771
Sederholm, Jakob Johannes (1863-1934, Finnish geologist), cartographer, 319, 321-22
Seib, J., cartographer, 1138,1168
Seiglstrup, Denmark, 178
Seismological Society of America, 1493
Seismotectonic study, New England, 633
Selamiyeh, Assyria, 374
Selden, Henry Bill (1886-1934, geologist), cartographer, 1232
Selkirk, Manitoba, Canada, 524; Magnetic variations, 524
Seller, John (fl. 1677-1700, hydrographer to Charles II and James II), cartographer, 858; publisher, 858, 975
Seller & Fisher, publishers, 858
Sellers, David (1757-1813), 43 (2)
Sellers, George Escol (1808-1899, engineer, inventor), 42; 43 (2); collection, 42
Sellers, John (1762-1847), cartographer, 43 (1), (5)
Sellers, Nathan (1751-1830, surveyor, scrivener, manufacturer), 43 (2)
Sellers family collection, 43
Sellier, L. M., cartographer, 1249, 1253
Selves fils, lithographers, 268; publishers, 431
Selwyn, Alfred Richard Cecil (1824-1902, Canadian geologist), cartographer, 471
Senécal, C. O., cartographer, 488, 496, 531, 542
Senegambia, 434
Senex, John (d. 1740, British cartographer, engraver, publisher, geographer to Queen Anne), cartographer, 122, 144, 186-87, 215; engraver, 186-87, 246, 248, 288, 1053; publisher, 214, 246-48, 288

Senex, John, & John Maxwell, publishers, 214, 246, 288
Sengteller, A., cartographer, 1182; engraver, 1138
Sentis, Switzerland, geology, 240-41
Serjali River, Peru, 1570
Serpentarius (constellation), 50 (2)
Sertzenich, 29 (2)
Service Géographique de l'Indochine, engraver, 402-403; publisher, 402-403
Seutter, Mattheus (1678-1756, Swiss geographer), cartographer, 218-19, 226
Seven Years War, 501
Seybert, Adam (1773-1825, physician, scientist, congressman, APS 1797), cartographer, 44 (1); collection, 44
Seymour, Joseph H (?) (fl. 1796-1818, engraver), engraver, 655, 664, 671
Shade, Theodor (fl. 1860-1875, German cartographer), 302
Shaefer, P. W., cartographer, 32 (83)
Shaeffer, M. B., cartographer, 1373
Shaffner, M. N., cartographer, 984
Shallus, Francis (ca. 1774-1821, engraver), engraver, 1070, 1330
Shallus, Jacob, 33 (5)
Shannon, John, 33 (10)
Shark River, NJ, 890
Shasta County, CA, geology, 1486
Shawanese Indians, 901
Sheaff, J. A., cartographer, 942
Sheboygan, WI, 1372; harbor, 1362
Sheerness, England, 125
Sheffield Island harbor, CT, 744
Shelburne, NH, 657
Shelby County, KY, 1249
Shepps, V. C., cartographer, 984
Sheridan, WY, 1443-44
Sherman, ——, publisher, 1209, 1474
Sherman, George E. (ca. 1810, engraver), 373, 684-85, 688, 741, 832, 1209, 1361, 1629; lithographer, 1474
Sherman, William Tecumseh (1820-1891, soldier), 1091
Sherman & Smith, engravers, 373, 741, 832, 1209, 1361; publishers, 1209, 1474
Shetland Islands, Scotland, 148, 152-54, 1662
Shiloh, TN, 1208
Ship Island harbor, MS, 1209
Shippen, Edward (1729-1806, Chief Justice of PA, APS 1768), 46 (1), (16-17)
Shippen, Joseph (1706-1793, merchant), 46, (1), (16-17)
Shippen, Joseph, Jr. (1732-1810, soldier, judge, APS 1768), 45, (1-4)
Shippen, William, Jr. (1736-1808, physician, anatomist, teacher, APS 1768), 46, (5), (16-17); estate, 46 (18)
Shippen collection, 45
Shippen estate, South Street, Phila., PA, 46 (4)
Shippen family collection, 46
Shippen Street, Phila., PA, 46 (14)
Shippens Lane, Phila., PA, 46 (5), (15)
Shippensburg, PA, 33 (4)
Shoalwater Bay, WA, 1467
Short, Charles Wilkins (1794-1863, physician, teacher, botanist, APS 1835), cartographer, 1 (2)
Short, William (1759-1849, diplomatist, APS 1804), donor, 213, 230, 255-56, 342-43, 346, 348-49, 356
A short account of Algiers . . . , M. Carey, 436
Shryock, Richard Harrison (1893-1972, historian, APS 1944), donor, 1040
Shuckburgh, Sir George, bart., 1646
Shutting, Julius, cartographer, 1206
Siaelland, Denmark, 167-69, 171
Sicily, Italy, 347
Sidney, J. C. (fl. 1847-1850, architect, author), cartographer, 1014
Sieb, John, cartographer, 1277
Siebert, Selmar, engraver, 642, 746, 848, 1204-205, 1283, 1466, 1479, 1482
Siebold, Philipp Franz von (1796-1866, German physician, botanist), cartographer, 1699; *Atlas von Land- und Seekarten vom Japanischen Reich,* 1699
Siegen, P. M., cartographer, 217
Sierra Leone, 435, 437, 439
Sierra Leone Company, *Substance of the report delivered by the Court of Directors,* 437
Silkeborg, Denmark, 176, 183
Silver Lake, PA, 926
Silverton County, CO, 1419
Sim, W., engraver, 1187
Simcoe, John G., 500
Simmern, Germany, 29 (12)
Simonel, ——, cartographer, 432
Simowitch, S., cartographer, 319
Sinclair, Sir John, 1st bart. (1754-1835, agriculturalist), 156
Sinclair, Thomas, (fl. 1846-1862, lithographer, publisher), donor, 1012; lithographer, 952, 1012, 1039
Sioux Indians, 48 (2)
Sipe, E. H., engraver, 1469

Sitgreaves, Lorenzo (d. 1888, soldier), cartographer, 1151-52
Sitka, Alaska, 1497
Skaane, Denmark, 171
Skanderborg, Denmark, 174, 176
Skaniadarâde Indians, 752-53
Skanke, H. (1766-1807, Danish cartographer), cartographer, 170, 171, 173
Skene, A. J. (fl. 1872-1876, Australian cartographer), cartographer, 1606
Skerchly, S. B. J., cartographer, 1588
Sketches of the life and correspondence of Major General Greene, William Johnson, 1089
Skinner, Andrew (fl. 1780-1791, British naval captain), cartographer, 32 (8), 328
Skivehuus, Denmark, 181
Slight, James (fl. 1875-1883, Australian engraver), cartographer, 1606
Sloane, Charles S., cartographer, 1728, 1732
Sloane, Sir Hans (1660-1753, physician), 32 (3)
Slough Creek, B.C., Canada, 537
Small, D., publisher, 927
Smedley, Samuel Lightfoot (b.1832, surveyor, author, publisher), 1037
Smith, Capt., ——, cartographer, 101; ship *William,* 101
Smith, ——, engraver, 373, 557, 678, 684-85, 688, 741, 1209, 1361, 1629; lithographer, 1474; publisher, 779, 1014, 1209, 1240, 1474
Smith, B., engraver, 564
Smith, Benjamin E., cartographer, 1719; *The Century atlas of the world,* 1719
Smith, Benjamin Hays (d.1918), cartographer, 1709; *Atlas of Delaware County, Pennsylvania,* 1709
Smith, Gen. Daniel (1748-1818, soldier, Tennessee official), cartographer, 32 (21)
Smith, E., cartographer, 1020-21, 1197
Smith, Elvino V., engraver, 1029
Smith, Erwin Frink (1854-1927, plant pathologist, APS 1916), cartographer, 47; collection, 47
Smith, George B., cartographer, 832-33
Smith, George Girdler (1795-1859, engraver), engraver, 313, 684-85, 688, 741, 832-33, 1209, 1361, 1629
Smith, George Otis (1871-1944, geologist), cartographer, 463, 610, 612-14, 616, 619, 668, 708, 1155, 1183, 1211, 1214, 1264, 1319, 1328, 1357, 1382, 1491, 1730
Smith, Glenn S., cartographer, 1371, 1420
Smith, J. L., publisher, 1023, 1025, 1207, 1030, 1033
Smith, James Perrin, cartographer, 1494
Smith, Lloyd Edwin (b. 1802), cartographer, 1743; *Commercial atlas of the world,* 1743
Smith, Murphy De Witt (b. 1920, librarian), donor, 1649
Smith, Richard A. (b. 1871, geologist), cartographer, 1351-53
Smith, Thomas Peters (1777-1802, chemist, mineralogist, APS 1799), donor, 264
Smith, W., engraver, 642, 692, 745-46, 687, 1153, 1167, 1203, 1279, 1462, 1468, 1480-81
Smith, W. F., cartographer, 1259, 1262, 1278
Smith, Warren D., cartographer, 420
Smith, William (1727-1803, educator, clergyman, first provost of the College of Philadelphia, APS original member), *An examination of the Connecticut claim to lands in Pennsylvania,* 719; *An historical account of the expedition under the command of Henry Bouquet,* 901, 1302
Smith, William, cartographer, 138
Smith, Zachariah Frederick (1827-1911, author), *History of Kentucky,* 1236
Smith and Stroup, publishers, 779
Smith and Wistar, publishers, 1014
Smith, Jones & T. Foot, engravers, 557
Smith, Reid & Wayland, publishers, 1240
Smith Sound, Canada, 100
Smither, B., engraver, 767
Smither, James (fl. 1768-1802, engraver), engraver, 87b, 755, 760-60a, 990, 992, 1652
Smither, John, engraver, 177
Smithsonian Institution, 96
Smock, John Conover (1842-1926, geologist, author), 886, 889-91
Smyth, Sir David William, bart. (1764-1837, English historian), cartographer, 500
Snake River, 1404-405; Chopunnish Indian's map of, 28 (10)
Sneed's ordinary, VA, to Williamsburg, VA, 1116
Sneffels, Red Mountain and Telluride districts, CO, 1418-19
Snider, Jacob, Jr., donor, 628, 1664, 1668, 1671
Snyder, Michael, 33 (5)
Sociedad Geográfica de Lima, Peru, 1570
Société des Arts de Genève, *Bulletin de la Classe d'Agriculture,* 239
Société Helvétique des Sciences Naturelles, 236, 243
Society for the Diffusion of Useful Knowl-

edge, publisher, 143, 1011
Society of Antiquaries, London, 129
Society of Arts and Commerce, London, 1535
Society of Colonial Wars in the Commonwealth of Pennsylvania, 1037
Society of Jesus (Outaovacs), 32 (74)
Soest, Westphalia, Germany, 299
Sokolov, N., cartographer, 319
Sokolow, A. A., cartographer, 984
Solar eclipse, Europe, 116
Solar system, 55
Solitude, Phila., PA, 46 (9)
Solodurensis, Switzerland, 229
Solothurn, Switzerland, 229
Somerset County, PA, Germany, 920
Somerville, NJ, 892
Soniat Plantation to New Orleans, LA, 1333
Sonis, I., cartographer, 1543
Sonne, Jörger (or Jeppe Jörgen, 1771-1833, Danish engraver), engraver, 184
Sonora, Mexico, 1454; Guimas, 32 (70)
Sorokin, A., cartographer, 319
Sotzmann, Daniel Friedrich (1754-1840, German cartographer), cartographer, 292
Soule, G. G., publisher, 1701
South Africa, 447-48; Kenhardt, 448
South America, 74a, 88-91, 1536, 1553-62, 1685, 1756; Guyana coast, 1549
South Australia, Australia, 1581-83; geology, 1582-83
South Black River, MI, 1348
South Carolina, 455, 1081-84, 1093, 1157-69a, 1175, 1504, 1661, 1689; Cape Romain to Cape Henry, VA, 1148; Charleston, 32 (5), 644, 576-77, 1081, 1163, 1165; harbor, 1166-68; Department of Agriculture, Commerce and Industries, 1169; Santee Canal, 1162
South Central U.S., 1215
South China Sea, physiographic diagram, 1635
South Dakota, 1391-97; Black Hills, 1391-93, geology, 1394
South Daniel Islands, 1619
South Keeling Island, Chile, 1620
South Shetland Islands, 101; Georges Bay, 101
South Street, Phila., PA, Shippen estate, 46 (4)
Southack, Cyprian (1662-1745, British cartographer), cartographer, 467
Southern Pacific Rail Road, 1234
Southern states, 1081-93
Southgate, Horatio (1812-1894, traveller), 373
Southwark, Phila., PA, 46
Sowle, Andrew, publisher, 987
Space, 53-61, 1644, 1646, 1653, 1672, 1762
Space and the Solar System, 53-61
Spain, 251, 325-41, 1654; Cadiz, 329, 332; *Canary Islands,* Fuerteventura, 336, Lanzarote Island, 337 Tenerife Island, 338; Galicia, 326; Leaõ, 329; Rande, 326; roads, 330; Viguo River, 326, 335
Spanish Armada, 10 (1)
Spanish, British and French empires in North America are indexed under Canada, U.S., Mexico and Florida
Special Papers, Geological Society of America, 1632
Speck, Frank Gouldsmith (1881-1950, ethnologist), cartographer, 48 (1-16); collection, 48
Speer, Joseph Smith (fl. 1766-1796, British sea captain, cartographer), cartographer, 326, 1145, 1512, 1535, 1537, 1547; publisher, 326, 1535, 1537; *West-India pilot,* 1145, 1170, 1503, 1509-13, 1520-34, 1565-67
Spencer & Co., publisher, 936
Spitsbergen, Norway, 100, 107
Spofford, Elmer P., cartographer, 650
Spotsylvania battlefield, VA, 774
Springfield, PA, 43 (4)
Spruce Street, Phila., PA 997
Spurr, J. E., cartographer, 1419
Stafford, Cornelius William, *The Philadelphia directory for 1797,* 1000
Standard Map Co., lithographer, 1373
Stanford, Edward (1827-1904, British geographer), cartographer, 117
Stanford, Edward, publisher, 391
Stanford University, Palo Alto, CA, 1493
Stanislas, Augustus Poniatowsky, king of Poland, 1732-98
Stansbury, M. Howard, cartographer, 740
Stapleton, Darwin Heilman (b. 1947, historian of technology), 33 (7)
Starling, William, cartographer, 1341
State House, Phila., PA. *See:* Independence Hall, Phila., PA
Statistical atlas of the U.S., U.S. Dept. of Commerce, 1728, 1732
Statistical atlas of the U.S., Francis A. Walker, 1704
Statistical view of the district of Maine, Moses Greenleaf, 641

Statistics, PA, 938, 965; U.S., 575, 592, 1704; Vermont, 666; World, 81
Stavelo, Belgium, 189
Steam boat navigation 2 (3)
Stedman, ——, publisher, 1312
Stedman, Brown & Lyon, publishers, 1312
Steele, J. Dutton, cartographer, 945
Steiler's [Adolf] *Handatlas über alle Theile der Erde,* 1720
Stellwagen, H. S., cartographer, 691
Stephens, Thomas, *Stephen's Philadelphia directory for 1796,* 999
Sterling, Scotland, 148; castle, 148-49
Stern, G., engraver, 1563
Sterner, ——, lithographer, 580
Steuart, W. M., cartographer, 1732
Steuben County, NY, 765, 815
Stevens, E., 32 (32-3), (35), (51)
Stevens Point, WI, 1372
Stevenson, J. J., donor, 1645
Stewart, George, cartographer, 1701
Stewart, J., & Co., publisher, 260
Stickney, Amos, cartographer, 1341
Stieler, Adolf Karl, cartographer, 1720, 1733, 1739; *Stieler's atlas of modern geography,* 1739; *Stieler's Hand-Atlas,* 1733
Stiernholms, Denmark, 174
Stiles, Arthur, cartographer, 1235
Stiles, Ezra (1727-1795, scholar, Congregational clergyman, president of Yale College, APS original member), *A history of three of the judges . . . ,* 711, 729-37
Stiles, S., engraver, 1629
Stiles, S., & Co., engraver, 846, 1311, 1324
Stockdale, John (1739-1814, English bookseller, publisher), cartographer, 188
Stockton, John, 1277
Stokes, Pringle, cartographer, 1575
Stone, C. H., cartographer, 417
Stone, David, 1150
Stone, Rolph Clark, Ltd., lithographer, 490
Stone, William J. (fl. 1822-1836, engraver), engraver, 632, 679, 682, 716, 740, 880, 1151-52, 1166, 1191-93, 1310, 1323, 1331-32, 1383
Story, Joseph (1779-1845, lawyer), *Address delivered on the dedication of the cemetery,* 704
Stout, James De Forest (b. 1783, engraver), engraver, 500
Stout, Peter F., cartographer, 1139
Strait of Gibraltar. *See:* Gibraltar, Strait
Strait of Mackinac, MI. See: Mackinac, Strait, MI.
The stranger in America, Charles William Janson, 1002
Stratford, CT, 721-27; to N.Y.C., NY, 795-97; to Poughkeepsie, NY, 810-12
Strauch, ——, cartographer, 953
Strauch & Cochran, cartographers, 953
Strausz, A., cartographer, 1138
Stretch, R. H., cartographer, 1769
Strickland, G., cartographer, 1005
Strickland, William (1787-1854, architect, engineer, engraver, APS 1820), cartographer, 32 (68), 1004, 1007, 1047; donor, 1047
Strobel, M.F.O., engraver, 1480
Strother, Daniel Hunter (1816-1888, soldier), cartographer, 1133
Strother, John (d. 1815(?), cartographer), cartographer, 1150
Stroup, ——, publisher, 779
Strum, G. P., cartographer, 1381
Struther, ——, cartographer, 1148
Stuart, John (ca. 1700-1799, cartographer), 32 (5)
Stuart, John, 3rd earl of Bute (1713-1792), 751
Stull, S. E., engraver, 693, 1230, 1467
"The Submarine Relief off the Norwegian Coast," Olaf Holtedahl, 165
Substance of the report delivered by the Court of Directors. Sierra Leone Company, 437
Sudbury, Ontario, Canada, 509
Sugar Creek, B.C., Canada, 537
Sullivan, J. A., cartographer, 1046
Sullivan, James (1774-1808, statesman, author), *The history of the district of Maine,* 638
Sullivan, John, publisher, 561
Sully, Thomas (1783-1872, artist, APS 1835), 1245
Sulpis, ——, engraver, 273-74
Sulu Sea, physiographic diagram, 1635
Superior, Lake, 32 (7), (74), 533, 1277, 1358; Agate Harbor, 1288; canal, 1298; Eagle harbor, 1282, 1289; Grand Island, 1297; to Gulf of the St. Lawrence, canal navigation, 32 (84)
Surinam, Guyana, 1549
A survey of the roads, Christopher Colles, 635, 721-27, 787-812, 863-74, 906-11, 1054-64, 1096-116
Surveyor's exercise (?), 24 (1)
Surville, ——, de, 1619
Susquehanna, Military Depot, U.S., 1132
Susquehanna County, PA, 926
Susquehanna River, 16 (2), 32 (3), (77), 45 (2), 1072; and Chesapeake bay, MD, 1068;

valley, central PA, 12 (2); West Branch, PA, 45 (1), (5)
Sutherland, A., cartographer, 32 (8)
Sutherland, Scotland, 155
Sutton, Frank, cartographer, 707, 837-38, 1024, 1026
Suzuki, K., cartographer, 382, 384
Sveriges Geologiska Undersökning, 163; donor, 161; *Nagra ord till Upplysning om Bladet . . .*, 161
Swanitz, A. W., cartographer, 1336
Sweden, 159-163, 177; geology, 161-63; Kongl. Vetenskaps Academien, 1538
Swift, William Henry (1800-1879, cartographer), cartographer, 740, 746, 1192
Switzerland, 218-44, 283; Appenzell, 228; Basel, geology, 238; Bern, 219; Brugge (Aargau), geology, 237; Chamouny valley, 244; Commission Géologique Fédérale, 242; Freiburg, 222; Geneva, 230, 239; geology, 32 (94), 236, 281; glaciers, 235, 243; Glarus, 225; Lake Constance, 221; Lepontina, 226; Lucern, 220; Mt. Blanc, glaciers, 243, model, 1774; Orten, 228; Rheinthal, 228; Schweiz, 223; Sentis, geology, 240-41; St. Gallen, 228; St. Gotthard, geology, 231; Solodurensis, 229; Solothurn, 229; Thurgau, 220-21, 228; Toggenburg, 228, 232; Urien, 226; Unterwalden, 224; Valais, geology, 242; Vallesia, 227; Vaudois Alps, geology, 242; Wallis le Valais, 227; Werdenberg, 225; Zugewandte, 228
Sword, James B., cartographer, 32 (88)
Sydney, J. C., cartographer, 1014
Sydney, N.S.W., Australia, geology, 1608
Sylva Mountain, Canada, 23 (2-3)
Syosset Bay, NY. *See:* Oyster Bay, NY
Syout, Egypt, 427
Syria, 325

Taasinge, Denmark, 173
Tableau des Etats Danois . . . , J. P. G. Catteau-Calleville, 182
Tacoma, WA, 1471
Taitt, David (fl. 1772-1773, British surveyor), cartographer, 32 (5)
Talbot, A.D., cartographer, 547
Talbot, William (fl. 1670-1693, surveyor), cartographer, 973
Talcott, George (d. 1862, soldier), cartographer, 1277
Talcott harbor, 1277
Tallis, F., publisher, 579
Tallis, John, publisher, 579
Tallis, J. & F., publishers, 579
Tambopata River, Peru, 1570
Tampa Bay, FL, 644, 1184, 1197
Tamura, E., cartographer, 384
Tanner, Benjamin (1775-1848, engraver), engraver, 461, 824, 930, 934, 1122, 1175, 1242-43; publisher, 934, 1190
Tanner, Henry Schenk (1786-1858, cartographer, statistical geographer, APS 1829), 32 (25-29), (32-53), (55-64), 1009, 1090; cartographer, 32 (73), 81, 140, 269, 573, 576, 935, 1045, 1089, 1257, 1687, 1693-94; donor 81, 334, 576, 824, 879, 930, 1009-10, 1165, 1687, 1692-94; engraver, 483, 501, 565, 569, 572, 676, 879, 930, 925, 935, 1006, 1045, 1071, 1086-87, 1089, 1122, 1159, 1165, 1177, 1309, 1681; publisher, 140, 269, 334, 573, 576, 935, 1009-10, 1045, 1165, 1687, 1690, 1693-94; *Atlas Classica,* 1694; *General atlas, new edition,* 1690; *A new American atlas,* 1687; *Universal atlas,* 140, 269, 334, 1693
Tanner, Vallance, Kearny & Co., publishers, 461
Tapscot, J.W., 1747
Tardieu, Pierre Antoine F. (1784-1869, French engraver), cartographer, 825, 1215; engraver, 258, 825, 1215; publisher, 1215
Tarpaulin Cove, MA, 685
Tasman, Abel (1602-1659, Dutch navigator, explorer), 1619
Taunton, England, 135
Taunton, MA, 680
Taunton, NJ, 890
Tawas Harbor, MI, 1286
Taylor, ——, cartographer, 32 (8)
Taylor, H. W. B., 1585, cartographer, 1585
Taylor, John (d. 1756, surveyor), cartographer, 973, 976
Taylor, Zachary (1784-1850, President of the U.S.), 1193
Taylor County, KY, 1249
Tehuacan, Mexico, 32 (18)
Tehuantepec, Golf of, 32 (16)
Telephone lines, U.S., 609
Telluride mining district, CO, 1421
Telluride, Red Mountain and Sneffels districts, CO, 1418-19
Temple, W. G., cartographer, 853
Temsewar, 29 (48-9)
Tenerife Island, Canary Islands, Spain, 338

Tennessee, 32 (21), 36 (1), 1134, 1179, 1254-62; Chattanooga, 1208; Franklin, battlefield, 1261; Great Smoky Mountains National Park, 1156; Indians, 32 (21); Knoxville, 1258; siege, 1260; Nashville, 1258; roads, 1258; Shiloh, 1208
Tennessee River, 1181; Muscle Shoals, 32 (66), canal, 1206
Tepeaca, Mexico, 32 (18)
Tepixi de la Seda, Mexico, 32 (18)
Terlingua quadrangle, Texas, 1235
Terrestrial globe, 1773
Tetela y Xonotla, Mexico, 32 (18)
Texas, 1225-35; Brewster County, 1234; El Paso County, 1234; Galveston bay, 644, harbor, 1230-31; geology, 1234; Jeff Davis County, 1234; railroads, 1233; Presidio County, 1234; Terlingua quadrangle, 1235
Texcoco, lake, Mexico, 32 (11)
Texel, The Netherlands, 206
Tezuitlan, Mexico, 32 (18)
Thackara, James (1767-1848, cartographer, engraver), cartographer, 1078; engraver, 354, 763, 1065, 1078
Thackara & Vallance, engravers, 1065
Thames River, CT, 644, 747; England, 132, 1662
Tharp, Marie, cartographer, 1632-33, 1635
Theatrum Orbis Terrarum, Abraham Ortelius, 1636
Thebes, Egypt, 427
Thom, George (d. 1891, soldier), cartographer, 32 (82)
Thomas, ——, publisher, 658, 1221, 1258, 1680
Thomas, C. M., cartographer, 1020-21
Thomas, Cowperthwait & Co., publishers, 658, 1221, 1258
Thomas, E. B., cartographer, 1137, 1197
Thomas, E. J., cartographer, 1336
Thomas, Isaiah (1749 o.s.-1831, printer, historian of the press, founder of the American Antiquarian Society, APS 1816), publisher, 368, 1540, 1557
Thomas & Andrews, publisher, 76, 368, 1540, 1557, 1680
Thompson, ——, cartographer, 1330; Lt., 1191; *New general atlas,* 1088
Thompson, Carl, cartographer, 775, 1370
Thompson, Henry A.(?) (d. 1880, soldier), 1191
Thompson, J. G., engraver, 645, 885, 1196
Thompson, W. A., engraver, 854, 885
Thompson, W. B., cartographer, 677
Thompson, William, cartographer, 1586
Thomson, A. R., cartographer, 490
Thomson, John Edgar (1808-1874, third president of the Pennsylvania Rail Road), cartographer, 1072
Thomson's new general atlas, 630, 1627
Thornton, John (fl. 1652-1704, British chartseller, engraver, publisher, hydrographer to Hudson Bay and East India Companies), donor, 897; publisher, 897, 975
Thornton, John, and John Seller, publishers, 975
Thorsinge, Denmark, 171
Thotmes IV, 3 (1)
Thoulet, ——, cartographer, 1613
Throop, D. S., engraver, 575
Throop, J. V. N., 32 (6); engraver, 642, 1204, 1464
Throop, O.H., cartographer, 575; engraver, 575
Thunder Bay, MI, Lake Huron, 1294
Thurgau, Switzerland, 220-21, 228
Thurso River, Scotland, 156
Thutmes I, 3 (2)
Thwaites, F. T., cartographer, 465
Tiburon, St. Domingue, 32 (37)
Tichnor, William D., publisher, 1695
Tiebout, Cornelius (1777-1832, line and stipple engraver), engraver, 656, 665, 674, 761, 1084
Tiernan, ——, publisher, 1233
Tierra del Fuego, Argentina, 1562, 1576; Chile, 1562, 1576
Tigre River, Peru, 1570
Tigress (ship), 6 (1)
Tigris River, 374
Tiiuxsoxrúntie Indians, 752-53
Tilghman, ——, 33 (5)
Tilghmann, F., engraver, 321-22; publisher, 321-22
Tiller, S., publisher, 324
The Times atlas of the world, John Bartholomew, ed., 1745
Times Publishing Co., publisher, 1745
Timiskaming, Ontario, Canada, 509
Tindal, Nicholas (1688-1774, English historical writer), *The history of England,* 450
Ting, V. K., cartographer, 395
Titicaca, Lake, 1570
Todd's Ordinary, VA, to Annapolis, MD, 1105
Tønder, Denmark, 300
Tofiño de San Miguel, Vincente (1732-1795, Spanish cartographer), cartographer, 332,

329
Togawa, T., cartographer, 382, 384
Toggenburg, Switzerland, 228, 232
Tokyo Bay, Japan, 389
Tollemer, ——, cartographer, 1613
Toluca, Mexico, to Vera Cruz, 32 (18)
Tonnawanta Creek, NY, 830
Tonopah, NV, 1460
Tootoomi oilfields, Japan, 30
Topeka, Kansas, Constitutional Hall, 1399
Topley, William, cartographer, 114
Topografich Atlas Danmärk. K. Dansk Geografishe Selskab, Copenhagen, 1759
Topographical and geological atlas of the district of the high plateau of Utah, 1708
Topographical and Geological Survey of PA, 966
A topographical description of Kentucky, Gilbert Imlay, 1238, 1270, 1308
A topographical description of Virginia. Thomas Hutchins, 1269, 1304, 1321
Toronto, Canada, 39
Toronto Lithographic Co., publisher, 473, 1724
Torreon, Mexico, 1507
Torrey, Jason, cartographer, 1681; *A map of Wayne and Pike Counties, Pa.,* 1681
Totopotomoy battlefield, VA, 774
Toulon, France, 262
Toulouse, France, 252
Touring Club Italiano, publisher, 1737
Toussaint l'Ouverture, Pierre Dominique (1746-1803, Negro revolutionary leader), 32 (47)
Tower of London, London, England, 143
Tracy, Gibbs & Co., 1366
Traité des Arbes et Arbustes, Du Hamel de Monceau, 14
Transactions. American Philosophical Society, 1(2), 2 (1), 32 (18a), (19), (23-24), (84), 66-7, 73, 92, 107, 128, 153, 160, 192, 195, 233, 308, 350, 361, 367, 400, 424, 576, 570, 755, 760, 1166, 1298, 1542, 1554-55, 1625-26, 1628
Transportation routes—world, 85
Travels through North & South Carolina, Georgia, East and West Florida, William Bartram, 1186
Travels through the United States, La Rochefoucald Liancourt, 557
Treaty of Ghent, 631
Treaty of Paris of 1783, 88, 90, 550, 552-53, 564, 632
Treaty of Vienna of 1815, 112
Treaty of Washington, D.C., of 1842, 632
Trebes, France, 252
Trego, Charles B. (1794-1874, geographer, APS 1843), 453, 772
Trenchard, James (fl. 1777-1793, engraver), engraver, 700, 912
Trent River, England, 136
Trenton, NJ, 576, 886, 890; to N.Y.C., NY, 867
Treuttel, ——, publisher, 182
Treuttel et Würtz, publishers, 182
Trierweiller, 29 (2)
Trigg County, KY, 1249
Trinity Fort, PA, 898
Triple Alliance, 282
Triple Entente, 282
Tripoli, Lybia, 436
Trowbridge, William Petit (d. 1892), cartographer, 717
Truando River, Colombia, 1568
Trumbauer, Horace, cartographer, 1028
Trumbull, ——, cartographer, 1506
Trumbull, John (?), cartographer 25 (1)
Trummer, C., engraver, 562
Truro, MA, 679, 682
Truxtun, Thomas (1755-1822, U.S. naval officer), cartographer, 78; donor, 78; *Remarks, Instructions . . . ,* 78
Tryon, County, NY, 16 (2)
Tschernyschev, Th., cartographer, 319
Tschirikow, Capt., 1496
Tucker, R. C., cartographer, 1141-43
Tuckerman, Edward, 32 (67)
Tuckernuck Island, MA, 681
Tucson, AZ, 1455-56
Tueur, T. W., Jr., engraver, 286
Tunisia, 436
Tupper, James, cartographer, 670
Turgot, Michel Etienne (1690-1751, French cartographer), 1643
Turgovie, 221
Turin, Italy, 267, 349
Turkestan, 375
Turkey, 325, 359-62, 373; Constantinople (Istanbul), 52 (362); Istanbul, 52, 362
Turkish-Austrian wars, 29
Turkish-Russian wars, 29
Turnbull, William (d. 1857, soldier), cartographer, 1191
Turnbull County, OH, coal, 1314
Turner, James (fl. 1744-1759, engraver), engraver, 750, 752, 757, 859-60, 900
Turner valley, Canada, 491
Turner's Lane, Phila., PA, 46 (12)

Turtle, Thomas (d. 1894, soldier), cartographer, 1341
Tuscaloosa River, AL, 1202
Tuscarora Valley, PA, 960
Tuscuco Lake, Mexico. *See:* Texcoco, Mexico
Tuttle, H. P., cartographer, 1394
Tutuila archipelago, Samoa, 608
Twenthe, The Netherlands, 207
Two Sicilies, king, donor, 358; kingdom, Italy, coastline, 358
Tyi, Queen, 3 (3)
Tyrrell, Joseph Burr (1858-1957, Canadian geologist, historian), cartographer, 547
Tyson, John R., donor, 1692

Ucayali River, Peru, 1570
Udden, Johan Augustus (1859-1932, geologist), cartographer, 1234
Ughi, K. Lodovico (fl. 1729-1787, Italian cartographer), cartographer, 348
Uinta Mountains, UT, 1449
Umpquah River, OR, to Canada, 1464; to San Francisco, CA, 1463
Ungava Bay, Quebec, Canada, 48 (15)
Union Canal, PA, 576
Union Canal, Leicestershire and Northamptonshire, England, 136
Union Lithographic Co., lithographer, 1494; publisher, 1494
Union of Soviet Socialistic Republics (U.S.S.R.). *See also:* Russia, 312-20, 587, 611; Bjelaja Gora, 317; Dvina River, 313; (European), geology, 319; Franz Josef Land, 100; Goktscha Lake, 320; hydrography, 318; Kamchatka Peninsula, 378, 456, 1647; Kuril Islands, 1699; Novaya Zemlya, 100; Ural Mountains, 317
United States of America (U.S.A. or U.S.), 85, 88, 90, 499-51, 453-55, 467, 469, 484, 550-623, 785, 965, 1223, 1427, 1655, 1687, 1696, 1707, 1726-27, 1734-35, 1736, 1753, 1768-69; altimetry, 591; Antarctic Service, 103; *Army,* 585, 1123, 1129, 1134, 1181, Adjutant General's Office, Military Information Division, 447; of the James, 774, of the Potomac, 774, 1125-28, 1131, Bureau of Topographical Engineers. *See also:* U.S. Army Corps of Engineers, U.S. Engineer Bureau, and, U.S. Engineer Department, 631, 740, 828-31, 834, 880, 1131, 1151-52, 1166, 1192-93, 1280, 1310, 1326, 1344-49, 1360, 1422, 1506, camps and centers, 617, 632, 677, 679, 682, 950, 970, 1123, 1391, 1393, Corps of Engineers, 1179, 1206, 1223, 1229, 1260, 1262, 1284, 1331, 1337, 1339-40, 1362, 1364, 1383, Department of the Cumberland, 1179, Engineering Department, 1135, 1721, Map Service, publisher, 1770, Military Intelligence Division, cartographer, 117; *Report of explorations and surveys for a railroad from the Mississippi River to the Pacific Coast,* 1700, Signal Service, 586; Survey of the Northern and Northwestern Lakes, 505, War College Division, General Staff, 282-84; Atlantic Coast, 32 (86), lighthouses, 624, to Rocky Mountains, glaciers, 625; base map, 600, 613; Board on Examination and Survey of Mississippi River, 1343; Boundary, Canada, 631-32; Boundary Survey Office, 1461; *Bulletin of American Ethnology.* Franz Boas, "The Central Eskimo." *Sixth Annual Report,* 486-87; Bureau of the Census, 1732; Bureau of Science, Division of Mines, 420; Bureau of Soils, 1076; canals, 570, 573, 576, 579, 587, 591, 607, 611, 615, 1298; capitol, Washington, D.C., 579; Census Office, 1704; Civil War, 774, 950, 956, 969-71, 1091, 1131-32, 1179-81, 1259-62, 1689; coal, 965; *Coast and Geodetic Survey,* 420, 581-82, 593, 595-97, 642-47, 684-88, 690-93, 705, 717, 741-47, 773, 832-33, 847-54, 881-85, 957, 1020-21, 1046, 1049-50, 1073-75, 1080, 1130, 1137-38, 1153, 1182, 1194, 1197, 1203-205, 1209, 1223, 1230-31, 1333, 1336, 1461-65, 1469-70, 1479-83, 1485, 1487-89, 1495, 1497, 1500, 1721 *Atlas of the Philippine Islands,* 1723, donor, 644-47, 705, 717, 854, 882-83, 885, 957, 1050, 1073, 1075, 1080, 1137-38, 1168, 1182, 1195-97, 1204, 1231, 1259, 1333, 1470, 1479, 1485; Coast Survey. *See:* Coast and Geodetic Survey; *Coastal Charts* 1783-1861. Peter J. Guthorn, 1768; coastline, 99; Congress, 679, 1236-37, *A Collection of Maps . . . ,* 1166, 1226-28, 1331, 1629, *Road from Washington to New Orleans,* 1090, House of Representatives, 631 Senate, 578; Constitutional Sesquicentennial Commission, 720, 1077; contour map, 601, 614, 619; *Department of Agriculture,* 1199, 1478, Bureau of Agricultural Economics, 620, General Land Office. *See:* General Land Office, Weather Bureau (*see also:* U.S. Weather Bureau), 605, 1092, 1553 Forest Service, 622; *Department of Commerce,* Bureau of the Census, 620 (and

Labor), Bureau of Statistics, 85, *Statistical Atlas of the U.S.,* 1728-1732; Department of State, 393, 453, 1497; *Department of the Interior* (*see also:* U.S. General Land Office, Geological Survey, and Geological Survey of the Territories), 607, 621, 782, 1156, 1198, 1207-8, 1210, 1213-14, 1222, 1224, 1263-64, 1266-67, 1313, 1316, 1318, 1320, 1327-28, 1350, 1355, 1367, 1369, 1381, 1388-90, 1394, 1396-98, 1400, 1404-405, 1407-408, 1410-17, 1420-21, 1423-25, 1428-29, 1430-33, 1435-36, 1438, 1440-56, 1458-60, 1471-72, 1475-76, 1484, 1490, 1495, 1500, 1708, donor, 1404-405, *The National atlas of the U.S.,* 1753, National forests, 611, *World atlas of commercial geology,* 1730; Department of the International Survey, 1371; Diplomatic Offices, world, 84; District of Columbia, 82; donor, 956; Engineer Bureau, 1402, 1461; Engineer Department, *Report of a special board of engineers, on survey of the Mississippi River,* 1343; Exploring Expedition, 1461, 1474; Exploring Expedition, 1839-40, 103; farming, 620; General Land Office, 587, 598-99, 607-608, 611, 615, 1198, 1201, 1207-208, 1210, 1213, 1222, 1224, 1263, 1267, 1313, 1316, 1318, 1320, 1327, 1340, 1350, 1355, 1367, 1369, 1381, 1388-90, 1396-98, 1400, 1412-13, 1417, 1423, 1428, 1443, 1446-47, 1451, 1453-54, 1456, 1458, 1460, 1471, 1475, 1490, 1495, 1500; Geographical and Geological Survey of the Rocky Mountain Region, 1394, 1708; Geological and Geographical Survey of the Territories, 1404-11, 1425, 1433-35, 1438, 1440-43, 1449-50, 1484, donor, 1441, *Report,* by Clarence Edward Dutton, 1450; Geological Expedition to the 40th Parallel, 1769, *Atlas,* accompanying vol. III on mining industry, 1769; *Geological Survey,* 393, 463, 473, 599, 601-604, 606, 610, 612-14, 616, 618, 668, 707-708, 720, 781-82, 837, 856, 888, 973-74, 977, 979-80, 1024, 1026, 1076-77, 1093, 1141, 1144, 1154-56, 1183, 1211, 1213-14, 1223, 1232, 1235, 1251, 1264, 1266, 1315, 1319, 1328, 1351, 1357, 1377, 1382, 1419, 1427, 1432, 1473, 1477, 1491, 1493, 1730, 1753, *Bulletin,* 1093, donor, 1441, engraver, 463, 600-601, 606, 613-14, 616, 622, 1155-56, 1315, 1319, 1328, 1382, 1415-16, 1429, 1432, 1445, 1473, 1477, 1491-92, lithographer, 603, publisher, 117, 463, 612, 616, 618, 622, 1183, 1235, 1319, 1328, 1382, 1415-16, 1429, 1432, 1445, 1473, 1477, 1491-92; geology, 570, 623, 1706; glaciers (east of the Rocky Mountains), 625; Government Printing Office, publisher, 84, 592, 1478, 1723, 1728, 1732; Guam Island, 608; Gulf of Mexico coast, 32 (86), lighthouses, 624, Homestead Act of 1909, 1414, 1424, 1431, 1444, 1448, 1451, 1455, 1459, 1472, 1476; hurricanes, 1092; Hydrographic Office, 99, 103, 117, 415, 1495, 1618, 1721; Inland Waterways Commission, 608; Indians, 587, 598, 607, 611, 615, 621-22, 638, 1375-76, 1378-80; isogenic map, 595-96; Land Office, 1461; lake coast, lighthouses, 1295; Light House Board, 624, 1295, 1465; Library of Congress, 32 (8), (54), 1077, 1721; loess, 465, 623; magnetic variations, 581-82, 593, 603; magnetism, 597; meteorology, 585-86, 604-05; National Aeronautics and Space Administration, 61; National Archives, 32 (82); National Forests, 607-08, 613, 622; National Parks, 607, 611, 615, 622; National Research Council; *See:* National Research Council; Navy Department, 716, 1553; Commissioners, 1629, hydrographic office, 1493, 1615, 1630; northeast, 627-633, Bouguer gravity map, 633; north East boundary, 631, dispute, 760-60a; northwest boundary survey, 1461; orography, 473; Pacific railroad explorations, 1461; Pension Bureau, 592; Porto Rico, 965, 1661; Post Office Department, 777; railroads, 578, 580, 583-84, 588-91, 594, 598, 607, 615, 617, 1700; Reclamation Service, 608; relief map, 602; roads, 452, 490, 555, 559, 568, 570, 1090; Sanitary Commission, 1259; statistics, 592, 1704, 1728, 1732; Susquehanna Military Department, 1132; telephone lines, 609, Treasury Department, 1723, donor, 642, 684, 687, 690, 692-93, 773, 832-33, 1074; Virginia Military Department, 1132; *War Department* (*see also:* U.S. Army), 769-70, 774-75, 956, 1090, 1129, 1132-33, 1177, 1180, 1259-62, 1392, 1402, 1516, Adjutant General's Office, Military Information Division, 416-17, 419, Bureau of Insular Affairs, 418, donor, 774, General Staff, 390, General Staff, Military Information Division, 379, Survey of the north and north western lakes, 504, 506, 1278-83, 1285-97, 1299-1301, Topographical Bureau, 1375-76, War College Division, General Staff, 1552;

Washington Military Department, 1132; Weather Bureau (*see also:* U.S. Department of Agriculture), 605, 608, 1092; Works Project Administration, 49
The United States coastal charts, 1783-1861, Peter J. Guthorn, 1768
The United States gazetteer, Joseph Scott, 637, 654, 662, 672, 713, 813, 875, 915, 1043, 1066, 1117, 1146, 1157, 1172, 1239, 1254, 1271
U.S. Gettysburg National Park Commission, 969-971
U.S. Grinnell Expedition, second, 32 (70)
U.S.S. *Tigress* (ship), 32 (85)
U.S.S. *Vincennes* (ship), 83
U.S.S. *Water Witch* (ship), 1563
Universal atlas, H. S. Tanner, 140, 269, 334, 1693
Universe, 1762
University of California, Berkeley, CA, 1493
University of Chicago Press, publisher, 1747
University of Nebraska Press, donor, 1765; publisher, 1765
University of Pennsylvania, 48
University of Texas Mineral Survey, 1234-35
University of the State of New York, 836, 839-42, 633
University of Virginia, 1121-22; Press, publisher, 1750
Unterwalden, Switzerland, 224
Upper Geyser Basin, Yellowstone National Park, WY, 1434, 1439
Urabá, Golfo de, Panamá, 1514
Ural Mountains, U.S.S.R., 317
Uranographia Britannica, John Bevis, 1646
Urien, Switzerland, 226
Urubamba River, Peru, 1570
Utah, 1406-407, 1427, 1449-53, 1708; Uinta Mountains, 1449
Utrecht, The Netherlands, 193, 195

Vache Island, Grande Baye de Mesle, to Abacou Point, St. Domingue, 32 (53)
Valais, Switzerland, geology, 242
Vallance, F., engraver, 921
Vallance (Valance, Valence, Vallence), John (?) (1770-1833, engraver), engraver, 309, 331, 386a, 387, 437, 569, 572, 1006, 1065, 1078, 1543; publisher, 461, 1065
Vallesia, Switzerland, 227
Valley of the Kings, Egypt, 3 (1-3)
Valltravers, Johann Rudolph (1723-1815, Swiss-American naturalist), donor, 74
Vambéry, Armin (1823-1913, traveller, linguist), 375
Vance, David H., cartographer, 574; donor, 574
Vancouver, George (1758-1798, British explorer), 80, 82-3, 370
Vancouver, Island, B.C., Canada, 541
Vandergucht, Gerard, engraver, 1016
Van Olden, A. E., cartographer, 984
Vargas, Federico, de, cartographer, 408
Varlé, Charles (1725-1795?, geographer, engraver), cartographer, 1120
Varlé, P. C. (fl. 1790-1820, French geographer, engineer), cartographer, 1001, 1548
Varley, Cromwell, J., cartographer, 50 (1-3); collection, 50
Vauban, Sébastien le Prestre de (1633-1707, French military engineer), 29 (6), 187
Vaudois Alps, Switzerland, geology, 242
Vaughan, A., lithographer, 1581, publisher, 1581
Vaughan, Benjamin (1751-1835, diplomat, political economist, agriculturist, APS 1786), collection, 51, 234
Vaughan, George Arthur, publisher, 1590
Vaughan, John (1755-1841, merchant, secretary, and librarian of the APS, APS 1784), 2 (4), 32 (65-6), (69), 139, 453, 628, 1192, 1194, 1664, 1668, 1671, 1688; donor, 386a-87, 436, 484, 671, 703, 761, 900, 1016, 1184-85, 1200, 1269, 1304, 1637, 1648, 1692
Vaughan, Samuel (1720-1802, merchant, philanthropist, horticulturalist, APS 1784), 51
Vaughan, William (1752-1850, London merchant, APS 1820), donor, 132, 141; *Plan of the London docks,* 132
Veluwe, The Netherlands, 208
Venango County, PA, 951
Venezuela, 1569, 1668; Amazonas Territory, 1569; Bolivar (state), 1569; geology, 1569; Maracaibo, 32 (69); orogeny, 1569
Venice, Italy, 348
Vendyssel (Vensyssel), Denmark, 178, 180
Vera Cruz, Mexico, 1503; intendency, 32 (16); to Orizaba and Xalapa, 32 (71); to Toluca, 32 (18)
Verdun, France, 13 (2)
Vermeule, Cornelius Clarkson (b. 1858, cartographer), cartographer, 884, 886, 888-93
Vermont, 658, 661-68, 694; Franklin, 666; Green Mountains, 32 (84), 1298; roads,

663, 780; statistics, 666
Vervest, A., engraver, 444
Vestervig, Denmark, 180
Vevelst, E., engraver, 290
Vicq, ——, (fl. 1804-1824, French engraver), engraver, 432
Victoria, Australia, 1606; Department of Lands and Surveys, 1606
Vidal, Alexander Thomas Emeric, cartographer, 338
Viegas Island, Peru, 35 (4)
View of South Carolina, as respects her natural and civil concerns, John Drayton, 1161-62
Vigo (Viguo) River, Spain, 326, 335
Viguier, Constant, cartographer, 113
Ville Marie, Quebec, Canada. *See:* Montreal, Quebec, Canada
Vinal, W. I., cartographer, 882, 885, 1050, 1197
Virginia, 32 (19), (99), 451, 455, 469, 752-53, 756, 758-59, 774-745, 777, 779, 1078, 1083-84, 1094-1138, 1661, 1750, 1761; Albermarle County, 32 (54); Appomattox Court House, 775, battlefield, 774; Ayletts Warehouse to Williamsburg, VA, 1110-11; Berkeley County, 1120; Bermuda Hundred, 774; Big Walker Mountain, 32 (80), Bowling Green ordinary to Annapolis, MD, 1096; Cape Henry to Cape Romain, SC, 1148; Chancellorsville, 774; Cold Harbor battlefield, 774; donor, 1122; Dumfries, to Annapolis, MD, 1097; Elizabeth River, 1137; Farmville, 774, battlefield, 774; Five Forks battlefield, 774; Fort Monroe, 1125; Frederick County, 1120; Fredericksburg, 774, to Annapolis, MD, 1098-99; Hanover and Newcastle to Annapolis, MD, 1100; Hanover Court House to Annapolis, MD, 1101; Head Lynchs ordinary to Annapolis, MD, 1102; High Bridge, 774, battlefield, 774; Hooe's Ferry to Williamsburg, VA, 1112-13; Indians, 90; Jefferson County, 1120; Jetersville battlefield, 774; Manassas battlefield, 1123; Monticello, 32 (54); New Kent Court-House to Annapolis, MD, 1103-104; Norfolk harbor, 644, 1137; North Anna battlefield, 774; NC boundary, 32, (91); Peak Mountain, 32 (80); Petersburg, 1125, siege, 774; Port Royal to Williamsburg VA, 1114-15; Richmond, 1122, 1125, 1134-35, siege, 774; roads, 777, 1096-1116; Sailor's Creek battlefield, 774; Sneed's ordinary to Williamsburg, VA, 1116; Spotsylvania battlefield, 774; Todd's ordinary to Annapolis, MD, 1105; Totopotomy, 774; battlefield, 774; U.S. Military Department, 1132; University of Virginia, 1121-22; Wilderness battlefield, 774; *Williamsburg,* to Annapolis, MD, 1106-107, to Ayletts Warehouse, VA, 1110-11, to Hooe's Ferry, VA, 1112-13, to Port Royal, VA, 1114-15, to Sneed's Ordinary, VA, 1116, to Washington, D.C., 1127; Winchester, 1120; Wythe County, 32 (80), geology, 32 (80); York, to Annapolis MD, 1108-09; *Yorktown,* 32 (9), 1136, battle of, 32 (9), siege, 1125, 1131, to Williamsburg, 1126
Viscayan Islands, Philippines, 417
Visscher Nikolaes Jansz (1649-1702, Dutch engraver, publisher), cartographer, 898
Voegelin, Charles Frederick (b. 1906, anthropologist), cartographer, 464, 468; donor, 468
Voegelin, E. W., cartographer, 646
Voegelin, Florence Marie Robinett (b. 1947, anthropologist), cartographer, 468
Voegelin, F. M., cartographer, 468
Vohsen, E., 115
Volcanoes, Japan, 383
Volney, Constantin François Chasseboeuf, comte de (1757-1820, French scholar, author, APS 1797), *The Ruins . . .* , 87a
Voyage à l'Est, Cailliaud, 428
Voyage à l'Ouest, Cailliaud, 427
Voyage de l'Ambassade . . . , A. E. van Braam Houckgeest, 386a, 387
Voysard, Étienne Claude (1746-1812, French engraver), engraver, 265

Waddell, James, 82
Wadden, The Netherlands, 209
Wadsworth, Alexander S., cartographer, 704, 716, 853
Wagner, O. G., cartographer, 1131
Wagner, Thomas S. (fl. 1840-1865, publisher, lithographer), lithographer, 852
Wagner, William (fl. 1820-1835, engraver, publisher), engraver, 927; publisher, 927
Wagner & McGuigan, lithographers, 852
Wainwright, R., cartographer, 853-54
Walcheren, The Netherlands, 210
Walcott, Charles Doolittle (1850-1927, geologist), cartographer, 707, 837-38, 856, 606, 1024, 1026, 1076, 1235, 1315
Waldo patent, Maine, 32 (20)
Wales, 1679; canals, 139; geology, 138-39,

142, railroads, 139, 141; roads, 124-25, 127-28, 133, 137
Walker, C., engraver, 95, 374, 1576-77; publisher, 93
Walker, Emery, Ltd., cartographer, 1617; engraver, 1617
Walker, Francis A., cartographer, 1704; *Statistical atlas of the United States,* 1704
Walker, George H., publisher, 649
Walker, James (1748-1808, English engraver, geographer to the East India Company), cartographer, 374
Walker, John (1759-1830, English man of science), cartographer, 79, 139; engraver, 79, 95, 1576-77; publisher, 93
Walker, J., & C., engraver, 95, 374, 1576-77; publishers, 93
Wall, J. Sutton, cartographer, 967
Walla Walla River, WA, 1426
Wallace, J., cartographer, 1282, 1288-90
Wallace, ID, 1447-48
Walling, Harry Francis (1825-1888, writer on railroads), *New topographical atlas of the state of Ohio,* 1312
Wallis, James, publisher, 1679
Wallis, John (fl. 1783-1792, surveyor), 1619; cartographer, 913
Wallis le Valais, Switzerland, 227
Wallis's new pocket edition of the English Counties, 1679
Walser (Walserum), Gabriel (1695-1776, Swiss cartographer, chronicler), cartographer, 220, 222-25, 227-29
Wampler, J. M., cartographer, 687, 743, 1230-31
Wanbesa Lake, WI, 1358
Wang, C. C., cartographer, 395
Wankijfs, S., publisher, 896, 898, 1048
Wansleben, Thomas O., cartographer, 1320, 1388, 1428, 1453
War, 295-99
War of 1821, 32 (68), 768, 824, 1071, 1220, 1274
War with Mexico. *See:* Mexican War
Warberg, O. (fl. 1776-1806. Danish publisher), cartographer, 174, 176
Wardle, J. M., cartographer, 480, 482
Warner, ——, publisher, 921-22
Warner, John, collection, 52
Warner, W. H., cartographer, 1279
Warner and Hanna, 1070
Warr, John II (fl. 1825-1845, engraver), engraver, 324
Warr, John, & William W., engravers, 324
Warr, William W., engraver, 324
Warren, ——, cartographer, 1393
Warren, Gouverneur Kemble (1830-1882, soldier), cartographer, 950
Warren, J. G., cartographer, 1336
Warren County, KY, 1249
Washington, George (1732-1799, patriot, first president of the U.S., APS 1780), 74, 579, 1078, 1236-37, 1767
Washington (state), 1466-1473; Admiralty Inlet, 1469; False Dungeness Harbor, 1468; Gray's Harbor to Admiralty Inlet, 1462; Mount Rainier National Park, 1473; Port Townsend, 85; Puget Sound, 644, 1469-70; Seattle, 1471; Shoalwater Bay, 1467; Tacoma, 1471
Washington, Adams County, MS, 1 (3)
Washington, D.C., 576-77, 644, 770, 1065, 1078, 1134; and Georgetown harbors, 1080; Capitol, 579; Hydrographic Office, 99; to New Orleans, LA, 1090; to Williamsburg, VA, 1127; Military Department, 1132; Washington Monument, 579
Washington College, Chestertown, MD, 1767; publisher, 1767
Washington College Presents: On the Map, Russell Morrison, 1767
Washington County, KY, 1249
Washington Monument, Washington, D.C., 579
Washington Square, Phila., PA, 1012, 1040
Washoe mining district, NV, 1457; Comstock Lode, 1457
The water masses of the North Atlantic Ocean, W. R. Wright and L. V. Worthington, 1754
Water power, Canada, 472
Waterford, Ireland, 145
Waterman, Henry Ely (d. 1898, soldier), 1337; cartographer, 1338
Watertown, WI, 1358
Watertown, Fond du Lac and Lake Superior Rail Road, 1365
Watmough, John G., donor, 716, 1639-40
Watson, John Frampton (b. ca. 1805, lithographer), lithographer, 937, 771
Waukesha, WI, 1372
Wausau, WI, 1372
Wauters, Alphonse Jules (1845-1916, Belgian cartographer), cartographer, 444-45
Wayland, ——, publisher, 1240
Wayles, John, 22 (1)
Wayne County, PA, 18, 1681; roads, 1681
Webber, F. P., cartographer, 717, 1182
Weber, E., & Co., lithographer, 1426; pub-

lisher, 1426
Weber, E. A., cartographer, 1075
Webster, J., publisher, 483
Webster County, KY, 1252
Weigel, Johann Christoph (d. 1746, German goldsmith, engraver), publisher, 91
Weigel-Schneider, Chr., publisher, 1665
Weinek, L., 58
Weissport Borough, PA, 33 (9)
Welch, William H., 1720
Welcome Society of Pennsylvania, 981
Weldon, Daniel, cartographer, 1094
Weller, E. (d. 1884, publisher, engraver), lithographer, 1631; publisher, 1631
Weller, E., and Grahams, Lith., lithographer, 1631; publisher, 1631
Wellfleet harbor, MA, 690
Wells, L. A., cartographer, 1581
Werdenberg, Switzerland, 225
Wersebe, Hermann Martin Christian von, cartographer, 401
Werner, T. W., cartographer, 849
Wessel, Caspar (1745-1818. Danish geographer), cartographer, 167-68, 171-72; engraver, 169
West Africa, 434-42
West Chester, PA, 43 (2), (4-5); PA-DE quadrangle, 973
West Florida (*See also:* Florida), 32 (5); Middle River, 17 (1); Yellow River, 17 (1)
The West India pilot, J. S. Speer, 1145, 1170, 1503, 1509-13, 1520-34, 1565-67
The West-India atlas, Thomas Jefferys, 1664
West Indies, 88-90, 182, 450, 456, 569, 572, 598, 1517-53, 1562, 1627, 1638, 1661, 1663-64, 1668, 1674-75, 1738
West Virginia, 973, 1139-44; Boone County, 1140; coal, 1140-41; coal fields (bituminous), 32 (83); Coal River, 1140; gas, 1141, 1143; Geological Survey, 1144; geology, 1139, 1141, 1143; Harper's Ferry, 774; iron ore, 1141, 1143; Kanawha County, 1140; oil, 1139, 1141, 1143; oil district, 32 (83); railroads, 1142; roads, 777
Western Australia, Australia, 1581, 1584-85; geology, 1585; gold mines, 1584; Meekatharra, 1585
Western hemisphere, 88-91
Western Printing and Lithographing Co., lithographers, 1372
Western Reserve Historical Society, 32 (22)
Western United States, 1426-29
Westerwolde, The Netherlands, 211
Westmoreland County, England, 120
Westphaliaprovinz, Germany, 303; Soest, 299
Wetherell, John P., donor, 1692
Weyl, C. G., cartographer, 1336
Weyss, ——, cartographer, 1707
Weyss, John E., cartographer, 775
Whales, 1618
Wheat, James (author). *See:* 74a-74d, 76, 78-83, 87a-87b, 109, 130-31, 147, 154, 177, 196, 259, 261-63, 295-99, 309, 315, 323, 331, 352, 354, 366, 368-69, 386a-87, 405-407, 425-26, 436-37, 457, 458-59, 467, 497, 555-56, 558, 626, 628-29, 635, 637-40, 652-55, 661-65, 671-74, 697-98, 700, 705, 711-15, 719, 721-39, 750-52, 755, 760-61, 763-65, 783, 785, 787-815, 837, 859-60, 863-77, 900-903, 905-11, 913-18, 989-93, 997-1000, 1042-44, 1051, 1053-69, 1077-78, 1082-85, 1095-119, 1146-48, 1157-60, 1172, 1174-76, 1184-86, 1200, 1236, 1238-43, 1254-56, 1269, 1271-72, 1302-808, 1539-40, 1543-44, 1546, 1556-59, 1616, 1621, 1623-26
Wheeler, George Montague (1842-1905, soldier), cartographer, 1427, 1707; donor, 1427, 1457
Wheeler, Junius Brutus (1830-1886, Army teacher), cartographer, 32 (82)
Wheeler, Samuel, 46 (3)
"Where is Franklin's Chart of the Gulf Stream?," Franklin Bache, APS *Proceedings,* 1622
Whitaker, E. A., 1747
White, George W., cartographer, 465
White, I. C., cartographer, 960, 1141, 1143
White, James (1863-1928, Canadian geographer, publisher), cartographer, 472, 474-76, 516, 518-19, 530, 532-34, 536, 539-41, 546-47, 1724; donor, 1724; *Atlas of Canada,* 1724
White Mountains, NH, 657
White River, Yukon Terr., Canada, 547
Whiteburn gold district, Queens County, Nova Scotia, 496
Whitelaw, James (1748-1829, surveyor-general of Vermont), cartographer, 661, 663
Whiting, Henry L., cartographer, 684, 686-87, 690-91, 693, 746, 850, 853-54, 1049, 1074-75, 1182
Whitman, Edmund Burke (d. 1883, cartographer), 1399
Whitney, Milton, cartographer, 1076
Whitney, Peter, *The history of the county of Worcester,* 671
Whitson, Andrew Robson (b. 1870, geolo-

gist), cartographer, 1370
Whittle, James (1757-1818, British publisher), publisher, 90, 193, 332, 370, 762, 1672, 1674
Widen, K. A., cartographer, 1336, 1340
Wies, Nicolas (1817-1789, French geologist), cartographer, 217; *Guide de la carte géologique du Grand-Duché de Luxembourg,* 217
Wilbur, D. C., cartographer, 1225
Wilderness battlefield, VA, 774
Wiley, ——, publisher, 1695
Wiley & Long, publishers, 1695
Wilhelms, ——, publisher, 607, 611, 615
Wilkes, Charles (1798-1877, Naval officer, explorer), cartographer, 716, 1470, 1474, 1629; donor, 1629
Wilkes, Charles C., 1688
Wilkins, Lt. ——, cartographer, 6 (1), 32 (85)
Wilkins, Sir Hubert (1888-1958, Australian Antarctic explorer), 103
Wilkins-Hearst Antarctic Expedition, 103
Wilkinson, J., cartographer, 32 (66)
Wilkinson, James, 39 (1)
Wilkinson, Robert (fl. 1748-1794, English publisher), 121, 133, 257, 289, 564, 1517, 1541
Wilkinson, Robert, and Robert Sayer, publishers, 121
Wilkinson, Santiago, cartographer, 32 (71)
Willett, James R., cartographer, 1261
William L. Clements Library, 32 (5)
Williams, ——, lithographer, 625, 481; publisher, 82, 1633, 1635
Williams, Archibald B., cartographer, 284, 1522
Williams, I., cartographer, 32 (8)
Williams, J. S., cartographer, 705, 1230-31
Williams, John, 1646
Williams, Jonathan, Jr. (1750-1815, merchant, soldier; APS 1787), 32 (68); donor, 32 (68), 1625-26; "Memoir on the Use of the Thermometer in Discovering Banks, Sounding, etc." APS, *Transactions,* 1625-26
Williams, Lemuel D., cartographer, 1138, 1461, 1469
Williams, Samuel (1743-1817, historian of Vermont), *The natural and civil history of Vermont,* 661
Williams, T. J., cartographer, 958
Williams, William (b. 1787, engraver), cartographer, 946; engraver, 82, 946
Williams, William George (1801-1846, soldier), cartographer, 828-31, 834, 1276, 1281; donor, 830
Williams, W. S., cartographer, 1336
Williams & Heintz, publisher, 481, 1569
Williams & Heintz Map Corp., publisher, 984, 1633, 1635; lithographer, 481, 625
Williams, Creek, B.C., Canada, 537
Williams (ship), 101
Williamsburg, VA, to Annapolis, MD, 1106-107; to Ayletts Warehouse, VA, 1110-11; to Hooe's Ferry, VA, 1112-13; to Port Royal, VA, 1114-15; to Sneed's Ordinary, VA, 1116; to Washington, D.C., 1127; to Yorktown, VA, 1126
Williamson, Charles (1757-1808, British officer, land promoter, secret agent), *Description of the Genesee Country,* 764-65; *Description of the settlement of the Genesee Country,* 815
Willing, Thomas, 19
Willing family, 19
Willink, John, 816, 819
Willink, William, 816, 819
Willis, Bailey (1857-1949, geologist), cartographer, 395, 466, 1493, 1725
Willman, H. G., cartographer, 625
Willow River, B.C., Canada, 537
Wilson, ——, publisher, 649
Wilson, A. D., cartographer, 1407-10, 1442
Wilson, H. Maclean, cartographer, 707, 837-38, 856, 1024, 1026
Wilson, John (1741-1793, cartographer), cartographer, 1164-65; donor, 1655
Wilson, John L. (governor of SC), donor, 1165
Wilster, F., cartographer, 184-85, 300
Wiltberger, C., Jr., cartographer, 324
Winchester, VA, 1120
Wind, current and thermal charts, M. F. Maury, 1697
Winds, oceans, world, 1612
Windward Islands, 463
Winlock, Herbert E., donor, 3
Winnecke, Charles (Australian geographer), 1582
Winnemucca, NV, 1460
Winnipeg, Lake, Manitoba, Canada, 525
Winslow, Arthur (1860-1938, geologist), cartographer, 1384
Winterbotham, William (1763-1829, author, geographer), *An historical view of the U.S.,* 1242-43
Winthrop, ME, 51 (1)
Wischnewsky, 317
Wisconsin, 1335, 1358-73; Appleton, 1372;

Beloit, 1372; Dairy and Food Commission, 1368; dairy products, 1368; Dane County, 1358; Eau Claire, 1372; Experiment Station, 1368; Fond du Lac, 1277, 1358, 1372; Fox River, 1281; geology, 1359, 1370; Green Bay, 1281, 1372; Janesville, 1372; Kegonsa Lake, 1358; Kenosha, 1372, harbor, 1364; La Crosse, 1372; Lac du Flambeau, quadrangle, 1371; Madison, 1358, 1365-66, 1372; Manitonoc, 1372, harbor, 1360; Memoma Lake, 1358; Mendota Lake, 1358; Milwaukee, 1361, 1372; Oshkosh, 1372; Outer Island, 1299; Racine, 1372; railroads, 1358, 1363, 1365; roads, 1372; Sheboygan, 1372, harbor, 1362; Stevens Point, 1372; Wanbesa Lake, 1358; Watertown, 1358; Waukesha, 1372, Wausau, 1372
Wise, G. D., cartographer, 1049, 1138
Wissahickon Valley, Fairmount Park, Phila., PA, roads, 1034
Wistar, ——, publisher, 1014
Wit, Frederick de (1610-1698, cartographer, publisher), *Atlas,* 1637; cartographer, 1637, publisher, 1637
Witzel, P., cartographer, 647
Wolbrecht, George H., cartographer, 1336
Wolfe, James (1727-1759, British general), 501
Wollaston, William Hyde (1766-1828, English physiologist, chemist, physicist), 1772
Wong, W. H., 395
Wood, George Bacon (physician, APS 1829), donor, 1247, 1629
Wood, G. H., Jr., cartographer, 984
Wood, H. O., cartographer, 1493
Woodbury, NJ, 890
Woodhull, M., cartographer, 692, 746, 850-51
Woodman, ——, publisher, 73, 128, 308, 350, 361, 367, 400, 424
Woodman, Richard, Sr. (English engraver), engraver, 160, 192, 195
Woodman & Mutlow, publishers, 73, 128, 160, 192, 308, 350, 361, 367, 400, 424
Woodruff, I. C., cartographer, 834, 1278-79
Woods Hole Oceanographic Institute, donor, 1754
Woodward, ——, publisher, 1233
Woodward & Tiernan, publishers, 1233
Woodward, E. F., engraver, 577, 1195, 1322, 1462, 1479, 1758
Worcester, MA, 671
World, 32 (93), 62-90, 1076, 1223, 1636-37, 1639-40, 1649, 1651, 1653, 1659, 1669, 1671-73, 1676-78, 1680-84, 1690, 1693, 1695, 1698, 1702, 1713-14, 1716-17, 1719-20, 1726-27, 1729, 1733, 1738-39, 1741-46, 1762; atlas, 1636-37; geology, 1730; meteorology, 1722; mineralogy, 1730; oceans, 1611-15; bathymetry, 1613-14; water power, 1730; zoogeography, 1722
World atlas and gazetteer, P. F. Collier & Son Corp., 1746
World atlas of commercial geology, U.S. Dept. of the Interior, 1730
(World ocean atlas), Sergei G. Gorshkov, 1755
World Syndicate Publishing Co., 1742; *New revised atlas of the world,* 1742
World War I, 282-84, 617
World's Fair, St. Louis, MO, 964, 1385, 1387
Worthington, L. V., cartographer, 1754; and W. R. Wright, *The water masses of the North Atlantic ocean,* 1754
Wragg, S., cartographer, 1166
Wreck chart, North Atlantic ocean, 1630
Wresina, 29 (27)
Wright, ——, cartographer, 1040
Wright, A. A., lithographer, 1586
Wright, Charles Cushing (1796-1857, engraver), engraver, 1188
Wright, John K., cartographer, 1735
Wright, W. R., cartographer, 1754; and L. V. Worthington, *The water masses of the North Atlantic ocean,* 1754
Wüllerstorf-Urbair, B. von, cartographer, 1634
Würtz, ——, publisher, 182
Wütquinàw, ——, 29 (58)
Wuhrer, L., engraver, 271, 273-74, 340; publisher, 339-40
Wurster, Johann Ulrich (1841-1880, Swiss topographical engineer, cartographer, publisher), publisher, 231, 236, 239-43, 281
Wurster, J., et Cie., publisher, 279
Wurster, Randegger, publishers, 231, 236, 239-43, 281
Wyandot Indians, 901
Wyld, James, & Son, publishers, 32 (93)
Wyman & Sons, publishers, 1588
Wynkoop, ——, publisher, 580
Wyoming, 1406-407, 1425, 1433-46; Cheyenne, 1443-44; Evanston, 1443-44; Fire Hole River, 1433; geology, 1403; Laramie, 1443-44; Sheridan, 1443-44; Lower Geyser Basin, Yellowstone National Park, 1433, 1438; Upper Geyser Basin, Yellowstone

National Park, 1434, 1439; Yellowstone Lake, 1435-36; Yellowstone National Park, 1433-36, 1438-39
Wyoming coal basin, PA, 958
Wythe County, VA, 32 (80); geology, 32 (80)

Xalapa, Mexico. *See:* Jalapa, Mexico
Xochimilco lake, Mexico, 32 (11)

Yale, Eli (1648-1721, governor of Madras), 186
Yale University, William Robertson Coe Collection of Western Americana, 560
Yampolsky, Mrs. Helene Boas, donor, 6 (1-6), 486-87
Yarmouth, England, 123, 125
Yeager, E., engraver, 577, 692, 1167, 1195, 1482
Yeats, Thomas (1768-1839, British orientalist), 79
Yellow River, FL, 17 (1)
Yellowstone Lake, Yellowstone National Park, WY, 1435-36
Yellowstone National Park, WY, 1437, 1445; Lower Geyser Basin, 1433, 1438; Upper Geyser Basin, 1434, 1439; Yellowstone Lake, 1435-36
Yellowstone River, 1402, 1404-405, 1425; fork. *See:* Clarks River
Yerkes Observatory, 1747
Yonge, Ena L., *A catalogue of early globes made prior to 1850 . . .*, 1773
York, Cape, Queensland, Australia, 1588; peninsula, 1600
York County, PA, 927
York Factory, Manitoba, Canada, 548
York Haven, PA, 923
York River, Hampton Roads, Chesapeake Bay, 1138
York River harbor, ME, 643
York, VA, to Annapolis, MD, 1108-109
Yorktown, VA, 32 (9), 1136; battle of, 32 (9); siege, 1125, 1131; to Williamsburg, VA, 1126
Yosemite National Park, CO, 1492
Yoshii, G., publisher, 382, 384
Young, ——, 948, publisher, 1161-62
Young, G. A., cartographer, 465
Young J., engraver, 643, 666, 692-93, 1466-67
Young,˙ James H. (fl. 1817-1866, cartographer, engraver), cartographer, 577, 1322; donor, 577; engraver, 574, 577, 687, 1073; lithographer, 1474; publisher, 1474
Young, Robert Evans (1861-1911, Canadian surveyor, civil servant), cartographer, 507, 541, 549
Young and Delleker, engravers, 666
Young's Ford on the Schuylkill to Darby Creek, PA, 32 (68)
Your Mexico, Winifred Duncan, 41
Yucatan, Mexico, 1505, 1507; intendency 32 (16)
Yugoslavia, Banjaluca, 29 (50); Belgrade, 29 (46), (57-59); Pancevo, 29 (47)
Yukon Territory, Canada, 539, 547; Alsek River, 547; Kluane River, 547; White River, 547
Yust, Walter, cartographer, 1744

Zab River, 374
Zabarah Mountain, 428
Zacatlan, Mexico, 32 (18)
Zaire, Katanga, 446
Zalapa, Mexico. *See:* Jalapa, Mexico
Zannoni, G. A. Rizzi. *See:* Rizzi Zannoni, G.A.
Zantzinger, D. C., cartographer, 1028
Zealand, The Netherlands, 188, 190, 193, 195
Zee Atlas tot het Gebruik van de Vlouten des Konigs van Groot Britain, 1640
Zehender Theil der Orientalische Indien . . ., De Bry, 1645
Zimmerman, ——, cartographer, 1032
Zimpel, Charles, cartographer, 1220
Zingg, A., publisher, 235
Zingg, George A., cartographer, 1336
Zodiac, 53
Zoogeography, world, 1722
Zugewandte, Switzerland, 228
Zuiderzee, The Netherlands, 212
Zuliani, Felice, engraver, 355
Zworneck, 29 (26)